Metal Ions in Neurological Systems

Wolfgang Linert · Henryk Kozlowski
Editors

Metal Ions in Neurological Systems

Springer

Editors
Wolfgang Linert
Vienna University of Technology
Institute of Applied Synthetic Chemistry
Vienna
Austria

Henryk Kozlowski
University of Wroclaw
Faculty of Chemistry
Wroclaw
Poland

Including contributions from the special issue
Metals in the Brain. Monatshefte für Chemie/Chemical Monthly,
Issue: Volume 142, Number 4, April 2011,
ISSN: 0026-9247 (paper) 1434-4475 (online)

ISBN 978-3-7091-1000-3 ISBN 978-3-7091-1001-0 (eBook)
DOI 10.1007/978-3-7091-1001-0
Springer Wien Heidelberg New York Dordrecht London

Library of Congress Control Number: 2012948964

Printed on acid-free paper

Springer is part of Springer Science+Business Media (www.springer.com)

Foreword: Metal Ions in Neurological Systems

At the beginning of the last century a newborn in Europe had a life expectancy of about 45 years. Now, approximately 100 years later, this number has increased in the West to more than 80 years [1]. The reasons for this extended life span are manyfold, the more important ones being improved hygiene, progress in medicine, a reduced mortality of mother and child, control of epidemics, and better food provisions [2]. As we all know, this has led to a tremendous growth of the world population: About 1 billion ($=10^9$) people inhabited the Earth in 1804; it took nearly 125 years to double this number, by 1974 it was 4 billion, and the estimate for 2025 is 8 billion [1].

However, this story of success has a seamy side: because of the increased life span, cancer and other age-related diseases, especially those connected with neurodegeneration, mainly manifested in the brain, are getting more and more prominent. To this category belong, among others, Alzheimer's disease (AD), Parkinson's disease (PD), amyotrophic lateral sclerosis (ALS), Huntington's disease (HD), transmissible spongiform encephalopathy (prion disease; Creutzfeldt–Jakob disease), neurodegeneration with brain iron accumulation (NBIA; formerly Hallervorden–Spatz disease), Friedreich's ataxia, progressive supranuclear palsy (PSP), and multisystem atrophy (MSA) [3].

The best known of these debilitating illnesses are certainly Alzheimer's [3] and Parkinson's [3, 4] diseases. About 29 million people worldwide suffered from Alzheimer's disease in 2009; this is about 60% of those being affected by dementia. Computer models predict that the number of patients with AD will increase to 106 million by 2050 [1], which means one per 85 persons—even if this number should be off by a factor of 2, one out of 170 persons is still truly frightening!

It is an old wisdom that metals are indispensable for life; in fact, several of them, like sodium, potassium, and calcium, are easily discovered in living matter [5]. However, the role of metals and their impact on life remained largely hidden until inorganic chemistry and coordination chemistry experienced a pronounced revival in the middle of the last century. The experimental and theoretical tools created and their application to biochemical problems led to the development of the field or discipline *Bioinorganic Chemistry*, which was established by 1970. It also became known as *Inorganic Biochemistry* and more recently as *Biological Inorganic Chemistry* [3].

At present there is substantial evidence that metal ions play a significant role in the pathogenesis and pathophysiology of neurodegenerative diseases, which reveal themselves mainly in the brain. This is not surprising because the brain utilizes metal ions for a myriad of biochemical reactions. The metal-related disorders are caused (1) by defects in metal ion transport or homeostasis, (2) by toxicological exposure to metals, and (3) they are associated with metalloprotein aggregation and/or misfolding and oxidative reactions [3]. This organizational frame is certainly helpful, yet to assign a given illness to a distinct category is not easy because of our lack in understanding the mechanisms involved in a given disease, though misfolded proteins are often implicated. Of course, toxic metal ions like mercury or cadmium are easily categorized, but with essential metal ions like copper, iron, zinc, or manganese, which may be toxic as well, matters are considerably more difficult and we are only at the brink of understanding the fine line between essentiality and toxicity. There is an urgent need to unravel the mechanisms involved in these neurodegenerative diseases; only once these are understood, is there hope for the development of effective drugs.

Jawaharlal Nehru, the first Prime Minister of free India, having in mind the difficulties India was facing, made a statement in 1961 [6] that does in a shortened form also hold for the difficulties in treating neurodegenerative disorders: "It is Science alone that can solve the problems... Who indeed could afford to ignore Science today? At every turn we have to seek its aid... The future belongs to Science and those who make friendship with Science." Therefore, the editors of this book, devoted to Metal Ions in Neurological Systems, must be congratulated for composing this volume and it is hoped that their and the authors' efforts have an impact on pertinent research, not only by established but also by younger scientists because new ideas are needed. It is evident that the mentioned diseases, which are increasingly affecting mankind, are closely linked to the role metal ions are playing in the brain. Clearly, huge efforts by devoted scientists will be needed to overcome the obstacles, and this also means that politicians need to provide the necessary financial support without a top-down policy, otherwise the disaster is programmed. It is our sincere hope that this book is a strong stimulus for future research!

Astrid Sigel
Helmut Sigel

References

1. Wikipedia (2010) http://www.wikipedia.org. Accessed 26 Oct 2010
2. Perutz MF (1982) Ging's ohne Forschung besser? Der Einfluss der Naturwissenschaften auf die Gesellschaft. Wissenschaftliche Verlagsanstalt, Stuttgart, pp 1–54
3. Sigel A, Sigel H, Sigel RKO (eds) (2006) Metal ions in life sciences, neurodegenerative diseases and metal ions, vol 1. Wiley, Chichester, pp 1–463
4. Grayson M, Schnabel J et al (2010) Outlook: Parkinson's disease. Nature 466(7310):S1–S19 (on p S1 under "Collection" five more pertinent articles from 2009 and 2010 are cited)
5. Sigel H, Sigel A (2000) J Indian Chem Soc 77:501
6. Ram Atma (1961) Proc Natl Inst Sci India 27A:531 (see also [2] on page 5)

Preface

Metal ions in the brain are a necessity as well as a poison. The presence of metal ions in the active sites of biological catalysts or metalloproteins and in the biological functioning of nucleic acids is very well documented, and they are required for brain activity. On the other hand, essential metals like copper or iron are very effective in generating oxidative stress which not only plays an important role in our immunology but also is the root of practically all neurodegenerative disorders, since it induces disease via the death of neurons.

The brain (2 % of the body weight) is a major consumer of oxygen in the human body (20 % of total oxygen), so production of reactive oxygen species (ROS) is believed to be inherent to the progression of many disorders. Because metals are involved in both ROS production and in the active centers of antioxidant enzymatic systems, the proper distribution and strictly controlled homeostasis of metal ions are critical for brain functioning. Indeed, metal ions that are highly concentrated in some regions of the brain are also able to promote protein aggregation and thus aid the formation of protein aggregates, e.g., fibrils, which are toxic for neurons. Therefore, managing metal ions in the brain could be an important strategy in the search for therapeutic agents to be used in the treatment of neurodegenerative diseases. The actual biological functions of many proteins involved in neurodegenerative disorders are not yet understood. Biochemical and biological data suggest that many of them could be metalloproteins. Metal ions such as Cu(II), when bound to protein, can act either as an antioxidant enzyme (e.g., when bound to prion protein) or as a Fenton-type catalyst producing ROS (when bound to beta-amyloid peptide).

Aluminum plays a very peculiar role in neurotoxicity and development of neurodegeneration; its distinct effect on brain activity was discovered during dialysis processes in humans. Three chapters will be devoted to the potential role of this metal in pathologies of the brain. The role of Al(III) ions in particular neurodegenerative disorders is still controversial, but the same could be said about other metal ions. It seems to be generally accepted that aluminum is certainly neurotoxic.

It can, therefore, be seen that much remains to be learnt about the chemistry and biology of proteins involved in neurodegeneration and about the metals bound to them before convincing mechanisms of neurological pathologies can be understood. There are also two other enigmatic functional structures closely related to the brain, the blood–brain barrier (BBB) and blood–cerebrospinal fluid barrier. These unusual structures formed by endothelial cells protect the brain from chemical species which could otherwise be potentially harmful. Both barriers are rather effective and also metal ion transport is strictly controlled; thus, entry into the brain is already a difficult challenge for metals. Therefore, as well as understanding the basic biology and various pathological disorders within the brain areas, we must learn how to design a drug molecule able to pass through these barriers.

Vienna, Austria — Wolfgang Linert
Wroclaw, Poland — Henryk Kozlowski

Contents

List of contributors

Bulent Acikyol School of Biomedical Sciences and Pharmacy MSB, University of Newcastle, Callaghan, NSW, Australia

Michael Aschner Division of Clinical Pharmacology and Pediatric Toxicology, Vanderbilt University Medical Center, Nashville, TN, USA; Center in Molecular Toxicology, Vanderbilt University Medical Center, Nashville, TN, USA; Center for Molecular Neuroscience, Vanderbilt University Medical Center, Nashville, TN, USA; The Kennedy Center for Research on Human Development, Vanderbilt University Medical Center, Nashville, TN, USA

Emmanuel L. Barbier Inserm, U836, Grenoble Institut des Neurosciences, Grenoble, France; Université Joseph Fourier 1, Grenoble Institut des Neurosciences, Grenoble, France

Brianna Bassett School of Biomedical Sciences and Pharmacy MSB, University of Newcastle, Callaghan, NSW, Australia

Valerio Bertolasi Dipartimento di Chimica, Università di Ferrara, Ferrara, Italy; Centro di Strutturistica Diffrattometrica, Università di Ferrara, Ferrara, Italy

Sylvain Bohic Inserm U-836, Grenoble Institut des Neurosciences, Team 6 : Rayonnement Synchrotron et Recherches Médicales, ESRF, Grenoble, France; European Synchrotron Radiation Facility, ESRF, Grenoble, France; Université Joseph Fourier 1, Grenoble Institut des Neurosciences, Grenoble, France

Silvia Bolognin Department of Biology, Padua "Metalloproteins" Unit, CNR-Institute for Biomedical Technologies, University of Padua, Padua, Italy

Samuel Caito Division of Clinical Pharmacology and Pediatric Toxicology, Vanderbilt University Medical Center, Nashville, TN, USA; Center in Molecular Toxicology, Vanderbilt University Medical Center, Nashville, TN, USA

David G. Churchill Department of Chemistry, Korea Advanced Institute of Science and Technology (KAIST), Yuseong–gu, Daejeon, Republic of Korea

Peter Cloetens European Synchrotron Radiation Facility, ESRF, Grenoble, France

Robert R. Crichton Institute of Condensed Material and Nanosciences, Université Catholique de Louvain, Place Louis Pasteur 1, Louvain-la-Neuve, Belgium

Guido Crisponi Dipartimento di Scienze Chimiche, Università di Cagliari, Monserrato, Italy

Alexia Daoust Inserm, U836, Grenoble Institut des Neurosciences, Grenoble, France; Université Joseph Fourier 1, Grenoble Institut des Neurosciences, Grenoble, France

Paolo deBona Department of Chemical Sciences, University of Catania, Catania, Italy

Annalisa Dean Department of Chemical Sciences, University of Padova, Padua, Italy

Roheeth Delima School of Medicine and Pharmacology, University of Western Australia, Fremantle, WA, Australia

David T. Dexter Department of Cellular and Molecular Neuroscience, Faculty of Medicine, Imperial College London, London, UK

Valerio B. DiMarco Department of Chemical Sciences, University of Padova, Padua, Italy

Christopher Exley The Birchall Centre, Lennard-Jones Laboratories, Keele University, Staffordshire, UK

Gavino Faa Dipartimento di Citomorfologia, Divisione di Anatomia Patologica, Università di Cagliari, Cagliari, Italy

Daniela Fanni Dipartimento di Citomorfologia, Divisione di Anatomia Patologica, Università di Cagliari, Cagliari, Italy

Elena Gaggelli Department of Chemistry, University of Siena, Siena, Italy

Clara Gerosa Dipartimento di Citomorfologia, Divisione di Anatomia Patologica, Università di Cagliari, Cagliari, Italy

Ross Graham School of Biomedical Sciences, Curtin University of Technology, Bentley, WA, Australia

Alberto Granzotto Metalloproteins Unit, Department of Biology, National Research Council (CNR), Institute for Biomedical Technologies, University of Padova, Padova, Italy; Centre for Excellence on Aging (Ce.S.I.), "G. D'Annunzio" University of Chieti–Pescara, Chieti, Italy

Yonghwang Ha Department of Chemistry, Korea Advanced Institute of Science and Technology (KAIST) 373–1 Guseong–dong, Yuseong–gu, Daejeon, Republic of Korea; Hydro–Technology Research Laboratory, Convergence Science & Technology Research Center (CSTRC), Jungwon University, Goesan–gun, Chungbuk, Republic of Korea

Dominic Hare Elemental Bio-imaging Facility, University of Technology, Sydney, Broadway, NSW, Australia

Pedro L. Hernandez Faculty of Agronomy, University of Concepción, Concepción, Chile

Emily R. House The Birchall Centre, Lennard-Jones Laboratories, Keele University, Staffordshire, UK

Tamás Jakusch Department of Inorganic and Analytical Chemistry, University of Szeged, Szeged, Hungary

Guy N. L. Jameson Department of Chemistry, Dunedin, New Zealand

Daniel Johnstone School of Biomedical Sciences and Pharmacy MSB, University of Newcastle, Callaghan, NSW, Australia

Csilla Kállay Department of Inorganic and Analytical Chemistry, University of Debrecen, Debrecen, Hungary

Kibong Kim Department of Chemistry, Korea Advanced Institute of Science and Technology (KAIST), Yuseong–gu, Daejeon, Republic of Korea

Tamás Kiss Department of Inorganic and Analytical Chemistry, University of Szeged, Szeged, Hungary; HAS-USZ Bioinorganic Research Group, Szeged, Hungary

Radmila Milacic Department of Environmental Sciences, Josef Stefan Institute, Ljubljana, Slovenia

Elizabeth Milward School of Biomedical Sciences and Pharmacy MSB, University of Newcastle, Callaghan, NSW, Australia

Zoltán Nagy Department of Colloid Chemistry, University of Debrecen, Debrecen, Hungary

Sonia Nemolato Dipartimento di Citomorfologia, Divisione di Anatomia Patologica, Università di Cagliari, Cagliari, Italy

Yuzo Nishida Medical Research Institute, Kanazawa Medical University, Kahoku, Ishikawa, Japan

Valeria Marina Nurchi Dipartimento di Scienze Chimiche, Università di Cagliari, Monserrato, Italy

John Olynyk Department of Gastroenterology, Fremantle Hospital, Fremantle, WA, Australia

Giuseppe Pappalardo CNR Institute of Biostructures and Bioimagining, Catania, Italy

Irmgard Paris Program of Molecular and Clinical Pharmacology, ICBM, Faculty of Medicine, Casilla, Santiago, Chile; University of Santo Tomás, Viña del Mar, Chile

Maurizio Remelli Dipartimento di Chimica, Università di Ferrara, Ferrara, Italy

Enrico Rizzarelli CNR Institute of Biostructures and Bioimagining, Catania, Italy; Department of Chemical Sciences, University of Catania, Catania, Italy

Daniele Sanna CNR Institute of Biomolecular Chemistry, Baldinca-Li Punti (Sassari), Italy

Janez Scancar Department of Environmental Sciences, Josef Stefan Institute, Ljubljana, Slovenia

Juan Segura-Aguilar Program of Molecular and Clinical Pharmacology, ICBM, Faculty of Medicine, Casilla, Santiago, Chile

Eleni Siakkou Department of Chemistry, Dunedin, New Zealand

Éva Sija Department of Inorganic and Analytical Chemistry, University of Szeged, Szeged, Hungary

Carlos P. Sotomayor Instituto de Química, Pontificia Universidad Católica de Valparaíso, Valparaíso, Chile

Imre Sóvágó Department of Inorganic and Analytical Chemistry, University of Debrecen, Debrecen, Hungary

Mario Suwalsky Faculty of Chemical Sciences, University of Concepcion, Concepcion, Chile

Sarolta Timári Department of Inorganic and Analytical Chemistry, University of Debrecen, Debrecen, Hungary

Deborah Trinder School of Medicine and Pharmacology, University of Western Australia, Fremantle, WA, Australia

Olga G. Tsay Department of Chemistry, Korea Advanced Institute of Science and Technology (KAIST), Yuseong–gu, Daejeon, Republic of Korea

Ildikó Turi Department of Inorganic and Analytical Chemistry, University of Debrecen, Debrecen, Hungary

Gianni Valensin Department of Chemistry, University of Siena, Siena, Italy

Alfonso Venzo Institute of Science and Molecular Technologies, CNR, Padua, Italy

Roberta J. Ward Department of Cellular and Molecular Neuroscience, Faculty of Medicine, Imperial College London, London, UK; Biologie du Comportement, Université Catholique de Louvain, Louvain-la-Neuve, Belgium

Evan Williams School of Biomedical Sciences and Pharmacy MSB, University of Newcastle, Callaghan, NSW, Australia

Paolo Zatta Department of Biology, Padua "Metalloproteins" Unit, CNR-Institute for Biomedical Technologies, University of Padua, Padua, Italy

Brain iron metabolism and its perturbation in neurological diseases

Robert R. Crichton, David T. Dexter, and Roberta J. Ward

Abstract

Enormous advances have been made in the last decade in understanding iron metabolism and iron homeostasis at both the cellular and the systemic level. This includes the identification of genes and proteins involved in iron transport, such as the ferric reductase DcytB, the proton-coupled ferrous (divalent) iron transporter DMT1, the iron exporter ferroportin and the membrane-bound ferroxidase hephaestin. The modulation of their translation by the iron regulatory protein (IRP) system has also been identified together with the impressive signalling cascades involved in regulating the *chef d'orchestre* of systemic iron homeostasis, hepcidin. However, exactly how the brain regulates fluxes and storage of iron between neurons, oligodendrocytes, astrocytes and microglial cells remains an enigma. In this review we discuss the possible mechanisms which may be involved in the transfer of iron across the blood–brain barrier(BBB), together with the possible role played by astrocytes. The consequences of iron deficiency and iron excess on brain function are described. Finally, various neurodegenerative diseases, where accumulation of iron may be important in the pathogenesis, are presented as well as the possible use of iron chelators to diminish disease progression.

Keywords

Iron • Neurodegeneration • Parkinson's disease • Alzheimer's disease

Introduction

Iron (Fe) is a necessary cofactor in many metabolic processes in the central nervous system (CNS), including oxidative phosphorylation, myelin synthesis, neurotransmitter production, nitric oxide metabolism and oxygen transport. It plays an important role in electron transfer and is a cofactor for a large number of enzymes [1], including a number of key enzymes of neurotransmitter biosynthesis in brain, e.g. tyrosine hydroxylase (involved in the synthesis of catecholamines, including dopamine), tryptophan hydroxylase (involved in the synthesis of serotonin) and monoamine oxidase (involved in the metabolism of dopamine). It is essential that iron fluxes and storage within the brain are controlled within very exact limits in order to have adequate supplies for such metabolic processes but not 'too much', which could exacerbate damage via Fenton chemistry.

A brief overview of iron metabolism and homeostasis

Iron is transported throughout the circulation bound to the iron transport protein transferrin (Tf), which binds two atoms of Fe^{3+} as diferric transferrin $(Fe^{3+}{}_2-Tf)$. Such iron is delivered to cells via the transferrin-to-cell cycle (Fig. 1) and binds to its receptor, and the complex is invaginated into clathrin-coated pits, which fuse with the target membranes

R.R. Crichton (✉)
Institute of Condensed Material and Nanosciences,
Université Catholique de Louvain, Place Louis Pasteur 1,
1348 Louvain-la-Neuve, Belgium
e-mail: robert.crichton@uclouvain.be

W. Linert and H. Kozlowski (eds.), *Metal Ions in Neurological Systems*,
DOI 10.1007/978-3-7091-1001-0_1, © Springer-Verlag Wien 2012

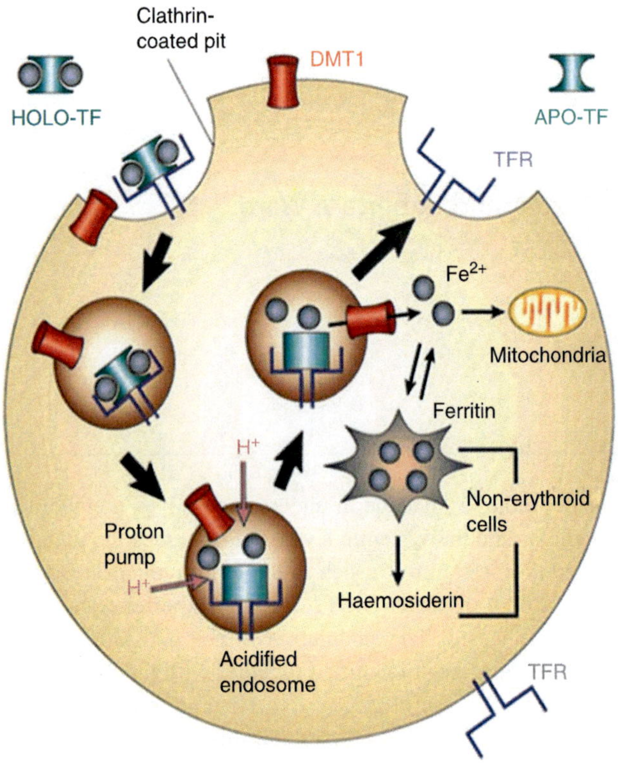

Fig. 1 The transferrin cycle. Holotransferrin (HOLO-TF) binds to transferrin receptors (TFR) at the cell surface. The complexes localize to clathrin-coated pits, which invaginate to initiate endocytosis. Specialized endosomes form, and are acidified by a proton pump. At the acidic pH, iron is released from transferrin and is co-transported with protons out of the endosomes by the divalent metal ion transporter DCT1. Apotransferrin (APO-TF) bound to TFR is returned to the cell membrane, where at neutral pH it dissociates to participate in further rounds of iron delivery. In non-erythroid cells, iron is stored as ferritin and haemosiderin [47]

of endosomes, delivering the vesicle contents into the interior of this organelle. The pH of the endosome is reduced to around 5–6 by the action of an adenosine triphosphate (ATP)-dependent proton pump. At this pH, iron is released from the transferrin–receptor complex as $Fe^{3+}Tf_2$, in a TfR-facilitated process (Fig. 1). It is proposed that TfR binding stimulates iron release from $\left(Fe^{3+}{}_2-Tf\right)$ at acidic pH by protonation of the bound carbonate, and by stabilization of the Tf molecule in the apo-Tf (iron-free) conformation. The divalent cation transporter DMT1 then ensures the transport of iron out of the endosome into the cytoplasm, presumably after reduction of Fe^{3+} to Fe^{2+}, by a member of the six transmembrane epithelial antigen of the prostate (STEAP) family of metalloreductases [2]. The cytoplasmic iron can then be transferred to the mitochondria for use in haem and iron–sulphur cluster synthesis, or stored in cytosolic ferritin. Unlike most other protein, a ligand taken up by receptor-mediated endocytosis, i.e. apotransferrin, retains high affinity for its receptor at acidic pH values, and is recycled back to the plasma membrane, where, at the slightly alkaline extracellular pH, it dissociates from its receptor and goes off into the circulation in search of further iron. This sequence of events constitutes the transferrin-to-cell cycle, which ensures iron uptake by cells that have transferrin receptors.

There is only one identified pathway for cellular iron export, from the basolateral membrane of duodenal enterocytes, from macrophages, hepatocytes and a number of other cell types. This involves ferroportin, which, together with hepcidin, plays a key role in systemic iron homeostasis. About a quarter of total body iron is stored in macrophages and hepatocytes in a readily mobilised form, mostly as ferritin for red blood cell formation (erythropoiesis). Mammalian ferritins are heteropolymers, made up of two subunit types, H and L. Whereas H-subunits have a ferroxidase activity, catalysing the oxidation of two Fe^{2+} atoms to Fe^{3+}, L subunits appear to be involved in the nucleation of the mineral iron core.

The regulation of cellular iron homeostasis is to a large degree controlled at the level of the translation of the messenger RNAs (mRNAs) of proteins involved in cellular iron metabolism. The key players in this post-transcriptional regulation are two iron regulatory proteins (IRP1 and IRP2), which function as cytosolic iron sensors. In conditions of iron deficiency, IRPs bind with high affinity ($K_D \approx 20$–100 pM) to stem loops, known as iron regulatory elements (IREs), in mRNAs encoding the regulated proteins (Fig. 2). When the IREs are in the 5′-untranslated region (UTR) of the mRNA, as is the case for ferritin and ferroportin,[1] binding to IRPs prevents initiation of translation. In contrast, in the case of the transferrin receptor and DMT1, where the IREs are in the 3′-UTR, binding of the IRPs to the mRNAs protects them against degradation by nucleases. This results in increased iron uptake and blockage of iron storage and export. When iron is abundant, the IRPs are no longer active in binding, allowing ferritin and ferroportin mRNAs to be translated and resulting in the down-regulation of transferrin receptor and DMT1 synthesis as a result of the nuclease-catalysed degradation of their mRNAs. Under these conditions, IRP1 acquires aconitase activity, associated with the incorporation of a 4Fe–4S cluster, whereas IRP2 is condemned, after ubiquitination, to degradation in the proteasome.

We now have an increasingly detailed understanding of how systemic iron homeostasis is regulated (Fig. 3). The first index of iron loading, increased transferrin saturation, leads to increased levels of $\left(Fe^{3+}{}_2-Tf\right)$, which is detected by the liver via a complex pathway involving HFE, TFR2 and haemojuvelin (HJV). Hepatocytes respond to this signal by

[1] Other proteins with functional IREs in their 5′-UTR include hypoxia inducible factor 2a [99], erythroid aminolaevulinate synthase and mitochondrial aconitase (reviewed in [100]), and more recently amyloid precursor protein [101].

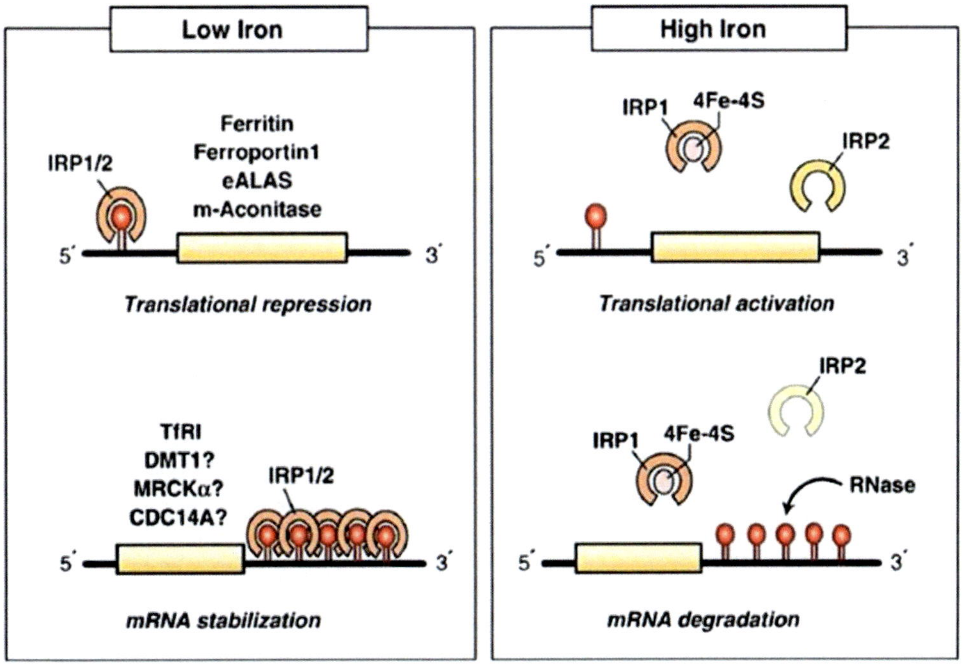

Fig. 2 Outline of translational regulation of mRNAs of a number of proteins involved in iron metabolism in low and high iron. IRPs bind to IREs located in either the 5′- or 3′-UTRs of specific mRNAs. During low-iron conditions, IRP1 and IRP2 bind with high affinity to 5′-IREs and to the five 3′-IREs in Tf R mRNA, resulting in the translational repression of 5′-IRE-containing mRNAs and the stabilization of the Tf R mRNA. During high-iron conditions, IRPs lose their affinity for IREs, increasing translation of 5′-IRE-containing mRNAs and mediating degradation of the Tf R mRNA. Increased iron levels result in the conversion of the IRP1 RNA binding form into the [4Fe–4S] cluster c-acon form, while increased iron and/or haem levels mediate IRP2 proteasomal degradation [1]

increased expression of the *HAMP* gene, resulting in increased secretion of the regulatory peptide hepcidin. Circulating hepcidin blocks dietary iron uptake by duodenal enterocytes and iron recycling from macrophages, in both cases through internalization of ferroportin, which therefore blocks iron export. The outcome of this is to decrease serum iron levels, leading logically to the feedback response of down-regulating hepcidin synthesis and secretion. This once again allows ferroportin to be displayed on the surfaces of enterocytes and macrophages, allowing them once again to export iron into the circulation.

The blood–brain barrier (BBB)

The brain is unique among all the organs of the body, being hidden behind a relatively poorly permeable vascular barrier, which limits its access to plasma nutrients, such as iron. It is generally accepted that iron transport into the brain mostly involves the transferrin-to-cell cycle, by the use of transferrin receptors within epithelial cells lining the blood–brain barrier (BBB), although the precise mechanism of iron transfer is still uncertain. The BBB proper (Fig. 4) is essentially composed of cerebral capillary endothelial cells, joined by tight junctions, a basal lamina, pericytes and astrocyte endfoot processes.

The types of cells found at the BBB and their associations are illustrated in Fig. 4. The endothelial cells form tight junctions, thereby sealing the paracellular pathway between the cells, such that substances which enter the brain must use dedicated endothelial cell transport systems, either passive or active transporters, which include the ATP binding cassette (ABC) transporters, as well as by transcytosis. Unlike other blood vessel epithelia, the BBB epithelia express different receptors at the luminal membrane (facing the circulation) compared with the abluminal membrane, which is surrounded by astrocyte endfeet, neuronal processes and interstitial fluid. Pericytes are distributed along the length of the cerebral capillaries, partially surrounding the endothelium. Both the cerebral endothelial cells and the pericytes are enclosed by the local basement membrane, forming a distinct perivascular extracellular matrix (basal lamina 1, BL1), different from the extracellular matrix of the glial endfeet bounding the brain parenchyma (BL2). Foot processes from astrocytes form a complex network surrounding the capillaries. Microglia, the resident immunocompetent cells of the brain, are also found in the vicinity of the BBB. Since no brain cell is further than about 25 μm from a capillary, once the BBB is crossed, diffusion distances for solutes to neurons, astrocytes and glial cells are short.

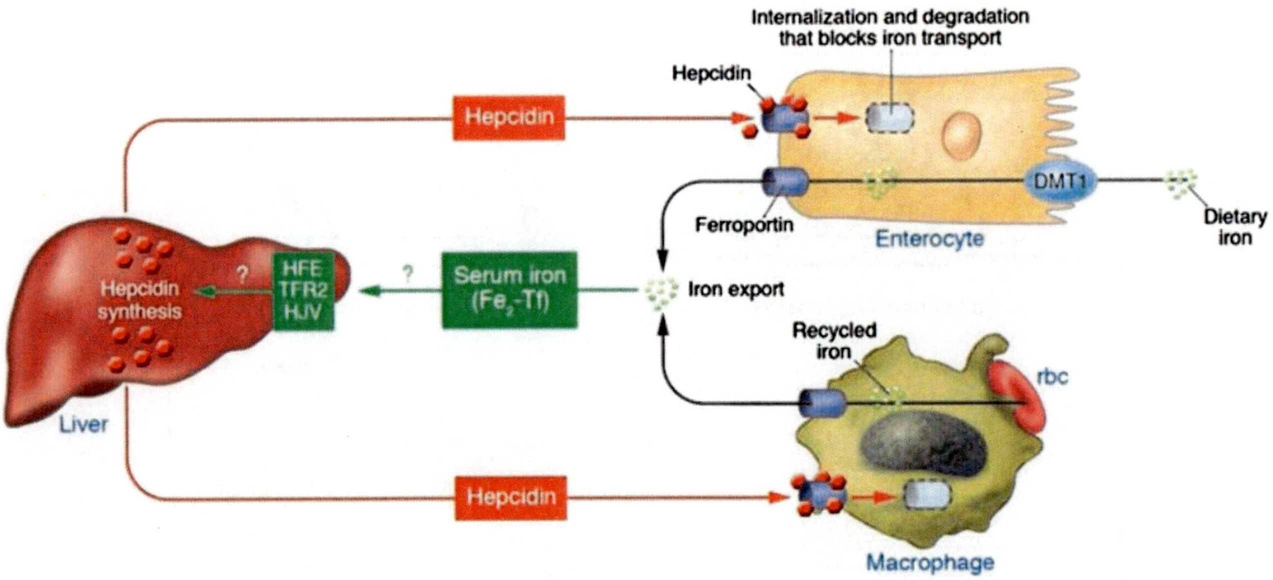

Fig. 3 Regulation of systemic iron homeostasis. Increased diferric transferrin, Fe_2–Tf, is detected by the liver via an as-yet unknown complex regulatory pathway involving HFE, TFR2 and HJV. Hepatocytes respond to this signal by inducing HAMP expression and hepcidin secretion. Circulating hepcidin acts in turn to diminish dietary iron absorption by enterocytes and iron recycling by macrophages through internalization of ferroportin, which blocks iron export. As a consequence, serum iron decreases. As a feedback response, hepcidin synthesis is down-regulated, which allows ferroportin molecules to be displayed on the surface of the target cells [1]

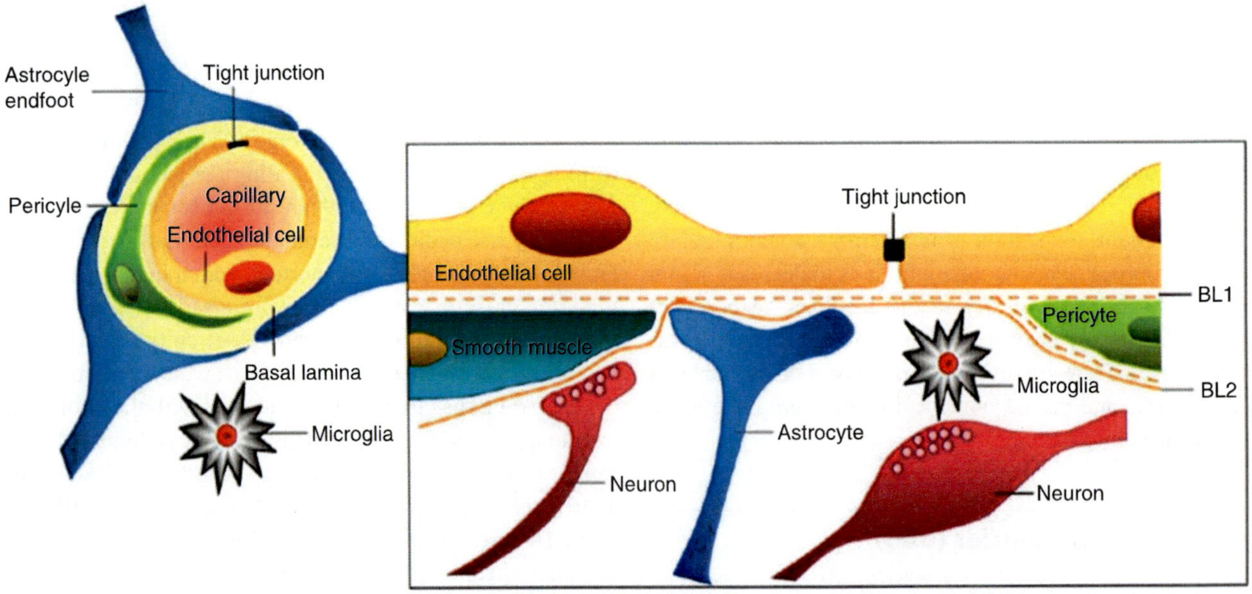

Fig. 4 Cell associations at the BBB [98]

Mechanisms for iron transport into and within the brain

Endothelial cells make up the blood–brain barrier(BBB) and express transferrin receptors (transferrin receptor 1) on the luminal side of the capillaries. Although iron deficiency does not increase its expression [3], it is suggested that, in the latter situation, perhaps the cycling rate of endosomes containing the diferric transferrin receptor is increased.

The first step in the transport of iron from the luminal to the abluminal side of the blood capillary endothelial cells (BCECs) involves diferric transferrin binding to transferrin receptors expressed at the luminal membrane of the BCECs, followed by receptor-mediated endocytosis, with iron released from transferrin within the endosome by the slightly

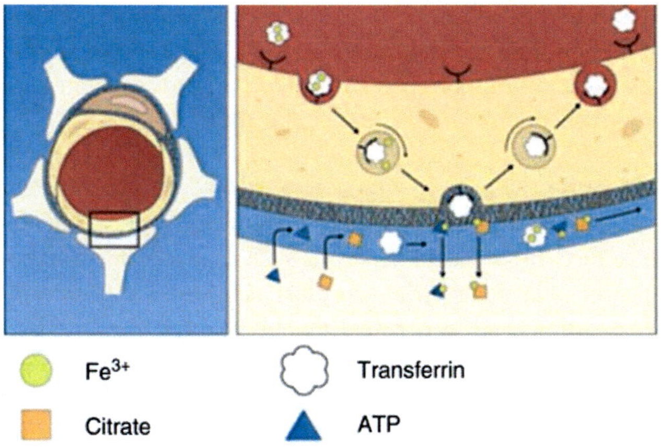

Fig. 5 Cross-section of a brain capillary demonstrating the close interaction between brain capillary endothelial cells (BCECs) and astrocytic endfeet (*left*). The area marked with a *rectangle* is shown on the *right* to demonstrate the possible interactions between BCECs and astrocytes which facilitate iron transport into the brain [11]

acidic pH. Iron could then be released from the endosomes, with recycling of apotransferrin to the luminal side of the endothelial cells and its ultimate release into the plasma [4, 5]. However, there is some controversy as to whether DMT1, which is required to transport Fe(II) out of the endosome, is present or not in BCECs (blood capillary endothelial cells) [6–8]. Therefore, it has been suggested that BCECs mediate iron transport into the brain by segregating iron from transferrin without the involvement of DMT1 [9]. On account of its ubiquitous expression pattern, transient receptor potential mucolipin 1 might also play an important role in the endolysosomal iron release in the BVEC [10]. It has been suggested that the interactions between BCECs and astrocytes might facilitate iron transport into the brain (Fig. 5) [11]. Subsequent to the binding of the iron-transferrin at the luminal surface of the BCEC, the $Fe^{3+}{}_2$–Tf–transferrin receptor complex would be internalized in an endosome, which is then possibly transported towards the abluminal side of the BCEC. However, the release of iron from such transferrin receptor complexes at the abluminal side remains very hypothetical. Iron bound to its receptor would then be exposed to the local microenvironment, e.g. hydrogen ions, ATP, other nucleotides and citrate [12, 13], which would bind the iron and lead to its release from transferrin. However, this would appear to be a very non-specific process. The apotransferrin would remain bound to its receptor, which has a high affinity for apotransferrin at acidic pH, and would be recycled back to the luminal cell surface, where it would be released from the transferrin receptor and returned to the circulation [11]. In addition, it is suggested that there may be a Tf-independent mechanism, since hypotransferrinaemic mice have normal amounts of iron in brain, as well as a mechanism for the entry of non-transferrin-bound iron. Since ferroportin is the only known iron exporter, iron may be exported into the brain interstitial fluid via ferroportin. This iron protein has been shown by immunohistochemistry to be present in the BBB, although its exact membrane localization is unclear [14]. Complexes of iron bound to citrate or ATP could then circulate in brain extracellular fluid to be taken up by other cell types or to bind to transferrin. Transferrin synthesised by the oligodendrocytes in the brain will bind the majority of iron that traverses the blood–brain barrier(BBB) after the oxidation of the iron, possibly by a glycophosphoinositide-linked caeruloplasmin found in astrocytic foot processes that surround brain endothelial cells. Since astrocytes have intimate contact with the abluminal side of the BBB, it is considered that they may play an important role as iron importers into the brain. Receptors for H ferritin receptors have been identified on the brain micro-vasculature, which might imply a role for this iron protein in iron transport. This could be an alternative pathway to that of transferrin [15].

Neuronal iron homeostasis

A model for neuronal uptake and export of iron is presented in Fig. 6. An astrocytic endfoot forming intimate contact with the neuron is also shown and as such could play a critical role as a gate-keeper in regulating brain iron absorption and metabolism at the junction of the BBB. Subsequent to binding of iron-transferrin to the transferrin receptor at the cell surface, iron is transported into the neuron bound to transferrin. The resulting endosome contains DMT1 which facilitates iron transport across the endosomal membrane into the cytosol. Ferroportin is present in neurons and could therefore represent the mechanism by which iron is exported from such cells [8]. It has been suggested that iron can also enter the neuron (the interstitium contains ascorbic acid which can maintain iron in the ferrous iron) as a low-molecular-weight form bound to citrate or ATP possibly via non-vesicular import mechanisms or voltage-gated Ca^{2+} channels. Transferrin receptors are expressed on central neurons, as well as DMT1 and TRPML-1. Since both of these latter proteins transport Fe^{2+}, a ferric reductase must exist within the endosomes and lysosomes of neurons. Under normal circumstances little iron is stored in the neurons, indicating that iron is taken up for rapid utilisation or secreted via ferroportin.

Glial cell iron homeostasis

Astrocytes

As already stated, astrocytes may play an important role in regulating iron absorption and metabolism at the BBB. It is

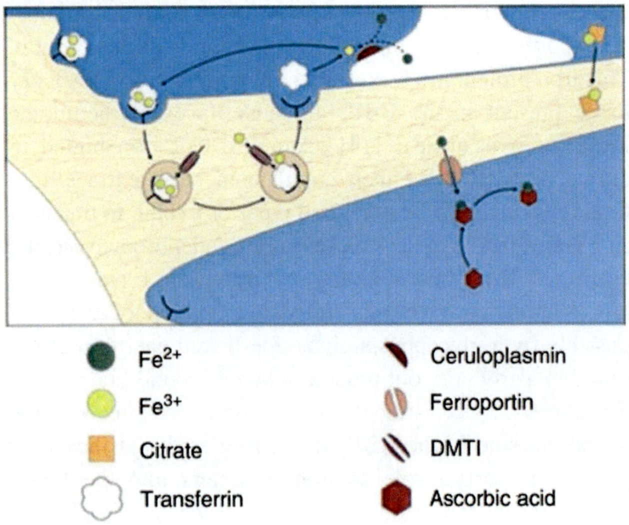

Fig. 6 A model for neuronal uptake and export of iron. An astrocytic endfoot forming intimate contact with the neuron is also shown. Subsequent to binding of iron-transferrin to the transferrin receptor at the cell surface, iron is transported into the neuron bound to transferrin. The resulting endosome contains divalent metal transporter 1 that facilitates iron transport across the endosomal membrane into the cytosol, while pumping protons into the endosome. Astrocytes contain caeruloplasmin that exhibits ferroxidase activity, which is capable of oxidising ferrous iron to ferric iron. The ferric iron can enter the neuron in a low-molecular-weight form such as iron bound to citrate or ATP. The neuron expresses the iron exporter ferroportin that transports ferrous iron out of the cell. The interstitium contains ascorbic acid that can bind and thereby neutralize the toxicity of ferrous iron [11]

unclear whether they express transferrin receptors or transferrin [16]. However, significant amounts of non-transferrin-bound iron may be taken up by these cells. In cultured astrocytes, Lane et al. [17] demonstrated that at least half of the accumulated iron is initially reduced by effluxed ascorbate and then imported via DMT1 in ascorbate-replete astrocytes, thereby confirming their role as an important contributor to iron homeostasis. Astrocytes contain caeruloplasmin (Cp) attached to their endfeet membranes by a glycophosphoinositide (GPI) linkage, which exhibits ferroxidase activity, capable of oxidising ferrous iron to ferric iron and thereby facilitating its binding to transferrin in the brain interstitial fluid. This GPI-Cp co-localizes on the astrocyte cell surface with the divalent metal transporter ferroportin and is physically associated with ferroportin. Ferroportin alone is unable to efflux iron from astrocytes in the absence of GPI-Cp or Cp [18]. The ferroxidase activity is required for the stability of cell surface ferroportin in cells expressing GPI-caeruloplasmin [19].

Oligodendrocytes

Since there is a high demand of iron for myelination, most of the histologically detectable iron in the brain is present in oligodendrocytes. Iron may be taken up into oligodendrocytes via the ferritin receptor Tim-2s as well as in its non-transferrin-bound form via DMT1 or other non-vesicular iron import mechanisms [16]. These cells are the predominant producers of transferrin [20].

Microglia

The control of iron homeostasis within microglia remains undefined–other phagocytic cells such as macrophages take up iron via transferrin receptors and release iron via ferroportin [21]. It is of interest that both microglia and iron deposits accumulate at the site of damage in many neurodegenerative diseases such as Alzheimer's and Parkinson's. Whether these accumulations are a cause or effect of the disease is currently unknown. In preliminary studies we have shown that the expression of the mRNAs of transferrin receptor 1 and ferroportin were significantly down-regulated in response to lipopolysaccharide (LPS) treatment in immortalised microglial cells (N9), while that of divalent metal ion transporter DMT1 showed no change in expression under control conditions and after LPS treatment [21]. However, the results of real-time polymerase chain reaction (RT-PCR) studies indicated that hepcidin was not produced by microglia either under control conditions or in response to LPS treatment after 24 h. This might suggest that iron homeostasis is under the control of alternative mechanisms in microglia, for instance cytokines such as interleukin (IL)-1 or IL-6. Of course, the lack of hepcidin production does not preclude the action of hepcidin from other sources, both in the brain and elsewhere. In addition, other studies that have detected hepcidin production in response to LPS treatment noted that hepcidin induction was transient and was often undetectable at 24–36 h post LPS treatment. Therefore, it cannot be ruled out that hepcidin was produced but was undetectable at 24 h post LPS exposure. Whether iron accumulates in activated microglia (which could explain the association between increased iron stores and microglia activation in specific brain regions of Parkinson's and Alzheimer's patients) remains unknown.

Importance of iron in the developing foetus

In pioneering studies, it was shown that iron deficiency could affect infant behaviour and development [22, 23]. Such a role for iron in neurocognitive and neurobehavioural development is highlighted by our growing understanding of the biology of developmental iron deficiency [24]. Iron requirements are expected to exceed iron intake during the first 6–18 months of postnatal life at a time of rapid neural development, during which morphological, biochemical and bioenergetic alterations may influence the way in which the brain functions in later life [25, 26]. Iron deficiency either in utero or in early postnatal life can result in abnormal cerebral development, because iron is essential for proper neurogenesis and differentiation of certain brain cells and brain regions: recent studies of iron-deficient rodents clearly identify altered morphology in the hippocampus and striatum [27–29]. This includes decreased arborization of dendrites and alteration in the location and functioning of oligodendrocytes, the cells responsible for myelin formation. Iron deficiency results in persistent alterations in both the composition and amounts of myelin in white matter, which do not return to normal levels later in life [30, 31]. Iron deficiency also affects dopamine and noradrenaline metabolism [32, 33].

In various epidemiological studies it is reported that children with iron-deficiency anaemia have poorer performance on tests of some specific cognitive functions. Animal experiments have identified some of the defects of reduced iron availability on brain function. These include post-translational changes (which result in a failure of iron incorporation into protein structures, which are subsequently degraded), vulnerability of the developing hippocampus (with loss of the neuronal metabolic marker cytochrome *c* oxidase) and altered dendritic structure. Iron deficiency will also have a direct effect on myelin, including a decrease in myelin lipids and proteins, as well as neurotransmitter systems, since iron is essential for a number of enzymes including tryptophan hydroxylase (serotonin) and tyrosine hydroxylase (norepinephrine and dopamine). Long-term follow-up studies of iron deficiency in the human infant brain indicate that such alterations in myelination result in slower conduction in both the auditory and visual systems. Both of these sensory systems are rapidly myelinating during the period of iron deficiency and are critical for learning and social interaction. Together with the reduced energy, impaired glial function and altered activation of monoamine circuits, this may alter experience-dependent processes, which are critical to brain structure and function during early development.

It is well known that there is specificity and selectivity for brain iron acquisition, which is emphasised by the fact that certain brain regions respond to iron deficiency and iron depletion more aggressively than other regions. For example, the densities of TfRs in different regions of human brain vary widely [34] and do not seem to correlate with the published iron content of these brain regions [35]. Of particular interest is the fact that substantia nigra (SN) iron content is not high in young rats, but that during iron repletion a dramatic increase of iron is observed in this brain region, thereby indicating that iron is targeted to this specific brain region [36]. High rates of iron accumulation occur in different brain regions at different ages of postnatal life, indicating that shortage at these important stages of development might be of significance in later life. The iron requirements by the brain are much greater than the iron uptake into the tissue, indicating that most brain iron is derived from recycling behind the blood–brain barrier(BBB) comparable to what occurs in the periphery. This would indicate that, following birth, iron recycling rather than uptake from the circulation may be the major source of iron. It is still under debate as to whether impaired iron homeostasis may be a primary cause of many neurological diseases. As yet, no major defects have been identified. One of the major points of debates concerns the amount of iron that might be transported from the systemic circulation into the brain. Earlier studies indicated that people with high circulating iron, saturated transferrin and high levels of low-molecular-weight iron, e.g. genetic haemochromatosis and thalassaemic patients, did not show increased brain iron. For example, in a study of Italian population, the most common HFE mutations, H63D and C282Y, were not associated with the individual risk to develop Parkinson's disease (PD), although no indications of their iron loading were given [37]. However, in another smaller study of only 66 people, the results indicated that these genetic variants in iron metabolism genes can influence brain iron levels in men [38]. In one further study, the H63D polymorphism was overrepresented in individuals with sporadic amyotrophic lateral sclerosis (ALS) [odds ratio 1.85, confidence interval (CI) 1.35–2.54] [39], but there was no association with cognitive decline [40]. Furthermore, a large, well-powered meta-analysis of eight studies comprising in total 758 cases and 626 controls failed to find a significant association between Alzheimer's disease and any haemochromatosis HFE genotype [41]. Such studies now need to be extended to include patients with high iron loading, e.g. thalassaemia, such that correlations can be made between systemic iron overload, brain iron content, cognitive impairment and neurodegeneration.

Ageing and metal-based neurodegeneration

It is becoming apparent that the function of the BBB is affected by a number of patho-physiological factors such as age and inflammation. Immune cells are able to penetrate the BBB, either at the endothelial blood–brain barrier(BBB) or the epithelial blood CFS barrier. Inflammation as well as reactive oxygen species (ROS) and reactive nitrogen species (RNS) can acutely disrupt BBB at tight junctions. 4-Hydroxynonenal (HNE), a second messenger of free radicals, is present in the BBB under pathological conditions and could make the endothelial part of the BBB permeable. As yet, little is known about the expression pattern of transferrin receptors in the capillarics of aging brains. Modulation of P-glycoproteins at the BBB may be an important factor to improve drug delivery, as well as to augment CNS protection from challengers from the peripheral system.

Multiple structural and functional changes occur in the ageing brain, often accompanied by changes in the immune function, as well as an elevation of brain iron in specific brain regions, e.g. in the putamen, motor cortex, prefrontal cortex, sensory cortex and thalamus, which is possibly localized within H- and L-ferritin and neuromelanin. Recent studies have identified various adverse effects of such iron accumulations on cognitive decline; for example, magnetic resonance imaging (MRI) quantified iron deposits in the basal ganglia of 143 non-demented subjects at mean ages of 11, 70 and 72 years and showed that higher iron deposits were associated with lower cognitive ability [42], higher iron burdens in the putamen and caudate nucleus were associated with lower scores on dementia rating scales and longer reaction times [43], while assessment of genomic integrity in a small number of brain samples from elderly individuals identified a higher number of single-strand and double-strand DNA breaks as the concentration of iron and copper increased in the hippocampus and frontal cortex [44]. In contrast, when brain iron deposition was reduced in the globus pallidus, substantia nigra, red nucleus and temporal cortex in old rhesus monkeys receiving a calorie-restricted diet [45], there were reductions in both age-related inflammation and oxidative damage. Recent studies have suggested that indices of iron homeostasis in the blood, e.g. serum iron and transferrin saturation, as well as hepatic iron content may be correlated with iron content of various brain regions, when the latter was assessed by MRI proton traverse relaxation time, R2 [46].

Oxidaive stress in Alzheimer's disease and Parkinson's disease

Ill-placed excessive amounts of iron, either in specific brain cellular constituents such as mitochondria or in specific brain regions, could lead to the generation of toxic free radicals leading to neurodegenerative diseases. Over the last decade, it has become widely accepted that inflammation, associated with dysfunction of metal ion homeostasis (Fe, Cu, Zn) accompanied by concomitant oxidative stress, is a key factor in a large number of neurodegenerative diseases such as Parkinson's disease, Alzheimer's disease, Huntington's disease, multiple sclerosis, Friedreich's ataxia and others [47]. Support comes from the observation that Alzheimer's disease (AD), Parkinson's disease (PD) and many other neurodegenerative diseases are characterised by increased levels of iron in specific regions of the brain. Furthermore, iron increases appear to predominate in motor areas of the brain, such as the basal ganglia, which may explain the motor deterioration observed in many neurodegenerative diseases [16].

The 'metal-based neurodegeneration' hypothesis can be described by the following postulates:

1. Redox-active metal ions (Fe, Cu), present within specific brain regions, can generate oxidative stress by production of ROS;
2. ROS then cause peroxidation of polyunsaturated fatty acids in membrane phospholipids;
3. This in turn leads to the formation of reactive aldehydes (Fig. 7);
4. The reactive aldehydes, together with other oxidative processes, react with proteins to generate carbonyl functions, which damage proteins (Fig. 8);
5. Damaged, misfolded proteins aggregate, overwhelming the cytosolic ubiquitin/proteasome protein degradation system, and accumulate within intracellular inclusion bodies;
6. Such intracellular inclusion bodies are found in a great many neurodegenerative diseases (Alzheimer's, Parkinson's, ALS, Huntington's etc.).

Although both ROS and RNS are involved in physiologically relevant signalling pathways, there is considerable evidence that, in situations of oxidative stress, they are associated with a number of neurodegenerative pathologies. Oxidative stress refers to a situation where elevated levels of ROS are observed, and can result from a variety of conditions that represent either increased ROS production or a decreased level of antioxidant defence. In the case of stimulation of ROS production by macrophages during the innate immune response to bacterial infection, the ROS so generated act in a protective manner. However, dysregulation of ROS levels in a variety of tissues, notably in the brain, has been linked to a growing number of inflammatory and age-associated diseases. During oxidative stress, the oxidation of cellular components results in the modification of DNA, proteins, lipids and carbohydrates, and the resulting oxidative damage is frequently associated with cell death either by necrosis or by apoptosis [48].

When ROS are generated by redox trace metals in the proximity of membrane phospholipids, they initiate the peroxidation of polyunsaturated acyl chains of phospholipids or *n*-6 polyunsaturated fatty acids (PUFA) (Fig. 7). The lipid hydroperoxides are highly susceptible to breakdown through non-enzymatic Hock cleavage, forming a variety of lipid-derived α,β-unsaturated 4-hydroxyaldehydes of which the most prominent is 4-hydroxynonenal (HNE) [49]. 4-hydroxynonenal (HNE) is the main aldehyde formed during lipid peroxidation of *n*-6 polyunsaturated fatty acids, such as linoleic acid C18:2 *n*-6 and arachidonic acid C20:4 *n*-6, whereas peroxidation of *n*-3 polyunsaturated fatty acids such as α-linolenic acid C18:3 *n*-3 and docosahexaenoic acid C22:6 *n*-3 generates a closely related compound, 4-hydroxy-2-hexenal (HHE) [50]. The mechanisms by which these 4-hydroxyalkenals might be formed from membrane phospholipid PUFAs have been recently reviewed [51].

Fig. 7 Schematic diagram of reactive hydroxy-alkenals generated during lipid peroxidation of *n*-3 and *n*-6 polyunsaturated fatty acids [50]

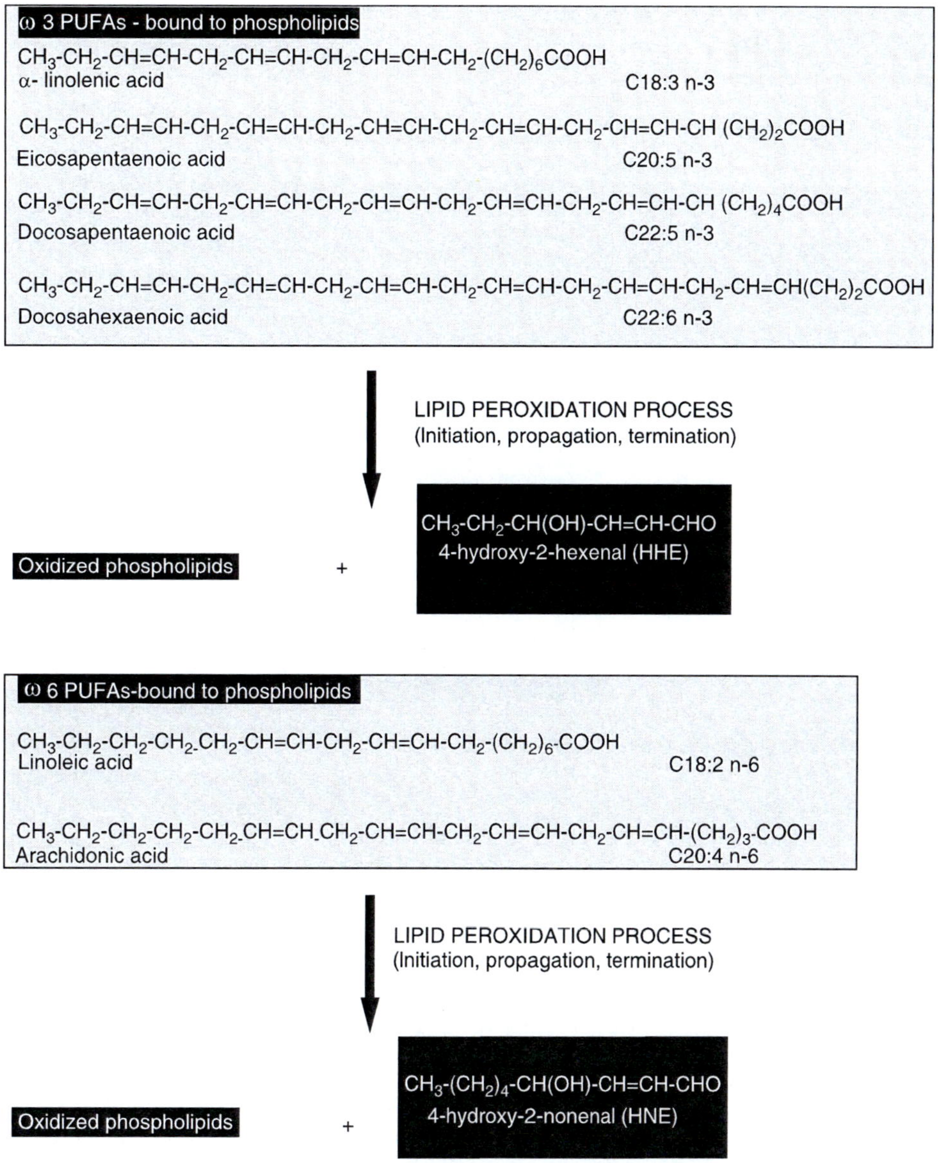

HNE was initially recognised as the product of lipid peroxidation with the greatest toxicological potential; it is subsequently considered to be one of the most reliable markers of oxidative stress: It can also trigger signalling events in a physiological context as well as acting as a growth modulating factor [52]. Immunohistochemical studies show the presence of HNE in neurofibrillary tangles and senile plaques in Alzheimer's disease (AD), in the cytoplasm of the residual motor neurons in sporadic amyotrophic lateral sclerosis (ALS), in Lewy bodies in neocortical and brain stem neurons in Parkinson's disease (PD) and in diffuse Lewy bodies disease (DLBD) [53]. 4-HNE is relatively stable in vivo and has been proposed to be one of the key mediators of the damage resulting from exposure to reactive oxygen and nitrogen species.

During oxidative stress, numerous post-translational modifications of proteins have been characterised resulting either from direct oxidation of amino acid residues by highly reactive oxygen species (that are formed during normal metabolism), or through the conversion of lipid and carbohydrate derivatives to compounds that react with functional groups on proteins [54]. A significant portion of these ROS-induced post-translational modifications result in the formation of reactive protein carbonyl derivatives, generically termed 'protein carbonylation'. The level of carbonyl groups in proteins is widely used as a marker of oxidative protein damage [55, 56].

Direct oxidation of certain amino acid side-chains in proteins (proline, arginine, lysine and threonine) or oxidative cleavage of the protein backbone [55] can lead to the formation of protein carbonyl derivatives (Fig. 8). Methionine and cysteine can be directly oxidised. Unlike other types of modification (except cysteine oxidation), oxidation of methionine residues to methionine sulphoxide is reversible;

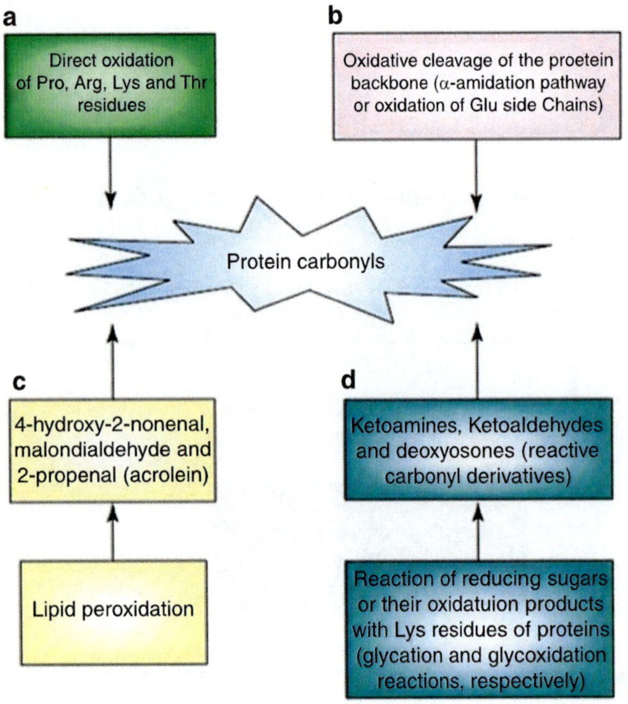

Fig. 8 Production of protein carbonyls (aldehydes and ketones) [47]

thus, cyclic oxidation and reduction of methionine residues leads to consumption of ROS and thereby increases the resistance of proteins to oxidation [57]. Carbonyl groups can also be introduced into proteins by addition of reactive carbonyl compounds (ketoamines, ketoaldehydes and deoxyosones) produced by a complex series of reactions between reducing sugars or their oxidation products with the amino groups of lysine residues in proteins, by mechanisms known as glycation and glyoxidation [57]. Because of their electron-withdrawing functional groups, the double bond of 4-HNE and other α,β-unsaturated aldehydes serves as a site for Michael addition with the sulphur atom of cysteine, the imidizole nitrogen of histidine and, to a lesser extent, the amine nitrogen of lysine [58]. After forming Michael adducts, the aldehyde moiety may in some cases undergo Schiff base formation with amines of adjacent lysines, producing intra- and/or intermolecular cross-linking [59, 60]. Recent studies have suggested that such protein carbonylation from lipid-derived aldehydes is more prevalent than that formed via direct amino acid side-chain oxidation [58].

Peroxynitrite is able to oxidise methionine residues and to nitrate tyrosine residues in proteins; however, this depends on the availability of CO_2, which stimulates the nitration of tyrosine residues but inhibits the oxidation of methionine residues [55]. Nitration of tyrosine residues may contribute significantly to peroxynitrite toxicity, since nitration will prevent the phosphorylation or nucleotidylation of key tyrosine residues in enzymes which are regulated by phosphorylation/adenylation, thereby seriously compromising one of the most important mechanisms of cellular regulation and signal transduction.

ROS can also readily attack DNA, generating a variety of DNA lesions, such as oxidised bases, abasic sites and single- and double-strand breaks. If not properly removed, DNA damage can be potentially dangerous, leading to mutagenesis and/or cell death, especially in the case of lesions that block the progression of DNA/RNA polymerases.

Iron involvement in Alzheimer's disease and Parkinson's disease

There is an increasing body of evidence that elevated iron levels are found in particular brain regions in specific-neurological diseases including Alzheimer's disease (AD), Parkinson's disease (PD), Huntington's disease (HD) and Friedreich's ataxia (FA). We summarize some of the findings below.

As life expectancy in the developed world increases, there is a significant increase in the development of mild cognitive impairment (dementia) which can progress to Alzheimer's disease (AD). Over 24 million people worldwide suffer from some form of dementia, and estimates are that, by 2040, 80 million people will be demented, with Alzheimer's disease accounting for some 60% of all dementia. AD is the most common cause of age-related neurodegeneration, affecting memory and behaviour centres of the brain. This progressive loss of cognitive and behavioural functions is associated with the temporal and frontal lobes of the brain [61]. The classical patho-physiological hallmarks are the presence of toxic insoluble aggregates of amyloid-β peptide (Aβ) in extracellular senile plaques and of neurofibrillary tangles (NFT) created by the hyperphosphorylation and subsequent aggregation of the microtubule-associated protein, tau, associated with the loss of cortical neurones [62]. Changes in the levels of iron, ferritin and transferrin have been reported in areas of the brain associated with centres of memory and thought processing which are lost in the clinical development of AD, notably the hippocampus and the cerebral cortex [63, 64], with iron accumulating more rapidly than ferritin in areas of particular neurodegeneration [65]. Another important pathological finding in AD is that the iron accumulation occurs in the same brain regions characterised by Aβ deposition [36]. It has been suggested therefore that abnormal deposition of iron in AD plaques associated with β-amyloid can mediate free-radical-related neurotoxicity [66].

Aβ is derived by the proteolytic cleavage of the amyloid precursor protein (APP), a type 1 transmembrane glycoprotein. APP is cleaved by three types of proteases, the α-, β- and γ-secretases. The majority of APP is processed in the non-

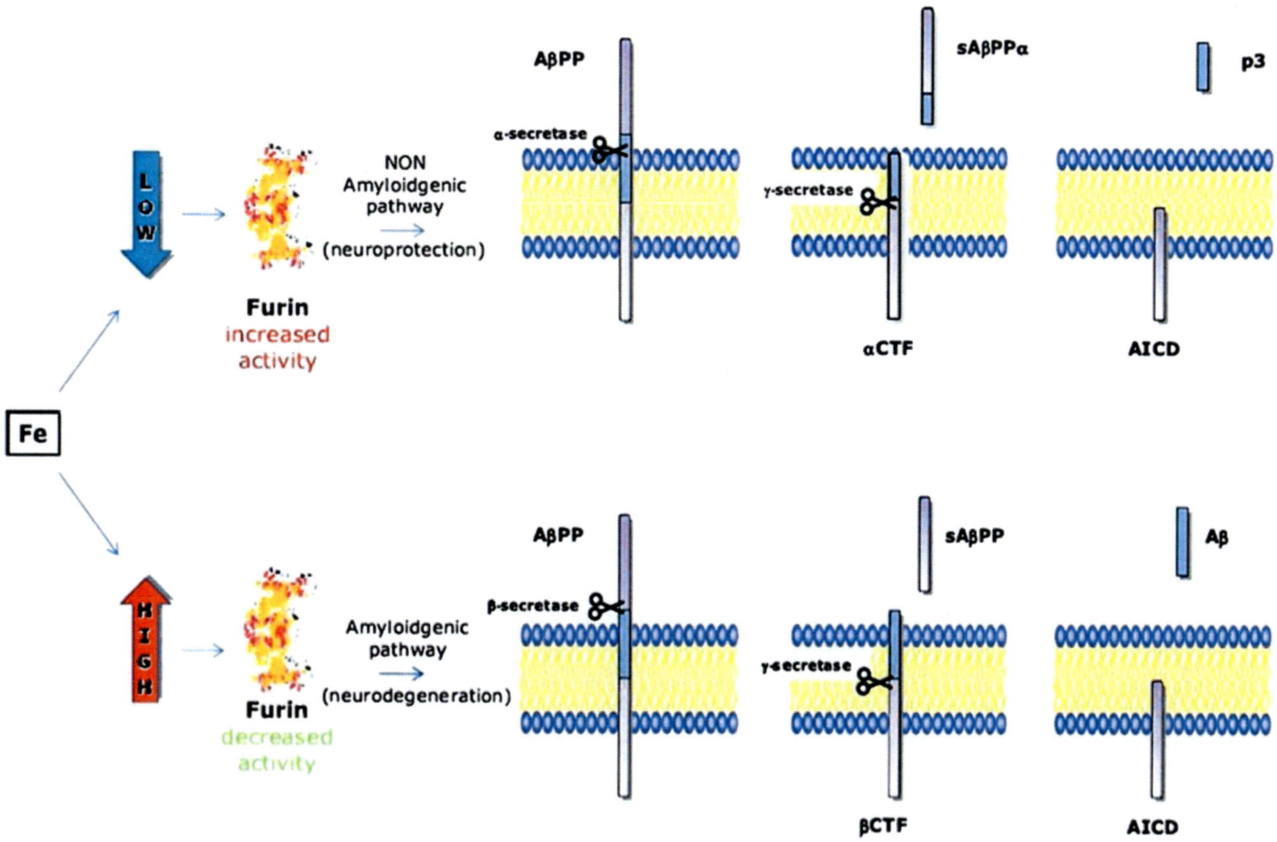

Fig. 9 Furin activity and the fate of AψPP cleavage by ψ- and ψ?-secretases. Low cellular iron levels are thought to increase furin activity, stimulating the non-amyloidogenic pathway. In contrast, high cellular iron levels decrease furin activity and may activate the amyloidogenic pathway [93]

amyloidogenic pathway (Fig. 9); APP is first cleaved by α-secretase within the amyloid-β protein (Aβ) domain, leading to release of the neuroprotective extracellular soluble sAβPPsα fragment, and precluding Aβ generation. The membrane-anchored α-carboxy terminal fragment (αCTF) is then cleaved by γ-secretase within the membrane, releasing the p3 peptide and the APP intracellular domain (AICD). Alternatively, amyloidogenesis takes place when APP is first cleaved by β-secretase, producing AβPP. Aβ and AICD are generated upon cleavage by γ-secretase of the βCTF fragment retained in the membrane. Therefore, stimulation of the α-secretase pathway attenuates Aβ accumulation in the brain and amyloid formation [67]. The processing of both α- and β-secretases is modulated by furin, a member of the subtilisin-like proprotein convertase family which catalyses the cleavage of precursor proteins into their biologically active forms [68]. Furin is also involved in modulation of systemic iron homeostasis through the production of soluble haemojuvelin (HJV) [69], an antagonist of bone morphogenic protein (BMP)-mediated activation of hepcidin [70], an important regulator of iron homeostasis. Furin transcription is modulated by cellular iron levels and by hypoxia [69, 71]. Excess iron decreases furin protein levels and therefore impairs the production of soluble HJV. In contrast, iron deficiency or hypoxia up-regulates furin activity, thereby increasing the production of soluble HJV, and blocking hepcidin activation [69]. This has led to the hypothesis, illustrated in Fig. 9, that iron regulation of furin may play a role in AD [72]. Increased levels of iron in the brain could down-regulate furin protein levels, thereby impairing the ability of α-secretase to generate the neuroprotective sAβPPsα fragment, thereby activating the amyloidogenic pathway, leading to Aβ production and ultimately neurodegeneration. In addition, the iron-dependent production of ROS could shift the IRP1 to its IRE-binding form [73], thereby increasing cellular iron uptake via the transferrin receptor, creating a vicious circle which would progressively increase the intracellular iron content, further down-regulating furin and shifting the secretase equilibrium in favour of Aβ production. Furin mRNA levels in the brains of AD patients and the AD animal model, Tg2576 mice, were significantly lower than those in controls. Moreover, injection of furin-adenovirus into Tg2576 mouse brains markedly increased α-secretase activity and reduced β-amyloid protein (Aβ) production in infected brain regions [74]. Further support for the connection between iron metabolism and AD comes from the identification of a functional IRE in the 5′-

UTR of the amyloid precursor protein mRNA. As is the case of ferritin, APP levels increase in the presence of iron and decrease upon addition of an iron chelator in neuroblastoma cells [75]. Increased APP formation in parallel with inhibition of α-secretase activity would favour Aβ deposition. Recently it has been shown that APP possesses ferroxidase activity, oxidising Fe^{2+} incorporating Fe^{3+} into transferrin and shows an interaction with ferroportin in specific cells which lack caeruloplasmin, HEK293T cells [76]. Further studies are clearly necessary to ascertain this important function of APP in health and disease in man.

PD is the second most common neurodegenerative disease after AD, affecting about 1 % of the population older than 60 years. Unlike AD, which affects memory and behaviour centres in the brain, PD is characterised by progressive loss of control over voluntary movement. The characteristic symptoms (bradykinesia, rigidity, tremor and loss of balance) arise from progressive loss of dopaminergic neurons (neurons which synthesise and release dopamine) in the substantia nigra pars compacta (SNPC), located in the midbrain [77].

In the brains of Parkinson's disease (PD) patients there is a specific elevation of iron in the substantia nigra and the lateral globus pallidus, by approximately two-fold in comparison with age-matched controls (reviewed in [35, 36]). This is in marked contrast to other iron storage diseases, such as untreated genetic haemochromatosis and thalassaemia patients, where 10–20-fold iron increases in iron stores must be attained before clinical abnormalities occur [78]. Increased iron levels are also found by phase microscopy in individual dopaminergic neurones of PD patients [79].

A second characteristic hallmark of PD is the presence, within dopaminergic neurons, axons and synapses of the substantia nigra, of intracellular, eosinophilic proteinaceous aggregates called Lewy bodies, which are composed mostly of α-synuclein [80], but also contain ubiquitin, tyrosine hydroxylase and IRP2 [35]. Many studies have shown that iron promotes the aggregation of α-synuclein [81–84] while chelation of free iron with desferrioxamine blocks α-synuclein aggregation [82–85]. Iron is also found to accumulate within Lewy bodies in the brains of PD patients [86–88].

The role of iron in PD is outlined in Fig. 10 [36]. In the brain interstitial fluid, iron is transported bound to transferrin (Tf), which is absorbed by neuronal cells via transferrin receptor (TfR1)-mediated endocytosis. Iron levels are elevated in the neurons of the substantia nigra in PD. Iron is normally stored in ferritin, but in PD, ferritin levels are found to be inappropriately low. If the capacity of the neurons to store iron is exceeded, potentially toxic free iron will accumulate. Similarly, neuromelanin, a dark pigment produced by dopaminergic neurons which binds free iron, is also found to be decreased in PD. Iron also promotes conformational changes within parkin and α-synuclein, which cause their aggregation. Iron is an important cofactor of tyrosine hydroxylase (TH), which is involved in dopamine biosynthesis, and of monoamine oxidase (MAO), an enzyme which is involved in dopamine metabolism. Hydrogen peroxide, generated in this reaction, can be converted to reaction oxygen species (ROS) by 'free' iron.

Therapeutic considerations

From the description of the pathologies of aged brains and neurodegenerative disease it can be observed that there are certain common factors between each of these pathologies which include inflammation, oxidative stress and increased iron accumulation. Therefore, therapeutic agents that could reduce (a) the inflammatory signalling process, (b) oxidative stress and (c) excessive iron accumulation could show efficacy in reducing the progression of the disease, for which there is currently no known cure.

Neuroinflammation is considered to be an important contributor to pathogenesis of neurological disorders, with microglial activation as a hallmark of neuroinflammation. Microglia, a subset of the glial cells, are the resident macrophage population in the CNS and are regarded as the resident immunocompetent effector cells of innate immunity in the brain. In the adult brain, under normal conditions, the blood–brain barrier(BBB) prevents molecules from gaining access to the vascular lumen. However, molecules of the systemic innate immune system are able to stimulate immune cells of the brain as well as the neuronal populations. Microglia are considered to be primary mediators of neuroinflammation. In the healthy adult brain they exist in a non-activated state displaying a ramified morphology and minimal expression of surface antigens. The activation of microglia is an early event in neurodegenerative process when they are rapidly transformed to a reactive phenotype and release cytotoxic pro-inflammatory molecules, including oxygen radicals, NO, glutamate, cytokines and prostaglandins which can have a detrimental effect on other neural cells. Their activation involves NFκB, which may involve neuronal–microglial interactions [89]. Therefore, inhibition of microglia-mediated neuroinflammation may present a promising therapeutic target for neurological disorders. Various compounds such as resveratrol, a non-flavonoid polyphenol rich in red wine and grapes [90], and taurine [91] have been shown to have antioxidant and anti-inflammatory properties, respectively, in animal models of neurodegeneration.

Disease-modifying therapies aimed at removing excess iron, without affecting iron-containing enzymes involved in neurotransmitter function, could be part of the therapeutic approaches utilised to prevent the progression of PD. There has been considerable speculation as to whether such

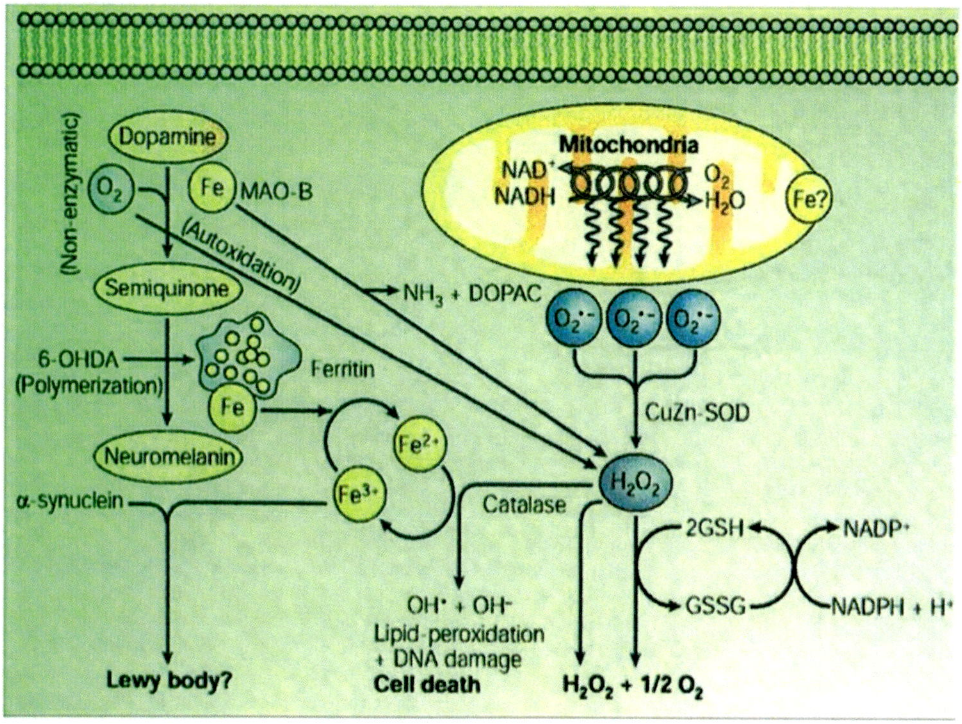

Fig. 10 Iron induces oxidative stress in PD [36]

chelators could be utilised in the treatment of neurodegenerative disease, where there are marginal increases in toxic iron in specific brain regions. However, in our earlier studies it was demonstrated that both DFO and deferiprone, a hydroxypyridone, were able to traverse the BBB and chelate iron in various brain regions in the ferrocene model of brain iron overload [92]. Subsequently, in more recent experiments where DFO was administered to experimental animals with intracranial haemorrhage, mortality was decreased [93, 94]. In addition, DFO administration to control animals had no significant effect on either iron homeostasis or haematological parameters. Several commercially available iron chelators, e.g. deferrioxamine (DFO–hexadentate), deferiprone (bidentate) and deferasirox (tridentate), which are clinically effective for the treatment of peripheral iron overload disorders, could potentially be utilised to treat PD. Although early studies indicated that the BBB is relatively impermeable to DFO [95], in our recent studies of these three chelators it was shown that each of them could penetrate the BBB and prevent the loss of tyrosine hydroxylase granules as well as improving dopamine levels in the 6-hydroxy dopamine model of Parkinson's disease [96].

Other therapies which incorporate iron chelators into drug delivery systems, e.g. liposomes and nanoparticles, may improve the targeting of these compounds to regions of the brain with excessive iron deposition. However, the toxicity associated with these approaches needs to be clarified. Chelators bound to transferrin receptors may also be a viable drug delivery system.

Conclusions

The cause of the enhanced brain iron content in PD remains unknown but may be attributable to a variety of factors which include changes in iron release mechanisms across the BBB, or perhaps more likely, a dysregulation of iron homeostatic control in the substantia nigra. In our recent studies [97], mRNA was isolated from two regions within the substantia nigra and from the cortex of Parkinson's brain, and the expression of a number of iron genes was compared with those from control post mortem material. Our global conclusions, in line with previous studies, are that, in human PD brain, cells of the SN behave phenotypically as if they were Fe deficient. Iron uptake systems are up-regulated, whereas iron storage in ferritin is down-regulated.

We conclude that excess iron accumulation in specific brain regions, albeit relatively slight when compared with other iron-accumulating diseases such as haemochromatosis and thalassaemia, is implicated in neurodegenerative disorders, notably in AD and PD as we have illustrated above. The mechanisms involved in these processes remain to be established, but the involvement of iron in PD and AD presents us with what could be an extremely precious therapeutic possibility of using iron chelators for its removal [47]. Clearly identifying mechanisms which are involved in brain iron accumulation will facilitate new therapeutic interventions to reduce the progression of age-related neurodegenerative diseases.

Acknowledgments Grateful thanks are due to COST D34 for facilitating the interaction between the authors.

References

1. Crichton RR (2008) Biological chemistry, an introduction. Elsevier, Amsterdam, p 369
2. Ohgami RS, Campagne DR, McDonald A, Fleming MD (2006) Blood 108:1388
3. Moos T, Oates PS, Morgan EH (1998) J Comp Neurol 31:420
4. Moos T, Morgan EH (2002) Dev Neurosci 24:99
5. Rouault TA, Cooperman S (2006) Semin Pediatr Neurol 13:142
6. Burdo JR, Menzies SL, Simpson IA, Garrick LM, Garrick MD, Dolan KG, Haile DJ, Beard JL, Connor JR (2001) J Neurosci Res 66:1198
7. Moos T, Morgan EH (2004) J Neurochem 88:233
8. Moos T, Rosengren Nielsen T (2006) Semin Pediatr Neurol 13:149
9. Moos T, Skjoerringe T, Gosk S, Morgan EH (2006) J Neurochem 98:1946
10. Dong XP, Cheng X, Mills E, Delling M, Wang F, Kurz T, Xu H (2008) Nature 455:992
11. Moos T, Rosengren Nielsen T, Skjørringe T, Morgan EH (2007) J Neurochem 103:1730
12. Morgan EH (1977) Biochim Biophys Acta 499:169
13. Morgan EH (1979) Biochim Biophys Acta 580:312
14. Wu LJ, Leenders AG, Cooperman S, Meyron-Holtz E, Smith S, Land W, Tsai RY, Berger UV, Sheng ZH, Rouault TA (2004) Brain Res 1001:108
15. Fischer J, Devraj K, Ingram J, Slagle-Webb B, Madhankumar AB, Liu X, Klinger M, Simpson IA, Connor JR (2007) Am J Physiol Cell Physiol 293:C641
16. Mills E, Dong X-P, Wang F, Xu H (2010) Future Med Chem 2:51
17. Lane DJR, Robinson SR, Czerwinska H, Bishop GM, Lawen A (2011) Biochem J 432:123–132
18. Jeong SY, David S (2003) J Biol Chem 278:27144
19. De Domenico I, Ward DM, di Patti MC, Jeong SY, David S, Musci G, Kaplan J (2007) EMBO J 26:2823
20. Todorich B, Pasquini JM, Garcia CI, Paez PM, Connor JR (2009) Glia 57:467
21. Ward RJ, Crichton RR, Taylor DL, Corte LD, Srai SK, Dexter DT (2010) J Neural Transm. doi:10.1007/s00702-010-0479-3
22. Dallman PR, Beutler E, Finch CA (1978) Br J Haematol 40:179
23. Dallman PR, Siimes MA, Stekel A (1980) Am J Clin Nutr 33:86
24. Beard JL (2008) J Nutr 138:2534
25. Lozoff B, Beard J, Connor J, Barbara F, Georgieff M, Schallert T (2006) Nutr Rev 64:S34
26. Rao R, Georgieff MK (2007) Semin Fetal Neonatal Med 12:54
27. Rao R, Tkac I, Townsend EL, Gruetter R, Georgieff MK (2003) J Nutr 133:3215
28. Rao R, Tkac I, Townsend EL, Ennis K, Gruetter R, Georgieff MK (2007) J Cerebr Blood Flow Metab 27:872
29. Ward KL, Tkac I, Jing Y, Felt B, Beard J, Connor J, Schallert T, Georgieff MK, Rao R (2007) J Nutr 137:1043
30. Beard JL, Wiesinger JA, Connor JR (2003) Dev Neurosci 25:308
31. Ortiz E, Pasquini JM, Thompson K, Felt B, Butkus G, Beard J, Connor JR (2004) J Neurosci Res 77:681
32. Beard J, Erikson KM, Jones BC (2003) J Nutr 133:1174
33. Burhans MS, Dailey C, Beard Z, Wiesinger J, Murray-Kolb L, Jones BC, Beard JL (2005) Nutr Neurosci 8:31
34. Bradbury MWB (1997) J Neurochem 67:443
35. Götz ME, Double K, Gerlach M, Youdim MB, Riederer P (2004) Ann N Y Acad Sci 1012:193
36. Zecca L, Youdim MB, Riederer P, Connor JR, Crichton RR (2004) Nat Rev Neurosci 5:863
37. Biasiotto G, Goldwurm S, Finazzi D, Tunesi S, Zecchinelli A, Sironi F, Pezzoli G, Arosio P (2008) Parkinsonism Relat Disord 14:426
38. Bartzokis G, Lu PH, Tishler TA, Peters DG, Kosenko A, Barrall KA, Finn JP, Villablanca P, Laub G, Altshuler LL, Geschwind DH, Mintz J, Neely E, Connor JR (2010) J Alzheimers Dis 20:333
39. Goodhall EF, Greenway MJ, van Marion I, Carrol CB, Hardiman O, Morrison KE (2005) Neurology 65:934
40. Johnstone D, Milward EA (2010) J Neurochem 113:1387
41. Ellervik C, Birgens H, Tybjaerg-Hansen A, Nordestgaard BG (2007) Hepatology 46:1071
42. Penke L, Valdés Hernandéz MC, Maniega SM, Gow AJ, Murray C, Starr JM, Bastin ME, Deary IJ, Wardlaw JM (2010) Neurobiol Aging. doi:10.1016/j.neurobiolaging.2010.04.032
43. Sullivan EV, Adalsteinsson E, Rohlfing T, Pfefferbaum A (2009) Brain Imaging Behav 3:167
44. Vasudevaraju P, Bharathi TJ, Shamasundar NM, Subba Rao K, Balaraj BM, Ksj R (2010) Indian J Psychiatry 52:140
45. Kastman EK, Willette AA, Coe CL, Bendlin BB, Kosmatka KJ, McLaren DG, Xu G, Canu E, Field AS, Alexander AL, Voytko ML, Beasley TM, Colman RJ, Weindruch RH, Johnson SC (2010) J Neurosci 30:7940
46. House MJ, St Pierre TG, Milward EA, Bruce DG, Olynyk JK (2010) Magn Reson Med 63:275
47. Crichton RR, Ward RJ (2006) Metal-based neurodegeneration. From molecular mechanisms to therapeutic strategies. Wiley, Chichester, p 227
48. Dalle-Donne I, Giustarini D, Colombo R, Rossi R, Milzani A (2003) Trends Mol Med 9:169
49. Benedetti A, Pmpella A, Fulceri R, Romani A, Comporti M (1986) Toxicol Pathol 14:457
50. Catala A (2009) Chem Phys Lipids 157:1
51. Schneider C, Porter NA, Brash AR (2008) J Biol Chem 283:15539
52. Zarkovic N (2003) Mol Aspects Med 24:281
53. Zarkovic K (2003) Mol Aspects Med 24:293
54. Grimsrud PA, Xie H, Griffin TJ, Bernlohr DA (2008) J Biol Chem 283:21837
55. Berlett BS, Stadtman ER (1997) J Biol Chem 272:20313
56. Beal MF (2002) Free Radic Biol Med 32:797
57. Stadtman ER (2006) Free Radic Res 40:1250
58. Sayre LM, Lin D, Yuan Q, Zhu X, Tang X (2006) Drug Metab Rev 38:651
59. Oe T, Arora JS, Lee SH, Blair IA (2003) J Biol Chem 27:42098
60. Zhang WH, Liu J, Xu G, Yuan Q, Sayre LM (2003) Chem Res Toxicol 16:512
61. Bush AL, Tanzi RE (2002) Proc Natl Acad Sci USA 99:97317
62. Honda K, Casadesus G, Peterson RB, Perry G, Smith MA (2004) Ann NY Acad Sci 1012:179
63. Honda K, Smith MA, Zhu X, Baus D, Merrick WC, Tartakoff AM, Hattier T, Harris PL, Siedlak SL, Fujioka H, Liu Q, Moreira PI, Miller FP, Nunomura A, Shimohama S, Perry G (2005) J Biol Chem 280:20978
64. Sipe JC, Lee P, Beutler E (2002) Dev Neurosci 24:188
65. Connor JR, Snyder BS, Beard JL, Fine RE, Mufson EJ (1992) J Neurosci Res 31:327

66. Smith MA, Harris PL, Sayre LM, Perry G (1997) Proc Natl Acad Sci USA 94:9866
67. Tachida Y, Nakagawa K, Saito T, Saido TC, Honda T, Saito Y, Murayama S, Endo T, Sakaguchi G, Kato A, Kitazume S, Hashimoto Y (2008) J Neurochem 104:1387
68. Seidah NG, Chrétien M, Day R (1994) Biochimie 76:197
69. Silvestri L, Pagani A, Camaschella C (2008) Blood 111:924
70. Babitt JL, Huang FW, Wrighting DM, Xia Y, Sidis Y, Samad TA, Campagna JA, Chung RT, Schneyer AL, Woolf CJ, Andrews NC, Lin HY (2006) Nat Genet 38:531
71. McMahon S, Grondin F, McDonald PP, Richard DE, Dubois CM (2005) J Biol Chem 280:6561
72. Silvestri L, Camaschella C (2008) J Cell Mol Med 12:1548
73. Mueller S (2005) Biofactors 24:171
74. Hwang EM, Kim SK, Sohn JH, Lee JY, Kim Y, Kim YS, Mook-Jung I (2006) Biochem Biophys Res Commun 349:654
75. Rogers JT, Randall JD, Cahill CM, Eder PS, Huang X, Gunshin H, Leiter L, McPhee J, Sarang SS, Utsuki T, Greig NH, Lahiri DK, Tanzi RE, Bush Al, Giordano T, Gullans SR (2002) J Biol Chem 277:45518
76. Duce JA, Tsatsanis A, Cater MA, Janes SA, Robb E, Wikhe K, Leong SL, Perez K, Johanssen T, Greenough MA, Cho H-H, Galatis D, Moir RD, Masters CL, McLean C, Tanzi RE, Cappai R, Barnham KJ, Ciccotosto GD, Rogers JT, Bush AI (2010) Cell 142:857
77. Schapira AH, Olanow CW (2004) JAMA 291:358
78. Crichton RR (2009) Inorganic biochemistry of iron metabolism. from molecular mechanisms to clinical consequences, 3rd edn. Wiley, Chichester, p 461
79. Oakley AE, Collingwood JF, Dobson L, Love G, Perrott HR, Edwardson JA, Elstner M, Morris CM (2007) Neurology 68:1820
80. Goedert M (2001) Nat Rev Neurosci 2:492
81. Golts N, Snyder H, Frasier M, Theisler C, Choi P, Wolozin B (2002) J Biol Chem 277:16116
82. Hashimoto M, Hsu LJ, Xia Y, Takeda A, Sisk A, Sundsmo M, Masliah E (1999) Neuroreport 10:717
83. Münch G, Lüth HJ, Wong A, Arendt T, Hirsch E, Ravid R, Riederer P (2000) J Chem Neuroanat 20:253
84. Ostrerova-Golts N, Petrucelli L, Hardy J, Lee JM, Farer M, Wolozin B (2000) J Neurosci 20:6048
85. Sangchot P, Sharma S, Chetsawang B, Porter J, Govitrapong P, Ebadi M (2002) Dev Neurosci 24:143
86. Takanashi M, Mochizuki H, Yokomizo K, Hattori N, Mori H, Yamamura Y, Mizuno Y (2001) Parkinsonism Relat Disord 7:311
87. Castellani RJ, Siedlak SL, Perry G, Smith MA (2000) Acta Neuropathol 100:111
88. Arawaka S, Saito Y, Murayama S, Mori H (1998) Neurology 51:887
89. Kaltschmidt C, Kaltschmidt B, Lannes-Vieira J, Kreutzberg GW, Wekerle H, Baeuerle PA, Gehrmann J (1994) J Neuroimmunol 55:99
90. Zhang W, Murao K, Zhange X, Matsumoto K, Diah S, Okada M, Miyake K, Kawai N, Tamiya T (2010) BMC Cancer 10:593
91. Ward RJ, Lallemand F, de Witte P, Crichton RR, Piette J, Tipton K, Hemmings K, Pitard A, Page M, Della Corte L, Taylor D, Dexter DT (2011) Biochem Pharmacol 81:743–751
92. Ward RJ, Dexter D, Florence A, Aouad F, Hider R, Jenner P, Crichton RR (1995) Biochem Pharmacol 49:1821
93. Hua Y, Keep RF, Hoff JT, Xi G (2008) Acta Neurochir Suppl 105:3
94. Selim M (2009) Stroke 40:S90
95. Shachar DB, Kahana N, Kampel V, Warshawsky A, Youdim MB (2004) Neuropharmacology 46:254
96. Dexter DT, Statton S, Whitmore C, Freinbichler W, Weinberger P, Tipton KT, Della Corte L, Ward RJ, Crichton RR (2011) J Neural Trans 118:223–231
97. Crichton RR, Dexter DT, Ward RJ (2008) Coord Chem Rev 252:1189
98. Abbott NJ, Patabendige AA, Dolman DE, Yusof SR, Begley DJ (2009) Neurobiol Dis 37:13
99. Zimmer M, Ebert BL, Neil C, Brenner K, Papaioannou I, Melas A, Tolliday N, Lamb J, Pantopoulos K, Golub T, Iliopoulos O (2008) Mol Cell 32:838
100. Hentze MW, Kühn LC (1996) Proc Natl Acad Sci USA 93:8175
101. Cho HH, Cahill CM, Vanderburg CR, Scherzer CR, Wang B, Huang X, Rogers JT (2010) J Biol Chem 285:31217

Brain changes in iron loading disorders

Elizabeth Milward, Bulent Acikyol, Brianna Bassett, Evan Williams, Ross Graham, Roheeth Delima, Deborah Trinder, John Olynyk, and Daniel Johnstone

Abstract

Abnormal iron accumulation within the brain is associated with various neurodegenerative diseases; however, there is debate about whether milder disorders of systemic iron loading, such as haemochromatosis, affect the brain. Arguments on both sides of the debate are often based on some common assumptions that have not been rigorously tested by appropriate experimentation. Recent research from our lab has applied high-throughput molecular techniques such as microarray to models of dietary and genetic iron loading to identify subtle but important effects on molecular systems in the brain that may go undetected by other methods commonly used in the field. In this chapter, we review the existing research in animal models and human patients and discuss the strengths and limitations of the different approaches commonly used. Using our findings as an example, we argue that transcriptomic methods can provide unique insights into how systemic iron loading can affect the brain and suggest some basic guidelines for extracting the most robust and reliable information from microarray studies.

Keywords

Iron overload • Brain • Microarray • Mouse model • HFE

Introduction

As reviewed elsewhere, severe iron dyshomeostasis can impair cognition, movement and behaviour in rare conditions such as neurodegeneration with brain iron accumulation (NBIA) disease [1]. In this chapter, we will review the evidence that even relatively mild iron disorders give rise to brain changes of potential functional significance. Although iron deficiency will be considered briefly, we will focus primarily on iron loading disorders. We will first give an overview of some of the past work in this field. We will then go on to discuss some new approaches to studying this issue and some of the new data being generated by our laboratory and others which challenge some common assumptions about the effects of iron overload on the brain.

This area is of current interest since patients with the common iron overload disorder haemochromatosis sometimes report neurological symptoms such as memory impairment, headaches or extreme fatigue, and it is not yet clear whether such symptoms are directly attributable to haemochromatosis or not. Haemochromatosis, most commonly caused by polymorphisms in the *HFE* gene [2, 3], is a disorder of iron metabolism in which there is excessive iron accumulation in a number of organs throughout the body, particularly in the liver and also in the pancreas, endocrine organs and heart [4–6]. Iron accumulation can cause liver fibrosis and cirrhosis, increasing the risk of hepatocellular carcinoma, as well as other disorders such as diabetes, cardiomyopathy and arthritis [4–6]; however, effects on the brain are still under debate.

Past animal models of iron overload have used iron supplementation and will be described in more detail below. More

E. Milward (✉)
School of Biomedical Sciences and Pharmacy MSB, University of Newcastle, Callaghan, NSW 2308, Australia
e-mail: Liz.Milward@newcastle.edu.au

W. Linert and H. Kozlowski (eds.), *Metal Ions in Neurological Systems*,
DOI 10.1007/978-3-7091-1001-0_2,

recently, new genetic models have been developed that may give more accurate pictures of human haemochromatosis. We have been studying brain gene expression changes in these models using high-throughput microarray technologies. Even relatively mild changes in (body) iron homeostasis appear to influence the expression of important genes within the brain, often in unexpected ways. In this chapter, we will discuss some common assumptions about iron and the brain and some of the key findings and limitations of past studies. We will go on to describe how high-throughput technologies such as microarrays and other methods of examining genome-wide gene expression can be used to expand our understanding of the effects of iron on brain and how the findings from such studies are starting to challenge many of the existing assumptions in this area.

A few common assumptions about iron and the brain

There have been relatively few experimental studies on the effects of iron perturbations in the CNS or CNS-derived models and these have often been performed under pharmacological conditions. For example, abnormally high levels of iron are often assumed to exert toxic effects primarily through the oxidative actions of free iron on neurons or other cells, but this is based primarily on studies using pharmacological iron doses. Surprisingly few studies provide in vivo evidence for iron-induced oxidative damage within the brain in conditions likely to be physiologically relevant.

Other common assumptions relate to the blood–brain barrier (BBB), which restricts the free passage of many substances between the brain and the rest of the body. Notably it is sometimes assumed that people with systemic iron overload conditions do not experience brain sequelae because the BBB limits the entry of iron into the brain. This in fact involves two assumptions, first that the BBB effectively protects the brain from iron loading in such conditions and second that if there is no iron loading in the brain, systemic iron loading in itself does not exert indirect effects on the brain.

With regard to these assumptions, as will be discussed in more detail further below, iron supplementation studies in rodents have demonstrated that brain iron levels can increase as a result of systemic iron increases, and that this can subsequently affect brain function [7–10]. This suggests that the BBB may not provide complete protection; however, these studies have generally used very high doses of iron and may not necessarily be relevant under more physiological conditions. Therefore, the validity of the first assumption remains uncertain.

Iron transport across the BBB is covered in detail in a chapter by Crichton and colleagues and will not be revisited in detail here. Another consideration that may be more relevant in this context is that iron overload often arises as a result of genetic mutations, as further discussed below. This could lead to perturbations in brain systems influenced by the gene in question, independent of systemic iron status. These brain systems could be related to iron homeostasis within the brain or could be involved in other brain functions unrelated to iron. Therefore, genetic disorders of iron dyshomeostasis may affect brain systems even if the BBB is limiting entry of iron per se or even if there is no gross iron loading in the brain.

Also brain iron loading per se is not necessarily the only mechanism by which altered brain function may occur in response to systemic iron dyshomeostasis. There can be circumstances in which systemic iron loading can transduce indirect effects on the brain. For example, extreme systemic iron loading can cause severe liver damage, which can result in the accumulation of toxic substances such as ammonia in the blood. This, in turn, can lead to perturbation of brain functions through hepatic encephalopathy, a condition characterised by symptoms such as impairment of speech and movement and, in severe cases, seizures and coma [11]. A less extreme example may involve systemic inflammatory and immune changes. Iron overload in the liver can lead to low-grade inflammation [12] and alterations in circulating cytokines and other peripheral immune responses could act on the brain without brain iron levels being affected in any way.

Past animal studies

Iron supplementation studies in animal models have usually employed one of two broad approaches. One approach is to use acute models involving direct injection of one or more high doses of iron into the CNS of adult mice, in order to bypass the BBB [13–15]. Most such studies have involved highly artificial, often pharmacological, conditions, in large excess of usual exposure levels, and their relevance to human iron disorders remains to be established.

It is sometimes assumed that the early phases of extreme pharmacological changes or changes in response to less extreme stimuli (acute or chronic) may be inferred from 'scaling down' observations in more extreme, pharmacological conditions. Yet often early changes in response to stimuli may involve completely different systems to those that become prominent under more extreme circumstances. One counter-example to this occurs for end-stage neurodegenerative disease. By the time patients die, the brain may be reduced in mass by as much as 30 % or more due to loss of particular cell populations. The original molecular profiles caused by the stimulus may have long since vanished, as cells move through early, possibly compensatory responses, through damage and repair profiles and into mechanistic profiles associated with cell death. In end-stage disease, the cells in which the first problems originally occurred may have long since died, to be replaced by cells expressing different sets of genes or proteins. This needs to be considered when investigating phenomena that may lead to neuronal cell death, such as approaches that use pharmacologically high iron loading.

The other broad approach used by some researchers has been to look at the effects of iron supplementation in early post-natal rodents, with the goal of examining the effects of abnormally high iron exposures on the brain before the BBB is fully formed [8, 16]. While this kind of approach has been informative, one limitation is that it is still not yet fully clear how the transport of iron into and out of the brain alters through embryogenesis into infancy and adulthood. Also, while younger animals may have a more iron-permeable BBB, supplementing iron in young rodents can be difficult. Supplementation during embryogenesis or in pre-weanling neonatal mice has primarily used maternal dietary iron supplementation, making it difficult to achieve high levels of iron loading [17]. It may also be difficult to determine the actual exposure levels of individual neonates, although a few studies have used gavage (force-feeding) to deliver high known doses of iron [18, 19].

Issues relating to the sensitivity of methods used to measure iron also need to be kept in mind. The brain is a heterogeneous organ comprising various different anatomical regions and cell types which have different iron usages and contents. Measuring iron levels in whole brain homogenate tells only part of the story, as it provides no information on regional variation. Even when brain structure is preserved using histology, histochemical stains such as Perls' or Turnbull's stains are often not sensitive enough to pick up subtle changes in iron levels. Ideally, brain iron would be measured using a technique that combines the sensitivity of inductively coupled plasma-mass spectrometry (ICP-MS) or inductively coupled plasma-atomic emission spectroscopy (ICP-AES) with maintenance of brain morphology. Recent innovations and emerging technologies, for example laser ablation inductively coupled plasma mass spectrometry (LA-ICP-MS), may be the solution. Such methods can provide detailed and quantitative mapping of regional distributions of iron and other metals in brain sections [20].

Yet, despite these limitations, various studies using such approaches have successfully demonstrated alterations in brain iron levels and other brain changes in response to iron supplementation, as will now be reviewed in more detail.

Animal studies of how iron supplementation affects brain iron levels

Several studies have assessed changes in brain iron levels in response to iron administration at various ages. Pinero and colleagues investigated the effects of early systemic iron loading on brain iron content using normal rat pups mothered by iron-supplemented dams [16]. This resulted in higher total iron content in the pups in most brain regions investigated. If weanling rats were instead given an iron-supplemented diet for two weeks only the pons and hippocampus showed increased total iron levels, suggesting that the brain may be less affected by increases in systemic iron at older ages. Yet, even at older ages, iron supplementation over longer periods can still accelerate accumulation of iron within the brain. Male weanling rats fed with a high iron diet for six or eight weeks showed significant increases of 15–30 % in total iron levels in the cortex, hippocampus, striatum and substantia nigra [21–23]. The degree of brain iron accumulation resulting from iron supplementation depends on the iron dose [7, 8, 18]. This supports the hypothesis that the BBB is capable of protecting the brain from excessive iron accumulation arising as a consequence of systemic iron overload, but only up to a certain threshold.

Effects of iron supplementation on brain expression of iron-related genes and proteins

1. Ferritin, transferrin receptor 1 and transferrin—Most studies of changes in brain gene and protein expression in response to systemic iron loading have focussed on specific molecules such as those controlling iron handling and homeostasis elsewhere in the body. In peripheral organs such as the liver, iron loading is often accompanied not only by increases in the iron storage protein ferritin but also by decreases in transferrin receptor 1 (TFR1), the main receptor for the iron transport protein transferrin. This has the effect of reducing the cellular uptake of iron, which is not only important in the liver but may be particularly important in organs such as the brain, which have not evolved the capacity to store large amounts of iron. The concerted inverse changes in ferritin and TFR1 occur in response to alterations in intracellular iron concentrations and this is coordinated at least in part by the iron-responsive element/iron regulatory protein (IRE/IRP) system [24].

Pinero and colleagues [16] observed decreased TFR1 protein levels in some brain regions in their iron-supplemented rat models, described in detail above, but no changes in ferritin levels in their model of early iron overload (rats mothered by dams iron-supplemented from postnatal day 10 to postnatal day 21). In contrast, their model of late iron overload (rats fed an iron-supplemented diet for two weeks following weaning) showed decreased TFR1 expression in different brain regions and increased ferritin expression in pons and cortex. There were no changes in transferrin levels in either model. Decreases in TFR1 and increases in ferritin are predicted to help maintain cellular iron homeostasis and protect against iron toxicity by restricting cellular iron uptake and increasing sequestration within ferritin. For regions such as the pons, which also showed increased iron levels, the changes are consistent with post-transcriptional regulation through the IRE/IRP system.

2. Divalent metal transporter 1—Another protein important in cellular and organelle iron uptake is divalent metal transporter 1 (DMT1). Ke and colleagues assessed DMT1 protein and mRNA in different brain regions by Western and Northern blotting in male weanling rats fed a high-iron or normal diet for eight weeks. As described in the preceding section, all brain regions investigated showed increases in brain iron content of 15–30 %. Yet there was no difference in brain DMT1 mRNA or protein levels relative to rats fed a normal diet, despite one DMT1 transcript variant containing an IRE. So, at least in adulthood, DMT1 expression may not be regulated by iron in the particular brain regions examined [22].
3. The ferroxidases hephaestin and caeruloplasmin—Movement of iron between different cells in the brain is influenced by ferroxidases in the brain interstitium which oxidise ferrous iron exported from neural cells or transported across the BBB. This facilitates binding to transferrin, which binds ferric iron with greater affinity than ferrous iron.

One ferroxidase essential for normal brain iron homeostasis is caeruloplasmin—patients with acaeruloplasminaemia due to loss-of-function mutations in the caeruloplasmin gene or with low levels of caeruloplasmin in Wilson's disease show brain MRI abnormalities suggestive of basal ganglia iron accumulation [25–27] and often display symptoms such as cerebellar ataxia and cognitive impairment [28–30].

In iron-supplemented rat models similar to those used in the Ke DMT1 study above, with increased brain iron content in all regions investigated, hephaestin protein is increased in cortex and hippocampus but decreased in striatum and substantia nigra [23], while caeruloplasmin protein is increased in the substantia nigra only [21]. This suggests that the ferroxidases hephaestin and caeruloplasmin are not regulated in the same way in response to alterations in iron status and that regulation of these proteins differs between different brain regions. This may relate to differences in iron usage between brain regions with different functions.

Recently the Alzheimer's disease amyloid precursor protein (APP) has been reported to have ferroxidase activity and proposed to be 'the neuronal ferroxidase' [31]. This protein has both membrane-bound and secreted forms and contains a region with sequence similarity to the ferroxidase domain of ferritin heavy chain [31]. Furthermore, the corresponding mRNA transcript contains an iron-responsive element in the 5′ untranslated region [32]. The contributions of APP to brain ferroxidase activity remain to be determined but are likely to be limited insofar as APP appears unable to protect adequately against caeruloplasmin deficiency and accompanying iron deposition in patients with acaeruloplasminaemia or Wilson's disease.

Iron supplementation and the brain microvasculature

Studies of changes in protein levels in brain microvasculature in response to iron supplementation have found no evidence for changes in ferritin, TFR1, transferrin or DMT1 [33]. This suggests any changes which occur in expression of these proteins in response to iron supplementation do not take place in brain capillary endothelial cells, influencing uptake of blood iron into the brain, but instead occur in neurons or glia and influence iron trafficking and distribution within neural tissue after uptake across the BBB. However, it is unclear what changes occur in brain microvasculature in response to more extreme circumstances of iron release, for example during cerebrovascular haemorrhage.

Functional changes

On the whole, the weight of current evidence suggests that iron supplementation affects the expression within the brain of genes and proteins important in iron homeostasis. Several investigations suggest there may also be functional phenotypic changes in response to iron supplementation, including effects on movement and behaviour. Fredriksson and colleagues found that oral administration of iron to mice (37 mg/kg body weight) at 10–12 days of age, a critical period for brain development, resulted in significantly increased total iron content (60 %) in the basal ganglia at three months of age [8], as assessed by atomic absorption spectroscopy. This was accompanied by signs of neurobehavioural dysfunction, including altered spontaneous motor behaviour, poor performance in radial arm maze learning tests and deficits in habituation.

Similarly, Sobotka and colleagues found that supplementing the diet of male weanling rats with a high iron dose (20,000 parts per million carbonyl iron) for 12 weeks caused increased brain non-haem iron levels in association with neurobehavioural dysfunctions. These included deficits in conditioned active avoidance response, which indicates changes in associative processes, impaired startle response, which indicates decreased ability to respond to environmental stimuli, and decreased motor activity [7].

Using an iron-supplementation protocol sufficient to cause regional brain iron increases, as described above [16], Pinero and colleagues showed evidence of altered motor functions in iron-supplemented rats, including decreased activity and stereotypic behaviour [9]. In addition, studies by Maaroufi and colleagues demonstrated that intraperitoneal injection of ferrous sulphate (3 mg/kg) in adult male rats for five days or more, not only increases iron levels in various brain regions (see above), but causes impaired emotional behaviour and spatial learning [10].

Some of these effects may be due to altered neurotransmitter levels in certain brain regions as a result of iron supplementation. Kaur and colleagues administered a dose of iron (120 mg/kg body weight) that was 40 times the usual daily iron intake to neonatal mice at 10–17 days of age, resulting in increased total iron levels in the substantia nigra by two months of age, as assessed by inductively coupled plasma mass spectrometry. This was accompanied by depletion of striatal dopamine levels and increased levels of oxidative stress markers [18]. Whether pharmacological effects have more generalised validity and applicability is always open to question but another study also found depletion of dopamine levels in rats fed an iron-supplemented diet (0.3 % w/w ferrous sulphate) for 10 weeks, in addition to decreased brain serotonin levels [34].

Neither of the two studies just described investigated higher level brain function such as movement, behaviour or cognition. Even so, taken together, the body of findings in the various animal models subjected to iron supplementation suggests that systemic iron levels can influence some brain functions, in addition to brain gene and protein expression.

However, most of these studies have used pharmacological doses of iron that may not accurately reproduce what occurs in most human iron overload conditions. Supplementation models are also constrained by the acute, short-term nature of typical experimental iron-supplementation protocols as well as the partial restriction of brain iron entry by the blood–brain barrier and the blood–cerebrospinal fluid (CSF) barrier. All these factors may limit the usefulness of such models in addressing how brain iron abnormalities cause neurologic dysfunction. Further insights can be gained by studying the effects on the brain of longstanding perturbations in brain iron homeostasis due to inherent genetic mutations. In addition, such mutations also have the potential to disrupt the brain's internal homeostasis by mechanisms which are partly or fully independent of systemic iron status.

Functional effects in haemochromatosis patients and animal models

There are several case reports [35–37] of movement disorders in patients with the iron loading disorder haemochromatosis, described earlier in this chapter. However, it has been argued that the movement impairment in these patients is coincidental and not attributable to haemochromatosis [38, 39]. There is a dearth of studies investigating *HFE* genotype in patients with movement disorders other than Parkinson's disease, but one case study does report a female with severe HFE haemochromatosis and cerebellar ataxia. This patient showed evidence of brain iron accumulation and symptoms of ataxia were stabilised following phlebotomy [37], suggesting that iron may have had a causative role.

There are virtually no papers on brain effects in animal models of HFE dysfunction, but Golub and colleagues investigated motor function in male $Hfe^{-/-}$ knockout mice at three months of age [40]. The $Hfe^{-/-}$ mice experienced more falls from a rotarod, in addition to having a wider forelimb landing footsplay, greater variability in stride length and hypersensitivity to proximal stimulation. These features are consistent with a motor co-ordination deficit, which did not appear to be attributable to deficits in motor learning or strength or generalised debility.

Interestingly, histochemical staining revealed no obvious accumulation of iron in the $Hfe^{-/-}$ brain [40], suggesting that movement disorders can occur in the absence of gross brain iron accumulation. As noted earlier, lack of gross brain iron accumulation does not rule out the possibility that systemic iron overload causes other brain changes. At least three different animal models—a rat model of dietary iron overload [7], $Hfe^{-/-}$ mice [40] and *Irp2* knockout mice, deficient for iron regulatory protein 2 [41, 42]—all show behavioural or motor abnormalities without apparent increases in brain iron. Such findings raise the possibility that some iron disorders, such as those due to *HFE* gene mutations, may disrupt brain functions by mechanisms that are at least partially independent of brain iron accumulation as well as through effects directly due to changes in brain iron homeostasis.

In summary, there is now considerable evidence that both iron deficiency and iron overload can cause changes in gene expression and protein levels in the brain. To date, most studies have focussed on changes in iron-related genes and proteins but we believe there may also be expression changes for important genes and proteins relating to key brain functions, although as yet few studies have investigated this.

Exploratory studies of brain gene expression changes in response to iron status

As exemplified by most of the studies discussed above, most research relating to the effects of iron on the brain has been restricted to investigations of a small number of molecules of interest. So far there have been very few studies of changes at the level of molecular systems, yet some studies have now begun to use exploratory, discovery driven approaches rather than focussing solely on select molecules.

Liu and colleagues measured gene expression changes in response to iron supplementation or deprivation in rat hippocampal cells by suppression subtractive hybridisation [43]. Expression changes for select genes were then confirmed in vivo by performing Northern blot analysis on hippocampal tissue isolated from rats fed either an iron-supplemented or iron-deficient diet for two months following weaning. This is one of the earliest examples in the brain iron field of combining both discovery-driven and hypothesis-based approaches to find new genes of interest. In total, 31 genes were identified by SSH as differentially expressed in comparisons of iron-supplemented and iron-deprived hippocampal cultures. Quantitative analysis of ten of these by Northern blot analysis revealed only modest expression changes (less than twofold). The genes that were identified as having higher expression in response to iron supplementation were endothelial-derived gene 1, ribophorin II, ubiquitin-specific protease 14, etoile Sam68-like protein and the partial coding sequence Ir22. These genes are generally poorly characterised and had few known functions.

Four of the genes with higher expression in response to iron deprivation (BS69, Pdcd5, Anapc8, Ruvbl2) related to cell cycle regulation and may encode proteins with functions in DNA damage and repair and the transition to apoptosis [43]. This suggests that iron deficiency in the hippocampus may be associated with cell damage or apoptosis.

This illustrates a common problem confronting researchers working with emerging high-throughput technologies. What can be made of a bevy of small changes in poorly characterised proteins? Such changes may simply be artefacts of the model or the technology. Yet even small changes may be of real relevance, not only if acting in concert but even when acting in opposing directions if this indicates systems under stress which are exhibiting both potentially pathogenic and compensatory changes.

When performed rigorously, gene expression microarray studies can provide unprecedented insights into the behaviour of molecular systems. However, some facets of rigorous array data analysis are still not well understood by many researchers. Array data often show a mass of changes, many involving molecules whose functions are unknown. Out of the sea of changes, often of relatively small magnitude even in grossly abnormal samples such as cancers, how do we decide, first, which are real and, second, which of these are important? Resources need to be focused on effects of real importance. Yet the volume of information generated is often so large that it is far beyond the capacity of individual groups to follow up and validate all important effects. The following sections address these issues in more detail.

High-throughput advantages and concerns

The emergence of microarray technology has facilitated high-throughput gene expression studies unimaginable only years ago, allowing differential gene expression to be examined across the entire genome in a single experiment. Microarray technology has several advantages over approaches traditionally used to assess gene expression. Researchers often studied molecules artificially isolated from their natural environment, making it difficult to assess how molecular phenomena were affected by interactions with other molecules within cells and tissues. Methods such as RT-PCR or Northern blot analysis require the user to select a small number of specific genes for investigation of expression. In contrast, the microarray technique measures the expression of thousands of genes in a single experiment, providing a cost-effective and time-efficient way of acquiring information on genome-wide gene expression. This allows researchers to explore biomolecular systems and the orchestrated behaviour of large groups of molecules in far greater depth than ever before. Microarray experiments can provide valuable insights into molecular mechanisms or pathways or ontologies that are perturbed due to a given experimental manipulation, as opposed to simply identifying changes in the expression of select individual genes.

But these powerful tools bring new challenges with regard to interpreting the mass of data now pouring out. Some deeply rooted assumptions about how best to progress bioscientific research may need to be re-assessed in the light of the new perspectives now starting to emerge. As just one obvious example, the high-throughput 'omics' approaches now being applied have already begun to raise awareness of the validity of conducting both discovery-driven research and hypothesis-driven research and indeed of the value of viewing these as complementary approaches in order to maximise yields of useful information.

Explanation of arrays

Microarrays are manufactured on small chips or slides, with an individual gene expression array containing many different oligonucleotide or cDNA probes, each targeting a specific gene transcript. Each probe is present in millions of copies, allowing transcript levels to be assessed quantitatively. With advances in current technology, microarrays are now commercially available that, in theory, contain a sufficient number of different probes to measure every transcript encoded by the genome, although the comprehensiveness and effectiveness of such investigations are reliant on probe quality.

One important barrier to uptake of any high-throughput technology, whether array-based or sequencing-based, is concerns about the validity of the output data. The difficulty lies in extracting reliable, relevant information from the mass of molecular data generated by high-throughput studies. This raises some important issues in analysing and interpreting array and other high-throughput datasets, particularly in circumstances where effect sizes are relatively small. These issues include assessment of background thresholds, minimising effects of technical variation, assessing whether or not an observed signal difference is likely to be real and, if this is done statistically, deciding whether or not it is appropriate to correct for multiple testing [44–46].

In addition, the cost of microarrays, although small in terms of the amount of data obtained, is still a barrier for many labs. In our experience, sample pooling, while superficially appealing on economic grounds, is problematic, since even a single outlier sample can make an entire experimental run useless, over-riding any potential cost efficiencies of pooling. We therefore recommend avoiding pooling if at all possible. But this does restrict the number of variables that can be examined effectively. For example, it is difficult to examine the effects of iron on gene transcript levels over a wide range of different ages, dietary regimens or brain regions. However, initial microarray results can guide validation studies over expanded experimental conditions using cheaper technologies such as PCR, Western analysis of protein or other more traditional approaches. Issues relating to the accuracy of high-throughput data are addressed in more detail below.

Normalisation and analysis issues when investigating small magnitude expression changes

Microarray intensity values are usually normalised by one or more transforming functions. Reasons for normalising can include forcing a normal data distribution or increasing comparability between probes, samples, chips, machines or platforms [47–49]. Even small technical variations (e.g. cRNA loading on arrays, scanning and hybridisation inconsistency) can sometimes cause considerable differences in signal intensities. The overarching aim of normalisation is to reduce differences due to technical variation (false positives) while conserving true biological effects (i.e. maximising true positives and minimising false negatives).

The particular normalisation methods which are most appropriate will vary depending on the type of experiments, the array platform and type of array (e.g. RNA or cDNA, one-colour or two-colour). It is often advisable to apply more than one normalisation approach to identify robust effects that are not artefacts of a particular normalisation method. These issues are particularly relevant for datasets where expression changes are expected to be modest but similar considerations also apply even for datasets where most expression changes are large, since these will also contain some genes of biological interest with small expression changes.

The challenge for any analytical approach lies in reducing false positives while avoiding false negatives. A statistical p-value approach allows estimation of false positive error probability, which can be considerable when conducting large numbers of comparisons, as occurs in arrays. However, the methods currently used to adjust for multiple comparisons [50] are often conservative, missing real changes. Such adjustments may be most useful for identifying restricted groups of target genes (e.g. class prediction aimed at identifying biomarkers for diagnosis or prognosis). For studies aimed at identifying complete sets of target genes (e.g. class comparison or class discovery aimed at understanding biological mechanisms), accepting non-informative false positives may be less problematic than omitting informative genes. Minimising false negatives by not applying a multiple testing correction has been recommended for such studies [51, 52].

One drawback of using statistical metrics is that important expression changes can be missed if there is high replicate variation due to chance outliers. To avoid this, fold-change metrics have been advocated by various groups including the Microarray Quality Control (MAQC) consortium, which has recommended cut-offs based on fold-change in combination with a relaxed p-value cut-off [53]. However, other groups disagree [46, 54]. Chen and colleagues note that if there is no treatment effect, the strategy of a fold-change cut-off with a non-stringent p-value cut-off will result in 100 % false-positive error selection [46]. Other problems may occur when analysing datasets with low fold-changes, where small but biologically significant alterations may be missed by fold-change filters (false negatives).

We recommend routinely using multiple normalisation and analytical approaches to decrease the likelihood of false-positive results due to artefacts of the analytical methodology. In addition, we suggest focusing most attention on molecular systems containing several genes with altered expression, as co-ordinated expression changes in multiple related genes may reflect perturbations of pathways or functions and may be more likely to be real findings of biological relevance than changes in individual genes with no common functions. This is achieved in part through the use of pathway and ontology tools and other bioinformatics approaches, as well as by more traditional investigative strategies.

Microarray studies of brain gene expression in models of iron disorders

Despite the potential opportunities for discovery using microarray, few researchers have utilised the technology to investigate brain molecular systems perturbed as a result of iron disorders. Clardy and colleagues used microarray to investigate brain expression of ~8,000 genes in a rat model of developmental iron deficiency [55]. There were 334 genes with altered expression, including decreased transferrin and increased transferrin receptor 1 gene expression. Down-regulation of myelin-related genes was also seen and may contribute to the hypomyelination that occurs in states of developmental iron deficiency [56–58]. Iron repletion following weaning corrected most of the alterations, with only five genes showing alterations in iron replete rats [55]. This suggests that the effects of iron deficiency early in development may be partially or even fully reversible, at least when deficiency is corrected at a relatively young age.

There are few other array studies examining genome-wide brain gene expression. However, recently our group has performed microarray and real-time RT-PCR analyses of brain gene expression in mouse models of iron loading. We believe these data may be painting a far richer picture of what iron is doing in the brain than has ever been captured before.

We first investigated genome-wide gene expression changes in a mouse model of dietary iron overload. In this model, male mice were maintained on normal chow from weaning until 7 weeks of age and then transferred to a high-iron diet comprising normal chow supplemented with 2 % carbonyl iron for 3 weeks. The control group consisted of mice maintained on normal chow from weaning until 10 weeks of age. These experiments utilised mice of the AKR strain, which exhibit a strong iron loading phenotype [59]. The high iron diet has previously been shown to increase serum iron indices and liver iron loading [59, 60], and this has been confirmed specifically in the AKR strain under the experimental conditions used in our study [61].

Yet, despite the peripheral iron loading phenotype seen in these mice and confirmed in our study, there was no increase in the amount of non-haem iron in the brain of mice on the high iron diet, as shown in Fig. 1. Therefore, in this model utilising short-term dietary iron loading, the brain appears to be at least partly protected from the increased iron present throughout the rest of the body, whether by the BBB or by other mechanisms.

However, even though there was no apparent increase in brain non-haem iron levels, we observed various other changes. Notably, small but significant increases in brain transcripts for ferritin light chain of approximately 20 % were detected by microarray and validated by real-time RT-PCR. Consistent with this, brain transcripts for iron regulatory protein 1, which negatively regulates ferritin translation, decreased 30 %.

These changes suggest brain gene expression can be influenced even by relatively mild systemic iron loading insufficient to cause significant gross increases in brain iron levels. However, the extent of the observed changes was small and levels of transcripts for the transferrin receptor and various other important iron-related proteins were unaltered. This is consistent with previous reports that, in general, average brain levels of transferrin, transferrin receptor 1 and divalent metal transporter 1 proteins often do not change substantially in response to high systemic iron [62, 63].

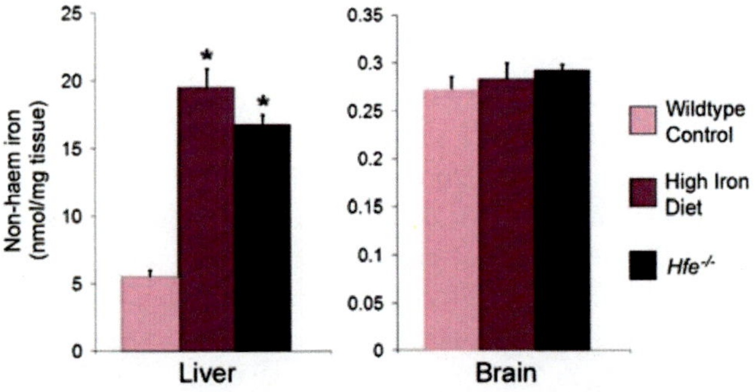

Fig. 1 Comparison between non-haem iron levels in the liver and the brain of mouse models of iron loading. All mice were male and of the AKR background strain. The levels of non-haem iron in the liver increased by similar amounts in both models of iron loading, while the levels of non-haem iron in the brain remained unchanged. $*p < 0.05$, $n \geq 4$ per group

So the most likely explanation for our findings is probably that on the one hand the BBB is indeed partly protecting the brain against the effects of peripheral iron loading in the short-term dietary model. On the other hand, the level of systemic iron loading is apparently sufficiently high to partially over-ride the protective effects enough to perturb brain iron homeostasis at the subcellular level and trigger early compensatory responses. These appear to involve initial increases in ferritin gene transcription, as well as transcriptional and probably also post-transcriptional changes involving the IRE/IRP system. However, these effects do not appear to be sufficient to lay down significantly detectable extra iron stores within ferritin or to cause gross brain iron loading.

One might therefore conclude that the BBB is essentially fully effective at protecting the brain against the effects of dietary iron overload in this short-term model. However, in addition to the changes observed for ferritin gene transcription, around 300 other genes showed significant expression changes.

Most of these changes were small in magnitude (less than twofold) and unable to be readily validated by real-time RT-PCR or other methods, suggesting short-term increases in dietary iron intake produce only subtle effects on brain gene expression. However, the alterations included sets of expression changes for groups of genes relating to particular pathways or functions already known to involve iron and likely to be biologically relevant.

We have observed similar effects in a model of genetic haemochromatosis, the $Hfe^{-/-}$ deletion mutant mouse model. This mouse was generated using a targeting vector to introduce a 1.7-kb PGKneo cassette to replace a 360-bp fragment of the *Hfe* gene encompassing a portion of exon 4 and intron 3 [64]. This produces a truncated dysfunctional protein. Two strains of mice commonly used as backgrounds

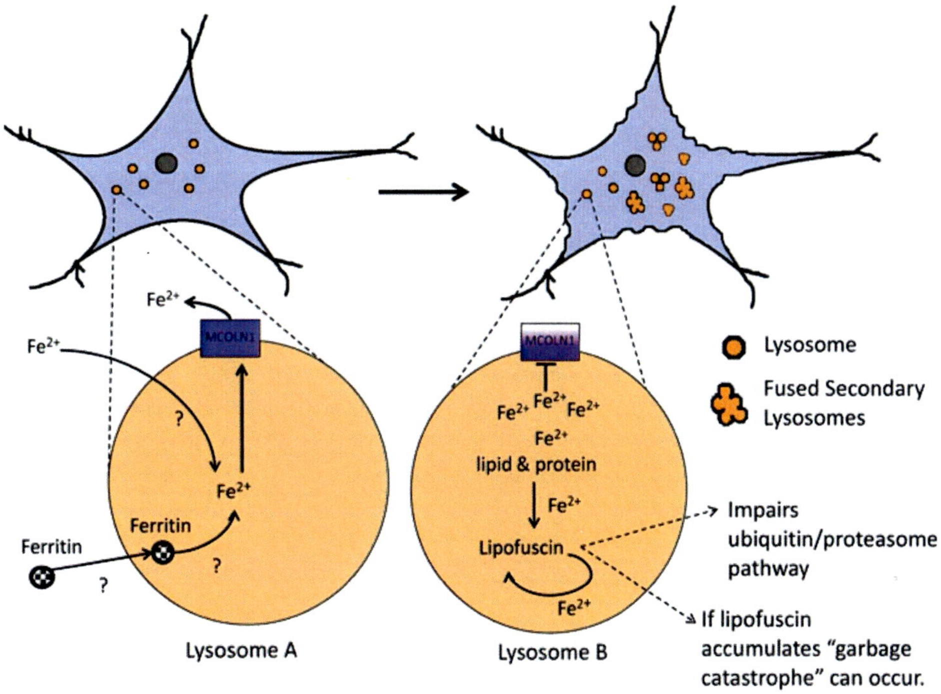

Fig. 2 The role of iron in the formation of cellular lipofuscin. Lysosome A shows the normal movement of iron through lysosomes in a healthy cell where MCOLN1 transports iron across the membrane of the lysosome. Lysosome B comes from a cell in which MCOLN1 function is impaired and iron is building up within the lysosomes. This iron acts as a catalyst in the formation of lipofuscin from degraded lipids and proteins. Some lysosomes are able to fuse together

in *Hfe*$^{-/-}$ deletion mutant mice studies are C57BL/6J and AKR [59]. Both models replicate the human disease with increased liver iron content (a gauge of total body content) [59, 64]; however, liver iron concentrations differ markedly between the two strains, with the AKR strain showing higher levels of hepatic iron [59]. We found that *Hfe*$^{-/-}$ deletion mutant mice on the AKR background, like the short-term dietary iron supplementation model, again showed relatively few changes in key genes associated systemic iron regulation yet still displayed many gene expression changes, across numerous important brain systems.

Overall, the dietary and genetic models of iron loading we have examined to date have shown very little evidence for changes in oxidative stress-related systems. There is virtually no evidence for inflammation (in strong contrast to other preliminary brain microarray studies we have analysed in adult mouse models of systemic infection, where there are numerous increases in expression of genes encoding cytokines and other inflammatory molecules, even after clearance of the systemic infection). We have also so far observed little if any evidence of changes relating to angiogenesis or vasculogenesis and few changes related to reactive oxygen species, although in the dietary iron model in particular we did observe some changes that may relate to nitric oxide signalling, which is dependent on iron, and nitrosative stress [65]. We have also seen few if any signs of gene expression changes clearly related to apoptosis or other forms of cell death in these models.

However, we do see effects on a range of other molecular systems. Notably expression changes were observed for genes causatively linked to neuronal ceroid lipofuscinosis and mucolipidosis. These diseases involve intralysosomal lipofuscin build-up that may reflect lysosomal iron accumulation (Fig. 2). The *Hfe*$^{-/-}$ brain showed most total changes and had expression changes for several genes causatively linked to neuronal ceroid lipofuscinosis [66]. However expression changes for genes causatively linked to lipofuscinosis diseases were also seen even in the short-term dietary iron model, which had fewer total changes [65].

Intriguingly, these included decreased levels of transcripts for the mucolipin 1 gene (MCOLN1), an iron channel protein and calcium transporter thought to play a role in iron release from late endosomes and lysosomes [67]. Mutations in MCOLN1 can impair the iron permeability of lysosomal membranes, preventing iron release and causing the disease Type IV mucolipidosis [67].

Lipofuscin is an intralysosomal substance that cannot be exocytosed from the cell or degraded by lysosomes [68]. It is hydrophobic and 'yellow-brownish' in colour [69]. It is considered to be a telltale sign of the ageing process because its rate of accumulation is proportional to the age of an individual [68].

The prevailing hypothesis for lipofuscin formation is that hydrogen peroxide produced as a by-product by mitochondria and other organelles permeates into the lumen of secondary lysosomes [70]. (Newly formed lysosomes are termed 'primary' lysosomes, becoming 'secondary' lysosomes once they acquire and start to degrade exogenous materials such as pathogens or membrane fragments.) Lysosomal degradation of iron-containing proteins such as ferritin and cytochromes releases iron, which can then catalyse the Fenton reaction with hydrogen peroxide [68]. The Fenton reaction produces hydroxyl radicals which cause protein and lipid peroxidation, which is followed by intermolecular cross-linking and lipofuscin formation [70].

Secondary lysosomes can fuse with other secondary lysosomes or with primary lysosomes [71]. When this occurs, materials within the lysosomes aggregate, so over time lipofuscin accumulates [71]. The buildup of lipofuscin can have strong effects on cellular functioning as it can impair the ubiquitin/proteasome pathway. Accumulation of lipofuscin has been proposed to lead to a phenomenon dubbed the 'garbage catastrophe', the cumulative effect of cellular waste not being properly eliminated due to concurrent inhibition of both proteasomes and lysosomes [72]. These changes are potentially pathogenic and therefore have the potential to affect brain function unless compensated.

We have also observed changes in transcripts of genes tied more directly to brain function. These included expression changes for genes involved in important brain functions such as neurotransmission. One prominent example of a gene showing consistent brain expression changes across both dietary and genetic models of iron overload is the gene encoding calcium/calmodulin-dependent protein kinase II α (Camk2a), a protein of considerable prominence in learning, memory and other cognitive processes. This gene shows reproducible decreases in transcripts across the models, in conjunction with a variety of other changes potentially affecting cognition [66].

As far as we are aware, there have not been any other reports examining changes in genome-wide brain gene expression in the whole brain in response to iron supplementation. The study by Liu and colleagues described above identified 31 genes with altered expression in the hippocampus of rats fed an iron-supplemented diet for 2 months after weaning [43]. None of these genes were identified as having altered expression in whole brains from iron-supplemented mice in our study.

The study by Clardy and colleagues described above, which investigated changes in 21 day-old pups of female rats fed an iron-deficient diet through gestation and lactation, identified 334 genes with altered expression [55]. Only one, stathmin-like 4, showed altered expression in our study of brains from mice following short-term (3 weeks) dietary iron supplementation [65]. The stathmin-like 4 gene is believed to be involved in microtubular organisation but is not well characterised and expression was decreased both in our study of iron supplementation and in the Clardy study of developmental iron deficiency, making it hard to interpret the findings. One possibility is that iron abnormalities of any kind, high or low, can perturb microtubular organisation.

None of the gene clusters highlighted in the Clardy iron deficiency study were found to be enriched in our study of short-term dietary iron supplementation, including myelin-related genes, although we did observe expression changes for two genes (Gdap1, Litaf) causatively linked to demyelinating forms of Charcot–Marie–Tooth disease [73]. Taken together, the above considerations suggest that gene expression changes in systemic iron loading in adults are not simply the opposite of changes in developmental iron deficiency.

Validation and the limitations of PCR

Any experimental technology can give rise to artefacts, making it important that experimental replication is carried out wherever possible using alternative approaches. Validation of RNA microarray data is often performed by real-time RT-PCR. This raises the question of phenotype and effect size. For genes with strong genetically dominant disease associations, an abnormality of one allele is sufficient to cause clinical disease. Yet this may represent a change of 50 % or less in gene function. For genes of this kind, capable of strong effects with relatively small changes, it is unreasonable to expect to observe big changes in experimental models that accurately represent the disease if a 50 % reduction or less in function causes a severe disease phenotype. Yet smaller changes are often below the limits of detection of relevant technologies such as real-time RT-PCR.

Although people often think of PCR as having molecular scale sensitivity, with the ability to amplify and detect a single molecule of DNA, sensitivity is often lower in practice, in particular for reverse transcription PCR, where RNA samples must first be reverse transcribed to complementary DNA (cDNA) before PCR. The exponential nature of PCR also imposes sensitivity limits in detecting differential expression when the relative gene expression of the test and control differs by less than twofold. This is because a twofold difference represents one cycle of a PCR reaction. While it may be possible to detect smaller fold changes that are statistically significant, basic principles of measurement suggest that it is not possible to determine effects of less than 0.5 cycles with certainty.

This is not a problem for genome-wide technologies such as microarrays, since the expression of genes of interest can be normalised relative to 'averaged' expression of large sets of genes, and typically the full background gene set (global gene expression). For changes of twofold or less microarrays may be more sensitive than real-time RT-PCR. This raises the question of the utility of using a less sensitive technique to validate the results from a more sensitive technology, at least for small changes, and suggests new approaches may be required for more accurate validation.

Other emerging technologies

New technologies now emerging include next-generation and real-time sequencing technologies. Although hybridisation-based approaches such as microarray can already provide fast and cost-effective large-scale transcriptomics for many applications, there are some applications for which sequence-based approaches are better suited [74–76]. These include determination of exact transcription boundaries as well as detection of transcripts that do not map to a known genomic sequence or are not specifically targeted by microarray probes. In addition, RNA-sequencing techniques may avoid certain problems that can lead to errors in microarray data. These include non-specific background signal or signal saturation for genes with very low or very high expression, as well as probe binding anomalies due to affinity differences, the existence of different transcript spliceoforms or other factors [74–77].

As yet, these technologies are rarely utilised in clinical practice, drug discovery or pharmacogenomics. More detailed molecular studies will be facilitated by new technologies such as real-time DNA and RNA sequencing. This will allow the tracking of a range of molecular changes over time to determine the mechanisms underlying disease progression and to help distinguish primary causative changes from secondary effects. These sensitive techniques may also overcome some of the limitations of currently available methods such as real-time RT-PCR, which has been demonstrated as not sufficiently sensitive to validate gene expression changes detected by microarray with a fold-change of less than approximately 1.4 [78], most likely due to the exponential nature of the amplification protocol.

Focussed studies of particular brain regions of interest could utilise cutting-edge RNA sequencing techniques, which provide information on novel transcripts and splice variants not specifically targeted by the arrays. This technology has already been used to determine region-specific gene expression changes in post-mortem brains from patients with Alzheimer's disease [79], although accurate quantification of transcript levels using these methods is currently limited due to issues relating to normalisation [76].

It will be valuable to clarify which gene expression changes result from effects dependent on systemic iron overload and which result from effects of deletion of *Hfe* or other iron-related genes independent of systemic iron overload. This could be investigated by performing regular venesection or iron chelation on $Hfe^{-/-}$ mice to maintain body iron levels within normal ranges or by using conditional knockout mice with brain-specific disruption of the *Hfe* gene. Rodent or human neural cell culture models may help further delineate which changes arise due to iron and which are due to dysfunction of HFE or other gene targets. This could be achieved by assessing genome-wide gene expression after either treating cells with iron or using RNA interference to knock down HFE expression in iron replete or iron-deficient cells.

Although there is clearly still a long way to go in understanding how iron acts, we believe the new high-throughput and other technologies now emerging will reveal that iron is not only a key player in assorted brain functions but that even relatively small, short-term changes in systemic iron levels may be having far more effect on everyday brain activities than is presently realised. This is particularly true for conditions such as haemochromatosis which involve effects accumulating over long time periods of decades or more, even in the juvenile forms of the disease. We therefore believe that it is important to report on systems undergoing large numbers of small changes in the literature, with the appropriate caveats on the limitations of this kind of data. Ultimately confirmation or negation will come from the cumulative weight of independent studies from different groups using a variety of different approaches. In the interim, we believe that the data now coming from high-throughput technologies is providing valuable signposts for guiding future research.

References

1. Johnstone D, Milward EA (2010) Molecular genetic approaches to understanding the roles and regulation of iron in brain health and disease. J Neurochem 113:1387–1402
2. Jazwinska EC, Cullen LM, Busfield F, Pyper WR, Webb SI, Powell LW et al (1996) Haemochromatosis and HLA-H. Nat Genet 14:249–251
3. Brissot P, Moirand R, Jouanolle AM, Guyader D, Le Gall JY, Deugnier Y et al (1999) A genotypic study of 217 unrelated probands diagnosed as "genetic hemochromatosis" on "classical" phenotypic criteria. J Hepatol 30:588–593
4. Pietrangelo A (2006) Hereditary hemochromatosis. Annu Rev Nutr 26:251–270
5. Adams PC, Barton JC (2007) Haemochromatosis. Lancet 370:1855–1860
6. Ayonrinde OT, Milward EA, Chua AC, Trinder D, Olynyk JK (2008) Clinical perspectives on hereditary hemochromatosis. Crit Rev Clin Lab Sci 45:451–484
7. Sobotka TJ, Whittaker P, Sobotka JM, Brodie RE, Quander DY, Robl M et al (1996) Neurobehavioral dysfunctions associated with dietary iron overload. Physiol Behav 59:213–219
8. Fredriksson A, Schroder N, Eriksson P, Izquierdo I, Archer T (1999) Neonatal iron exposure induces neurobehavioural dysfunctions in adult mice. Toxicol Appl Pharmacol 159:25–30
9. Pinero D, Jones B, Beard J (2001) Variations in dietary iron alter behavior in developing rats. J Nutr 131:311–318
10. Maaroufi K, Ammari M, Jeljeli M, Roy V, Sakly M, Abdelmelek H (2009) Impairment of emotional behavior and spatial learning in adult Wistar rats by ferrous sulfate. Physiol Behav 96:343–349
11. Eroglu Y, Byrne WJ (2009) Hepatic encephalopathy. Emerg Med Clin N Am 27:401–414
12. Bridle KR, Crawford DH, Fletcher LM, Smith JL, Powell LW, Ramm GA (2003) Evidence for a sub-morphological inflammatory process in the liver in haemochromatosis. J Hepatol 38:426–433
13. Demougeot C, Methy D, Prigent-Tessier A, Garnier P, Bertrand N, Guilland JC et al (2003) Effects of a direct injection of liposoluble iron into rat striatum. Importance of the rate of iron delivery to cells. Free Radic Res 37:59–67
14. Junxia X, Hong J, Wenfang C, Ming Q (2003) Dopamine release rather than content in the caudate putamen is associated with

behavioral changes in the iron rat model of Parkinson's disease. Exp Neurol 182:483–489
15. Bueno-Nava A, Gonzalez-Pina R, Alfaro-Rodriguez A (2010) Iron-dextran injection into the substantia nigra in rats decreases striatal dopamine content ipsilateral to the injury site and impairs motor function. Metab Brain Dis 25:235–239
16. Pinero DJ, Li NQ, Connor JR, Beard JL (2000) Variations in dietary iron alter brain iron metabolism in developing rats. J Nutr 130:254–263
17. Moos T, Oates PS, Morgan EH (1999) Iron-independent neuronal expression of transferrin receptor mRNA in the rat. Brain Res Mol Brain Res 72:231–234
18. Kaur D, Peng J, Chinta SJ, Rajagopalan S, Di Monte DA, Cherny RA et al (2007) Increased murine neonatal iron intake results in Parkinson-like neurodegeneration with age. Neurobiol Aging 28:907–913
19. Wang Q, Luo W, Zheng W, Liu Y, Xu H, Zheng G et al (2007) Iron supplement prevents lead-induced disruption of the blood-brain barrier during rat development. Toxicol Appl Pharmacol 219:33–41
20. Becker JS, Zoriy M, Matusch A, Wu B, Salber D, Palm C (2010) Bioimaging of metals by laser ablation inductively coupled plasma mass spectrometry (LA-ICP-MS). Mass Spectrom Rev 29:156–175
21. Chang YZ, Qian ZM, Wang K, Zhu L, Yang XD, Du JR et al (2005) Effects of development and iron status on ceruloplasmin expression in rat brain. J Cell Physiol 204:623–631
22. Ke Y, Chang YZ, Duan XL, Du JR, Zhu L, Wang K et al (2005) Age-dependent and iron-independent expression of two mRNA isoforms of divalent metal transporter 1 in rat brain. Neurobiol Aging 26:739–748
23. Qian ZM, Chang YZ, Zhu L, Yang L, Du JR, Ho KP et al (2007) Development and iron-dependent expression of hephaestin in different brain regions of rats. J Cell Biochem 102:1225–1233
24. Crichton RR, Dexter DT, Ward RJ (2011) Brain iron metabolism and its perturbation in neurological diseases. J Neural Transm Suppl 118:301–314
25. Miyajima H, Nishimura Y, Mizoguchi K, Sakamoto M, Shimizu T, Honda N (1987) Familial apoceruloplasmin deficiency associated with blepharospasm and retinal degeneration. Neurology 37:761–767
26. Takahashi Y, Miyajima H, Shirabe S, Nagataki S, Suenaga A, Gitlin JD (1996) Characterization of a nonsense mutation in the ceruloplasmin gene resulting in diabetes and neurodegenerative disease. Hum Mol Genet 5:81–84
27. Lucato LT, Otaduy MC, Barbosa ER, Machado AA, McKinney A, Bacheschi LA et al (2005) Proton MR spectroscopy in Wilson disease: analysis of 36 cases. AJNR Am J Neuroradiol 26:1066–1071
28. Hellman NE, Gitlin JD (2002) Ceruloplasmin metabolism and function. Annu Rev Nutr 22:439–458
29. de Bie P, Muller P, Wijmenga C, Klomp LW (2007) Molecular pathogenesis of Wilson and Menkes disease: correlation of mutations with molecular defects and disease phenotypes. J Med Genet 44:673–688
30. McNeill A, Pandolfo M, Kuhn J, Shang H, Miyajima H (2008) The neurological presentation of ceruloplasmin gene mutations. Eur Neurol 60:200–205
31. Duce JA, Tsatsanis A, Cater MA, James SA, Robb E, Wikhe K et al (2010) Iron-export ferroxidase activity of beta-amyloid precursor protein is inhibited by zinc in Alzheimer's disease. Cell 142:857–867
32. Rogers JT, Randall JD, Cahill CM, Eder PS, Huang X, Gunshin H et al (2002) An iron-responsive element type II in the 5'-untranslated region of the Alzheimer's amyloid precursor protein transcript. J Biol Chem 277:45518–45528
33. Burdo JR, Simpson IA, Menzies S, Beard J, Connor JR (2004) Regulation of the profile of iron-management proteins in brain microvasculature. J Cereb Blood Flow Metab 24:67–74
34. Elseweidy MM, Abd El-Baky AE (2008) Effect of dietary iron overload in rat brain: oxidative stress, neurotransmitter level and serum metal ion in relation to neurodegenerative disorders. Indian J Exp Biol 46:855–858
35. Demarquay G, Setiey A, Morel Y, Trepo C, Chazot G, Broussolle E (2000) Clinical report of three patients with hereditary hemochromatosis and movement disorders. Mov Disord 15:1204–1209
36. Demarquay G, Thobois S, Latour P, Broussolle E (2006) Hereditary hemochromatosis and movement disorders: the still controversial relationship. Response to Russo et al. in J Neurol (2004) 251:849–852. J Neurol 253:261–262
37. Rutgers MP, Pielen A, Gille M (2007) Chronic cerebellar ataxia and hereditary hemochromatosis: causal or coincidental association? J Neurol 254:1296–1297
38. Russo N, Edwards M, Andrews T, O'Brien M, Bhatia KP (2004) Hereditary haemochromatosis is unlikely to cause movement disorders–a critical review. J Neurol 251:849–852
39. Fasano A, Bentivoglio AR, Colosimo C (2007) Movement disorder due to aceruloplasminemia and incorrect diagnosis of hereditary hemochromatosis. J Neurol 254:113–114
40. Golub MS, Germann SL, Araiza RS, Reader JR, Griffey SM, Lloyd KC (2005) Movement disorders in the Hfe knockout mouse. Nutr Neurosci 8:239–244
41. LaVaute T, Smith S, Cooperman S, Iwai K, Land W, Meyron-Holtz E et al (2001) Targeted deletion of the gene encoding iron regulatory protein-2 causes misregulation of iron metabolism and neurodegenerative disease in mice. Nat Genet 27:209–214
42. Galy B, Holter SM, Klopstock T, Ferring D, Becker L, Kaden S et al (2006) Iron homeostasis in the brain: complete iron regulatory protein 2 deficiency without symptomatic neurodegeneration in the mouse. Nat Genet 38:967–969
43. Liu M, Xiao DS, Qian ZM (2007) Identification of transcriptionally regulated genes in response to cellular iron availability in rat hippocampus. Mol Cell Biochem 300:139–147
44. Tefferi A, Bolander ME, Ansell SM, Wieben ED, Spelsberg TC (2002) Primer on medical genomics. Part III: Microarray experiments and data analysis. Mayo Clin Proc 77:927–940
45. Shi L, Reid LH, Jones WD, Shippy R, Warrington JA, Baker SC et al (2006) The MicroArray Quality Control (MAQC) project shows inter- and intraplatform reproducibility of gene expression measurements. Nat Biotechnol 24:1151–1161
46. Chen JJ, Hsueh HM, Delongchamp RR, Lin CJ, Tsai CA (2007) Reproducibility of microarray data: a further analysis of microarray quality control (MAQC) data. BMC Bioinforma 8:412
47. Quackenbush J (2001) Computational analysis of microarray data. Nat Rev Genet 2:418–427
48. Workman C, Jensen LJ, Jarmer H, Berka R, Gautier L, Nielser HB et al (2002) A new non-linear normalization method for reducing variability in DNA microarray experiments. Genome Biol 3: research0048
49. Shippy R, Fulmer-Smentek S, Jensen RV, Jones WD, Wolber PK, Johnson CD et al (2006) Using RNA sample titrations to assess microarray platform performance and normalization techniques. Nat Biotechnol 24:1123–1131
50. Reiner A, Yekutieli D, Benjamini Y (2003) Identifying differentially expressed genes using false discovery rate controlling procedures. Bioinformatics 19:368–375
51. Rothman KJ (1990) No adjustments are needed for multiple comparisons. Epidemiology 1:43–46
52. Bender R, Lange S (2001) Adjusting for multiple testing–when and how? J Clin Epidemiol 54:343–349
53. Shi L, Perkins RG, Fang H, Tong W (2008) Reproducible and reliable microarray results through quality control: good laboratory proficiency and appropriate data analysis practices are essential. Curr Opin Biotechnol 19:10–18
54. Maouche S, Poirier O, Godefroy T, Olaso R, Gut I, Collet JP et al (2008) Performance comparison of two microarray platforms to

assess differential gene expression in human monocyte and macrophage cells. BMC Genomics 9:302

55. Clardy SL, Wang X, Zhao W, Liu W, Chase GA, Beard JL et al (2006) Acute and chronic effects of developmental iron deficiency on mRNA expression patterns in the brain. J Neural Transm Suppl 71:173–196
56. Beard JL, Connor JR (2003) Iron status and neural functioning. Annu Rev Nutr 23:41–58
57. Ortiz E, Pasquini JM, Thompson K, Felt B, Butkus G, Beard J et al (2004) Effect of manipulation of iron storage, transport, or availability on myelin composition and brain iron content in three different animal models. J Neurosci Res 77:681–689
58. Beard J, Han O (2009) Systemic iron status. Biochim Biophys Acta 1790:584–588
59. Fleming RE, Holden CC, Tomatsu S, Waheed A, Brunt EM, Britton RS et al (2001) Mouse strain differences determine severity of iron accumulation in Hfe knockout model of hereditary hemochromatosis. Proc Natl Acad Sci USA 98:2707–2711
60. Dupic F, Fruchon S, Bensaid M, Loreal O, Brissot P, Borot N et al (2002) Duodenal mRNA expression of iron related genes in response to iron loading and iron deficiency in four strains of mice. Gut 51:648–653
61. Drake SF, Morgan EH, Herbison CE, Delima R, Graham RM, Chua AC et al (2007) Iron absorption and hepatic iron uptake are increased in a transferrin receptor 2 (Y245X) mutant mouse model of hemochromatosis type 3. Am J Physiol Gastrointest Liver Physiol 292:G323–G328
62. Moos T, Trinder D, Morgan EH (2000) Cellular distribution of ferric iron, ferritin, transferrin and divalent metal transporter 1 (DMT1) in substantia nigra and basal ganglia of normal and beta2-microglobulin deficient mouse brain. Cell Mol Biol (Noisy-le-grand) 46:549–561
63. Lykkesfeldt J, Morgan E, Christen S, Skovgaard LT, Moos T (2007) Oxidative stress and damage in liver, but not in brain, of Fischer 344 rats subjected to dietary iron supplementation with lipid-soluble [(3,5,5-trimethylhexanoyl)ferrocene]. J Biochem Mol Toxicol 21:145–155
64. Zhou XY, Tomatsu S, Fleming RE, Parkkila S, Waheed A, Jiang J et al (1998) HFE gene knockout produces mouse model of hereditary hemochromatosis. Proc Natl Acad Sci USA 95:2492–2497
65. Johnstone D, Milward EA (2010) Genome-wide microarray analysis of brain gene expression in mice on a short-term high iron diet. Neurochem Int 56:856–863
66. Johnstone D, Acikyol B, Graham RM, Trinder D, Scott RJ, Olynyk JK et al (2010) Gene expression studies in three different mouse models support the case for neurologic sequelae in iron overload disorders and provide new insights into mechanism. Society for Neuroscience Meeting, San Diego, USA
67. Dong XP, Cheng X, Mills E, Delling M, Wang F, Kurz T et al (2008) The type IV mucolipidosis-associated protein TRPML1 is an endolysosomal iron release channel. Nature 455:992–996
68. Terman A, Brunk UT (2004) Lipofuscin. Int J Biochem Cell Biol 36:1400–1404
69. Hohn A, Jung T, Grimm S, Grune T (2010) Lipofuscin-bound iron is a major intracellular source of oxidants: role in senescent cells. Free Radic Biol Med 48:1100–1108
70. De Meyer GR, De Keulenaer GW, Martinet W (2010) Role of autophagy in heart failure associated with aging. Hear Fail Rev 15:423–430
71. Brunk UT, Jones CB, Sohal RS (1992) A novel hypothesis of lipofuscinogenesis and cellular aging based on interactions between oxidative stress and autophagocytosis. Mutat Res 275:395–403
72. Gray DA, Woulfe J (2005) Lipofuscin and aging: a matter of toxic waste. Sci Aging Knowledge Environ 2005:re1
73. Street VA, Bennett CL, Goldy JD, Shirk AJ, Kleopa KA, Tempel BL et al (2003) Mutation of a putative protein degradation gene LITAF/SIMPLE in Charcot-Marie-Tooth disease 1C. Neurology 60:22–26
74. Mortazavi A, Williams BA, McCue K, Schaeffer L, Wold B (2008) Mapping and quantifying mammalian transcriptomes by RNA-Seq. Nat Methods 5:621–628
75. Wilhelm BT, Marguerat S, Watt S, Schubert F, Wood V, Goodhead I et al (2008) Dynamic repertoire of a eukaryotic transcriptome surveyed at single-nucleotide resolution. Nature 453:1239–1243
76. Wang Z, Gerstein M, Snyder M (2009) RNA-Seq: a revolutionary tool for transcriptomics. Nat Rev Genet 10:57–63
77. Nagalakshmi U, Waern K, Snyder M (2010) RNA-Seq: a method for comprehensive transcriptome analysis. Curr Protoc Mol Biol Chapter 4:Unit 4.11.11–13
78. Morey JS, Ryan JC, Van Dolah FM (2006) Microarray validation: factors influencing correlation between oligonucleotide microarrays and real-time PCR. Biol Proced Online 8:175–193
79. Twine NA, Janitz K, Wilkins MR, Janitz M (2011) Whole transcriptome sequencing reveals gene expression and splicing differences in brain regions affected by Alzheimer's disease. PLoS One 6:e16266

The role of metal ions in dopaminergic neuron degeneration in Parkinsonism and Parkinson's disease

Irmgard Paris and Juan Segura-Aguilar

Abstract

Parkinson's disease is characterized by the selective degeneration of neuromelanin-containing dopaminergic neurons in the substantia nigra and locus coeruleus. Although the cause of this disease remains unknown, several transition metals, including manganese and copper, have been associated with the development of the atypical form of Parkinsonism, and iron accumulation has been associated with the development of Parkinson's disease. Manganese^{3+} is a strong oxidizing agent, which oxidizes dopamine to aminochrome (dopaminochrome), the precursor of neuromelanin. Aminochrome formation in cell culture medium induces acute cell death in cells that uptake aminochrome, explaining the role of manganese in the development of atypical Parkinsonism. Copper accumulation in Wilson's disease also induces Parkinsonism as one of the main symptoms, and an atypical Parkinsonism has also been observed in young copper miners. Interestingly, copper is able to complex with dopamine, which can be taken up by cells expressing the dopamine transporter, inducing caspase-independent cell death with formation of autophagic vacuoles. Iron is also able to form a complex with dopamine, the neurotoxic action of which also depends on the cellular expression of the dopamine transporter. The neurotoxicities of these transition metals to cells expressing the dopamine transporter all involve dopamine oxidation to quinones and require the inhibition of DT-diaphorase.

Keywords

Dopamine • Aminochrome • Neurodegeneration • Manganese • Copper • Iron • Parkinsonism • Neurotoxicity • Metals

Introduction

The possible role of the ions of the metals manganese, copper, and iron in the neurodegenerative process has been associated with the formation of reactive oxygen species. Manganese^{2+} is able to react with superoxide radicals generating hydrogen peroxide, the precursor of hydroxyl radicals, and both Fe^{2+} and Cu^{+} are able to catalyze the Fenton reaction with the formation of hydroxyl radicals in the presence of hydrogen peroxide. This review discusses other mechanisms of dopamine neuron degeneration induced by the ions of the metals copper, manganese, and iron, where the formation of reactive oxygen species is only a part of the mechanism that induces cell death. The degeneration of dopamine neurons is associated with the motoric symptoms observed in Parkinsonism induced by manganese and copper. The association of iron with Parkinson's disease is based on the accumulation of iron in the affected regions of the brain.

Manganese

Exposure to excessive levels of manganese (Mn) can induce severe psychiatric and extrapyramidal motor dysfunction closely resembling Parkinson's disease. The clinical

J. Segura-Aguilar (✉)
Program of Molecular and Clinical Pharmacology, ICBM, Faculty of Medicine, Independencia 1027, 70000 Casilla, Santiago, Chile
e-mail: jsegura@med.uchile.cl

W. Linert and H. Kozlowski (eds.), *Metal Ions in Neurological Systems*,
DOI 10.1007/978-3-7091-1001-0_3,

manifestations of manganese toxicity arise from focal injury to the basal ganglia. This region, characterized by intense oxygen consumption and significant dopamine content, can undergo mitochondrial dysfunction, depletion of peroxidase and catalase, and imbalances in catecholamine levels following manganese exposure [1]. Welding fumes have been associated with the development of Parkinsonism, and the evaluation of rats exposed to gas metal arc-mild steel with low manganese, or to manual metal arc-hard surfacing with high manganese, revealed a loss of tyrosine hydroxylase in the striatum and midbrain [2].

A brief exposure to Mn^{2+} increases reactive oxygen species levels and glutathione production, decreases oxygen consumption affecting mitochondrial membrane potential, and induces dopaminergic neuronal death [3]. Manganese chloride (Mn^{2+})- and LPS-induced reactive oxygen species and nitric oxide, cytokine release, and dopaminergic neurotoxicity in microglia were significantly attenuated by pre-treatment with the potential anti-inflammatory agents minocycline and naloxone [4, 5]. A sub-chronic exposure to Mn during fetal development leads to temporally distinct patterns of glial activation, which result in elevated nitric oxide stress in distinct populations of basal ganglia neurons [6]. Treatment with minocycline, an inhibitor of microglial activation, prior to manganese exposure attenuated microglial activation and mitigated IL-1beta, TNF-alpha, and iNOS production as well as dopaminergic neurotoxicity [7].

Tyrosine hydroxylase is the rate-limiting enzyme in dopamine synthesis and is regulated acutely by phosphorylation at Ser40 and chronically by protein synthesis. Sustained phosphorylation of tyrosine hydroxylase at Ser40 and the consequent stimulation of enzyme activity both occurred at low concentrations of Mn^{2+}, and this effect represents a potential mechanism for Mn^{2+}-induced neuronal toxicity that does not involve H_2O_2-mediated cell death [8]. Acute Mn^{2+} exposure induces cytoskeleton dysfunction prior to degeneration, and chronic Mn^{2+} exposure results in neurochemical dysfunction with features overlapping those of Parkinson's disease [9]. Induction of NF-kappaB and activation of nitric oxide synthase through reactive oxygen species was suggested to be involved in Mn^{2+}-induced apoptosis in primary mesencephalic cells exposed to dopamine prior to Mn^{2+} [10].

Weanling rats chronically exposed to Mn exhibited significant Mn accumulation in several brain regions. However, rats receiving the selective dopamine transporter (DAT) inhibitor GBR12909 had significantly lower Mn levels only in the globus pallidus compared with saline-treated rats. These data show that inhibition of the DAT exclusively inhibits Mn accumulation in the globus pallidus during chronic exposure to the metal [11]. Postnatal Mn exposure caused persistent declines in DAT protein expression and [^{3}H]dopamine uptake in the striatum and nucleus accumbens, as well as long-term reductions in striatal dopamine efflux [12]. However, Mn accumulated to similar levels in PC12 cells overexpressing the DAT and control PC12 cells following incubation with manganese chloride, suggesting that the DAT is not sufficient for manganese cytotoxicity [13]. In general, Mn^{2+} has been used to study the mechanism of manganese neurotoxicity, and the possible involvement of the DAT in Mn uptake was found to be restricted to the globus pallidus region [11]. However, it is worth noting that the divalent metal transporter 1 (DMT1) is able to take up Mn^{2+} into various tissues given its wide expression. DMT1 has been reported to be expressed in the striatum, ventral mesencephalon, substantia nigra, hippocampus, caudate nucleus, putamen, corpus callosum, and cerebellum; in neurons in the cortex, striatum, cerebellum, and thalamus; in ependymal cells lining the third ventricle; and in vascular cells throughout the brain. DMT1 is also expressed in several cell lines, such as MES23.5, SH-SY5Y, PC12, RCSN-3 (derived from rat substantia nigra), rat C6 astrocytoma, human U87 glioblastoma, and hippocampal cells [14–24]. Therefore, manganese does not accumulate selectively in dopaminergic neurons, and its neurotoxic effect will differentially affect various neuronal systems. The wide expression of DMT1 may explain why manganese induces an atypical Parkinsonism. Manganese induces a consistent pattern characterized by damage to the globus pallidus, sparing of the substantia nigra pars compacta, and absence of Lewy bodies. This finding contrasts what is seen in Parkinson's disease, in which there is preferential degeneration of dopaminergic neurons in the substantia nigra pars compacta coupled with Lewy bodies and preservation of the pallidum [25]. The uptake of Mn^{2+} mediated by DMT1 is probably not the most neurotoxic event because manganese can be oxidized to Mn^{3+} in the presence of superoxide (Fig. 1). Auto-oxidation of dopamine to aminochrome was found to be considerably potentiated by Mn^{2+}, suggesting that Mn^{2+} is contributing dopamine oxidation as a consequence of its oxidation to Mn^{3+} [26].

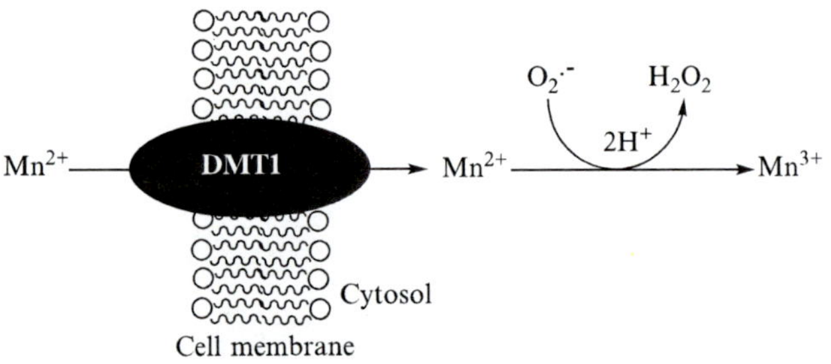

Fig. 1 Manganese uptake and intracellular manganese oxidation. Mediated by DMT1 transporter manganese^{2+} is taken up into the cytosol of the cell where it can be oxidized to manganese^{3+} catalyzed by superoxide radicals with concomitant formation of hydrogen peroxide

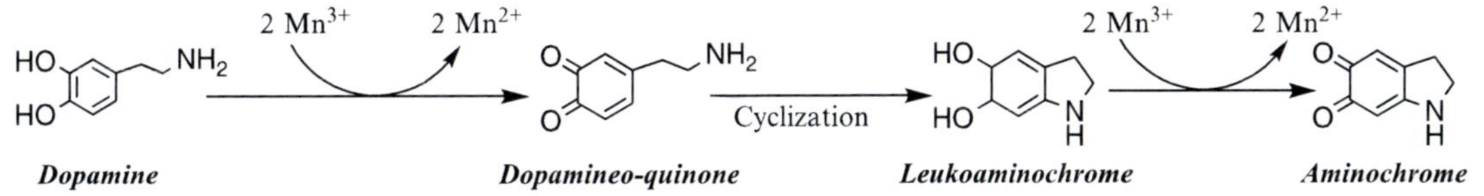

Fig. 2 Dopamine oxidation to aminochrome catalyzed by manganese^{3+}. Manganese^{3+} oxidizes dopamine to dopamine *o*-quinone that at physiological pH spontaneously cyclizes to leukoaminochrome that doesn't require oxygen to form aminochrome since manganese^{3+} is able to catalyze this reaction both in aerobic and anaerobic conditions [28]

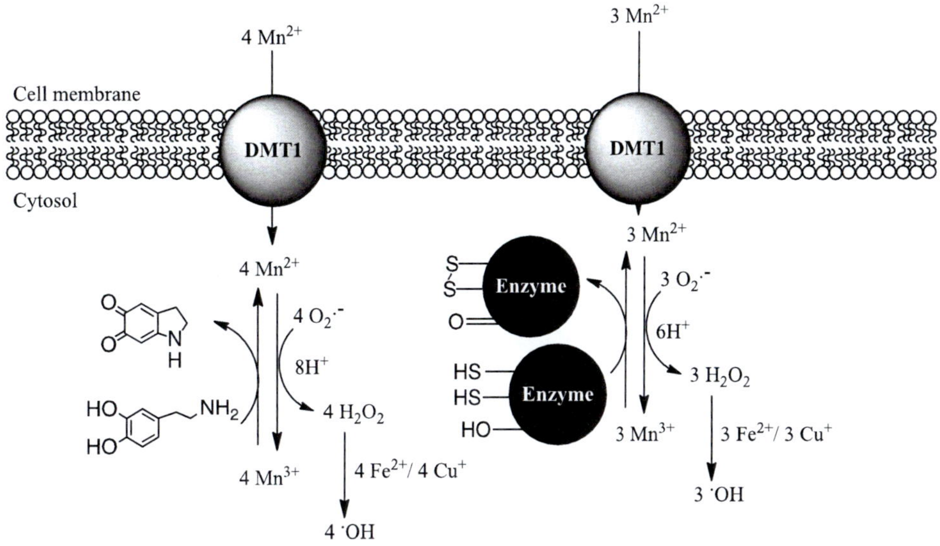

Fig. 3 Intracellular effect of manganese^{2+}. Manganese^{2+} uptake induces intracellular dopamine oxidation and protein oxidation by oxidizing the thiol and hydroxyl groups of cysteine and serine

Manganese^{3+} is a potent oxidizing agent that is able to oxidize dopamine, and dopamine oxidation catalyzed by Mn^{3+} does not require oxygen (Fig. 2). In addition, Mn^{3+} was far more efficient than Mn^{2+}, Mn^{4+} (MnO_2), O_2^-, or H_2O_2 in oxidizing the catecholamine [27].

Dopamine is oxidized by Mn^{3+} to yield dopamine *o*-quinone, which is the precursor of aminochrome. This is a transient compound at physiological pH because its amino chain undergoes cyclization to leukoaminochrome, which is oxidized by Mn^{3+} to aminochrome in the absence of oxygen. Stable dopamine *o*-quinone can only be found when the pH is lower than 2.0 [28]. However, Mn^{3+} can also oxidize thiol groups of proteins or hydroxyl groups of amino acids, thereby inactivating enzymes not only in dopaminergic neurons, but also in other neurons that express DMT1. Therefore, manganese is not a selective neurotoxic agent for dopaminergic neurons, explaining the atypical Parkinsonism resulting from the neurotoxic effects of manganese in different neuronal systems. This non-specific action of manganese explains why depletion of dopamine does not prevent manganese neurotoxicity and why, in the presence of 5 mM GSH and 10 mM *N*-acetylcysteine, cells otherwise sensitive to manganese are protected [29]. Intracellular dopamine oxidation and the oxidation of the thiol and hydroxyl groups of cysteine and serine, respectively, can participate in a dopamine-independent mechanism of manganese neurotoxicity, although both mechanisms result in the concomitant formation of reactive oxygen species, generating a redox cycling between Mn^{2+} and Mn^{3+} that potentiates the neurotoxic effects of manganese (Fig. 3).

One mechanism of manganese neurotoxicity involves Mn^{3+}-dependent oxidation of thiol groups in cysteine-containing enzymes, which leads to a loss of activity leading to loss of function. The total thiol content was lowered 40 % in cultures of the neuroblastoma clone N1E 115 treated with manganese. This decreased thiol content was also reflected by the reduced activity of the thiolenzyme glyceraldehyde-3-phosphate dehydrogenase in manganese-exposed cells [30]. Mn^{3+}, a powerful oxidizing agent, is able to oxidize hydroxyl groups of amino acids such as serine, threonine, and tyrosine present in the polypeptide chain of enzymes, which may result in a loss of enzymatic activity. Manganese treatment of PC12 cells induces inhibition of proteasome activity with significant accumulation of protein carbonyls arising from damage to proteins [31].

The dopamine-dependent mechanism of manganese neurotoxicity involves Mn^{3+}-catalyzed dopamine oxidation to aminochrome (Fig. 4). Studies in RCSN-3 cells of dopamine oxidation catalyzed by Mn^{3+} in the cell culture medium prior to addition to the cells showed that aminochrome

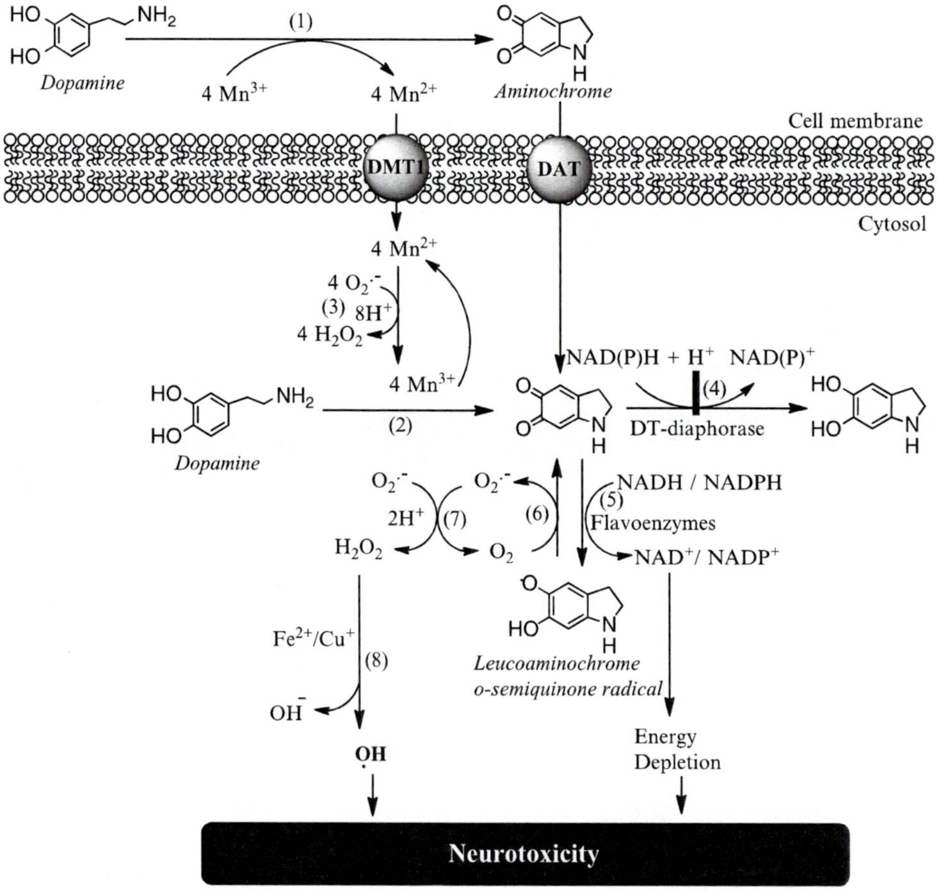

Fig. 4 The dopamine-dependent mechanism of manganese neurotoxicity

neurotoxicity was dependent on DT-diaphorase inhibition and resulted in the formation of a leukoaminochrome-*o*-semiquinone radical. The Mn^{3+} neurotoxicity induced the formation of intracellular hydroperoxides and hydroxyl radicals measured by electron spin resonance with an absence of apoptosis, a strong decrease of mitochondrial membrane potential, and a swelling and disruption of the outer and inner mitochondrial membranes (assessed by transmission electron microscopy) [32]. These results support the proposed role of leukoaminochrome-*o*-semiquinone radical as a neurotoxin in dopaminergic neurons and of DT-diaphorase as a neuroprotective enzyme [33–44].

Copper

Several important enzymes within the brain, such as dopamine *β*-mono-oxygenase, peptidyl-α-amidating mono-oxygenase, Cu/Zn superoxide dismutase, cytochrome c oxidase, and lysine oxidase, require a copper cofactor for their enzymatic activity. On the other hand, accumulation or deficiency of copper is associated with neurodegenerative disorders, such as Wilson's and Menke's diseases [45].

Copper induces an increase in oxidative stress, resulting in DNA damage and activation of p53-dependent cell death [46]. Neuronal cell death induced by GSH depletion is dependent on trace levels of extracellular copper in the culture medium [47]. Cultured astrocytes and neurons treated with copper reveal that mitochondrial membrane permeability transition and oxidative and nitrosative stress represent major factors in copper-induced toxicity, whereas oxidative and nitrosative stress appear to play a major role in neuronal injury [48]. In vivo, copper treatment alone caused no significant learning and memory impairments in behavioral tasks [49]. However, trace amounts of copper in cholesterol-fed mice induced significant learning and memory impairments accompanied by the upregulation of amyloid precursor protein, which contributes to amyloid beta-peptide deposition, increased protein expression of tumor necrosis factor-alpha, and degradation of IkappaB protein. Furthermore, an increased production of inducible nitric oxide synthase and cyclo-oxygenase-2 was observed in the hippocampus and cerebral cortex [50]. Copper concentrations are elevated in amyloid plaques, and copper binds with high affinity to the Abeta peptide, promoting its oligomerization and neurotoxicity [51].

Wilson's disease is a copper deposition disorder caused by a genetic mutation in a copper transporter ATPase (ATP7B), which is essential for the maintenance of intracellular copper concentration. This mutation prevents the excretion of excess copper from the liver to the bile, resulting in the hepatic accumulation of copper. Interestingly, the major impact of this copper overload is not in the liver but in the brain, leading to neurologic Wilson's disease. A study on the clinical presentations of Wilson's disease in 282 patients showed that 69 % of the patients were classified as having neurologic Wilson's disease, the predominant neurologic feature of which was Parkinsonism (62.3 %) [52]. The fact that Parkinsonism is one of the major neurologic symptoms of Wilson's disease opens the question as to why copper overload in the brain affects dopaminergic neurons so extensively. Young workers in the copper refining industry exposed to high concentrations of copper developed an irreversible atypical Parkinsonism [53]. The ability of reduced copper to catalyze the Fenton reaction is a non-specific mechanism and cannot explain the selective action of copper on dopaminergic neurons. One possible explanation for the selectivity of copper resulting in extensive neurodegeneration of dopaminergic neurons is the ability of Cu^{2+} to form a complex with dopamine (reaction 1) based on the formation of ionic binding between the positive charge of Cu^{2+} and the negative charges of dissociated hydroxyl groups [34].

The DAT is able to take up the Cu–dopamine complex into cells (reaction 2) that express the protein, conferring specificity to dopaminergic neurons or cells expressing DAT. However, uptake of the Cu–dopamine complex is not restricted to dopaminergic neurons, which may explain why Parkinsonism is the main symptom of Wilson's disease. Dopamine is oxidized to aminochrome (reaction 3) inside cells when Cu^{+} is released from the Cu–dopamine complex and chelated by proteins (reaction 4). Aminochrome can be reduced with two electrons to yield leukoaminochrome via a reaction catalyzed by DT-diaphorase (reaction 5), thus preventing aminochrome from participating in such neurotoxic reactions as one-electron reduction to leukoaminochrome *o*-semiquinone radical (reaction 6), which is extremely reactive with oxygen [54]. This reaction is catalyzed by flavoenzymes that reduce quinones with one electron using NADH or NADPH. Leukoaminochrome *o*-semiquinone radical reduces oxygen, generating superoxide radicals (reaction 7) that spontaneously or enzymatically (via superoxide dismutase) generate hydrogen peroxide (reaction 8). When NADH is used in this redox cycling, both oxygen and NADH are depleted, generating an energy collapse because of ATP depletion, which induces cell death. However, when the flavoenzymes use NADPH in this redox cycling, there results both a NADPH depletion affecting the reduction of oxidized glutathione and an oxygen depletion, which also affects the electron transport chain in the mitochondria and leads to ATP depletion (Fig. 5). Under these conditions, the Cu–dopamine complex induces caspase-independent cell death with formation of autophagic vacuoles [39]. Animals subjected to a unilateral intranigral injection of $CuSO_4$ and dicoumarol, an inhibitor of DT-diaphorase, presented a significant and characteristic contralateral rotational behavior and a loss of tyrosine hydroxylase-positive staining [55]. Interestingly, the formation of Cu–dopamine complex requires DT-diaphorase inhibition both in cell culture and in vivo, supporting the proposed neuroprotective role of DT-diaphorase [39, 55]. Copper inhibits the vesicular H(+)-ATPase required for DAT into monoaminergic vesicles mediated by VMAT-2, resulting in an increase of free cytosolic dopamine and its subsequent oxidation to aminochrome [56], supporting the role of dopamine in copper neurotoxicity.

Iron

Iron is essential for life because it is an essential component of the heme group of hemoglobin, cytochromes, and enzymes, but it is also involved in neurodegenerative mechanisms, such as protein aggregation, free radical generation, and oxidative stress. The role of iron in the pathophysiology of Parkinson's disease has been extensively studied and supported by the finding that iron accumulates in the brain regions affected by this disease [57]. The iron concentration in the substantia nigra correlates with UPDRS motor score, indicating that iron concentration can function as an in vivo biomarker to objectively evaluate the status of Parkinson's disease [58]. Iron accumulation in dopaminergic neurons and glial cells in the substantia nigra of Parkinson's disease patients may contribute to the neurodegenerative process in this disease, but the mechanisms involved in iron accumulation remain unclear. The major transport protein responsible for uptake of iron is the DMT1, and it is expressed in rat substantia nigra both with and without the iron regulatory element (IRE) in neurons, astrocytes, and microglia, but not in oligodendrocytes [59]. Recent studies demonstrate that the 1B isoform is regulated post-translationally by degradation via the proteasome pathway, and overexpression of parkin results in a decrease in the 1B isoform of DMT1 [60].

The role of transferrin and its receptor in iron accumulation was discarded, and the overexpression of lactoferrin receptors and DMT1 has been reported. A mutation in DMT1, which impairs iron transport, protected rodents against the Parkinsonism-inducing neurotoxins 1-methyl-4-phenyl-1,2,3,6-tetrahydropyridine hydrochloride (MPTP) and 6-hydroxydopamine [61, 62]. Ferroportin1 and

Fig. 5 The dopamine-dependent mechanism of copper neurotoxicity

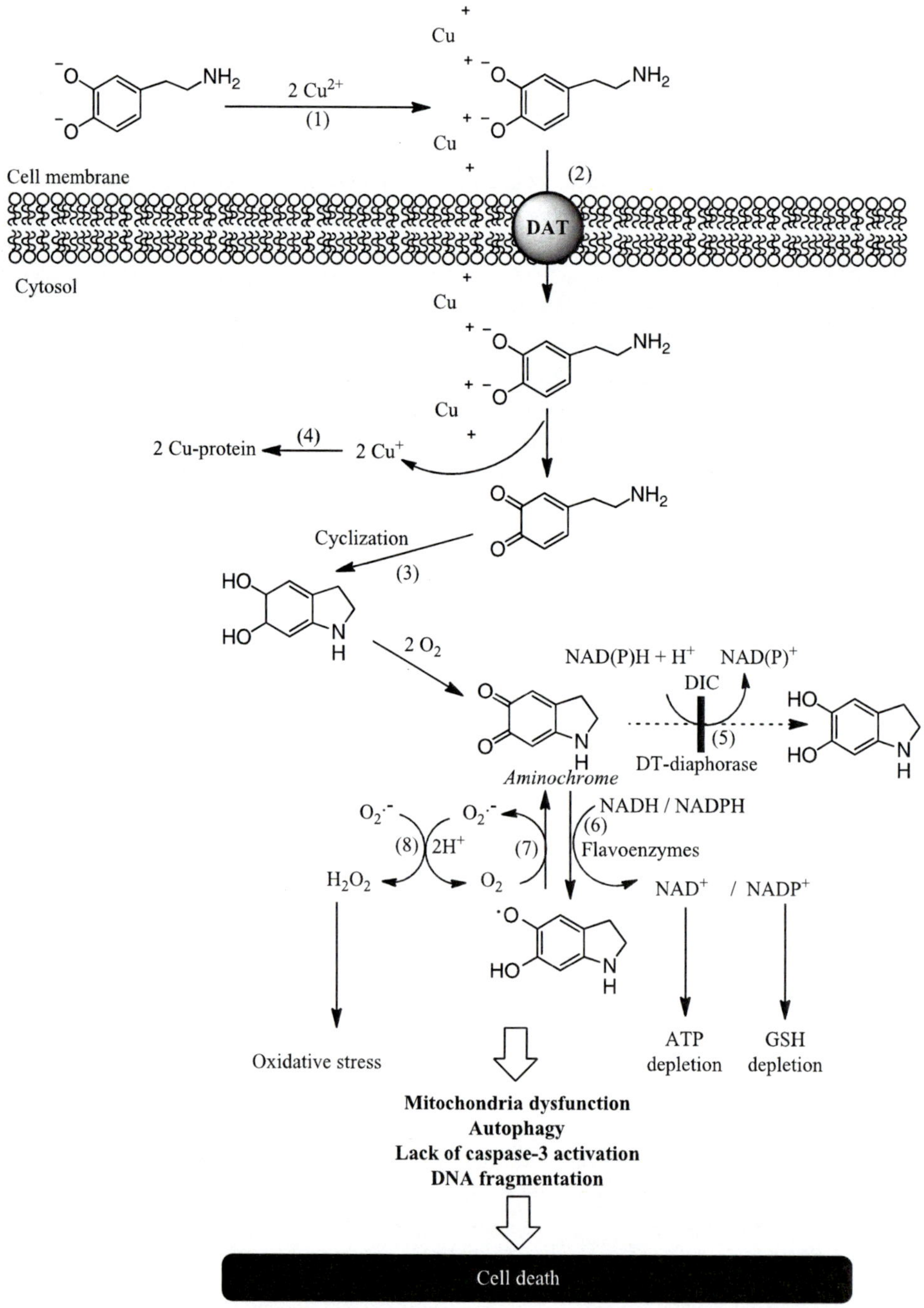

hephaestin, two iron export proteins, regulate iron export in the gut and are expressed by astrocytes, microglia, oligodendrocytes, and neurons in the substantia nigra [63]. It was reported that 6-hydroxydopamine induced the downregulation of the iron transporters ferroportin 1 and hephaestin, resulting in the decreased iron efflux and iron accumulation in primary ventral mesencephalic neurons and in MES23.5 dopaminergic cells. However, under iron-overload conditions, ferroportin 1 showed dose-dependent upregulation, whereas hephaestin showed no response, indicating that 6-hydroxydopamine-induced downregulation was not caused by increased intracellular iron content [64]. Overexpression of the human ferritin heavy chain (H-ferritin) in dopamine neurons exerts significant neuroprotection, possibly by modulating iron homeostasis and restoring ubiquitin proteasome activity [65]. Reduction in the levels of glutathione in immortalized midbrain-derived dopaminergic neurons leads to increases in the cellular pool of labile iron that are independent of the induction of either iron regulatory protein/iron regulatory element (IRP/IRE) or hypoxia inducible factor, but are dependent on hydrogen peroxide and protein synthesis. A link between glutathione depletion in dopaminergic neurons with an increase in iron levels based on translational activation of transferrin receptor-1 has been suggested [66]. L-Ferritin was found to be localized in neuromelanin granules [67].

In MPTP- and 6-hydroxydopamine-induced dopaminergic neurotoxicity and in Parkinson's disease, iron accumulates in the substantia nigra pars compacta. Pretreatment with iron chelators such as desferal, clioquinol, VK-28, and M30 is neuroprotective in both neurotoxin models [68]. The naturally occurring iron chelator phytic acid decreased 1-methyl-4-phenylpyridinium [MPP(+)]-induced caspase-3 activation and DNA fragmentation, and increased cell survival in immortalized rat mesencephalic/dopaminergic cells exposed to excess iron [69]. 6-Hydroxymelatonin reduces Fe^{2+}-induced neurotoxicity in the rat hippocampus [70]. Dopamine and Fe^{2+} in combination induce apoptosis with significantly increased concentrations of hydroxyl radicals and malondialdehyde in SH-SY5Y cells [71].

Neuromelanin blocks hydroxyl radical production by Fenton's reaction and also inhibits the iron-mediated oxidation of ascorbic acid. The sequestration of iron into a stable iron–neuromelanin complex prevents dopamine oxidation, inhibiting the formation of neurotoxic dopamine quinones [72]. L-Ferritin was found to be localized in neuromelanin granules [67]. Lisuride, a dopamine agonist, protects against iron-induced lesions [73].

The formation of a Fe–dopamine complex was reported to be a mechanism for specific neurotoxicity in cells expressing DAT such as dopamine, norepinephrine(NET), and serotonin (SERT) transporters [35, 74]. It was reported that iron forms a complex with dopamine [75] and that the Fe–dopamine complex protected isolated hepatocytes against hypoxia-re-oxygenation-induced injury [76] because these cells do not express DAT. It was also reported that iron accumulates into dopamine neurovesicles, and the inhibition of dopamine synthesis results in a decreased vesicular storage of iron [77]. These results suggest that VMAT-2 transports the Fe–dopamine complex into monoaminergic vesicles, supporting the idea that the Fe–dopamine complex can be formed in cells that synthesize dopamine. The precursor of dopamine, L-dopa, also forms a complex with Fe^{3+} [78].

Fe–dopamine complex formation occurs in the extracellular space (Fig. 6, reaction 1), and the uptake of the Fe–dopamine complex into cells is mediated by DAT, NAT and SERT. Once inside the cell, the Fe–dopamine complex undergoes dopamine oxidation to dopamine *o*-quinone (Fig. 6, reaction 2), which at physiological pH spontaneously cyclizes to aminochrome in a two-step reaction (reaction 3). Aminochrome can be reduced to leukoaminochrome in a reaction catalyzed by DT-diaphorase, and the product prevents aminochrome from participating in neurotoxic reactions, such as the one-electron reduction catalyzed by flavoenzymes that use NADH or NADPH as an electron donor (reaction 4). The one-electron reduction of aminochrome results in the formation of leukoaminochrome *o*-semiquinone radical (reaction 5), which is extremely reactive with oxygen [54] and neurotoxic [32, 34–40, 43, 79]. The auto-oxidation of leukoaminochrome *o*-semiquinone radical in the presence of oxygen generates a redox cycling (reaction 6) with concomitant formation of a superoxide radical, which spontaneously or enzymatically dismutates to H_2O_2. This redox cycling will proceed until oxygen, NADH, or NADPH is depleted. The depletion of NADH and oxygen will affect the mitochondrial electron transport chain, thereby inhibiting ATP formation, which induces an energy collapse and cell death. The depletion of NADPH prevents the reduction of oxidized glutathione and a subsequent depletion of GSH, which is an important antioxidant. Upon oxidation of the Fe–dopamine complex, Fe^{3+} is reduced to Fe^{2+} (reaction 2), which is able to catalyze the formation of a hydroxyl radical (reaction 8) from the hydrogen peroxide generated by the dismutation of superoxide radicals (reaction 7). The Fenton reaction oxidizes Fe^{2+} to Fe^{3+}, which is able to form a new complex with intracellular dopamine, potentiating the neurotoxic effect of the Fe–dopamine complex. The inhibition of ATP formation resulting from NADH and oxygen depletion generating an energy collapse and the oxidative stress caused by hydroxyl radicals and depletion of reduced glutathione eventually lead to cell death [35, 74]. It is important to note that Fe–dopamine complex neurotoxicity requires the expression of DAT and the inhibition of DT-diaphorase. Interestingly, norepinephrine is able to compete with dopamine to form a complex with Fe^{3+} and prevents Fe–dopamine complex neurotoxicity because the Fe–norepinephrine complex is not neurotoxic, despite the fact that cells are able to take up the Fe–norepinephrine complex.

Conclusions

The transition metals manganese and copper have been involved in neurodegenerative processes in dopaminergic neurons related to atypical Parkinson's disease and iron accumulation has been associated with Parkinson's disease. Interestingly, these metals are able to induce neurotoxicity by reacting with the neurotransmitter dopamine. Mn^{3+} is able to oxidize dopamine to aminochrome and potentiate aminochrome neurotoxicity. Cu^{2+} and Fe^{3+} are able to form complexes with dopamine and are able to induce neurotoxicity in cells expressing DAT, resulting in the selective action of copper on dopaminergic neurons.

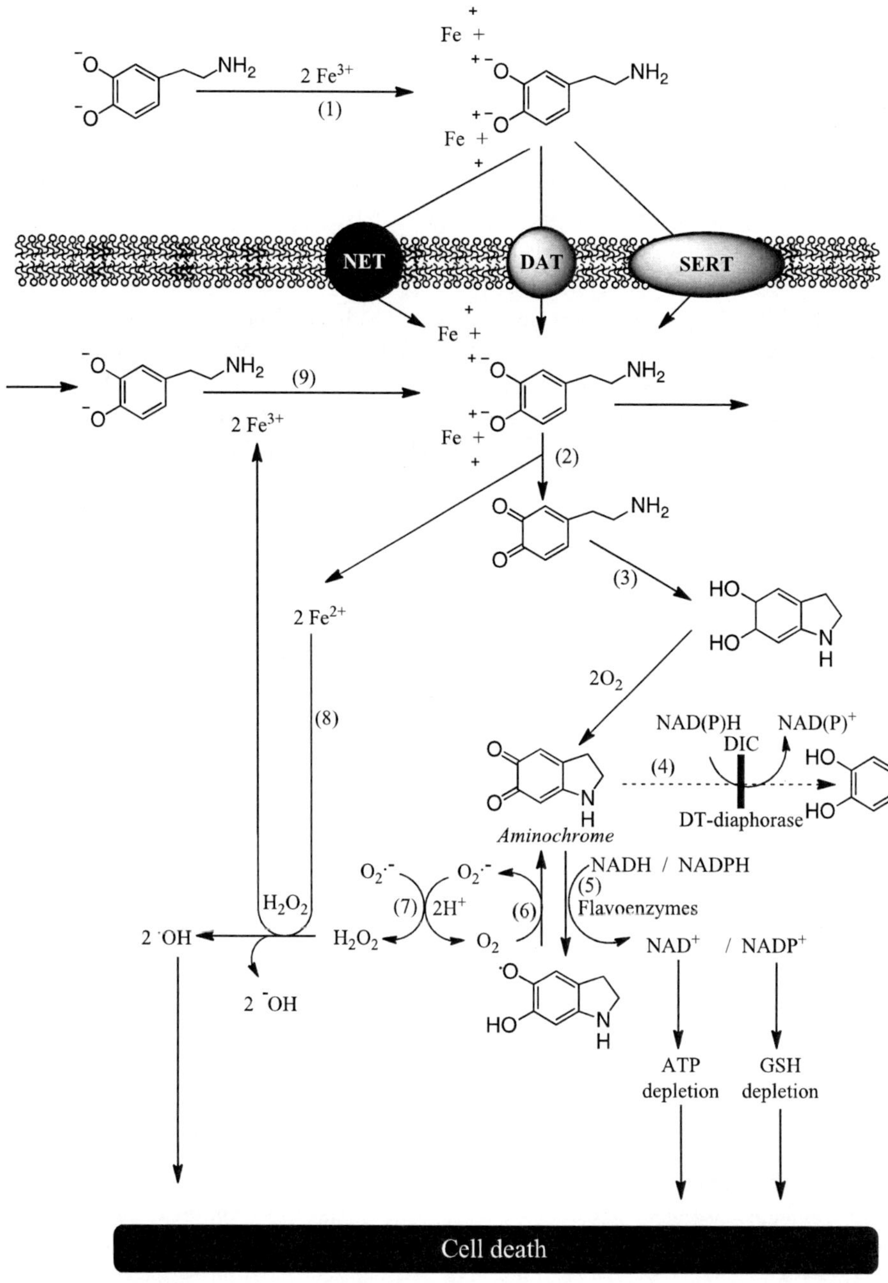

Fig. 6 The dopamine-dependent mechanism of iron neurotoxicity

However, the neurotoxicity induced by these metals was dependent on the one-electron reduction of aminochrome in conditions where DT-diaphorase was inhibited, supporting the proposed neuroprotective role of DT-diaphorase in dopaminergic neurons.

Acknowledgments Supported by FONDECYT 1100165.

References

1. HaMai D, Bondy SC (2004) Ann N Y Acad Sci 1012:129
2. Sriram K, Lin GX, Jefferson AM, Roberts JR, Chapman RS, Chen BT, Soukup JM, Ghio AJ, Antonini JM (2010) Arch Toxicol 84:521
3. Settivari R, Levora J, Nass R (2009) J Biol Chem 284:35758
4. Zhang P, Lokuta KM, Turner DE, Liu B (2010) J Neurochem 112:434
5. Zhang P, Wong TA, Lokuta KM, Turner DE, Vujisic K, Liu B (2009) Exp Neurol 217:219
6. Moreno JA, Streifel KM, Sullivan KA, Legare ME, Tjalkens RB (2009) Toxicol Sci 112:405
7. Zhao F, Cai T, Liu M, Zheng G, Luo W, Chen J (2009) Toxicol Sci 107:156
8. Posser T, Franco JL, Bobrovskaya L, Leal RB, Dickson PW, Dunkley PR (2009) J Neurochem 110:848
9. Stanwood GD, Leitch DB, Savchenko V, Wu J, Fitsanakis VA, Anderson DJ, Stankowski JN, Aschner M, McLaughlin B (2009) J Neurochem 110:378
10. Prabhakaran K, Ghosh D, Chapman GD, Gunasekar PG (2008) Brain Res Bull 76:361

11. Anderson JG, Cooney PT, Erikson KM (2007) Environ Toxicol Pharmacol 3:179
12. McDougall SA, Reichel CM, Farley CM, Flesher MM, Der-Ghazarian T, Cortez AM, Wacan JJ, Martinez CE, Varela FA, Butt AE, Crawford CA (2008) Neuroscience 154:848
13. Hirata Y, Suzuno H, Tsuruta T, Oh-hashi K, Kiuchi K (2008) Toxicology 244:249
14. Williams BB, Kwakye GF, Wegrzynowicz M, Li D, Aschner M, Erikson KM, Bowman AB (2010) Toxicol Sci 117:169
15. Hirsch EC (2009) Parkinsonism Relat Disord 15:S209
16. Du T, Ciccotosto GD, Cranston GA, Kocak G, Masters CL, Crouch PJ, Cappai R, White AR (2008) Free Radic Biol Med 44:44
17. Zhang S, Wang J, Song N, Xie J, Jiang H (2009) Neurobiol Aging 30:1466
18. Huang E, Ong WY, Go ML, Connor JR (2006) Exp Brain Res 170:376
19. Huang E, Ong WY, Connor JR (2004) Neuroscience 128:487
20. Aguirre P, Mena N, Tapia V, Arredondo M, Núñez MT (2005) BMC Neurosci 6:3
21. Lis A, Barone TA, Paradkar PN, Plunkett RJ, Roth JA (2004) Brain Res Mol Brain Res 122:62
22. Knutson M, Menzies S, Connor J (2004) J Neurosci Res 76:633
23. Wang XS, Ong WY, Connor JR (2003) 120:21
24. Burdo JR, Menzies SL, Simpson IA, Garrick LM, Garrick MD, Dolan KG, Haile DJ (2001) J Neurosci Res 66:1198
25. Perl DP, Olanow CW (2007) J Neuropathol Exp Neurol 66:675
26. Donaldson J, LaBella FS, Gesser D (1981) Neurotoxicology 2:53
27. Archibald FS, Tyree C (1987) Arch Biochem Biophys 256:638
28. Segura-Aguilar J, Lind C (1989) Chem Biol Interact 72:309
29. Stredrick DL, Stokes AH, Worst TJ, Freeman WM, Johnson EA, Lash LH
30. Eriksson H, Heilbronn E (1983) Arch Toxicol 54:53
31. Cai T, Yao T, Li Y, Chen Y, Du K, Chen J, Luo W (2007) Brain Res 1185:359
32. Arriagada C, Paris I, Sanchez de las Matas MJ, Martinez-Alvarado P, Cardenas S, Castañeda P, Graumann R, Perez-Pastene C, Olea-Azar C, Couve E, Herrero MT, Caviedes P, Segura-Aguilar J (2004) Neurobiol Dis 16:468
33. Segura-Aguilar J, Metodiewa D, Baez S (2001) Neurotox Res 3:157
34. Paris I, Dagnino-Subiabre A, Marcelain K, Bennett LB, Caviedes P, Caviedes R, Azar CO, Segura-Aguilar J (2001) J Neurochem 77:519
35. Paris I, Martinez-Alvarado P, Perez-Pastene C, Vieira MN, Olea-Azar C, Raisman-Vozari R, Cardenas S, Graumann R, Caviedes P, Segura-Aguilar J (2005) J Neurochem 92:1021
36. Paris I, Cardenas S, Lozano J, Perez-Pastene C, Graumann R, Riveros A, Caviedes P, Segura-Aguilar J (2007) Neurotox Res 12:125
37. Paris I, Lozano J, Perez-Pastene C, Muñoz P, Segura-Aguilar J (2009) Neurotox Res 16:271
38. Paris I, Perez-Pastene C, Cardenas S, Iturriaga-Vasquez P, Muñoz P, Couve E, Caviedes P, Segura-Aguilar J (2010) Neurotox Res 18:82
39. Paris I, Perez-Pastene C, Couve E, Caviedes P, Ledoux S, Segura-Aguilar J (2009) J Biol Chem 284:13306
40. Fuentes P, Paris I, Nassif M, Caviedes P, Segura-Aguilar J (2007) Chem Res Toxicol 20:776
41. Segura-Aguilar J, Diaz-Veliz G, Mora S, Herrera-Marschitz M (2002) Neurotox Res 4:127
42. Paris I, Lozano J, Cardenas S, Perez-Pastene C, Saud K, Fuentes P, Caviedes P, Dagnino-Subiabre A, Raisman-Vozari R, Shimahara T, Kostrzewa JP, Chi D, Kostrzewa RM, Caviedes R, Segura-Aguilar J (2008) Neurotox Res 13:221
43. Díaz-Véliz G, Mora S, Gómez P, Dossi MT, Montiel J, Arriagada C, Aboitiz F, Segura-Aguilar J (2004) Pharmacol Biochem Behav 77:245
44. Díaz-Véliz G, Mora S, Dossi MT, Gómez P, Arriagada C, Montiel J, Aboitiz F, Segura-Aguilar J (2002) Pharmacol Biochem Behav 73:843
45. Strausak D, Mercer JF, Dieter HH, Stremmel W, Multhaup G (2001) Brain Res Bull 55:175
46. Du F, Qian ZM, Zhu L, Wu XM, Yung WH, Tsim TY, Ke Y (2009) PLoS One 4:e4593
47. White AR, Cappai R (2003) J Neurosci Res 71:889
48. Reddy PV, Rao KV, Norenberg MD (2008) Lab Invest 88:816
49. Lu J, Zheng YL, Wu DM, Sun DX, Shan Q, Fan SH (2006) FEBS Lett 580:6730
50. Lu J, Wu DM, Zheng YL, Sun DX, Hu B, Shan Q, Zhang ZF, Fan SH (2009) Brain Behav Immun 23:193
51. Zheng Z, White C, Lee J, Peterson TS, Bush AI, Sun GY, Weisman GA, Petris MJ (2010) J Neurochem 114:1630
52. Taly AB, Meenakshi-Sundaram S, Sinha S, Swamy HS, Arunodaya GR (2007) Medicine (Baltimore) 86:112
53. Caviedes P, Segura-Aguilar J (2001) Neuroreport 12:A25
54. Segura-Aguilar J, Metodiewa D, Welch CJ (1998) Biochim Biophys Acta 1381:1
55. Díaz-Véliz G, Paris I, Mora S, Raisman-Vozari R, Segura-Aguilar J (2008) Chem Res Toxicol 21:1180
56. Wimalasena DS, Wiese TJ, Wimalasena K (2009) J Neurochem 101:313
57. Kume A, Shiratori M, Takahashi A, Kato T, Ito K, Tadokoro M, Sakuma S (1992) J Neurol Sci 110:37
58. Zhang J, Zhang Y, Wang J, Cai P, Luo C, Qian Z, Dai Y, Feng H (2010) Brain Res 1330:124
59. Song N, Jiang H, Wang J, Xie JX (2007) J Neurosci Res 85:3118
60. Roth JA, Singleton S, Feng J, Garrick M, Paradkar PN (2010) J Neurochem 113:454
61. Hirsch EC (2009) Parkinsonism Relat Disord 15:S209
62. Salazar J, Mena N, Hunot S, Prigent A, Alvarez-Fischer D, Arredondo M, Duyckaerts C, Sazdovitch V, Zhao L, Garrick LM, Nuñez MT, Garrick MD, Raisman-Vozari R, Hirsch EC (2008) Proc Natl Acad Sci USA 105:18578
63. Wang J, Jiang H, Xie JX (2007) Eur J Neurosci 25:2766
64. Song N, Wang J, Jiang H, Xie J (2010) Free Radic Biol Med 48:332
65. Zhu W, Li X, Xie W, Luo F, Kaur D, Andersen JK, Jankovic J, Le W (2010) Neurobiol Dis 37:307
66. Kaur D, Lee D, Ragapolan S, Andersen JK (2009) Free Radic Biol Med 46:593
67. Tribl F, Asan E, Arzberger T, Tatschner T, Langenfeld E, Meyer HE, Bringmann G, Riederer P, Gerlach M, Marcus K (2009) Mol Cell Proteomics 8:1832
68. Youdim MB, Grünblatt E, Mandel S (2007) J Neural Transm 114:205
69. Xu Q, Kanthasamy AG, Reddy MB (2008) Toxicology 245:101
70. Maharaj DS, Maharaj H, Daya S, Glass BD (2006) J Neurochem 96:78
71. Wang R, Qing H, Liu XQ, Zheng XL, Deng YL (2008) Neurosci Bull 24:125
72. Zecca L, Casella L, Albertini A, Bellei C, Zucca FA, Engelen M, Zadlo A, Szewczyk G, Zareba M, Sarna T (2008) J Neurochem 106:1866
73. Double KL, Halliday GM, Henderson J, Griffiths FM, Heinemann T, Riederer P, Gerlach M (2003) Exp Neurol 184:530
74. Paris I, Martinez-Alvarado P, Cárdenas S, Perez-Pastene C, Graumann R, Fuentes P, Olea-Azar C, Caviedes P, Segura-Aguilar J (2005) Chem Res Toxicol 18:415
75. Arreguin S, Nelson P, Padway S, Shirazi M, Pierpont C (2009) J Inorg Biochem 103:87
76. Siraki AG, Smythies J, O'Brien PJ (2000) Neurosci Lett 296:37
77. Ortega R, Cloetens P, Devès G, Carmona A, Bohic S (2007) PLoS One 2:e925
78. Mohamed GG, Zayed MA, El-Dien FA, El-Nahas RG (2004) Spectrochim Acta A 60:1775
79. Arriagada C, Dagnino-Subiabre A, Caviedes P, Armero JM, Caviedes R, Segura-Aguilar J (2000) Amino Acids 18:363

Iron, cysteine and neurodegeneration during Parkinson's disease

Eleni Siakkou and Guy N.L. Jameson

Abstract

A short review of the role of cysteine and iron in the progression of Parkinson's disease. The complex chemistry of iron and cysteine and its interactions are discussed and put into context of their probable role in oxidative stress during neurodegeneration.

Keywords

Radicals • Quinones • Metal complexes • Amino acids • Bioinorganic chemistry

Introduction

Iron in the brain: a short introduction

Well-defined levels of iron (in its ferrous and ferric form) are essential for survival, replication and differentiation of cells in almost all living organisms. In the brain, iron is involved in functions like neurotransmission, myelination and cell division [1]. The concentration of iron within the brain naturally varies depending on the region, and regions of the brain associated with motor function tend to contain higher levels of iron [2].

Iron enters the brain via the blood–brain barrier (BBB) and, to a lesser extent, the blood–cerebrospinal fluid barrier (Fig. 1). The uptake of iron at the BBB occurs through the interaction between iron(III)-loaded plasma transferrin (Tf) and Tf-receptors expressed by endothelial cells of the BBB [3]. The subsequent release and transport of iron within the endothelial cells is still disputed and is thought to involve either transcytosis or endocytosis of iron-loaded Tf. The mechanisms involved in further iron redistribution within the brain and iron efflux from the brain are also not very well understood yet, but may include involvement of brain-native Tf [1, 4].

Cellular iron storage calls for the presence of iron-storage proteins and other, smaller, iron-ligating molecules because free iron ions are very reactive species. Within the brain, the most important iron storage molecules are ferritin and neuromelanin [5]. The failure to properly ligate free iron has been associated with oxidative stress and the consequent occurrence of various vascular and other progressive inflammatory and degenerative diseases [6, 7]. This includes Parkinson's disease, which is the focus of this chapter.

On Parkinson's disease, and the involvement of cysteine and iron

Parkinson's disease (PD) is the most common neurodegenerative disease and affects approximately 2 % of all people over 65 in Europe [8]. It is characterised by loss of dopaminergic neurons in the brain, particularly in the *substantia nigra*, and, in the vast majority of cases, the formation of Lewy bodies, which are abnormal aggregates of protein comprised mainly of α-synuclein.

Cysteine is a non-essential sulfur amino acid that is important for redox balance within the cell, signalling and glutathione synthesis. The brain is likely to be supplied with cysteine in the form of the disulfide, cystine, which has been shown to cross the BBB, although a direct uptake of cysteine at the BBB from the peripheral blood has not been ruled out [9]. Depending on the cell type (e.g. neurons, astrocytes) and

Based on chapter published in: Chemical Monthly, Vol 142, No 4, 2011

G.N.L. Jameson (✉)
Department of Chemistry, PO Box 56, Dunedin 9054, New Zealand
e-mail: gjameson@chemistry.otago.ac.nz

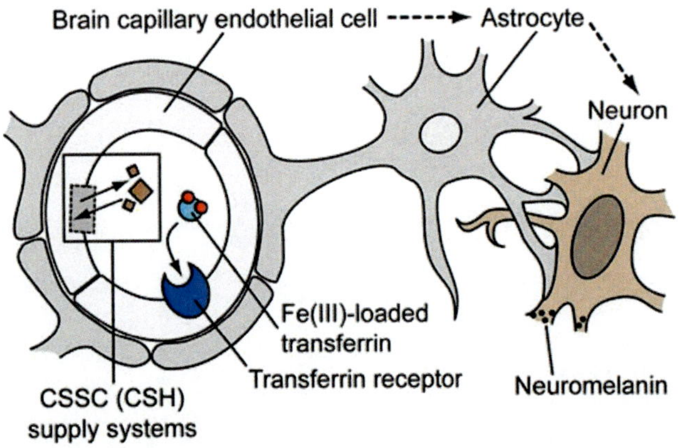

Fig. 1 Scheme showing how iron and cysteine cross the blood–brain barrier (BBB) into the brain. Iron is transported via the transferrin receptors, while cysteine is transported either as cysteine (CSH) or cystine (CSSC), depending upon cell type. Once across the BBB, the mechanism of flux of iron is unknown but both cysteine and iron must travel via the astrocytes to the neurons (*dashed arrows*) where among other places they are combined in neuromelanin

also the maturation state of the cell, brain cells can import cysteine either directly or as cystine from the extracellular space [10, 11]. Analogous to iron, elevated concentrations of cysteine can have toxic effects and brain cysteine levels must therefore be tightly regulated, which involves the concerted action of different cell types [12, 13]. Cysteine/iron interactions are themselves complex and have important roles to play in the progression of neurodegeneration.

There is substantial evidence that the impairment of Complex I of the mitochondrial respiratory chain has an important role in the pathogenesis of PD [14]. Results from genetic studies [15] have shown that a threonine to alanine polymorphism of dehydrogenase 3 of Complex I significantly reduces risk of PD. Furthermore, administration of Complex I inhibitors [16] (i.e. MPTP, paraquat and rotenone) replicates many of the features of PD. This impairment of Complex I suggests that oxidative stress is important in PD [17] and this view is supported by oxidative damage to DNA and protein that occurs in the brain during disease progression.

Oxidative stress describes a large number of different reactions involving many different participants. The brain is particularly vulnerable to oxidative stress due to relatively weak anti-oxidative defence systems [18]. Oxidative stress can be initiated by catecholamines in the presence of iron and oxygen or peroxide leading to formation of 6-hydroxydopamine [19, 20]. Cysteine can also participate in oxidative stress through its complicated chemistry involving redox, coordination and nucleophilic reactions, which are exacerbated by its interactions with the redox active transition metal iron. Indeed, combination of cysteine and iron has been reported to potentiate lipid peroxidation through formation of hydroxyl radicals [21].

Concentrations and location of cysteine and iron

In vivo both iron and cysteine levels are strongly controlled because of their potentially toxic effects. Iron is well known to catalyse Fenton-type reactions when suitably coordinated and a significant amount of work has been carried out to investigate these types of reactions [22]. Likewise, the role of elevated levels of cysteine in neurodegeneration has been extensively discussed [23–30]. Cysteine is neuroexcitory and is believed to interact with the *N*-methyl-D-aspartate (NMDA) receptor [26], allowing Ca^{2+} ions to enter the cell leading to apoptosis. Since cysteine's toxicity is attenuated by zinc cations, it has been suggested that ZnCys complexes block the NMDA receptors. Indeed, zinc binding to cysteine is strong ($\log K_1 = 9.04$, $\log K_2 = 8.50$) and the stability constants show zinc can compete successfully with iron (Table 2) for the formation of a cysteine complex. Concentrations of both transition metals are increased during PD [31]; however, iron is in higher concentrations and reacts with cysteine to form cysteine.

Table 1 collates representative data obtained from patients with PD and controls. One can see that cysteine levels are approximately 250 μM in plasma and this increases after the onset of PD. Interestingly, this increase appears to be higher after Levodopa treatment, still the main treatment of PD. Iron levels are also seen to increase dramatically after onset of PD but only significantly in the *substantia nigra* [32]. As can be seen from Table 1, although there is an increase, the absolute values depend upon how the samples are treated before measurement.

Chemical reactivity of cysteine and iron

Iron is present mainly as iron(II) and iron(III), although it can reach higher oxidation states in highly oxidising environments during the Fenton reaction [22] or during the catalytic cycles of certain enzymes. Unless it is complexed, iron(III) tends to form insoluble hydroxides and oxides and iron(II) is prone to oxidation at physiological pHs. The chemistry of iron in vivo is therefore dominated by coordination compounds and its accessible redox potential allows it to react with many different compounds. For example, iron (II) tends to bind oxygen while iron(III) can oxidise organic compounds or initiate radical reactions.

Cysteine's reactivity is dominated by the thiol group, which can act as an excellent nucleophile when deprotonated and is easily oxidised. Reactions include the formation of disulfides, thioethers, thioacetals and thioesters. A selection of representative physical data is given in Table 2.

Table 1 Concentrations of cysteine and iron in both controls and patients with Parkinson's disease

	Location	Control	Patients with Parkinson's disease
Cysteine [63]	Plasma	220 μM	260 μM Levodopa treated
Cysteine [64]	Plasma	280 μM	490 μM Levodopa treated
			340 μM before Levodopa treatment
Total iron [31]	*Substantia nigra*	10,000 nmols/g dry weight	13,000 nmols/g dry weight
Total iron [65]	*Substantia nigra*	860 nmols/g fresh weight	1,522 nmols/g fresh weight

Table 2 Physical data for cysteine and its interactions with iron

Protonation constants	log K_1	10.11[a]	log K_2	8.13[a]	log K_3	~2[b]
Stability constants	Iron(II) log K_1		6.66[c]	Iron(III) log K_1		13.70[d]

To aid comparison, constants measured under similar temperatures and the same ionic strength have been chosen
[a]25 °C, 0.1M KNO_3 [66]
[b]Published values vary greatly; glass electrodes do not function properly at very low pHs
[c]20 °C, 0.1M $NaClO_4$ [36]
[d]25 °C, 0.1M KNO_3, determined through kinetic experiments as iron(III) oxidises cysteine anaerobically [67]

Cysteine forms complexes with both iron(II) and iron(III), binding stronger to iron(III). Binding appears to occur through the amine nitrogen and the thiolate sulfur judging by the structure of the non-heme mono-iron enzyme cysteine dioxygenase with substrate bound [33]. The iron (III) complexes are, however, not stable and undergo internal electron transfer [34, 35] to form the thiyl radical, which unless otherwise trapped, combines with another thiyl radical to form cysteine. It is this radical chemistry that explains why the combination of cysteine and iron has been reported to potentiate lipid peroxidation [21].

Speciation

Iron is present mainly as iron(III) at pH = 7 and with suitable coordination is stabilised against precipitation as hydroxo species. Iron(II) is also present and in this case is stabilised in vivo usually by strong-field nitrogen donors.

At physiological pH the carboxyl group of cysteine is deprotonated and the amine group is protonated. The value of pK_{SH} can largely be associated with log K_2 (8.13) due to the large separation of the log K values although micro-constants are available [36]. This value of pK_{SH} means that at physiological pHs the thiol is present in both protonated and deprotonated forms and this fact dominates the reactivity of cysteine.

Cysteine as a nucleophile

The thiolate group acts as an extremely effective nucleophile, able to attack quinones and other electrophiles. In dopaminergic neurons (Fig. 2), cysteine is able to attack the 5 position of dopamino-quinone to form 5-cysteinyl-dopamine [37], which is an important component of neuromelanin

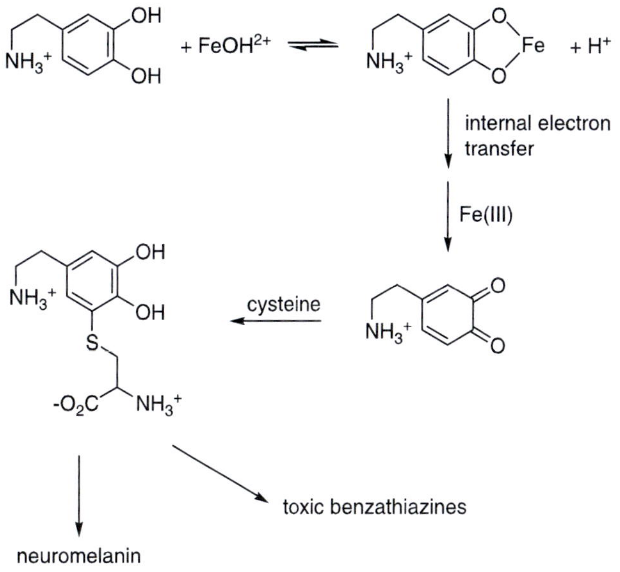

Fig. 2 Reaction pathway by which 5-cysteinyl-dopamine can be formed from iron(III), dopamine and cysteine. In healthy cells, 5-cysteinyl-dopamine is used to form neuromelanin but during neurodegeneration it may also further react to form toxic benzathiazines

[38–41]. Neuromelanin is formed in dopaminergic neurons and is a polymer formed from oxidation products of catecholamines. It forms spherical particles, granules, of approximately 30 nm in diameter with a pheomelanin core suggesting a kinetic model in which cysteine is initially used up to form 5-cysteinyl-dopamine before eumelanin is formed [40]. However, although cysteine reacts rapidly with quinones it seems to attenuate melanin formation in vitro [37].

It is not known, however, whether cysteine addition to dopaminoquinone is the source of 5-cysteinyl-dopamine in vivo or whether initially glutathione attacks to form

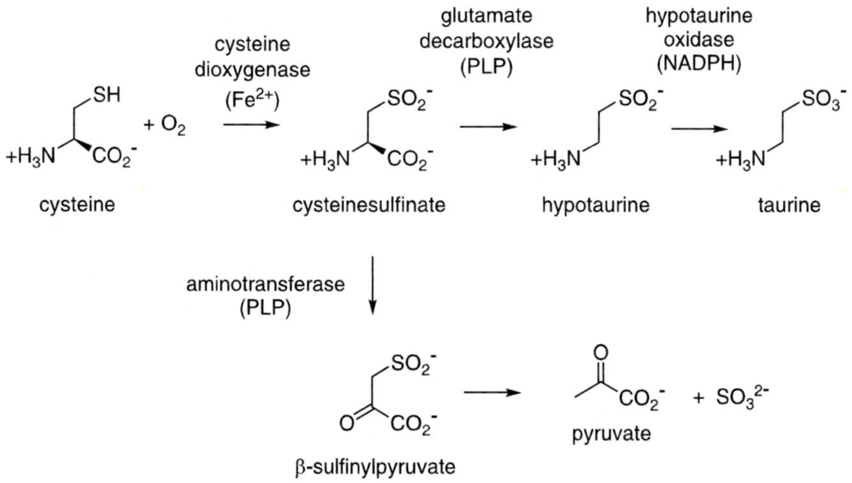

Fig. 3 The enzymatic pathway by which cysteine is broken down via the cysteine sulfinate in the brain. The cofactor for each enzyme is given in brackets: Fe^{2+} = high spin iron(II), PLP = pyridoxal-5′-phosphate, NADPH = reduced nicotinamide adenine dinucleotide phosphate. The last step from β-sulfinylpyruvate to pyruvate is believed to be spontaneous

5-glutathionyl-dopamine followed by cleavage by gamma-glutamyl transpeptidase and peptidase [24]. Work by Dryhurst [42–45] has shown that 5-cysteinyl-dopamine is not the final product and the reduced redox potential of subsequent species leads to a cascade of toxic benzathiazines that have been shown to produce Parkinson type symptoms in rats [44, 45].

Dopamino-quinone itself is formed by oxidation of dopamine by iron(III) [46] or oxygen [47] or both [20] and, indeed, all catecholamines undergo similar chemistry [24]. Iron concentrations increase during the onset of PD (see Table 1) and this could be caused by 6-hydroxydopamine, formed from dopamine, releasing iron from the iron storage protein ferritin [48]. The iron that is released can bind to neuromelanin [41, 49] where it can promote oxidative stress. There are two types [49] of binding sites to neuromelanin, high affinity non-reactive sites where the iron is bound to catechol groups and low affinity reactive sites where iron is bound as iron oxo-hydroxo clusters. Neuromelanin, thus works as both an anti-oxidant and as a pro-oxidant.

Cysteine oxidation

Cysteine can be oxidised in a number of different ways. One electron oxidation can occur anaerobically after complexation with iron(III) resulting in the thiyl radical [34, 35]. The redox potential for the formation of this cysteinyl radical is ~0.80 V at pH = 7 [68] and at pH = 7.4 the midpoint potential E_m for glutathione (GS·,H^+/GSH) is 0.92 V vs. NHE [50]. This is the same midpoint potential as hydrogen sulfide and it appears all thiols have similar values [69, 70].

The cysteinyl radical, in most cases, combines very rapidly with another cysteinyl radical to form cysteine ($k = 1.0 \times 10^{10}\ M^{-1}\ s^{-1}$ [51]) but can also combine with dioxygen to form the thiol peroxyl radical [52]. Dioxygen is well known to oxidise cysteine to form the disulfide. This reaction is metal-ion catalysed by many transition metals including copper and iron [53]. The kinetics are complicated by the fact that the peroxide formed can also react. The main product is disulfide except at low cysteine concentrations when sulfenate, sulfinate and sulfonate are also formed [54, 55].

Cysteine can also be oxidised by hypohalous acids [56] and at high pH the sulfenate can be observed transiently (although ultimately the main product is the disulfide with small amounts of oxygenated products including the sulfinate). Since hypohalous acids are produced during inflammatory response by myeloperoxidase a redox cascade may start with the oxidation of cysteine [57] and therefore these reactions may well be important in the progression of oxidative stress. Indeed, myeloperoxidase levels increase in patients suffering from PD [58].

Enzymatic breakdown of cysteine

The main enzymatic route for cysteine breakdown is shown in Fig. 3. The initial step involves the oxidation of the thiol group to the sulfinate [59]. This reaction is catalysed by the non-heme mono-iron enzyme cysteine dioxygenase [60] (CDO, [EC 1.13.11.20]). In rats it has been shown that CDO is degraded by the 26-S proteasome system [61]. The ubiquitin-proteasome system is used to degrade misfolded and damaged proteins. The 26-S proteasomes are multi-subunit proteases found in the cytosol. A characteristic feature of PD is the presence of Lewy bodies containing normal and damaged proteins including ubiquitin, neurofilaments, parkin, α-synuclein and others. It is not yet fully understood how Lewy bodies are formed but this may be linked to the ubiquitin-proteasome system failing [62] leading to subsequent failing of the initial step of cysteine oxidation by CDO.

Conclusions

During the progression of PD the concentrations of cysteine and iron are allowed to significantly increase leading to oxidative stress through a number of different mechanisms. These processes include complex formation, redox reactions and enzymatic reactions. Although a large amount of the basic chemistry of cysteine and iron is well characterised there is still a lot that is not yet fully understood about their role in vivo. For example, the underlying reason for the lack of control of the concentrations of cysteine and iron remains a serious unresolved problem. This area is proving to be a very fertile ground for further research.

Acknowledgements

I am grateful to Prof. R. F. Jameson and Dr. V. J. A. Jameson for useful discussions and to the Marsden Fund of the Royal Society of New Zealand for financial support.

References

1. Moos T, Nielsen TR, Skjørringe T, Morgan EH (2007) J Neurochem 103(5):1730–1740
2. Götz ME, Double KAY, Gerlach M, Youdim MBH, Riederere P (2004) Ann N Y Acad Sci 1012(1):193–208
3. Morris CM, Keith AB, Edwardson JA, Pullen RGL (1992) J Neurochem 59(1):300–306
4. Rouault TA, Cooperman S (2006) Semin Pediatr Neurol 13 (3):142–148
5. Gerlach M, Double KL, Youdim MBH, Riederer P (2006) Potential sources of increased iron in the substantia nigra of parkinsonian patients Parkinson's disease and related disorders. In: Riederer P, Reichmann H, Youdim MBH, Gerlach M (eds) vol 70. Journal of Neural Transmission. Supplementa. Springer, Vienna, pp 133–142
6. Kell D (2009) BMC Med Genomics 2(1):2–81
7. Kell D (2010) Arch Toxicol 84:825–889
8. de Rijk MC, Tzourio C, Breteler MM, Dartigues JF, Amaducci L, Lopez-Pousa S, Manubens-Bertran JM, Alperovitch A, Rocca WA (1997) J Neurol Neurosurg Psychiatry 62(1):10–15
9. Wang XF, Cynader MS (2000) J Neurochem 74(4):1434–1442
10. Dringen R, Hirrlinger J (2003) Biol Chem 384(4):505–516
11. McBean GJ, Flynn J (2001) Biochem Soc Trans 29(6):717–722
12. Escartin C, Joon Won S, Malgorn C, Auregan G, Berman AE, Chen P-C, Déglon N, Johnson JA, Won Suh S, Swanson RA (2011) J Neurosci 31(20):7392–7401
13. Guebel DV, Torres NV (2004) Biochim Biophys Acta 1674 (1):12–28, Gen Subj
14. Dawson TM, Dawson VL (2003) Science 302(5646):819–822
15. van der Walt JM, Nicodemus KK, Martin ER, Scott WK, Nance MA, Watts RL, Hubble JP, Haines JL, Koller WC, Lyons K, Pahwa R, Stern MB, Colcher A, Hiner BC, Jankovic J, Ondo WG, Allen FH, Goetz CG, Small GW, Mastaglia F, Stajich JM, McLaurin AC, Middleton LT, Scott BL, Schmechel DE, Pericak-Vance MA, Vance JM (2003) Am J Hum Genet 72(4):804–811
16. Sherer TB, Betarbet R, Greenamyre JT (2002) Neuroscientist 8 (3):192–197
17. Sayre LM, Perry G, Smith MA (2008) Chem Res Toxicol 21 (1):172–188
18. Crichton RR, Wilmet S, Legssyer R, Ward RJ (2002) J Inorg Biochem 91(1):9–18
19. Linert W, Herlinger E, Jameson RF, Kienzl E, Jellinger K, Youdim MBH (1996) Biochim Biophys Acta 1316(3):160–168
20. Napolitano A, Pezzella A, Prota G (1999) Chem Res Toxicol 12 (11):1090–1097
21. Searle AJ, Willson RL (1983) Biochem J 212(3):549–554
22. Groves JT (2006) J Inorg Biochem 100(4):434–447
23. Gazit V, Ben-Abraham R, Coleman R, Weizman A, Katz Y (2004) Amino Acids 26(2):163–168
24. Linert W, Jameson GNL (2000) J Inorg Biochem 79(1–4):319–326
25. McBean G (2007) 7 Sulfur-containing amino acids. In: Lajtha A, Oja S, Schousboe A, Saransaari P (eds) Handbook of neurochemistry and molecular neurobiology. Springer, Berlin, pp 133–154
26. Olney JW, Zorumski C, Price MT, Labruyere J (1990) Science 248 (4955):596–599
27. Puka-Sundvall M, Eriksson P, Nilsson M, Sandberg M, Lehmann A (1995) Brain Res 705(1–2):65–70
28. Stipanuk MH, Dominy JE Jr, Lee JI, Coloso RM (2006) J Nutr 136 (6 Suppl):1652S–1659S
29. Townsend DM, Tew KD, Tapiero H (2004) Biomed Pharmacother 58(1):47–55
30. Janaky R, Varga V, Hermann A, Saransaari P, Oja SS (2000) Neurochem Res 25(9–10):1397–1405
31. Dexter DT, Wells FR, Lees AJ, Agid F, Agid Y, Jenner P, Marsden CD (1989) J Neurochem 52(6):1830–1836
32. Gerlach M, Benshachar D, Riederer P, Youdim MBH (1994) J Neurochem 63(3):793–807
33. Ye S, Wu X, Wei L, Tang DM, Sun P, Bartlam M, Rao ZH (2007) J Biol Chem 282(5):3391–3402
34. Jameson RF, Linert W, Tschinkowitz A (1988) J Chem Soc Dalton Trans 8:2109–2112
35. Jameson RF, Linert W, Tschinkowitz A, Gutmann V (1988) J Chem Soc Dalton Trans 4:943–946
36. Berthon G (1995) Pure Appl Chem 67(7):1117–1240
37. Jameson GNL, Zhang J, Jameson RF, Linert W (2004) Org Biomol Chem 2(5):777–782
38. Wakamatsu K, Fujikawa K, Zucca FA, Zecca L, Ito S (2003) J Neurochem 86(4):1015–1023
39. Zecca L, Zucca FA, Wilms H, Sulzer D (2003) Trends Neurosci 26 (11):578–580
40. Bush WD, Garguilo J, Zucca FA, Albertini A, Zecca L, Edwards GS, Nemanich RJ, Simon JD (2006) Proc Natl Acad Sci USA 103 (40):14785–14789
41. Zecca L, Mecacci C, Seraglia R, Parati E (1992) Biochim Biophys Acta 1138(1):6–10
42. Shen XM, Dryhurst G (1996) Chem Res Toxicol 9(4):751–763
43. Shen XM, Dryhurst G (1998) Chem Res Toxicol 11(7):824–837
44. Shen XM, Zhang F, Dryhurst G (1997) Chem Res Toxicol 10 (2):147–155
45. Zhang F, Dryhurst G (1994) J Med Chem 37(8):1084–1098
46. ElAyaan U, Herlinger E, Jameson RF, Linert W (1997) J Chem Soc Dalton Trans 16:2813–2818
47. Herlinger E, Jameson RF, Linert W (1995) J Chem Soc Perkin 2 (2):259–263
48. Jameson GNL, Jameson RF, Linert W (2004) Org Biomol Chem 2 (16):2346–2351
49. Double KL, Gerlach M, Schunemann V, Trautwein AX, Zecca L, Gallorini M, Youdim MBH, Riederer P, Ben-Shachar D (2003) Biochem Pharmacol 66(3):489–494
50. Madej E, Wardman P (2007) Arch Biochem Biophys 462 (1):94–102

51. Hoffman MZ, Hayon E (1972) J Am Chem Soc 94(23):7950–7957
52. Sevilla MD, Yan MY, Becker D (1988) Biochem Biophys Res Commun 155(1):405–410
53. Ehrenberg L, Harmsringdahl M, Fedorcsak I, Granath F (1989) Acta Chem Scand 43(2):177–187
54. Claiborne A, Miller H, Parsonage D, Ross RP (1993) FASEB J 7 (15):1483–1490
55. Enami S, Hoffmann MR, Colussi AJ (2009) J Phys Chem B 113 (28):9356–9358
56. Nagy P, Ashby MT (2007) J Am Chem Soc 129(45):14082–14091
57. Hawkins CL, Pattison DI, Davies MJ (2003) Amino Acids 25 (3–4):259–274
58. Choi D-K, Pennathur S, Perier C, Tieu K, Teismann P, Wu D-C, Jackson-Lewis V, Vila M, Vonsattel J-P, Heinecke JW, Przedborski S (2005) J Neurosci 25(28):6594–6600
59. Siakkou E, Wilbanks SM, Jameson GNL (2010) Anal Biochem 405 (1):127–131
60. Kleffmann T, Jongkees SAK, Fairweather G, Wilbanks SM, Jameson GNL (2009) J Biol Inorg Chem 14(6):913–921
61. Dominy JE, Hirschberger LL, Coloso RM, Stipanuk MH (2006) Biochem J 394:267–273
62. McNaught KSP, Olanow CW, Halliwell B, Isacson O, Jenner P (2001) Nat Rev Neurosci 2(8):589–594
63. Muller T, Kuhn W (2009) Mov Disord 24(6):929–932
64. Heafield MT, Fearn S, Steventon GB, Waring RH, Williams AC, Sturman SG (1990) Neurosci Lett 110(1–2):216–220
65. Sofic E, Riederer P, Heinsen H, Beckmann H, Reynolds GP, Hebenstreit G, Youdim MBH (1988) J Neural Transm 74(3):199–205
66. Lenz GR, Martell AE (1964) Biochemistry 3(6):745–750
67. Jameson RF, Linert W (1991) Monatsh Chem 122(11):887–906
68. Prütz WA, Butler J, Land EJ, Swallow AJ (1986) Free Radical Res. 2, 69–75
69. Surdhar P, Armstrong DA (1986) J Phys Chem 90, 5915–5917
70. Surdhar PS, Armstrong DA (1987) J Phys Chem 91, 6532–6537

Copper uptake and trafficking in the brain

Guido Crisponi, Valeria Marina Nurchi, Clara Gerosa, Daniela Fanni, Sonia Nemolato, and Gavino Faa

Abstract

The aim of this chapter is to give a general view on the current status of the scientific basis for the role of copper in human health and disease, outlining the roles of copper in human metabolism and bioenergetics, its coordination chemistry as well as the biological ligands involved in the multiple steps of metal incorporation. In particular, our attention has been focused towards the interaction of copper status and brain function in health and disease, with particular consideration to the role of copper in the pathogenesis of Wilson's, of Menkes, and of human neurodegenerative diseases. Data on interactions between essential trace elements and copper, from the level of absorption in the gut to other systems in the body, are also presented. Particular attention is paid to copper-dependent enzymes in the central nervous system and to copper uptake and trafficking in brain cells.

Keywords

Copper • Copper overload • Bioinorganic chemistry • Neurodegenerative diseases • Copper-related diseases

Introduction

Copper is a transition metal characterized by a low redox potential between Cu(II) and Cu(I) 0.158 V in water [1] that accounts for its aptitude to exchange electrons with other chemical compounds. For this reason copper is utilized by a large number of enzymes, probably more than 300 in humans, involved in oxidative reactions. The cellular and biochemical mechanisms by which copper storage may induce cell damage and cell necrosis is still debated [2]. ROS formation during acute exposure of human cells to copper overload is generally considered the crucial event leading to cell death [3]. It has been hypothesized that excess copper could trigger peroxidative cell damage by ROS production and consequent oxidative injury [4]. In fact, oxidative stress is often cited as a possible cause of damage in cells exposed to high copper concentrations, even if no supporting quantitative data are presented. In hepatocytes incubated with high concentrations of Cu(II), lysosomes were identified as the major site of endogenous cytotoxic ROS formation [5]. In another study, the origin of ROS generated in hepatocytes was identified in mitochondria by confocal laser scanning microscopy, suggesting an important role for mitochondria in copper-induced apoptosis [6]. In the same study, co-incubation of Cu-treated hepatocytes with the iron chelator deferoxamine significantly inhibited ROS production preventing hepatocytic cell death. Recently, copper exposure was found to activate extracellular signal-regulated kinase (ERK) in trout hepatocytes, stimulating its phosphorylation to pERK. Activation of ERK, as well as of p38, was responsible for enhanced ROS formation and for apoptotic cell death [7]. Liver cells necrosis induced by Cu(II) has been also partly attributed to membrane lipid peroxidation, and to the increase of intracellular calcium levels observed when the hepatocytes were incubated with high Cu(II) levels [8]. Exposure of HepG2 cells, an in vitro system typical of liver parenchymal cells, to high copper dosages, was constantly associated with

G. Faa (✉)
Dipartimento di Citomorfologia, Divisione di Anatomia Patologica, Università di Cagliari, Cagliari, Italy
e-mail: gavfaa@tiscali.it

W. Linert and H. Kozlowski (eds.), *Metal Ions in Neurological Systems*,
DOI 10.1007/978-3-7091-1001-0_5, © Springer-Verlag Wien 2012

the induction of heat shock protein 70 (hsp70) [9] and of metallothioneins [10], providing a protective mechanism against copper toxicity. In patients affected by Wilson's disease and in Bedlington terriers, an animal model of WD [2], the presence of oxidant injury to hepatic mitochondria has been reported [11]. These findings confirm previous reports [12] indicating mitochondria as the main target of copper-related peroxidative damage in Wilson's disease, a metabolic human disease characterized by copper accumulation in different organs, including liver [13] and brain [14]. A severe mitochondrial involvement could be at the basis of the development of a cellular energy deficit and of increased intracellular free calcium levels, both leading to apoptotic cell death [6]. Recently, a transmission electron microscope study carried out in the liver of ATP7B rats, an animal model of Wilson's disease, evidenced relevant structural mitochondrial changes at an early stage, coinciding with massive copper accumulation, before the appearance of deficits of oxidative phosphorylation and of signs of oxidative damage, suggesting that mitochondrial changes could not be the result of copper-induced oxidative stress but a direct effect of copper overload [15]. In a comment on this study, this hypothesis was considered as an innovative theory of copper overload-related mitochondrial membrane cross-linking, allowing a new view of the pathogenesis of Wilson's disease [16].

The aim of this paper is to give a general view on the current status of the scientific basis for the role of copper in human health and disease, outlining the roles of copper in human metabolism and bioenergetics, its coordination chemistry as well as its biological ligands involved in the multiple steps of metal assimilation. In particular, our attention will be focused towards the interaction of copper status and brain function in health and disease, with particular attention to the role of copper in the pathogenesis of Wilson's disease, of Menkes disease and of human neurodegenerative diseases. Data on interactions between essential trace elements and copper, from the level of absorption in the gut to other systems in the body, are also presented, with particular attention to copper uptake and trafficking in brain cells.

Copper in human physiology

Copper is an essential trace element for living organisms, being an indispensable structural part of a huge number of enzymes which take part in all aspects of metabolism [17], including mitochondrial oxidative phosphorylation, free radical detoxification, neurotransmitter synthesis and denaturation, pigment formation, connective tissue formation and iron metabolism [18]. The essential role of copper in brain development is shown by the presence of demyelination and by neurodegeneration in patients affected by Menkes disease, the best documented congenital copper deficiency in humans [19, 20]. The biological utility of copper is mainly derived, as stated above, from its aptitude to cycle between oxidized Cu(II) and reduced Cu(I). The effects of metals characterized by variable oxidation states, such as copper and iron, depend markedly on the predominant form of the metal, which in turns depends on redox potential and coordinating environment [21]. As a consequence, the copper redox state is one of the critical variables that affect ion transport pathways. Copper uptake and its intracellular oxidative effects are mainly caused by the cupric ion, whereas the effects on membrane transport are largely induced by cuprous ion, likely formed from the cupric one at the membrane surface, where sulphydryl groups act as possible reducing agent [22]. Copper ion redox state is critical for methaemoglobin formation in trout erythrocytes [23]. Cu(I) is supposed to be found more often into the cell membrane and inside the cell, where the local redox potential is considerably lower than in blood [24].

Copper speciation study in humans has been scarcely developed: Cu(I) has a great affinity for thiol group (–SH), which is its most important coordinating group in biological environment [11]. The great part of copper in the human body, in physiological conditions, is probably functional, being copper atoms involved as cofactor of different redox enzymes. The best known among these are ceruloplasmin, a ferroxidase with an essential role in iron metabolism [25]; cytochrome *c* oxidase, the terminal enzyme of electron transport and oxidative phosphorylation; superoxide dismutase, an antioxidant enzyme able to remove superoxide radicals from tissues [26]; lysyl oxidase, essential for cross-linking of collagen and elastic fibres; monophenol monooxygenase, involved in the synthesis of melanin; dopamine-beta-monooxygenase, required for the synthesis of dopamine; peptide-alpha-amidating mono-oxygenase, essential for the synthesis of pituitary hormones; coagulation factors V and VIII; tyrosinase/laccase; hephaestin; nitrous oxide reductase and copper amine oxidase [23, 27, 28]. The diversity of potential copper ligands along a polypeptide primary structure makes prediction of the entire copper proteome a challenging area for future studies [28]. Copper plays further additional non-enzymatic functions: fundamental biological processes, such as angiogenesis [29], nerve myelination and activity of endorphin, need its presence [30].

Copper is present in the human body in relatively high amounts: a healthy 70 kg adult contains about 110 mg of copper, the major part (46 mg) in skeleton and bone marrow, 26 mg in skeletal muscles, 10 mg in liver , 8.8 mg in brain and 6 mg in blood [27]. Copper is essential for reproduction, regulation of gene expression, and for normal growth and development [31]. The interest on copper metabolism in humans begun in the late 1920s, when the role of copper in haemoglobin synthesis was revealed [32]. The essential role of copper in humans was first reported in 1956 in a paper on malnourished children presenting with anaemia refractory to iron therapy [33]. Further studies allowed a clear definition of pathological findings related to copper deficiency in children [34] and, in particular, in newborns [35]. A significant amount of insights into the kinetics of copper metabolism has been reached about 80 years after the first discovery of its fundamental role in human metabolism, even if the intimate mechanisms of copper homeostasis are not yet completely understood.

The need of copper in adults is generally reported as 1 mg/day [36]. Pipe water, because of the widespread use of copper pipes in household plumbing, can be an important dietary source of copper, differently from unpolluted fresh water that contains no or very little copper [27, 37]. The threshold value for copper in water for humans use is 2 mg/L in the Directives of most countries [38]. The principal dietary sources of copper are chocolate, animal liver, crustacean, shellfish, green vegetables, dried fruits and nuts. Copper concentration in these foods ranges from 20 up to 50 mg/kg, about 500 times higher than in human milk, which represents one of the poorest dietary sources of copper [39, 40]. Moreover, copper content in human milk shows a 50 % decrease over the first 3 months of lactation [41, 42], so that prolonging too much breast feeding could lead to copper deficiency [43]. The recommended dietary allowance (RDA) for copper in infancy is 80 μg per kg/day [44, 45]. Among newborns, premature and low birth weight infants are at higher risk in developing copper deficiency. In clinical practice, copper deficiency should be suspected in every infant with prolonged or recurrent diarrhoea [46].

After the first report of the association between diarrhoea and copper deficiency in malnourished infants in a children hospital in Lima [47], several studies have underlined the risk for copper deficiency [48] and the relevance of the intestinal mucosa in maintenance of normal copper status [49].

Relevant insights into the hierarchy of copper in vertebrates have recently been gained through the administration of molecules that interfere with copper metabolism in zebra fish [50]. The most severe copper-deficient phenotype was obtained by treating zebra fish embryos with the copper chelator neocuproine: it was characterized by impaired neurogenesis, deformed notochord, block in cartilage maturation, and hematopoietic defects probably due to disarrangement in hematopoietic stem cell differentiation.

Copper bioavailability

Copper bioavailability depends on four main factors: the absorption of copper from the gastrointestinal tract, the copper transport in blood, the extraction of copper by hepatocytes from the portal blood supply, and the copper transport into the brain. Many factors may influence copper bioavailability:

1. Ageing: it decreases the efficiency of copper homeostasis, resulting in higher serum copper concentrations in the elderly [48];
2. Sex: higher mean serum copper levels are detected in females [51];
3. Hormonal factors: women on contraceptive pill have increased serum copper levels with respect to control women [52].

The gastrointestinal tract does not receive only copper from the diet (1 mg/day) [36, 53], but much higher amounts, probably more than 4.5 mg/day, are released in the intestinal tube through saliva, bile, gastric secretion and through apoptotic intestinal cells shedding from the intestinal mucosa [54]. Copper is absorbed in the stomach, in duodenum and in the whole small intestine [27]. From a clinical point of view, absorption of copper in humans predominantly occurs in the ileum [53]. The first step in copper absorption is the uptake of the metal present in the diet by intestinal epithelial cells. The intimate mechanisms by which copper is transported across the brush border of the intestinal cells are not yet well known. The chemical speciation of copper contained in vegetables or in animal foodstuffs represents a key factor in copper absorption: copper salts, including carbonate, acetate, sulphate and chloride are well absorbed [55]. On the contrary, copper oxide shows a much lower ability to cross the intestinal barrier [56]. Food treatment may also influence copper absorption: salt for food preservation may modify the solubility of copper bound to proteins and its bioavailability [57]. Intestinal pH is probably the most important physiological factor affecting copper absorption: the acid environment is essential for freeing copper ions from complexes formed in foods and in mucosal secretions, setting the conditions for their absorption [58]. The presence in the diet of citric and lactic acids, present in large amounts in fruits, may enhance copper absorption by increasing its solubilization [59]. Many trace elements present in the diet, in particular divalent cations, may also act competitively on intestinal copper absorption [54]. The ability of zinc to halt copper absorption was first described in rats [60] and subsequently confirmed in humans: 3 mg of zinc added to the diet are sufficient to cause copper deficiency [61], resulting in the development of severe anaemia and leucopenia [62]. The mechanisms by which zinc interferes on copper metabolism are multiple, probably zinc acting at different sites of copper metabolism: (1) displacing copper ions from the specific carrier on the intestinal cells [63]; (2) increasing metallothionein content in the intestinal epithelium that leads to block copper trafficking, and favours copper loss in faeces through apoptosis of intestinal cells [64]. The molecular pathways followed by copper ions to enter the enterocytes remain poorly understood: little is known about the molecules and biochemical mechanisms responsible for copper uptake at the apical plasma membrane of intestinal cells in mammals [65].

Over the past few decades, critical progress has been made in the identification of genes encoding proteins that function in the regulation of copper homeostasis. The most important copper pump from the intestinal lumen into the enterocytes is probably the high affinity copper transporter "human cation transporter 1" (hCtr1), that has been clearly identified as the major source, though not the only source, mediating the entry of copper into human cells [66]. hCtr1 is a 190 amino acid integral membrane protein, predicted to have three transmembrane-spanning domains and existing in the plasma membrane as a homo-trimer [67]. Cryoelectron-microscopy analysis confirmed a trimer formation and revealed the presence of a hole at the threefold axis, consistent with the proposal that the copper translocation pathway is located at the centre of the trimer [68]. According with these data, the overall architecture of hCtr1 lies somewhere between that of a traditional monomeric transporter and that of a channel, which is generally tetrameric [69]. hCtr1 is considered an essential protein for early embryonic development and for intestinal copper uptake [70]. The transporter hCtr1, a fundamental component of the copper transport machinery at intestinal level, is localized on the plasma membrane of enterocytes where it acts as a conventional transporter, normally providing a permeation pathway for copper ions. In presence of elevated copper levels, hCtr1 has been shown to be internalized [64]. hCtr1 has been characterized as a transporter of copper with high affinity that is time-dependent and saturable, and it is stimulated by extracellular acidic pH and high potassium concentrations [64]. Recently, experimental evidences in Caco-2 cells showed that the basolateral copper uptake greatly exceeds the apical uptake of the metal in enterocytes [71]. In the same study, a basolateral localization of hCtr1 was also found in T84 cells, a common model for intestinal crypt cells. The basolateral transport was reported to be mediated by hCtr1 [70]. The role of hCtr1 as the major driving factor for intestinal copper absorption in mammals has been shown in Ctr1 knockout mouse, which exhibited severe growth and viability defects, systemic copper deficiency, iron overload, cardiac hypertrophy and severe growth deficit due to intestinal block of copper absorption [72]. Recently, the presence of an elevated intraluminal sodium gradient was shown to stimulate Ctr1-mediated copper absorption by intestinal epithelial cells, by increasing the H(+) concentration [73]. The finding of Ctr1 expression on endothelial cells of the blood–brain barrier suggests that copper could be transported from the plasma into the central nervous system (CNS) via Ctr1 [72]. The essentiality of Ctr1 during development has been evidenced in CTR1 mutant embryos, which showed growth retardation, poorly developed neural ectoderm and mesoderm, and defect in neural tube closure [74].

A role in apical Cu(I) transport in intestinal cells is probably played by the dication metal transporter 1 (DMT1), also known as Nramp 2 or DCT1. DMT1 is well known for uptake of Fe(II) at the apical pole of enterocytes [75]. Recent experimental data suggest that DTM1 is also a physiologically relevant Cu(I) carrier in intestinal cells, and indicate that intestinal absorption of Cu(I) and Fe(II) are intertwined [76]. Copper supplementation in Caco-2 cell culture is able to up-regulate the expression of DMT1 as well as that of the ferroxidase hephaestin (Heph) and of the iron transporter ferroportin-1 (Fpn1). This supports a major role of copper status in the modulation of iron uptake in intestinal cells [77]. A competition between iron and copper for the same transporter was also hypothesized, showing that the addition of Fe(II) to the diet reduces copper absorption in rats [78]. The molecular bases for these data have been recently clarified, with the demonstration that copper could compete with iron for uptake via DMT1 [79]. The copper–iron competition for uptake via DMT1 should induce, in clinical practice, to verify copper status in patients suffering from chronic anaemia and in women with hyposideremia during gestation, in which iron enters as a therapeutic agent in abnormally high amounts. Copper uptake in the intestinal lumen via DMT1 has been shown to be sensitive also to luminal pH: its decrease at the brush order of intestinal epithelial cells can create a suitable environment for copper uptake and transmembrane transport [73].

Copper transport across enterocytes

The second step in copper transport across the intestinal epithelium is the passage through their cytoplasm. The trip across the cytoplasm is mediated by metallothioneins (MTs), a group of proteins identified in the late 1950s as cadmium and zinc proteins [80]. The incorporation of copper atoms into MTs is a crucial step indispensable to protect cell structure from copper toxicity, preventing the oxidative damage [81]. MTs are the most important copper transporter from the intestinal cell luminal pole towards the basal pole [63]. Some competition among different trace elements occurs for MTs: cadmium and zinc are the most important ions involved in competition with copper for the binding site of MTs [27]. The relevance of zinc in influencing the binding of copper to MTs was first demonstrated in rat [82]. Zinc was shown to increase MTs content in intestinal mucosa cells up to 25-fold, by inducing their synthesis [83]. Subsequently, zinc supplementation was shown to easily produce copper deficiency in man [84].

A part of copper absorbed by hCtr1 is transferred to the chaperone ATOX1 to reach ATP7A localized at the trans-Golgi network [85]. Some copper is also most likely delivered by chaperone COX17 to cytochrome oxidase in mitochondria, and by the chaperone CCS to superoxide dismutase [86].

When arriving to the basolateral pole of enterocytes, copper ions are transferred to the main copper transporter across the basal cell membrane: the copper-translocating P-type ATPase ATP7A, also known as Menkes protein (MNK) [87], which has been proposed as the copper transporter that could play a relevant role in regulating the basolateral transport of copper from the enterocyte into the portal blood [88]. ATP7A is a member of the copper transporting ATPases, that play a fundamental role in copper physiology and, in particular, in copper trafficking inside cells, copper uptake at the cell membrane and copper efflux. At the intestinal level, a possible role of ATP7A in the regulation of copper absorption by intestinal cells raises several questions on ATP7A behaviour in the intestinal epithelium [89]. In rat, iron deprivation resulted in a strong induction of the ATP7A gene in the duodenal epithelium, followed by a significant copper overload in the liver and in the intestine, suggesting an increase in intestinal copper transport during iron deficiency [90]. In the same study, a translocation of ATP7A from the physiological site at the apical domain of enterocytes to the brush border and the basolateral membrane domains was observed after iron deprivation. In the structure of ATP7A, three putative domains may be identified: (1) six or eight transmembrane loops that presumably form a channel for metal ions passage; (2) an ATP-binding region; (3) six copper-binding sites [91]. The role of the six copper-binding domains in ATP7A is not fully understood, since two are considered sufficient for copper transport [88, 92]. An intriguing possible role for the other four metal-binding sites is that of sensors of the intracellular copper concentration [93]. Under copper-limiting conditions, ATP7A is retained in the trans-Golgi network, but delocalized in proximity of the basolateral membrane [89]. Bremner [63] presented a study on the dynamics of ATP7A in intestinal epithelial cells in conditions of copper depletion and copper overload. A prevalent localization of the protein in the trans-Golgi network in copper-depleted cells was observed, while in condition of copper overload, ATP7A re-localized from the Golgi apparatus to the cell periphery. Only about 10 % was detected at the basolateral cell surface, raising several questions on the exact role of ATP7A in copper trafficking.

Copper transport in blood

On entering portal blood, copper atoms are immediately bound by two main copper carriers in the way from gut towards liver: albumin, the most abundant plasma protein and transcuprein [94, 95]. Copper may be bound, in portal flow, even to some amino acids, to different peptides, to fatty acids and to tryptophan, but many evidences suggest that amino acid chelates are not relevant, in physiological conditions, for copper transport in blood [96]. Human serum albumin, the most abundant protein of blood serum, is a versatile carrier protein, involved in the transport of fatty acids, vitamins, hormones, xenobiotics and multiple metal ions, including Cu(II), Ca(II), Zn(II), Co(II), Ni(II) and Cd(II) [97]. Albumin has been shown to undergo conformational change upon contact with cell surfaces, originating two subpopulations of albumin which bind to cell surfaces with different affinities [98]. Albumin is considered the major copper binder of the exchangeable copper in human plasma, binding about 17 % of copper, thanks to the sequence Asp-Ala(Thr)-His with high Cu(II) affinity at the amino terminus of the protein [99]. A second binding site, highly conserved during evolution, contains two/three imidazole and one/two carboxylate coordination groups that, also being a relevant binding site for some divalent cations (Zn, Ni,Cd), does not play a significant role in copper transport (1 % of nonceruloplasmin plasma copper) [100, 101]. Albumin-bound Cu(II) is rapidly reduced by ascorbate to Cu(I)-Albumin that is reoxidized by molecular oxygen [102]. The redox activity of albumin-bound copper is regulated by multiple factors: at Cu/albumin ratios below 1:1, copper is virtually redox inactive, as long as cysteine-34 is in reduced state. The binding with fatty acids facilitates cysteine oxidation and converts copper–albumin complexes from antioxidants to prooxidants [103]. Complexation of Cu(II) causes conformational changes near the metal-binding site of human albumin [95]. Although it has been calculated that albumin normally present in human plasma could bind as much as 30 mg Cu(II)/mL, in fact it actually binds only 180 ng Cu(II)/mL [95]. This finding may induce to speculate on the existence of other plasma components with an even higher affinity for copper than albumin, such as transcuprein or transcuprein-like proteins, or some specific peptides. The observation that analbuminemic rats do not show relevant changes in the distribution of dietary copper in the liver supports the existence of other factors involved in copper transport in plasma [104]. On the other hand, the presence of an unsaturated copper-binding site on albumin might provide some protection against a sudden release in blood of copper ions, as can happen in patients with Wilson's disease [105], causing acute intravascular haemolysis [106].

Some amino acids, including histidine, have a role in copper transport in blood: a small percentage of human plasma copper appears attached to histidine or in a ternary complex of CuAlbHis [107]. A kinetic study on the distribution of copper ions between human albumin and transferrin, which has two sites capable of complexing Cu(II), gave evidence that transferrin binds copper more strongly than albumin [108]. The considerably lower concentrations of transferrin with respect to albumin in human blood prevent its competition with albumin for nonceruloplasmin cupric ions. Nevertheless, we may speculate that, in the absence of albumin, transferrin could play a relevant role in Cu(II) transport in plasma.

Transcuprein is an alpha glycoprotein with a molecular weight of 270 kDa, discovered in the rat, characterized by a higher affinity for Cu(II) than albumin, accounting for about 12 % of the exchangeable copper pool in rat serum [109]. It seems likely that humans do not have transcuprein, but employ for copper transport other transcuprein-like macroglobulins, carrying about 10 % of human plasma copper [31]. Ceruloplasmin, akin to Pirandello's *Six characters in search of an author*, has long been a protein in search of a function [22]. It, known as the major copper-containing protein in human plasma, comprising about 65 % of serum copper [110], is a glycoprotein, formed by a single polypeptide of 1,046 amino acids with several carbohydrate chains attached, molecular weight of 132 kDa, that carries six copper atoms per molecule [111].

Cu(II) ions also enter mammalian erythrocytes, via the band 3-mediated anion exchange pathway in the form of negatively charged complexes with chloride and carbonate [112]. Cupric ions are reduced to cuprous ions already on the cell membrane surface, probably at an externally facing sulphydryl group [113]. Copper uptake across the hepatocyte cell membrane occurs at the sinusoidal pole, where it arrives bound to albumin-specific and nonspecific binding sites as Cu(II) [114]. Whichever is the substrate for uptake, the hepatocyte has to remove copper ions from a high affinity complex, represented by albumin, histidine and transcuprein-like proteins [114]. The process of copper uptake by hepatocytes is probably initiated by the binding of copper in either a CuHis2 complex or as a CuHisAlb ternary complex, followed by reduction of cupric to cuprous ions by NADH oxidase localized at the plasma membrane [115]. It has been calculated that one copper ion is transported for each molecule of NADH oxidized [114]. The reducing agent vitamin C has been shown to stimulate copper uptake in hepatocytes, probably by accelerating reduction of copper ions [116]. NADH oxidase reduces both Cu(II) and Fe(III): this datum may explain why increased iron levels in the hepatocytes may halt copper uptake, perhaps by down-regulating the metalloreductase activity on the plasma membrane [117]. The reduction of copper would lower the stability of the complexes during their passage in the sinusoidal lumen, freeing part of copper atoms and allowing their uptake by hepatocytes.

Copper-dependent enzymes in the central nervous system

Copper plays an indispensable role in the development and in physiology of the human CNS, being a cofactor of numerous enzymes, including dopamine-β-hydroxylase (DBH), peptidyl-α-mono-oxygenase, superoxide dismutase and many other enzymes. These enzymes play fundamental roles in catecholamine biosynthesis, activation of neuropeptides and hormones, protection against reactive oxygen species, respectively, and in other processes essential for normal CNS function [118].

DBH catalyses the hydroxylation of the important neurotransmitter dopamine into norepinephrine, in the presence of both molecular oxygen and of reducing co-substrates, among which *N*-hydroxyguanidines probably play a relevant role in reducing Cu(II) to Cu(I) [119]. Recently, a very elegant experiment of molecular therapy carried out in a Menkes disease mouse model clearly showed the ability of ATP7A addition to the choroid plexus to enhance activity of DBH, leading to correction of brain pathology. This study first confirmed the role played by DBH in the pathogenesis of brain pathology in Menkes disease, giving the first evidence that gene therapy may be of clinical utility in the therapy of inborn errors of copper metabolism [120].

Superoxide dismutase 1 (SOD1) is a copper enzyme, highly represented in the CNS cells, mainly in cortical pyramidal neurons, Purkinje cells, deep cerebellar neurons and anterior horn cells in the spinal cord. SOD catalyses the conversion of reactive free radicals that can be toxic to cells and cause neuronal injury, to hydrogen peroxide, which can be converted to water by catalase or glutathione peroxidase [121]. The mitochondria function and protection against reactive oxygen species in the cytosol of neural cells mainly rely on the activity of cytochrome *c* oxidase and SOD1 [118]. Superoxides in the intercellular spaces and at the cell surface are neutralized by SOD3, which also maintains cerebral vascular tone and regulates neurogenesis [122].

Human cytochrome *c* oxidase (*CCO*) is a terminal enzyme of the mitochondrial respiratory chain, consisting of 13 subunits.

Another class of copper-dependent enzymes is represented by *lysyl oxidase* (LOX) protein family. LOX catalyses the oxidation of the side chain of a lysine giving rise to the cross-linking process that leads to collagen and elastin formation. In the rat brain, LOX has been found to be highly expressed in neurons, in the brain matrix, in the choroid plexus, and in the wall of blood vessels [123]. LOX activity is required during brain development, for the formation and remodelling of the extracellular matrix [124]. Since abnormalities in connective tissue cross-linking have been reported in patients affected by amyotrophic lateral sclerosis (ALS), in recent years an enzyme of possible relevance has been hypothesized to be LOX, the copper-containing enzyme which catalyses the cross-linking of collagens and elastin. The hypothesis that allelic variants or mutants of LOX gene could result in altered function of LOX in ALS patients has not been confirmed by studies on LOX polymorphisms, in which no association was found between any of these polymorphisms and ALS or its phenotype [125]. On the other hand, LOX immunoreactivity has been reported to increase in neurons of the spinal cord, brain stem and cortex, and the Purkinje cells of the cerebellum in transgenic G93A SOD1 (mSOD1) mouse, a model of ALS. The increase in the amount of LOX mRNA, protein and activity was coincidental with late-stage ALS, suggesting that LOX might be associated with the progression of the neurodegenerative disease [123].

Ceruloplasmin, a multicopper oxidase that plays a fundamental role in iron metabolism, utilizes copper atoms for electron transfer: by oxidizing ferrous iron into ferric ion, ceruloplasmin facilitates iron efflux from tissues and cells, including nervous cells [126]. Ceruloplasmin has been for a long time proposed as the principal copper transporter to peripheral tissues [127], and a role in copper transport for ceruloplasmin has been proposed even in recent years [128]. Copper chaperoned to the trans-Golgi network is transported into the Golgi apparatus by ATP7B and then incorporated into ceruloplasmin in vesicles that travel to the plasma membrane and release ceruloplasmin-bound copper into the plasma [129]. Copper which is used for ceruloplasmin synthesis is Cu(II), or goes from Cu(I) to Cu(II) at some stage prior to incorporation into the protein [130]. Intrahepatocytic chloride channels are involved in copper metabolism, by promoting copper incorporation into ceruloplasmin, probably by improving the efficiency of ATP7B [131]. The major part of researchers involved in the study of copper metabolism, on the contrary, think that this protein plays no essential role in the transport or, in general sense, in metabolism of copper [132–134]. The mechanism of copper transport into the brain is unclear. The expression of Cu transporters has been shown to be higher in brain barriers than in brain parenchyma; as a consequence, copper transport through the blood–brain barrier into the brain is mainly achieved as a free Cu ion [135]. New functions have been recently proposed for ceruloplasmin, in the CNS: it could play an important role in neuropathological conditions, by stimulating various neurotoxic molecules including nitric oxide (NO) in microglial cells [136]. The important role of ceruloplasmin in the striatal metabolism of catecholamines is well underlined by the observation that the lowest content of serum ceruloplasmin is observed in the most grave neurological forms of Wilson's disease [137]. The loss of ceruloplasmin function in the CNS cells, normally observed with age, has been recently associated with development of neural degeneration. This hypothesis has been reinforced by the observation that in patients affected by congenital aceruloplasminemia, astrocytes show a marked accumulation of haemosiderin, the storage toxic cytoplasmic evidence of a derangement in iron metabolism, probably due to the absence of ceruloplasmin, the most important iron-oxidizing copper enzyme [138]. In these patients, even cerebellar neurons, the Purkinje cells, have been reported to be reduced in number, and affected by iron storage. Similar lesions have been confirmed in ceruloplasmin-deficient mice: interestingly, whereas astrocytes show pathological changes very similar to those found in humans with aceruloplasminemia, neurons in experimental animals appear iron-deprived, suggesting a prevalent role in copper as well in iron metabolism for astrocytes [139]. It appears evident that the function of copper-dependent ceruloplasmin is strictly bound to copper bioavailability. As a consequence, we may hypothesize that every acquired pathological change in copper metabolism, including inadequate copper supply, malabsorption, blood transport and distribution, might cause iron overload in CNS cells, similar to those reported in patients with congenital aceruloplasminemia.

The blood coagulation *factor VIII,* a noncovalent heterodimer comprised of a heavy chain (A1–A2–B domains) and light chain (A3–C1–C2 domains), is an important component of the blood clotting cascade: it needs copper as a structural element, for reaching a stable binding between its protein chains [140]. A model where copper in factor VIII likely functions to increase specific activity of the hetero dimer rather than directly mediating the inter subunit interaction has been also proposed [141]. Factor VIII circulates as an inactive profactor in complex with von Willebrand factor, and recently metal ions have been confirmed to be critical to the integrity of factor VIII, with copper and calcium ions stabilizing the hetero dimer and generating the active conformation, respectively [142].

The normal cellular *Prion protein* (PrP(C)) is a copper-binding protein, highly conserved in mammals and expressed predominantly in the brain. The exact function of the normal prion protein in the brain has not been fully elucidated. Prion proteins have been hypothesized to function as metal-binding proteins, and show a superoxide dismutase-like activity, physiologically involved in the cell response to oxidative stress [143]. The conversion of the normal PrP(C) into other abnormal species, including the Scrapie prion (PrP(Sc)), has been associated in Scrapie-infected mice with brain metal imbalance and perturbations in the divalent metal content, especially copper, resulting in aberrant cellular responses to oxidative stress, predisposing CNS cells to oxidative impairment and eventual degeneration [144]. In humans, prion disease is characterized by a rapidly progressive dementia and cerebellar ataxia, resulting from neuronal spongiform degeneration and astrocytic gliosis [145]. The intimate mechanisms of neurotoxicity in prion disease have not been well clarified yet. Recent reports suggest that imbalance of brain metal homeostasis, due to loss of the physiological function of normal PrP(C) might represent a significant cause of neurotoxicity in prion disease. Other views suggest gain of toxic function by PrP(Sc), due to sequestration of metals within the aggregates of the pathological prion protein, resulting in the generation of redox-active PrP(Sc) complexes [146]. Involvement of oxidative stress in prion disease pathogenesis, associated with imbalance of copper and iron homeostasis, is generally considered an important trigger in neurotoxicity in prion disorders: whether neurotoxicity depends on loss of normal function of PrP(C) or because of gain of toxic function by PrP(Sc) remains unclear [147].

Amyloid precursor protein (APP) is a transmembrane glycoprotein, belonging to a multigene family that contains two other homologues known as amyloid precursor-like protein 1 (APLP1) and 2 (APLP2) [148]. Only APP contains the Aβ region and can be cleaved by β- and γ-secretase and incorrectly metabolized to generate Beta amyloid (Aβ), a 39–45 amino acid peptide that gives rise to extracellular plaques and intracellular deposits with neuronal death, leading to Alzheimer's disease. Thus APLPs cannot contribute to Aβ deposition in Alzheimer's disease [121]. The normal function of APP in brain cells has not been clarified yet. APP has been shown to bind zinc ions [149] and copper ions within cysteine-rich domain (residues 135–155), which contains three histidine residues at position 147, 149 and 151, the potentially very effective binding sites for Cu(II) [150]. APP has also been shown to reduce Cu(II) to Cu(I), potentially leading to increase oxidative stress in neurons [151]. In APP knockout mice, copper levels were significantly elevated in the cerebral cortex and in the liver [152]. Taken all together, these data indicate that the association between APP and copper is well established, and suggest that APP could have an important role in copper homeostasis. The presence in the amino acid sequence of the β-amyloid peptide of three histidine residues, able to anchor Cu(II) ions, could indicate possible involvement of copper ions in the aggregation process of Aβ [121].

Copper uptake and trafficking in brain

The mechanism of copper transport into the brain is unclear. The expression of Cu transporters has been shown to be higher in brain barriers than in brain parenchyma; as a consequence, copper transport through the blood–brain barrier into the brain is mainly achieved as a free copper ion [135]. The blood–brain barrier is a key structure in copper trafficking from general circulation into the brain. In the macular mouse, an animal model of Menkes disease, characterized by a defective intestinal absorption of copper and by copper deficiency in the brain, the administration of copper by intramuscular injections cannot improve the brain status because administered copper is not transported across the blood–brain barrier. The association of diethyldithiocarbamate with injected copper has been shown to facilitate the passage of copper ions across the blood–brain barrier, opening a new field of research on copper carriers for the brain which could solve the problem of finding a therapy for children affected by Menkes disease [153]. Copper is required to maintain the human brain biochemistry, copper deficiency or excess resulting in central nervous system disease [18]. However, understanding the intimate mechanisms regulating copper uptake and transport, including molecules involved in maintaining copper homeostasis in the human brain, is still in its infancy. Little is known about the molecules and biochemical mechanisms responsible for copper uptake at the plasma membrane of neurons and glial cells in mammals.

The *blood–brain barrier* represents the first and most important limit to blood copper access to the CNS. The marked differences in copper content between blood and the cerebrospinal fluid, with about 50-fold lower levels in the latter, suggest that effectively the blood–brain barrier limits copper entry into the brain [154]. Studies carried out utilizing radioactive copper have demonstrated that the major accumulation of copper ions was in the choroid plexus, entry of copper into cells of the choroid plexus being three to four times faster than that into brain capillaries, whereas copper entry into the cerebrospinal fluid was negligible [135]. Taken together, all these data suggest that brain cells receive copper from blood and not from the cerebrospinal fluid, and that choroid plexus cells play a key role in regulating copper entry into brain parenchyma.

Among brain cells, *astrocytes* probably play a central role in copper uptake and distribution to other cell types. Similar uptake rates and copper content of astrocytes and hepatocytes suggest that copper handling in astrocytes and in liver cells could be similar [155]. Another similarity, as for copper metabolism, between astrocytes and other cell types, including enterocytes and hepatocytes, is represented by the strong inhibition of copper uptake by zinc in cultured astrocytes [64].

New insights into brain cell copper metabolism have been acquired through the identification of the gene family SLC31 (solute-linked carrier 31) [156]. There are at least two members of the SLC31 gene family, the human cation transporter 1 gene (hCtr1) [157] and hCtr2 [22].

CTR1 is a high-affinity membrane copper permease, conserved from yeast to humans, that mediates the physiological uptake of Cu(I) from the extracellular environment, following reduction of Cu(II) by a cell surface metalloreductase [158]. CTR1 binds four Cu(I) ions as a cuprous-thiolate polynuclear cluster [159], is a component of the copper transport machinery at the cell membrane, which transports copper across the plasma membrane with high affinity in an energy-independent, time-dependent and saturable manner [64]. Two methionine-rich domains of hCtr1 have been shown to regulate copper-stimulated endocytosis:

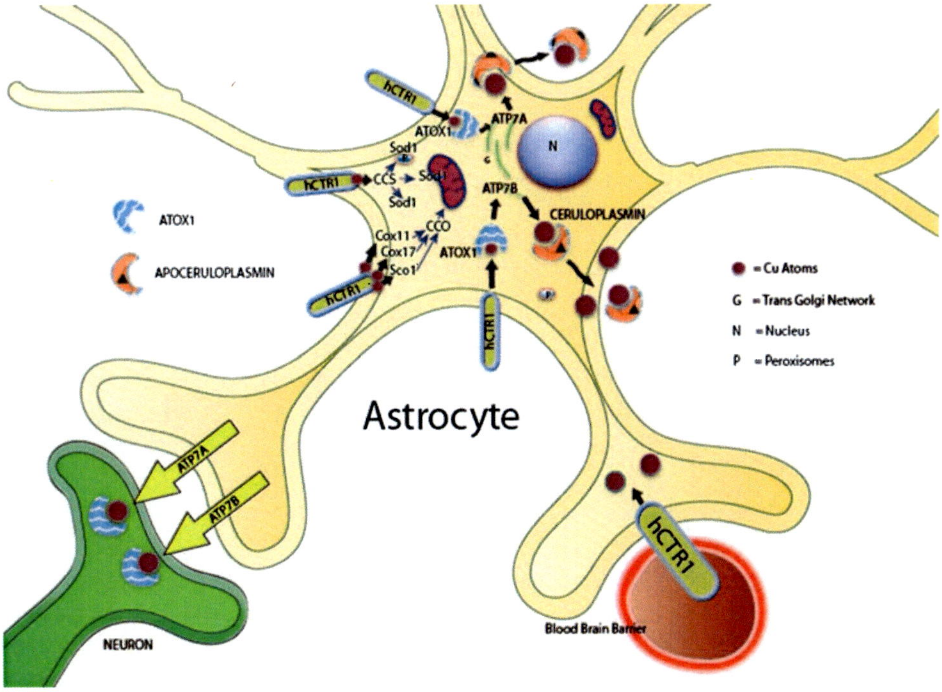

Fig. 1 Copper pumps and principal copper chaperones involved in copper trafficking between astrocytes, neurons and the blood–brain barrier

the response to low copper concentrations requires the amino-terminal methionine cluster MMMMPM, while the transmembrane MXXXM motif is required for the endocytic response to high copper concentrations [160]. In conditions of copper overload in an in vitro system, CTR1 is rapidly internalized by endocytosis, is ubiquitinated and degraded by vacuolar proteases [161].

In zebra fish, neural tissue resulted very sensitive to CTR1 down-regulation, which resulted in neuronal cell death [162]. In mice, the whole body knockout of CTR1 was associated with impaired neural tube closure and embryonic lethality [74]. In mammals, CTR1 is highly expressed in the whole brain parenchyma, reaching the highest levels in the choroid plexus cells [163]. In culture systems, astrocytes and neurons both showed high CTR1 expression levels, with similar uptake of copper ions [72].

CTR1 is not specific for copper: it may facilitate the transport of other substances, including the three platinum-based anticancer drugs cisplatin, carboplatin and oxaliplatin [164, 165]. Zinc and other divalent metal ions do not affect CTR1-mediated copper transport in brain cells [166].

The role played by *hCtr2*, identified in a database search, remains unknown: it is expressed in all human tissues examined and its gene is located in 9q31/32 [167]. In saccharomyces cerevisiae, CTR2 is localized at the vacuole membrane, where it mobilizes vacuolar copper stores towards the cytosol [158, 168]. A study on CTR2 immunolocalization in different human cell lines demonstrated its localization in late endosomes and lysosomes, with a putative role in maintaining copper homeostasis by stimulating copper delivery to the cytosol [169].

Copper chaperones

Copper imported by the cell membrane via CTR1 rapidly binds to intracellular carrier proteins (copper chaperones), which deliver copper ions to specific sites within the cell. Once entered into the astrocytic cytoplasm, copper ions exiting CTR1 are thought to bind to cytosolic copper chaperones that act as specific shuttles, able to carry copper ions towards different cell compartments (Fig. 1).

Atox1 is a 68 amino acid peptide, which exists as monomer that, in the presence of copper, can form dimers [170], playing a critical role in copper homeostasis [171]. Copper ions bind to Atox1 within the surface-exposed loop, containing a motif MxCxxC, in which two cystein coordinate copper [172]. Several possibilities have been proposed, regarding the mechanism by which copper ions are acquired by Atox1: the chaperone might directly interact with CTR1 or, exchange copper ions with a copper–glutathione complex [173]. In adult rat brain, Atox1 has been reported to be expressed in all brain compartments, reaching the highest levels in the pyramidal neurons of cerebral cortex, of the hyppocampus, in neurons of the locus coeruleus, and in Purkinje cells in cerebellum [174]. The main function of Atox1 is the delivery of copper ions to the copper-transporting ATPases, ATP7A and ATP7B (Fig. 1), through a direct interaction with the metal-binding sites located in the N-terminal region of both ATPases [175], followed by delivering of copper ions to P-type ATPases dislocated in the trans-Golgi network [176].

Moreover, in brain cells Atox1 probably functions as a copper-sensor, controlling the intracellular trafficking and distribution of copper ions throughout the cytoplasm and their proper delivery to the secretory pathway [177]. Atox1 may also have antioxidant properties and play a role in neuronal cell survival. In culture systems, neuronal cells overexpressing Atox1 showed higher resistance under conditions of serum deprivation, suggesting a role to Atox1 in cell protection against stress [178]. Recently, a role for Atox1 was assigned in cell growth, with the hypothesis that Atox1 could also act as a transcription factor [179]. The relevant role of Atox1 in development has been well evidenced in knockout mice, which are characterized by growth retard, reduced copper concentration in brain, seizure and high mortality [171].

Human *Cox17*, a 62-residue protein with a Cu(I)-binding site formed by two consecutive cysteine residues [180], is a mitochondrial copper chaperone and the only protein known so far to exhibit significant primary sequence homology to metallothioneins [181, 182]. In humans, mutations in Sco2 have been associated with encephalopathy and cardiomyopathy presenting in the perinatal period [183].

Mitochondria could also act as copper deposit, storing copper ions in a distinct ceruloplasmin and in metallothionein, and chaperoning copper to the trans-Golgi compartment and to the secretory pathways when too much copper is accumulated inside the cell [129].

The copper chaperone *CCS* activates the target essential antioxidant enzyme SOD through direct insertion of copper, prefolded dimers of apoSOD1 serving as substrate for CCS [184], and apparently functions to protect the metal ions from binding to intracellular scavengers of free copper, which is normally limited to less than one free copper ion per cell [185]. CCS1 is the necessary metallochaperone for delivering Cu(I) to Cu/Zn-superoxide dismutase 1 (SOD1) within the mitochondria, in the intermembrane space [184]. The turnover of CCS in fibroblasts, a cell type characterized by the abundance of this chaperone, has been shown to be related to copper content of the cell, revealing an unique post-translational component of the intracellular copper homeostasis [186]. Mammalian SOD1 may receive copper ions even by an alternative CCS-independent pathway of activation, which involves the reduced form of glutathione or GSH [187].

Conflicting results have been reported on the role of *metallothionein* (MT), a cysteine-rich cytoplasmic protein that chelates copper ions, in liver cells. MTs localize in the cytosol, in the nucleus as well as in the intermembrane spaces of mitochondria. They protect hepatocytes, as well other cells including duodenal and renal epithelium [188], against toxicity caused by excess copper, directly binding the metal ions through their cystein-rich domains alpha and beta and acting as antioxidants [189]. On the other hand, in the presence of H_2O_2, during oxidative stress, the degree of protection by MTs from copper-induced cell death may disappear, and MTs could release free copper and enhance the formation of reactive oxygen species, potentiating cellular damage [190].

MURR1, also called COMMD1, is a gene recently identified, which has been hypothesized to be essential for copper excretion at the biliary pole of hepatocytes, presumably acting downstream ATP7B. Its mutation is responsible for canine copper toxicosis, one of the animal models of Wilson's disease [191]. MURR1/COMMD1 encodes a protein without detectable sequence homology to known proteins, which is readily detectable in all tissues, suggesting that it could exhibit a pleiotropic function, in copper metabolism, in different organs [192]. XIAP is a potent suppressor of apoptosis which is found to interact with MURR1. In XIAP-deficient mice, cells and tissues were found to contain reduced copper levels, suggesting a role for XIAP in copper homeostasis through a negative regulation of MURR1 [193]. In situ hybridization experiments carried out in the brain of adult mice showed that MURR1 gene was expressed in neuronal cells in all regions of the brain [194].

Two P-type ATPases, *ATP7A* and *ATP7B*, are key molecules for the regulation of copper homeostasis in mammals [195]. In the liver, they have a dual role: when located in the trans-Golgi network, they provide copper ions to essential cuproenzymes; under copper stress, ATP7A and ATP7B sense the increased copper levels and translocate to the cell membrane in order to mediate the excretion of excess intracellular copper [196]. In human hepatocytes, the specific copper transporter Atp7B is localized to the trans-Golgi network when extracellular copper concentration is low (below 1 mmol/L); at increased Cu levels, it redistributes to vesicular structures and to biliary canaliculi [197]. A new intracellular transporter of ATP7B has been recently identified: it is the Niemann-Pick disease type C protein (NPC1), which is a late endosome protein which regulates intracellular vesicle traffic of ATP7B [198]. A recent study on ATP7B immunolocalization in hepatoma (HepG2) cells evidenced that the final trafficking destination of ATP7B is a paricanalicular vesicular compartment, rather than the canalicular membrane of the hepatocyte as previously reported. According to this report, ATP7B could be responsible of copper accumulation in vesicles, which subsequently undergo exocytosis, releasing excess copper across the plasma membrane [199]. The final destination of ATP7B in hepatocytes during the copper-induced trafficking process is still under debate [200]. Although the ATP7B translocation is conserved among nonhepatic cell lines, no co-localization with MURR1/COMMD1 was detected, suggesting that the translocation of ATP7B could take place independently [201]. ATP7B transports copper into cells against the gradient: to enter this transport pathway, cupric ions have to be reduced to cuprous ions: in fact, copper ATPase7B transports cuprous ions or complexes Cu^{+} -histidine. Increased expression of ATP7A and ATP7B genes has been recently observed in some human cancer specimens, and may be associated with tumorigenesis and chemotherapy resistance [202]. In neuroplastoma cells, ATP7A expression has been shown to be regulated by retinoic acid receptor β and it has effects on intracellular copper levels, revealing a link between the anticancer action of retinoids and copper metabolism. [203]. ATP7B stability is partially regulated by *COMMD1* (formerly MURR1) [204], the protein deleted in Bedlington terriers, an animal model of copper toxicosis [205]. COMMD1 exerts its regulatory role in copper homeostasis specifically interacting with newly synthesized ATP7B and decreasing its stability [206]. COMMD1 specifically binds copper as Cu(II), and may cooperate with ATP7B to facilitate biliary copper excretion [207, 208]. Glutaredoxin (GRX1) has been also proposed as an essential factor for ATP7A and ATP7B function; it catalyses the reduction of disulphide bridges and may reverse the glutathionylation of the cysteine residues within the six copper-binding motifs MxCxxC, facilitating copper binding for subsequent transport [209]. An additional protein has been shown to interact with ATP7B, and not with ATP7A, the dynactin subunit *p62*. The ATP7B/p62 interaction is a putatively key component of the pathway that delivers ATP7B-bound copper to subapical vesicles of the hepatocyte for the removal of excess copper into bile [209]. The function of ATP7B in tissues other than in liver is unclear [19].

The role of ATP7A in copper physiology has been well studied in transgenic mice overexpressing the human Menkes protein. The protein was hyper-expressed in heart, smooth muscle of the lung, distal tubules of the kidney, intestinal enterocytes, hepatocytes, as well as in the hippocampus, cerebellum and choroid plexus of the brain [210]. In mice overexpressing human ATP7A, copper concentration was reduced in most tissues, particularly in heart and brain, suggesting a relevant role of ATP7A in copper efflux from cells and tissues.

As for ATPases distribution in the CNS, although ATP7A and ATP7B have been often reported to be co-expressed in same cells, many authors have reported distinct expression patterns, suggesting different roles and different locations for these copper pumps. ATP7A is probably the more represented in the brain, being expressed through the entire brain during development and after birth, till adulthood [211]. In the foetal brain, ATP7A is mainly expressed in the choroid plexus cells, and in neuronal and glial precursors located around the lateral ventricles, whereas in the postnatal period, it reaches the highest values in the cerebellum and in the cortical regions [211]. Age-related differences have been also found as for ATP7A distribution in different brain compartments: this copper pump increases its expression with age in the cerebellum, where it localizes to Bergman glia and to Purkinje cells [212], and the hippocampal pyramidal layer [211]. On the contrary, ATP7B might be more expressed after birth. The neurons of the hippocampus, olfactory bulbs, cerebellum, neurons and glial cells in the cerebral cortex, and nuclei in the brainstem are generally considered the site of highest expression in the brain of ATP7B [213]. ATP7B is also expressed and functionally active in the pineal gland [214].

The principal intracellular localization of ATPases in glial cells and in neurons is in the trans-Golgi network, where they probably have a role in incorporating copper ions into the several copper enzymes present in these cells. Both ATPases have the ability to change their intracellular position, according to the different copper status of each cell. Differences have been reported about this ability between ATP7A and ATP7B: the former, when copper levels increase, disperses from the TGN in small vesicles, which move towards the basolateral membrane of polarized cells, where it re-localizes, functioning as an excretion copper pump, able to remove excess toxic Cu(I) from the cell [215]. ATP7B disperses from the TGN in larger vesicles, moving towards the apical membrane in polarized cells, re-localizing to the apical membrane. Glutathionilation of both ATPases has been proposed as the mechanism regulating the activity and trafficking of these proteins which play a fundamental, and largely unknown, role in copper metabolism in human cells, including neurons and glial cells.

Kidneys play a relevant role in copper metabolism and regulate their copper content very effectively. ATP7A and ATP7B are probably the most important copper transporters that contribute to this regulation. They are co-expressed in epithelial cells of the proximal and distal renal tubules, with ATP7A playing a major role in exporting excess copper and protecting the renal parenchima against copper overload [216]. ATP7A and ATP7B are also expressed in the glomeruli, suggesting that glomeruli may be responsible for regulating copper levels in the filtrate. ATP7B has been also immunodetected in the epithelial cell lining the loops of Henle, in the renal medulla, where it may have a role in copper reabsorption [217]. In HEK 293 cells, a model of human embryonic kidney cells, hCTR1 was shown to be expressed at the plasma membrane, acting as a conventional transporter providing a permeation pathway for copper [67]. In patients affected by Wilson's disease, copper content was increased even in kidneys, up to 70 times the normal values. Copper distribution, in carriers of Wilson's disease, was uneven, mainly concentrated in the inner cortical zones [218].

Conclusions

The study of copper trafficking in the adult as well as in the developing human brain will be the focus of many researchers in the next future, due to mounting evidences that copper homeostasis is disturbed in patients affected by Alzheimer's disease, leading to oxidative stress and neurodegeneration [219]. Two proteins related to neurodegeneration, the APP and the Prion protein are, as previously stated, copper-binding proteins and, contemporary, are major regulators of neuronal homeostasis. The relevance of metallothionein in neurophysiological and in neuromodulatory functions has been stressed by the observations of very high levels of MTs in the CNS and by the identification of a brain-specific isoform, called Mt-III [220]. Recent studies carried out in the macular mouse, an animal model of Menkes disease, revealed that the association of diethyldithiocarbamate with injected copper facilitates the passage of copper ions across the blood–brain barrier, opening a new field of research on copper carriers for the brain which could solve the problem of finding a therapy for children affected by Menkes disease [153].

Taken all together, these data encourage further studies on copper trafficking in the brain. Our knowledge of molecules regulating copper transport in the brain cells is rudimentary. In particular, very few studies are available on the expression of the known copper transporters in the human brain, during development and in the postnatal period, and the vast majority of studies here reviewed regard experimental animals. Given the amount of data available on copper trafficking, further studies should be encouraged on their expression in the different compartments of the human brain, in health and in subjects affected by neurodegenerative disorders, in order to better understand, at the molecular level, the intimate relationship between copper status and neuronal and glial cell function.

References

1. Weast RC (ed) (1970) Handbook of chemistry and physics. Chemical Rubber Co., Cleveland
2. Seymour CA, Howell JM, Gawthorne JM (eds) (1987) Copper in animals and man. CRC, Boca Raton, FL
3. Manzl C, Enrich J, Ebner H, Dallinger R, Krumschnabel G (2004) Copper-induced formation of reactive oxygen species causes cell death and disruption of calcium homeostasis in trout hepatocytes. Toxicology 196:57
4. Farinati F, Cardin R, D'Inca R, Naccarato R, Sturniolo GC (2003) Zinc treatment prevents lipid peroxidation and increases glutathione availability in Wilson's disease. J Lab Clin Med 141:372
5. Pourahmad J, O'Brien PJ, Jokar F, Daraei B (2003) Carcinogenic metal induced sites of reactive oxygen species formation in hepatocytes. Toxicol In Vitro 17:803
6. Krumschnabel G, Manzl C, Berger C, Hofer B (2005) Oxidative stress, mitochondrial permeability transition, and cell death in Cu-exposed trout hepatocytes. Toxicol Appl Pharmacol 209:62
7. Nawaz M, Manzl C, Lacher V, Krumschnabel G (2006) Copper-induced stimulation of extracellular signal-regulated kinase in trout hepatocytes: the role of reactive oxygen species, Ca2+, and cell energetics and the impact of extracellular signal-regulated kinase signaling on apoptosis and necrosis. Toxicol Sci 92:464–475
8. Pourahmad J, O'Brien PJ (2000) Contrasting role of Na(+) ions in modulating Cu(+2) or Cd(+2) induced hepatocyte toxicity. Chem Biol Interact 126:159
9. Urani C, Melchioretto P, Morazzoni F, Canevali C, Camatini M (2001) Copper and zinc uptake and hsp70 expression in HepG2 cells. Toxicol In Vitro 15:497
10. Balamurugan K, Schaffner W (2006) Copper homeostasis in eukaryotes: teetering on a tightrope. Biochim Biophys Acta 1763:737
11. Sokol RJ, Twedt D, McKim JM Jr, Devereaux MW, Karrer FM, Kam I, von Steigman G, Narkewicz MR, Bacon BR, Britton RS et al (1994) Oxidant injury to hepatic mitochondria in patients with Wilson's disease and Bedlington terriers with copper toxicosis. Gastroenterology 107:1788
12. Sternlieb I (1968) Mitochondral and fatty changes in hepatocytes of patients with Wilson's disease. Gastroenterology 55:354
13. Faa G, Nurchi V, Demelia L, Ambu R, Parodo G, Congiu T, Sciot R, Van Eyken P, Silvagni R, Crisponi G (1995) Uneven hepatic copper distribution in Wilson's disease. J Hepatol 22:303
14. Faa G, Lisci M, Caria MP, Ambu R, Sciot R, Nurchi VM, Silvagni R, Diaz A, Crisponi G (2001) Brain copper, iron, magnesium, zinc, calcium, sulfur and phosphorus storage in Wilson disease. J Trace Elem Med Biol 15:155
15. Zischka H, Lichtmannegger J, Schmitt S, Jagemann N, Schulz S, Wartini D et al (2011) Liver mitochondrial membrane crosslinking and destruction in a rat model of Wilson disease. J Clin Invest 121:1508
16. Merle U, Stremmel W (2011) Copper toxicity in Wilson disease explained in a new way. Hepatology 54:358
17. Pena MM, Lee J, Thiele DJ (1999) A delicate balance: homeostatic control of copper uptake and distribution. J Nutr 129:1251–1260
18. Madsen E, Gitlin JD (2007) Copper and iron disorders of the brain. Annu Rev Neurosci 30:317
19. Kodama H, Fujisawa C (2009) Copper metabolism and inherited copper transport disorders: molecular mechanisms, screening, and treatment. Metallomics 1:42
20. Kodama H, Murata Y, Kobayashi M (1999) Clinical manifestations and treatment of Menkes disease and its variants. Pediatr Int 41:423
21. Crisponi G, Remelli M (2008) Iron chelating agents for the treatment of iron overload. Coord Chem Rev 252:1225
22. Crichton RR (2008) Biological inorganic chemistry, chapter 7, Metal assimilation pathways. Elsevier Science, Amsterdam, pp 117–129
23. Bogdanova AY, Gassmann M, Nikinmaa K (2002) Copper ion redox state is critical for its effects on ion transport pathways and methaemoglobin formation in trout erythrocytes. Chem Biol Interact 139:43
24. Osterberg R (1980) Therapeutic uses of copper-chelating agents. CIBA Found Symp 79:283
25. Musci G, Polticelli F, Calabrese L (1999) Structure, function relationships in ceruloplasmin. Copper transport and its disorders. Adv Exp Med Biol 448:175–182

26. Kato S, Sumi-Akamuru H, Fujimura H, Sakoda S, Kato M, Hirano A, Takikawa M, Ohama E (2001) Copper chaperone for superoxide dismutase co-aggregates with SOD1 in neuronal Lewy body-like hyaline inclusions: an immunohistochemical study on familial amyotrophic lateral sclerosis with SOD1 gene mutation. Acta Neuropathol 102:233
27. Linder MC, Hazeg-Azam M (1996) Copper biochemistry and molecular biology. Am J Clin Nutr 63:797S
28. Turski ML, Thiele DJ (2009) New roles for copper metabolism in cell proliferation, signaling, and disease. J Biol Chem 284:717
29. Lane TF, Iruela-Arispe ML, Johnson RS, Sage EH (1994) SPARC is a source of copper-binding peptides that stimulate angiogenesis. J Cell Biol 125:929–943
30. Models OR (1974) for copper protein interaction based on solution and crystal structure studies. Coord Chem Rev 12:309–347
31. Linder MC (1991) The biochemistry of copper. New York, Plenum
32. Hart EB, Steenbock J, Waddell J, Elvehjem CA (1928) Iron in nutrition. Copper as a supplement to iron for hemoglobin bilding in the rat. J Biol Chem 77:797–81
33. Sturgeon P, Brubaker C (1956) Copper deficiency in infants. Am J Dis Child 92:254–265
34. Cordano A, Baertl J, Graham JJ (1964) Copper deficiency in infants. Pediatrics 34:324–336
35. al-Rashid RA, Spangler J (1971) Neonatal copper deficiency. N Engl J Med 285:841–843
36. Marsh MN, Riley SA (1998) Digestion and absorption of nutrients and vitamins, chapter 87. In: Feldman M, Sleisenger MH, Scharschmidt BF (eds) Sleisenger and Fordtran's gastrointestinal and liver disease. WB Saunders Co., Philadelphia, PA, pp 1471–1500
37. Lonnerdal B (1998) Copper nutrition during infancy and childhood. Am J Clin Nutr 67(Suppl):1046S–1053S
38. World Health Organization (1993) Guidelines for drinking water quality, vol 1. Recommendations. World Health Organization, Geneva
39. Delves HT (1980) Dietary sources of copper. In: Biological roles of copper, Ciba Foundation Symposium 79, Excerpta Medica, pp 5–22
40. Picciano MF, Guthrie HA (1976) Copper, iron and zinc contents of mature human milk. Am J Clin Nutr 29:242–254
41. Vuori E, Kuitunen P (1979) The concentration of copper and zinc in human milk. Acta Pediatr Scand 68:33–37
42. Krachler M, Shi F, Rossipal E, Irgolic KJ (1998) Changes in the concentration of trace elements in human milk during lactation. J Trace Elements Med Biol 12:159–176
43. Vuori E (1979) Intake of copper, zinc, and manganese by healthy exclusively-breast-fed infants during the firsy three months of life. Br J Nutr 42:407–411
44. Griscon NT, Craigh JN, Neuhauser EBD (1971) Systemic bone disease developing in small premature infants. Pediatrics 48:883–895
45. Yuen P, Lin HJ, Hutchinson JH (1979) Copper deficiency in a low birth weight infant. Arch Dis Child 54:553–555
46. Rodriguez A, Soto G, Torres S, Venegas G, Castillo-Duran C (1985) Zinc and copper in hair and plasma of children with chronic diarrhea. Acta Paediatr Scand 74:770–774
47. Graham GG, Cordano A (1969) Copper depletion and deficiency in the malnourished infant. John Hopkins Med J 124:139–150
48. Madsen E, Gitlin JD (2007) Copper deficiency. Curr Opin Gastroenterol 23:187–192
49. Johnson PE, Milne DB, Lykken GI (1992) Effects of age and sex on copper absorption, biological half-life, and status in humans. Am J Clin Nutr 56:917–925
50. Mendelsohn BA, Yin C, Johnson SL, Wilm TP, Solnica-Krezel L, Gitlin JD (2006) Atp7a determines a hierarchy of copper metabolism essential for notochord development. Cell Metab 4:155–162
51. Johson PE, Milne DB, Lykken GI (1992) Effects of age and sex on copper absorption, biological half-life, and status in humans. Am J Clin Nutr 56:917–925
52. Carruthers ME, Hobbs CB, Warren RL (1966) Raised serum copper and ceruloplasmin levels in subjects taking oral contraceptives. J Clin Pathol 19:498–501
53. Turnlund JR, Keyes WR, Anderson HL, Acord LL (1989) Copper absorption and retention in young men at three levels of dietary copper by the use of the stable isotope 65Cu. Am J Clin Nutr 49:870–878
54. Wapnir RA (1998) Copper absorption and bioavailability. Am J Clin Nutr 67(Suppl):1054–1060
55. Ledoux DR, Pott EB, Henry PR, Ammerman CB, Merritt AM, Madison JB (1995) Estimation of the relative bioavailability of inorganic copper sources for sheep. Nutr Res 15:1803–1813
56. Ledoux DR, Henry PR, Ammerman CB, Rao PV, Miles RD (1991) Estimation of the relative bioavailability of inorganic copper sources for chicks using tissue uptake of copper. J Anim Sci 69:215–222
57. Clydesdale FM (1988) Minerals: their chemistry and fate in food. Dekker, New York, pp 57–94
58. Gollan JL (1975) Studies on the nature of complexes formed by copper with human alimentary secretions and their influence on copper absorption. Clin Sci Mol Med 49:237–245
59. Shah BG (1981) Chelating agents and bioavailability of minerals. Nutr Res 1:617–622
60. Van Reen R (1953) Effects of excessive dietary zinc in the rat and the interrelationship with copper. Arch Biochem Biophys 46:337–344
61. Gregor JL, Snedeker SM (1980) Effect of dietary protein and phosphorus levels on the utilization of zinc, copper and manganese by adult males. J Nutr 110:2243–2253
62. Hoffman HN, Phyliki RL, Fleming CR (1988) Zinc-induced copper deficiency. Gastroenetrology 94:508–512
63. Bremner I (1980) Absorption, transport and distribution of copper. In: Biological roles of copper, Ciba Foundation Symposium 79, Excerpta Medica, pp 23–48
64. Lee J, Pena MM, Nose Y, Thiele DJ (2002) Biochemical characterization of the human copper transporter Ctr1. J Biol Chem 277:4380–4387
65. Sharp PA (2003) Ctr1 and its role in body copper homeostasis. Int J Biochem Cell Biol 35:288–291
66. Maryon EB, Molloy SA, Zimmicka AM, Kaplan JH (2007) Copper entry into human cells: progress and unanswered questions. Biometals 20:355–364
67. Eisses JF, Chi Y, Kaplan JH (2005) Stable plasma membrane levels of hCtr1 mediate cellular copper uptake. J Biol Chem 280:9635–9639
68. Aller SG, Unger VM (2006) Projection structure of the human copper transporter CTR1 at 6-A resolution reveals a compact trimer with a novel channel-like architecture. Proc Natl Acad Sci USA 103:3627
69. Boal AK, Rosenzweig AC (2009) Structural biology of copper trafficking. Chem Rev 109:4760–4779
70. Nose Y, Kim BE, Thiele DJ (2006) Ctr1 drives intestinal copper absorption and is essential for growth, iron metabolism, and neonatal cardiac function. Cell Metab 4:235–244
71. Zimnicka AM, Maryon FB, Kaplan JH (2007) Human copper transporter hCTR1 mediates basolateral uptake of copper into enterocytes: implications for copper homeostasis. J Biol Chem 7:26471–26480

72. Kuo YM, Gybina AA, Pyatskowit JW, Gitschier J, Prohaska JR (2006) Copper transport protein (Ctr1) levels in mice are tissue specific and dependent on copper status. J Nutr 136:21–26
73. Nadella SR, Grosell M, Wood CM (2007) Mechanisms of dietary Cu uptake in freshwater rainbow trout: evidence for Na-assisted Cu transport and a specific metal carrier in the intestine. J Comp Physiol 177:433–446
74. Lee J, Prohaska JR, Thiele DJ (2001) Essential role for mammalian copper transporter Ctr1 in copper homeostasis and embryonic development. Proc Natl Acad Sci USA 98:6842–6847
75. Arredondo M, Munoz P, Nùnez MT (2003) DMT1, a physiologically relevant apical Cu1+ transporter of intestinal cells. Am J Physiol Cell Physiol 284:C1525–C1530
76. Han O, Wessling-Resnick M (2002) Copper repletion enhances apical iron uptake and transepithelial iron transport by Caco-2 cells. Am J Physiol Gastrointest Liver Physiol 282:G527–G533
77. Wapnir RA, Devas G, Solans CV (1993) Inhibition of intestinal copper absorption by divalent cations and low molecular weight ligands in the rat. Biol Trace Elem Res 36:291–305
78. Yu S, West CE, Beynen AC (1994) Increasing intakes of iron reduce status, absorption and biliary excretion of copper in rats. Br J Nutr 71:887–895
79. Tennant J, Stansfield M, Yamaji S, Srai SK, Sharp P (2002) Effects of copper on the expression of metal transporters in human intestinal Caco-2 cells. FEBS Lett 11:239–244
80. Kagi JHR, Vallee BL (1960) Metallothionein: a cadmium and zinc-containing protein from equine renal cortex. J Biol Chem 235:3460–3467
81. Sturniolo GC, Mestriner C, Irato P, Albergoni V, Longo G, D'Incà R (1999) Zinc therapy increases duodenal concentrations of metallothionein and iron in Wilson's disease patients. Am J Gastroenterol 94:334–338
82. Hall AC, Young BW, Bremner I (1979) Intestinal metallothionein and the mutual antagonism between copper and zinc in the rat. J Inorg Biochem 11:57–66
83. Cousins RJ (1979) Synthesis and degradation of liver metallothionein. In: Kagi JHR, Nordberg M (eds) Metallothionein. Birkhauser, Basel, pp 273–280
84. Botash AS, Nasca J, Dubowy R, Weenberger HL, Oliphant M (1992) Zinc-induced copper deficiency in an infant. Am Dis Child 164:709–711
85. Lonnerdal B (2008) Intestinal regulation of copper homeostasis: a developmental approach. Am J Clin Nutr 88:846S–850S
86. Field LS, Luk E, Culotta VC (2002) Copper chaperones: personal escorts for metal ions. J Bioenerg Biomembr 34:373–379
87. Bankier A (1995) Menkes disease. J Med Genet 32:213–215
88. Mercer JFB (1998) Menkes syndrome and animal models. Am J Clin Nutr 67(suppl):1022S–1028S
89. Nyasae L, Bustos R, Braiterman L, Eipper B, Hubbard A (2007) Dynamics of endogenous ATP7A (Menkes protein) in intestinal epithelial cells: copper-dependent redistribution between two intracellular sites. Am J Physiol Gastrointest Liver Physiol 292: G1181–G1194
90. Ravia JJ, Stephen RM, Ghishan FK, Collins JF (2005) Menkes copper ATPase (Atp7a) is a novel metal-responsive gene in rat duodenum, and immunoreactive protein is present on brush-border and basolateral membrane domains. J Biol Chem 280:36221–36227
91. Vulpe C, Levinson B, Whitney S, Packman S, Gitschier J (1993) Isolation of a candidate gene for Menkes disease and evidence that it encodes a copper-transporting ATPase. Nat Genet 3:7–13
92. Solioz M, Odermatt A, Krapf R (1994) Copper pumping ATPases: common concepts in bacteria and in man. FEBS Lett 346:44–47
93. Monty JF, Llanos RM, Mercer JF, Kramer DR (2005) Copper exposure induces trafficking of the menkes protein in intestinal epithelium of ATP7A transgenic mouse. J Nutr 135:2762–2766
94. Linder MC, Weiss KC, Vu HM, Rucker RB (1987) Structure and function of transcuprein in transport of copper by mammalian blood plasma. In: Hurley LC, Lonnerdal B, Keen C (eds) Trace elements in man and animals. Plenum, New York, pp 141–144
95. Linder MC, Wooten L, Cerveza P (1998) Copper transport. Am J Clin Nutr 67(suppl):965S–971S
96. Carter DC, Ho JX (1994) Structure of serum albumin. Adv Protein Chem 45:153–203
97. Breslow E (1964) Comparison of cupric ion-binding site in myoglobin derivatives and serum albumin. J Biol Chem 239:3252–3259
98. Reed RG, Burrington CM (1989) The albumin receptor effect may be due to a surface-induced conformational change in albumin. J Biol Chem 264:9867–9872
99. Bal W, Christodoulou J, Sadler PJ, Tucker A (1998) Multi-metal binding site of serum albumin. J Inorg Biochem 70:33–39
100. Lovstad RA (2002) A kinetic study on the copper-albumin catalyzed oxidation of ascorbate. Biometals 15:351–355
101. Laussac JP, Sarkar B (1984) Characterization of the copper(II) and nickel(II)-transport site of human serum albumin. Studies of copper(II) and nickel(II) binding to peptide 1-24 of human serum albumin by 13C and 1H NMR spectroscopy. Biochemistry 23:2832–2838
102. Linder MC, Lomeli NA, Donley S, Mehrbod F, Cerveza P, Cotton S, Wooten L (1999) Copper transport in mammals. In: Leone A, Mercer JMC (eds) Copper transport and its disorders. Kluwer/Plenum, New York, pp 1–16
103. Gryzumov YA, Arroyo A, Vigne JL, Zhao Q, Tyurin VA, Hubel CA, Gandley RE, Vladimirov YA, Taylor RN, Kagan VE (2003) Binding with fatty acids facilitates oxidation of cysteine-34 and converts copper-albumin complexes from antioxidants to prooxidants. Arch Biochem Biophys 413:53–56
104. Vargas EJ, Shoho AR, Linder MC (1994) Copper transport in the Nagase analbuminemic rat. Am J Physiol 267:G259–G269
105. Scheimberg IH, Sternlieb I (1984) Wilson's disease. In: Smith LH (ed) Major problems in internal medicine. WB Saunders Co., Philadelphia
106. Roche-Sicot J, Benhamou JP (1977) Acute intravascular hemolysis and acute liver failure as a first manifestation of Wilson's disease. Ann Intern Med 86:301–303
107. Lau SY, Kruck TPA, Sarker B (1974) Peptide molecule mimicking the copper (II) transport site of human erum albumin. J Biol Chem 246:5878–5884
108. Lovstad RA (2004) A kinetic study on the distribution of Cu(II)-ions between albumin and transferrin. Biometals 17:111–113
109. Weiss KC, Linder MC (1985) Copper transport in rats involving a new plasma protein. Am J Physiol 249:E77–E88
110. Wirth PL, Linder MC (1985) Distribution of copper among components of human serum. J Natl Cancer Inst 75:277–284
111. Zaitseva I, Zaitsev V, Card G, Moshkov K, Box B, Ralph A, Lindley P (1996) The nature of the copper celtres in human ceruloplasmin. J Biol Inorg Chem 1:15–23
112. Alda JO, Garay R (1990) Chloride (or bicarbonate)-dependent copper uptake through the anion exchanger in human red blood cells. Am J Physiol 259:C570–C576
113. Bogdanova AY, Virkki LV, Gusev M, Nikinmaa M (1999) Copper effects on ion transport across lamprey erythrocyte membrane: Cl(-)/OH(-) exchange induced by cuprous ions. Toxicol Appl Pharmacol 159:204–213
114. McArdle HJ, Bingham MJ, Summer K, Ong TJ (1999) Cu metabolism in the liver. In: Leone A, Mercer JFB (eds) Copper transport and its disorders. Kluwer/Plenum, New York, pp 29–37
115. Van den Berg GJ, McArdle HJ (1994) A plasma membrane NADH oxidase is involved in copper uptake by plasma membrane vesicles isolated from rat liver. Biochim Biophys Acta 1195:276–280

116. Bingham MJ, McArdle HJ (1994) A comparison of copper uptake by liver plasma membrane vesicles and uptake by isolated cultured hepatocytes. Hepatology 20:1024–1031
117. Whitaker P, McArdle HJ (1997) Iron inhibits copper uptake by down-regulating the plasma membrane NADH oxidase. Trace Elem Man Anim 9:237–239
118. Lutsenko S, Bhattacharjee A, Hubbard AL (2010) Copper handling machinery of the brain. Metallomics 2:596–608
119. Slama P, Boucher JL, Règlier M (2009) N-hydroxyguanidines oxidation by N3S copper complex mimicking the reactivity of dopamine beta-hydroxylase. J Inorg Biochem 103:455–462
120. Donsante A, Zerfas PM, Brinster LR, Sullivan P, Goldstein DS, Prohaska J, Centeno JA, Rushing E, Kaler SG (2011) ATP7A addition to the choroid plexus results in long-term rescue of the lethal copper transport defect in Menkes disease. Mol Ther 19(12):2114–2123. doi:10.1038/mt.2011.143
121. Brown DR, Kozlowski H (2004) Biological inorganic and bioinorganic chemistry of neurodegeneration based on prion and Alzheimer diseases. Dalton Trans 1907–1917
122. Rola R, Zou Y, Huang TT, Fishman K, Baure J, Rosi S, Milliken H, Limoli CL, Fike JR (2007) Lack of extracellular superoxide dismutase (EC-SOD) in the microenvironment impacts radiation-induced changes in neurogenesis. Free Radic Biol Med 42:1133–1145
123. Li PA, He Q, Cao T, Yong G, Szauter KM, Fong KS, Karlsson J, Keep MF, Csiszar K (2004) Up-regulation and altered distribution of lysyul oxidase in the central nervous system of mutant SOD1 transgenic mouse model of amyoprophic lateral sclerosis. Mol Brain Res 120:115–122
124. Bronson NW, Hamilton JS, Han M, Li PA, Hornstra I, Horowitz JM, Horwitz A (2005) LOXL null mice demonstrate selective dentate structural changes but maintain dentate granule cell and CA1 pyramidal cell potentiation in the hippocampus. Neurosci Lett 390:118–122
125. Chioza BA, Ujfalusi A, Csiszar K, Leigh PN, Powell JF, Radunovic A (2001) Mutations in the lysyl oxidase gene are not associated with amyotrophic lateral sclerosis. Amyotroph Lateral Scler Other Motor Neuron Disord 2:93–97
126. Harris ZL, Durley AP, Man TK, Gitlin JD (1999) Targeted gene disruption reveals an essential role for eruloplasmin in cellular iron efflux. Proc Natl Acad Sci USA 96:10812–10817
127. Martin F, Linden T, Katskinski DM, Oehme F, Flamme I, Mukhopadhyay CK, Eckardt K, Troger J, arth S, Camenisch G, Wenger RH (2005) Copper-dependent activation of hypoxia-inducible factor (HIF)-1: implications for ceruloplasmin regulation. Blood 105:4613–4619
128. Healy J, Tipton K (2007) Ceruloplasmin and what it might to do. J Neural Transm 114:777–781
129. Mehta R, Templeton DM, O'brien PJ (2006) Mitochondrial involvement in genetically determined transition metal toxicity II. Copper toxicity. Chem Biol Interact 163:77–85
130. Bingham MJ, Sargeson AM, McArdle HG (1997) Identification and characterization of intracellular copper pools in rat hepatocytes. Am J Physiol 272:G1400–G1407
131. Wang T, Weinman SA (2004) Involvement of chloride channels in hepatic copper metabolism: CIC-4 promotes copper incorporation into ceruloplasmin. Gastroenterology 126:1157–1166
132. Meyer LA, Durley AP, Prohaska JR, Harris ZL (2001) Copper transport and metabolism are normal in aceruloplasminemic mice. J Biol Chem 276:36857–36861
133. Hellman NE, Gitlin JD (2002) Ceruloplasmin metabolism and function. Annu Rev Nutr 22:439–458
134. Crichton RR (2007) Biol Inorg Chem
135. Choi B-S, Zheng W (2009) Copper transport to the brain by the blood-brain barrier and blood-CSF barrier. Brain Res 1248:14–21
136. Lee KH, Yun SJ, Nam KN, Gho YS, Lee EH (2007) Activation of microglial cells by ceruloplasmin. Brain Res 1171:1–8
137. Mzhelskaya TI (2000) Biological functions of ceruloplasmin and their deficiency caused by mutation in genes regulating copper and iron metabolism. Bull Exp Biol Med 130:719–727
138. Miyajima H, Kono S, Takahashi Y, Sugimoto M, Sakamoto M, Sakai N (2006) Cerebellar ataxia associated with heteroallelic ceruloplasmin gene mutation. Neurology 57:2205–2210
139. Jeong SY, David S (2006) Age-related changes in iron homeostasis and cell death in the cerebellum of ceruloplasmin-deficient mice. J Neurosci 26:9810–9819
140. Bihoreau N, Pin S, Kersabiec AM, Vidot F, Fontaine-Aupart MP (1994) Copper-atom identification in the active and inactive forms of plasma-derived FVIII and recombinant FVIII-delta II. Eur J Biochem 222:41–48
141. Sudhakar K, Fay PJ (1998) Effects of copper on the structure and function of factor VIII subunits: evidence for an auxiliary role for copper ions in cofactor activity. Biochemistry 37:6874–6882
142. Fay PG (2006) Factor VIII structure and function. Int J Hematol 83:103–108
143. Choi CJ, Kanthasamy A, Anantharam V, Kanthasamy AG (2006) Interaction of metals with prion protein: possible role of divalent cations in the pathogenesis of prion disease. Neurotoxicology 27:777–787
144. Wong BS, Brown DR, Pan T, Whiteman M, Liu T, Bu X, Li R, Gambetti P, Olesik J, Rubenstein R, Sy MS (2001) Oxidative impairment in scrapie-infected mice is associated with brain metals perturbations and altered antioxidant activities. J Neurochem 79:689–698
145. Waggoner D, Bartnikas TB, Gitlin J (1999) The role of copper in neurodegenerative disease. Neurobiol Dis 6:221–230
146. Singh N, Das D, Singh A, Mohan ML (2010) Prion protein and metal interaction: physiological and pathological implications. Curr Issues Mol Biol 12:99–107
147. Singh N, Singh A, Das D, Mohan ML (2010) Redox control of prion and disease pathogenesis. Antioxid Redox Signal 12:1271–1294
148. Bayer TA, Cappai R, Masters CL, Beyreuther K, Multhaup G (1999) It all sticks together–the APP-related family of proteins and Alzheimer's disease. Mol Psychiatry 4:524–531
149. Multhaup G, Bush AI, Pollwein P, Masters CL (1994) Interaction between the zinc (II) and the heparin binding site of the Alzheimer's disease beta A4 amyloid precursor protein (APP). FEBS Lett 355:151–154
150. Hesse L, Beher D, Masters CL, Multhaup G (1994) The beta A4 amyloid precursor protein binding to copper. FEBS Lett 349:109–116
151. Multhaup G, Schlicksupp A, Hesse L, Beher T, Ruppert T, Masters CL, Beyreuther K (1996) The amyloid precursor protein of Alzheimer's disease in the reduction of copper(II) to copper(I). Science 271:1406–1409
152. White AR, Reyes R, Mercer JF, Camakaris J, Zheng H, Bush AI, Multhaup G, Beyreuther K, Masters CL, Cappai R (1999) Copper levels are increased in the cerebral cortex and liver of APP and APLP2 knockout mice. Brain Res 842:439–444
153. Kodama H, Sato E, Gu YH, Shiga K, Fujisawa C, Kozuma T (2005) Effect of copper and diethylthiocarbamate combination therapy on the macular mouse, an animal model of Menkes disease. J Inherit Metab Dis 28:971–978
154. Forte G, Bocca B, Senofonte O, Petrucci F, Brusa L, Stanzione P, Zannino S, Violante N, Alimonti A, Sancesario G (2004) Trace and major elements in whole blood, serum, cerebrospinal fluid and urine of patients with Parkinson's disease. J Neural Transm 111:1031–1040
155. Van Den Berg GJ, de Goeij JJ, Bock I, Gijbels MJ, Brouwer A, Lei KY, Hendriks HF (1991) Copper uptake and retention in liver

parenchymal cells isolated from nutritionally copper-deficient rats. J Nutr 121:1228–1235

156. Petris MJ (2004) The SLC31 (Ctr) copper transport family. Pflugers Arch Eur J Physiol 447:796–800
157. Eisses JF, Kaplan JH (2002) Molecular characterization of hCTR1, the human copper uptake protein. J Biol Chem 277:29162–29171
158. Rees EM, Thiele DJ (2007) Identification of a vacuole-associated metalloreductase and its role in Ctr2-mediated intracellular copper mobilization. J Biol Chem 282:21629–21638
159. Xiao Z, Loughlin F, George GN, Howlett GJ, Wedd AG (2004) C-terminal domain of the membrane copper transporter Ctr1 from *Saccharomyces cerevisiae* binds four Cu(I) ions as a cuprous-thiolate polynuclear cluster: subfentomolar Cu(I) affinity of three proteins involved in copper trafficking. J Am Chem Soc 126:3081–3090
160. Guo Y, Smith K, Lee J, Thiele DJ, Petris MJ (2004) Identification of methionine-rich clusters that regulate copper-stimulated endocytosis of the human Ctr1 copper transporter. J Biol Chem 279:17428–17433
161. Liu J, Sitaram A, Burd CG (2007) Regulation of copper-dependent endocytosis and vacuolar degradation of the yeast copper transporter, Ctr1p, by the Rsp5 ubiquitin ligase. Traffic 8:1375–1384
162. Mackenzie NC, Brito M, Reyes AE, Allende ML (2004) Cloning, expression pattern and essentiality of the igh-affinity copper transporter 1 (ctr1) gene in zebrafish. Gene 328:113–120
163. Platonova NA, Barabanova SA, Povalikhin RG, Tsymbalenko NV, Nodzdrachev AD, Puchkova LV (2005) Expression of Menkes AtPase and Wilson ATPpase in different regions of the adult rat brain. Dokl Biol Sci 401:88–91
164. Sinani D, Adle DJ, Kim H, Lee J (2007) Distinct mechanisms for Ctr1-mediated copper and cisplatin transport. J Biol Chem 282:26775–26785
165. Holzer AK, Manorek GH, Howell SB (2006) Contribution of the major copper influx transporter CTR1 to the cellular accumulation of cisplatin, carboplatin, and oxaliplatin. Mol Pharmacol 70:1390–1394
166. Kuo YM, Zhou B, Cosco D, Gitschier J (2001) The copper transporter CTR1 provides an essential function in mammalian embryonic development. Proc Natl Acad Sci USA 98:6836–6841
167. Zhou B, Gitschier J (1997) hCTR1: a human gene for copper uptake identified by complementation in yeast. PNAS 94:7481–7486
168. Rees EM, Lee J, Thiele DJ (2004) Mobilization of intracellular copper stores by the ctr2 vacuolar copper transporter. J Biol Chem 279:54221–54229
169. van den Berghe PV, Folmer DE, Malingré HE, van Beurden E, Klomp AE, van den Sluis B, Merkx M, Berger R, Klomp LW (2007) Human copper transporter 2 is localized in late endosomes and lysosomes and facilitates cellular copper uptake. Biochem J 407:49–59
170. Tanchou V, Gas F, Urvoas A, Cougouluegne F, Ruat S, Averseng O, Quemeneur E (2004) Copper-mediated homo-dimerisation for the HAH1 metallochaperone. Biochem Biophys Res Commun 325:388–394
171. Hamza I, Faisst A, Prohaska J, Chen J, Gruss P, Gitlin JD (2001) The metallochaperone Atox1 plays a critical role in perinatal copper homeostasis. PNAS 98:6848–6852
172. Rodriguez-Granillo A, Wittung-Stafshede P (2009) Differential roles of Met10, Thr11, and Lys60 in structural dynamics of human copper chaperone Atox1. Biochemistry 48:960–972
173. Walker JM, Tsivkovskii R, Lutsenko S (2002) Metallochaperone Atox1 transfers copper to the NH2-terminal domain of the Wilson's disease protein and regulates its catalytic activity. J Biol Chem 277:27953–27959
174. Naeve GS, Vana AM, Eggold JR, Kelner GS, Maki R, Desouza EB, Foster AC (1999) Expression profile of the copper homeostasis gene, rAtox1, in the rat brain. Neuroscience 93:1179–1187
175. Hussain F, Rodriguez-Granillo A, Wittung-Stafshede P (2009) Lysine-60 in copper chaperone Atox1 plays an essential role in adduct formation with a target Wilson disease domain. J Am Chem Soc 131:16371–16373
176. Hamza I, Prohaska J, Gitlin JD (2003) Essential role for Atox1 in the copper-mediated intracellular rafficking of the Menkes ATPase. Proc Natl Acad Sci USA 100:1215–1220
177. McRae R, Lai B, Fahrni CJ (2010) Copper redistribution in Atox1-deficient mouse fibroblast cells. J Biol Inorg Chem 15:99–105
178. Kelner GS, Lee M, Clark ME, Maciejewski D, McGrath D, Rabizadeh S, Lyons T, Bredesen D, Jenner P, Maki RA (2000) The copper transport protein Atox1 promotes neuronal survival. J Biol Chem 275:580–584
179. Itoh S, Kim HW, Nakagawa O, Ozumi S, Lessner SM, Aoki H, Akram K, McKinney RD, Ushio-Fukai M, Fukai T (2008) Novel role of antioxidant-1 (Atox1) as a copper-dependent transcription factor involved in cell proliferation. J Biol Chem 283:9157–9167
180. Banci L, Bertini I, Ciofi-Baffoni S, Janicka A, Martinelli M, Kozlowski H, Palumaa P (2008) A structural-dynamical characterization of human Cox17. J Biol Chem 283:7912–7920
181. Leary SC, Cobine PA, Kaufman BA, Guercin GH, Mattman A, Palaty J, Lockitch G, Winge DR, Rustin P, Horvath R, Shoubridge EA (2007) The human cytochrome c oxidase assembly factors SCO1 And SCO2 have regulatory roles in the maintenance of cellular copper homeostasis. Cell Metab 5:9–20
182. Briere JJ, Tzagoloff A (2007) The scoop on sco. Mol Cell 25:176–178
183. Mobley BC, Enns GM, Wong LJ, Vogel H (2009) A novel homozygous SCO2 mutation, p.G193S, causing fatal infantile cardioencephalomyopathy. Clin Neuropathol 28:143–149
184. Schmidt PJ, Kunst C, Culotta VC (2000) Copper activation of superoxide dismutase 1 (SOD1) in vivo. Role for protein-protein interactions with the copper chaperone for SOD1. J Biol Chem 275:33771–33776
185. Rae TD, Schmidt PJ, Pufahl RA, Culotta VC, O'Halloran TV (1999) Undetectable intracellular free copper: the requirement of a copper chaperone for superoxide dismutase. Science 284: 805–808
186. Caruano-Yzermans AL, Bartnikas TB, Gitlin JD (2006) Mechanisms of the copper-dependent turnover of the copper chaperone for superoxide dismutase. J Biol Chem 281:13581–13587
187. Carroll MC, Girouard JB, Ulloa JL, Subramanian JR, Wong PC, Valentine JS, Culotta VC (2004) Mechanisms for activating Cu- and Zn-containing superoxide dismutase in the absence of the CCS Cu chaperone. Proc Natl Acad Sci USA 101:5964–5969
188. Mullins JE, Fuentalba IC (1998) Immunohistochemical detection of metallothionein in liver, duodenum and kidney after dietary copper-overload in rats. Histol Histopathol 13:627–633
189. Shishido N, Nakayama K, Takazawa A, Ohyama T, Nakamura M (2001) Cu-metallothioneins (Cu(I)8-MTs) in EC rat livers 13 weeks after birth still act as antioxidants. Arch Biochem Biophys 387:216–222
190. Fabisiak JP, Tyurin VA, Tyurina YY, Borisenko GG, Korotaeva A, Pitt BR, Lazo JS, Kagan VE (1999) Redox regulation of copper-metallothionein. Arch Biochem Biophys 363:171–181
191. Fuentalba IC, Aburto EM (2003) Animal models of copper-associated liver disease. Comp Hepatol 2:1–12
192. Wijmenga C, Klomp LW (2004) Molecular regulation of copper excretion in the liver. Proc Nutr Soc 63:31–39
193. Burstein E, Ganesh L, Dick RD, van De Sluis B, Wilkinson JC, Klomp WJ, Cisca W, Brewer GJ, Nabel GJ, Duckett CS (2004) A novel role for XIAP in coper homeostasis through regulation of MURR1. EMBO J 23:244–254

194. Wang Y, Joh K, Masuko S, Yatsuki H, Soejima H, Nabetani A, Beechey CV, Okinami S, Mukai T (2004) The mouse Murr1 gene is imprinted in the adult brain, presumably due to transcriptional interference by the antisense-oriented U2af1-rs1 gene. Mol Cell Biol 24:270–279
195. Lutsenko S, Barnes NL, Bartee MY, Dmitriev OY (2007) Function and regulation of human copper-transporting ATPases. Physiol Rev 87:1011–1046
196. La Fontaine S, Mercer JF (2007) Trafficking of the copper-ATPases, ATP7A and ATP7B: role in copper homeostasis. Arch Biochem Biophys 463:149–167
197. Roelofsen H, Wolters H, Van Luyn MJ, Miura N, Kuipers F, Vonk RJ (2000) Copper-induced apical trafficking of ATP7B in polarized hepatoma cells provides a mechanism for biliary copper excretion. Gastroenterology 119:782–793
198. Yanagimoto C, Harada M, Kumemura H, Koga H, Kawaguchi T, Terada K, Hanada S, Taniguchi E, Koizumi Y, Koyota S, Ninomiya H, Ueno T, Sugiyama T, Sata M (2009) Niemann-Pick C1 protein transports copper to the secretory compartment from late endosomes where ATP7B resides. Exp Cell Res 315:119–126
199. Cater MA, La Fontaine S, Shield K, Deal Y, Mercer JF (2006) ATP7B mediates vesicular sequestration of copper: insight into biliary copper excretion. Gastroenterology 130:493–506
200. Bartee MY, Lutsenko S (2007) Hepatic copper-transporting ATPase ATP7B: function and inactivation at the molecular and cellular level. Biometals 20:627–637
201. Weiss KH, Carbajo Lozoya J, Tuma S, Gotthardt D, Reichert J, Ehehalt R, Stremmel W, Fullekrug J (2008) Copper-induced translocation of the Wilson Disease protein ATP7B independent of Murr1/COMMD1and Rab7. Am J Pathol 173:1783–1794
202. Zhang Y, Li M, Yao Q, Chen C (2009) Roles and mechanisms of copper transporting ATPases in cancer pathogenesis. Med Sci Monit 15:1–5
203. Bohlken A, Cheung BB, Bell JL, Koach J, Smith S, Sekyere E, Thomas W, Norris M, Haber M, Lovejoy DB, Richardson DR, Marshall GM (2009) ATP7A is a novel target of retinoic acid receptor β2 in neuroblastoma cells. Br J Cancer 100:96–105
204. de Bie P, van den Sluis B, Klomp L, Wijmenga C (2005) The many faces of the copper metabolism protein MURR1/COMMD1. J Hered 96:803–811
205. van Den Sluis B, Rothuizen J, Pearson PL, van Oost A, Wijmenga C (2002) Identification of a new copper metabolism gene by positional cloning in a purebred dog population. Hum Mol Genet 11:165–173
206. de Bie P, van de Sluis B, Burstein E, van den Berghe PV, Muller P, Berger R, Gitlin JD, Wijmenga C, Klomp LW (2007) Distinct Wilson's disease mutations in ATP7B are associated with enhanced binding to COMMD1 and reduced stability of ATP7B. Gastroenterology 133:1316–1326
207. Tao TY, Liu F, Klomp L, Wijmenga C, Gitlin JD (2003) The copper toxicosis product Murr1 directly interacts with the Wilson disease protein. J Biol Chem 278:41593–41596
208. Narindrasorasak S, Kulkarni P, Deschamps P, She YM, Sarkar B (2007) Characterization and copper binding properties of human COMMD1 MERR1. Biochemistry 46:3116–3128
209. Lim CM, Cater MA, Mercer JF, La Fontaine S (2006) Copper-dependent interaction of glutaredoxin with the N termini of the copper-ATPases (ATP7A and ATP7B) defective in Menkes and Wilson diseases. Biochem Biophys Res Commun 348: 428–436
210. Ke BX, Llanos RM, Wright M, Deal Y, Mercer JF (2006) Alterations of copper physiology in mice overexpressing the human Menkes protein ATP7A. Am J Physiol 290:R1460–R1467
211. Niciu MJ, Ma XM, El Meskini R, Ronnett GV, Mains ER, Eipper BA (2006) Developmental changes in the expression of ATP7A during a critical period in postnatal neurodevelopment. Neuroscience 139:947–964
212. Barnes N, Tsivkovskii R, Tsivkovskaia N, Lutsenko S (2005) The copper-transporting ATPases, Menkes and Wilson disease proteins, have distinct roles in adult and developing cerebellum. J Biol Chem 280:9640–9645
213. Saito T, Okabe M, Hosokawa T, Kurasaki M, Hata A, Endo F, Nagano K, Matsuda I, Urakami K, Saito K (1999) Immunohistochemical determination of the Wilson Copper-transporting P-type ATPase in the brain tissues of the rat. Neurosci Lett 266:13–16
214. Kitzemberg R, Madl C, Ferenci P (2005) Wilson disease. Metab Brain Dis 20:295–302
215. Hardman B, Michalczyk A, Greenough M, Camakaris J, Mercer JF, Ackland L (2007) Distinct functional roles for the Menkes and Wilson copper translocating P-type ATPases in human placental cells. Cell Physiol Biochem 20:1073–1084
216. Linz R, Barnes NL, Zimnicka AM, Kaplan JH, Eipper B, Lutsenko S (2008) Intracellular targeting of copper-transporting ATPase ATP7A in a normal and Atp7b-/- kidney. Am J Physiol Renal Physiol 294:F53–F61
217. Moore SD, Cox DW (2002) Expression in mouse kidney of membrane copper transporters Atp7a and Atp7b. Nephron 92:629–634
218. Crisponi G, Ambu R, Caria MP, Lisci M, Cristiani F, Nurchi VM, Pinna R (2001) Renal copper content and distribution in Wilson's disease. J Urol Pathol 13:23–30
219. Rossi L, Squitti R, Calabrese L, Rotilio G, Rossini PM (2007) Alteration of peripheral markers of copper homeostasis in Alzheimer's disease patients: implications inj aetiology and therapy. J NutrHealth Aging 11:408–417
220. Aschner M, Syversen T, Souza DO, Rocha JB (2006) Metallothioneins: mercury species-specific induction and their potential role in attenuating neurotoxicity. Exp Biol Med 231:1468–1473

Prion diseases and manganism

Yuzo Nishida

Abstract

Recent studies on mice experimentally infected with scrapie suggested that large increase in the levels of manganese ion occurs in blood and brain prior to the onset of symptoms of the prion disease, and the observed elevated manganese ion in several central nervous systems implies that the prion diseases should be considered to be one of the manganism. We have observed that oxidation of Mn(III) ion in several manganese chelates occurs in the presence of apo-transferrin, giving a di-μ-oxo bridged Mn(III/IV) species (hereafter we will call these Mn(III) and Mn(IV) ions to be labile plasma manganese ions), and at the same time facile uptake of manganese ions by apo-transferrin proceeds. This clearly shows that most manganese ions can be transported to the brain in a facile manner by transferrin under certain conditions. There are many iron-containing enzymes in the brain, and substitution of iron ion in these enzymes with other metal ions such as manganese ion results in complete or partial loss of enzymatic activity, and this is because the reactivity of the iron ion towards oxygen molecule is quite different from that of the manganese ions. Thus, the excess accumulation of the manganese ion in the brain should lead to (a) abnormality in iron metabolism, i.e., the increase of the labile plasma iron (or non-transferrin-bound iron, NTBI), which is in fact observed for the certain regions of the brain of scrapie strain infected mice; these iron ions are not transferred to transferrin, giving to the iron-deficiency state in the brain which leads to the defect of neurotransmitters such as dopamine and serotonin and (b) the abnormalities of the brain functions due to the toxicity of the labile plasma iron ions, which leads to neural cell death. Based on the above facts, and that (1) the labile plasma iron can in a facile manner produce the hydrogen peroxide and (2) the prion diseases can be elucidated by the "gain-of-function" of the prion proteins as copper(II)-containing enzyme in the presence of excess hydrogen peroxide, we have concluded that the prion diseases including both the sporadic and infected types should be elucidated by the combined toxicities due to the both labile plasma manganese and iron ions. Very recently we have succeeded in obtaining the chelate which captures both the labile plasma iron and manganese ions effectively and removes these ions without toxicity from the solution in vitro. Thus, we can hope that our new chelates should make notable contribution

Y. Nishida (✉)
Medical Research Institute, Kanazawa Medical University, 1-1
Daigaku, Uchinada, Kahoku, Ishikawa 920-0263, Japan
e-mail: nsd-2210@kanazawa-med.ac.jp

W. Linert and H. Kozlowski (eds.), *Metal Ions in Neurological Systems*,
DOI 10.1007/978-3-7091-1001-0_6, © Springer-Verlag Wien 2012

to the prevention and therapeutics for the prion disease, Parkinson's disease, amyotrophic lateral sclerosis (ALS), Alzheimer's disease, schizophrenia, and dementia, which are now in progress in Japan.

Keywords
Prion disease • Manganism • Labile plasma manganese ion • Labile plasma iron ion • Iron-overload syndrome

Prion diseases and transition metal ions

Prion diseases and prion protein

Between 1980 and roughly 1996, about 750,000 cattle infected with *BSE* (bovine spongiform encephalopathy) were slaughtered for human consumption in Great Britain, and it is now clear that BSE, also known as "*mad cow disease*" is neither merely a UK phenomenon nor is it merely an economic nuisance. In fact, it may be an impending worldwide health crisis, and in recent months several other European countries have found BSE in their cattle herds, and over the past few years about 100 mostly young individuals have fallen victim to a fatal condition known as new *variant Creutzfeldt–Jacob disease* (vJCD) [1–3]. BSE and vJCD are one of the transmissible spongiform encephalopathies (TSEs, or *prion disease*) which are a group of fatal neurodegenerative disorders that include BSE, vJCD, *scrapie* of sheep, chronic wasting diseases (CWD) of mule deer and elk, as well as Gestmann–Straussler–Scheinker disease (GSS) and fatal familial insomnia (FFI) of humans [4, 5].

Approximately 250 years ago, a sheep disease that presented with excitability, itching, ataxia, and finally paralysis and death was recognized and this is known today as *scrapie* in English-speaking countries, "the trembles" in France, "trotting disease" in Germany, and "itching disease" in Japan, reflecting the gamut of its symptoms. The first major advance in scrapie research took place in 1936 when Cuille and Chelle succeeded in transmitting the disease to sheep and goats by inoculating them with lumbar cord of diseased animals. Subsequently, transmission to mice and hamsters provided more-convenient experimental models. It was soon recognized that the transmissible agent had quite extraordinary properties, such as unusually long incubation periods, measured in months to years, and uncommon resistance to high temperature, formaldehyde treatment, and UV irradiation.

Enriching fractions from Syrian hamster (SHa) brain for scrapie infectivity led to the discovery of the *prion protein* (PrP), and at present it is generally accepted that the central event in TSEs is the post-translational conversion of the normal cellular prion protein (PrP^C) into a abnormal isoform called *scrapie PrP* (PrP^{Sc}) that *has a high-β-sheet content* and *is associated with the transmissible disease* [6]. These misfolded prions (PrP^{Sc}) ultimately kill neurons and leave the brain riddled with holes, like a sponge, and the 1997 Nobel Prize in Physiology and Medicine was awarded to Professor *S. Prusiner* of the University of California, San Francisco, for his contributions towards the identification of the infectious agent that causes TSEs.

The normal cellular prion protein PrP^C is a glycoprotein expressed on the surface of many cell types (see Fig. 1) and the fact that the protein is *expressed in neurons at higher levels* than in any other cell types suggests that PrP^C has special importance for neurons; PrP^C is highly concentrated at the *synapse* and there is evidence for intense localization not only at central nerves synapse but also at endplates. PrP^C is linked to the cell membrane by glycosylphosphatidylinositol (GPI) anchor (Fig. 1) [7], and has one or two sugar chains that are closely linked to the C terminus and also exists in a non-glycosylated form. Recent studies have showed that PrP^C not only binds *copper* (Cu) within the octarepeat region located in the unstructured N terminus, but under certain specific circumstances may bind along the C-terminal structured domain of protein fragments. It has been demonstrated that both recombinant and brain-derived PrP have *superoxide dismutase* (SOD)-like activity when Cu is bound to the octarepeat region [8].

The misfolded prion protein, PrP^{Sc} is extracted from affected brains as highly aggregated, detergent-insoluble materials that is not amenable to high-resolution structural technique and is covalently indistinguishable from PrP^C. During infection, the underlying molecular events that lead to the conversion of PrP^C to the scrapie agent remain ill defined.

Prion protein and transition metal ions

At present it is generally recognized that BSE may have originated from a *scrapie agent* infecting small ruminants, which have been recycled through cattle and disseminated through the use of contaminated meat and bone meal. Although BSE may spread among cattle by the feeding of infected offal, the majority of the cases of naturally occurring prion diseases arise sporadically with no known cause. Thus, the most important problem to be solved is to elucidate the intrinsic origin, i.e., the precise chemical mechanism of the prion diseases which arise sporadically.

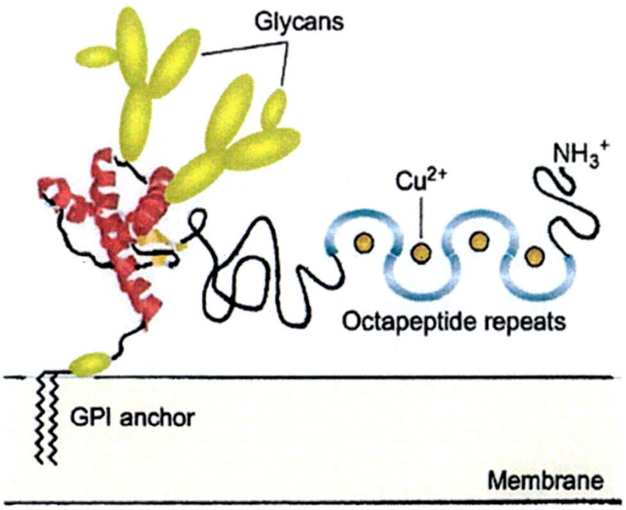

Fig. 1 Model of PrP^C structural domains. The folded C-terminal portion of PrP^C t contains the short β-sheet strands and the α-helix is based on a model derived from NMR-based coordinates of residues of hamster PrP [7]

Investigations of scrapie, CJD, and chronic wasting disease clusters in Iceland, Slovakia, and Colorado, respectively, have indicated that the soil in these regions is low in copper and *high in manganese*, and Brown et al. observed striking elevation of manganese ion accompanied by significant reduction of copper ion bound to purified PrP in all sCJD (sCJD = *sporadic CJD*) variants [9]. Brown et al. have reported that it loses the SOD-like activity when Cu is replaced with Mn in recombinant PrP, and Cu binding to PrP purified from sporadic CJD was significantly decreased while the binding of Mn and Zn was markedly increased [9]. These results suggest that altered metal-ion occupancy of PrP plays a pivotal role in the pathogenesis of prion diseases.

Elevated occupational exposures to manganese are known to cause significant neurotoxicity, and epidemiologic studies have suggested a relationship between elevated manganese exposure and an increased risk for parkinsonian disturbances, called *manganism* [10, 11], although the exact mechanisms underlying the neurotoxic effects of manganese remain unclear. Recent studies on mice experimentally infected with scrapie suggested that large increase in the levels of manganese ion occurs in blood and in several central nervous systems prior to the onset of symptoms of the prion disease [12]. It is quite likely that the excess accumulation of the manganese ion in the brain may lead to (a) abnormalities in iron metabolism, i.e., increase in the labile plasma iron (or non-transferrin-bound iron, NTBI), which is in fact observed in certain regions of the brain of scrapie strain infected mice [13, 14]; these iron ions are not transferred to transferrin, giving to a iron-deficiency state in the brain which leads to the defect of the neurotransmitters such as dopamine and serotonin and (b) abnormalities of the brain functions due to the toxicity by the labile plasma iron ions, which lead to neural cell death [15–17].

The sporadic neurodegenerative diseases are in general *endemic*; many years ago, amyotrophic lateral sclerosis (ALS) patients were collectively found in the New Guinea and Papua islands, and its origin has been attributed to drinking subterranean water, which contains much Al^{3+} and Mn^{2+} ions, and in these regions many patients of Alzheimer's and Parkinson's diseases were found [18], and increased aluminum levels were reported in the hippocampus of patients with Alzheimer's disease [19, 20]. In Alzheimer's disease, specific regions such as the hippocampus and the motor cortex contain elevated iron levels relative to normal, and abnormalities in brain iron metabolism have been described for several neurodegenerative disorders, including Alzheimer's diseases, Parkinson's disease, Huntington's, and prion diseases [17].

The above facts clearly indicate that the prion diseases should be considered to be one of manganism, and the sporadic neurodegenerative diseases are closely related to the abnormal functions of metal ions such as manganese and iron. Based on these facts we will show the new mechanism for pathogenesis of prion diseases including both sporadic and infected types and would like to postulate the new countermeasures to prevent the prion diseases, Parkinson's diseases, Alzheimer's diseases, schizophrenia, and other neurodegenerative disorders.

Manganism and labile plasma manganese ions

Manganism

As a nutrient, manganese is an essential component of several enzymes; a deficiency can lead to heart and bone problems and in children, stunted growth; however, since 1937 elevated occupational exposures to manganese are known to cause significant neurotoxicity, and epidemiologic studies have suggested a relationship between elevated manganese exposure and an increased risk for parkinsonian disturbances, called *manganism* [10, 11]. It has been reported that rats exposed to very elevated manganese levels via drinking water from an early age displayed increased brain manganese levels and altered copper and iron levels in the striatum and in the basal ganglia; when manganese (this should be manganese oxide) is inhaled, blood ferries it from the lungs to the brain, where it can readily cross the blood–brain barrier.

Transport of manganese ions into the central nervous system has been directly investigated in a limited number of studies. It is generally believed that iron and manganese ions are able to be complexed and carried by transferrin/ transferrin receptor, with iron being far more prevalent under normal circumstances [21]. Several authors suggest transport of trivalent manganese complexed to transferrin into the brain capillary endothelium, but the exact mechanism underlying the transport of manganese by transferrin to the brain has been elucidated very recently by us, which will be described below.

Transport of manganese ion by transferrin to brain

We have investigated the uptake of manganese ion by apo-transferrin in terms of the capillary electrophoresis method [22] and obtained clear evidence that (1) Mn(II) ion in several chelates is not transported to transferrin, (2) oxidation state, +3, and binuclear unit of a manganese chelate are critical factors for the facile uptake by apo-transferrin, similar to that observed for Fe(III) species [23], and (3) the facile oxidation of Mn(III) ion to a Mn(IV) state with simultaneous formation of a di-μ-oxo-bridged Mn(III/IV) species in several manganese chelates occurs in the reaction mixture containing Mn(III) chelates and apo-transferrin.

In our study the manganese compounds shown below were used: $Mn^{II}(dpea)Cl_2$, $Mn^{II}(dpa)ClClO_4$, $Mn^{II}(dpal)Cl_2$, $Mn^{II}(Me\text{-}en\text{-}py)Cl_2$, $Mn_2^{III/IV}O_2(dpa)_2(ClO_4)$, $Mn_2^{III/IV}O_2(Me\text{-}en\text{-}py)_2(ClO_4)_3$, and μ-oxo-μ-acetato-bridged binuclear manganese(III) complex, Mn_2^{III} (Me-en-py)$_2$(O)(CH_3COO)(PF_6)$_2$; the chemical structures of the ligands are illustrated in Fig. 2 [24].

All the manganese(II) complex solutions used in our study are colorless, and solutions of these Mn(II) complexes with (dpa), (dpal), and (Me-en-py) are stable in the air under our experimental conditions (in tris-buffer solution, pH 7.3). The capillary electrophoresis method indicated that all the manganese(II) ions in the Mn(II) chelates used in this study are not transported to transferrin, because no change in the electrophoresis of the apo-transferrin occurs by the addition of Mn(II)(dpal) chelate. However, the shift and increase in the intensity of peak due to the apo-transferrin were observed when the solution of $Mn_2^{III/IV}O_2(dpa)_2(ClO_4)$ was added to the solution of apo-transferin as illustrated in Fig. 3. These changes demonstrate that the Mn(III) ion is introduced to transferrin [22, 23], and a similar uptake of Mn (III) ion by apo-tremasferrin was observed for other di-μ-oxo-bridged binuclear Mn(III/IV) compounds.

The colorless $Mn(II)(dpea)Cl_2$ in the buffer solution became pale brown when the solution was kept standing for more than 1 day, demonstrating that Mn(II) ion in the Mn(II) (dpea) chelate is readily oxidized to a Mn(III) ion in the atmosphere [22]. The addition of colorless apo-transferrin to the aged-pale-brown Mn(dpea) complex solution immediately induced the color change to dark brown, and uptake of the manganese ion by apo-transferrin proceeded.

In order to obtain more information regarding the above process, we have measured the ESR spectra of the solutions.

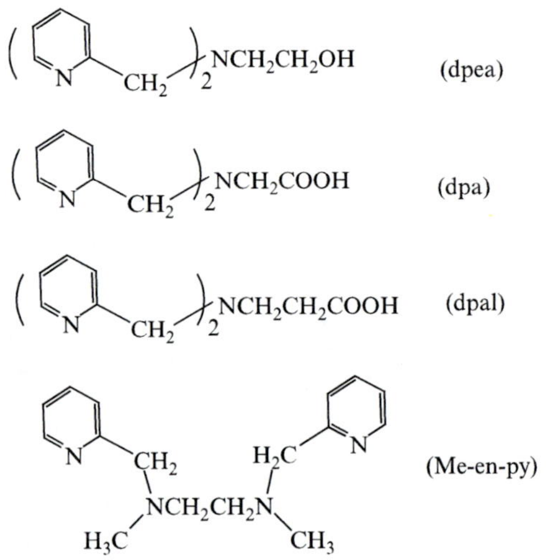

Fig. 2 Chemical structures of the ligands

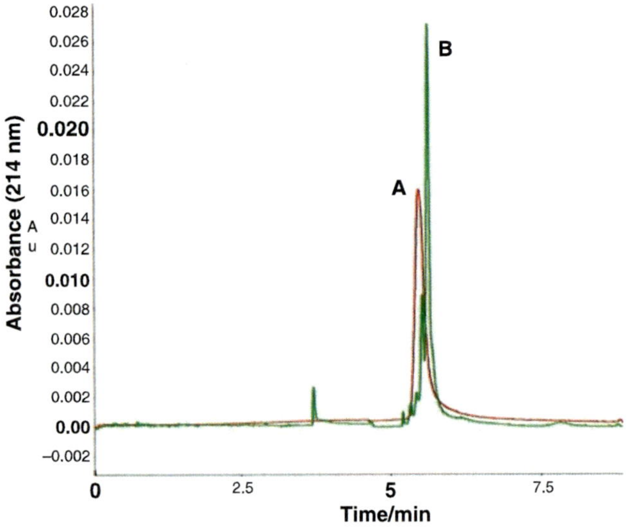

Fig. 3 CE profile of the solution. A (*red*); apo-transferrin, and B (*green*); di-μ-oxo Mn_2(III/IV))-(dpa) complex solution was added to the solution A

The freshly prepared coloress Mn(II)(dpea) solution is ESR detectable, exhibiting the multiline ESR signal characteristic for the high-spin type Mn(II) ion. However, no ESR signal was observed for the aged-pale-brown Mn(dpea) complex solution (see Fig. 4(A)), consistent with the general concept that almost all Mn(III) species are ESR-non-detectable [25]. On the other hand, we have found 16 lines in the ESR spectrum of the dark-brown solution described above, as illustrated in Fig. 4(B). The same 16-line ESR signals was also observed in the solution containing apo-transferrin and the binuclear Mn(III) complex with (Me-en-py), Mn_2^{III}(Me-en-py)$_2$(O)(CH_3COO)(PF_6)$_2$, but no 16-line ESR signal was detected in the solutions containing apo-transferrin and Mn(II) chelates with (dpa), (dpal), and (Me-en-py) under the same conditions.

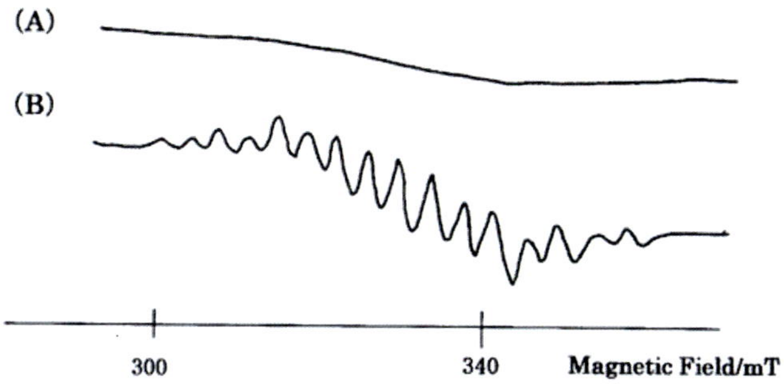

Fig. 4 ESR spectra of the solutions. A: Aged-pale-brown solution of Mn(II)(dpea), B: Apo-transferrin was added to the solution A

As it has been pointed out that 16-line ESR signal observed in the figure is diagnostic for the binuclear Mn (III)/Mn(IV) species with di-μ-oxo bridge [26], the present results are clearly implying that in the solution containing apo-transferrim, the Mn(III) ion of the chelates is readily oxidized to a Mn(IV) state to give a binuclear di-μ-oxo bridged Mn(III)/Mn(IV) species. Because the facile uptake of manganese ion by apo-transferrin was observed in the solution containing the di-μ-oxo-bridged binuclear Mn(III)/ Mn(IV) complexes with (dpa) and (Me-en-py) [22], it seems very reasonable to assume that formation of a di-μ-oxo-bridged Mn(III/IV) species in the solution plays a key role in the transport of manganese ion in the manganese chelates to apo-transferrin.

Labile plasma manganese ion

It is generally believed that iron and manganese ions are able to be complexed and carried by transferrin/transferrin receptor, with iron being far more prevalent under normal circumstances [21]. This is only true under the circumstance that the manganese ion exists as manganese(II) ion, and in fact uptake of manganese ion by apo-transferrin does not proceed in the solutions containing manganese(II) compounds with (dpa), (dpal), and (Me-en-py). The prevalence of the iron ion over the manganese ion is not valid when the Mn(III) or Mn(IV) ions are present in the plasma, and such condition may be induced in a facile manner as described above. Thus our results may give important information to understand the mechanism of the manganese ion transport to the brain, which should induce manganism, and we will call these Mn(III) and Mn(IV) ions to be *labile plasma manganese ions*.

The solutions of Mn(II) complexes with (dpa), (dpal) and (Me-en-py) are stable in the air, but these are converted to di-μ-oxo-bridged binuclear Mn(III)/Mn(IV) complexes by hydrogen peroxide [26]. Thus, it seems quite likely that the presence of hydrogen peroxide is closely related to the formation of the labile plasma manganese ions and thus with the pathogenesis of manganism, and it has been pointed out that formation of hydrogen peroxide is readily induced by the labile plasma iron ions (see sections "Hydrogen peroxide formation by binuclear iron(III) species", and also "New chelation therapy for the neurodegeneration"), which are observed in a certain region of the brain of scrapie strain infected mice [13, 14].

Recent studies on mice experimentally infected with scrapie suggested that large increase in the levels of manganese ion occurs in blood and brain prior to the onset of symptoms of the prion disease and the observed elevated manganese and iron ions in several central nervous systems of scrapie strain infected mice imply that prion diseases should be considered to be one of manganism [12–14]. Thus, it seems reasonable to point out that the best way to prevent the prion diseases is to remove both the labile plasma manganese and iron ions from the plasma (see section "New chelation therapy for the neurodegeneration").

Difference between Fe(II) and Mn(II) ions for oxidation reaction

The α-keto acid-dependent enzymes are distinguished from other non-heme iron enzymes by the requirement of an α-keto acid cofactor as well as Fe(II) and O_2 for reactivity (see Fig. 5) [27]; the reactions catalyzed by these enzymes involve the oxidation of an unactivated C–H bond to give either of the hydroxylated products. In general, these enzymes require 1 equiv of Fe(II), and α-keto acid and ascorbate for full activity, and α-keto acid and ascorbate may behave as pterin of reduced form in tyrosine hydroxylase (TH) as illustrated in Fig. 6.

Substitution with other divalent metal ions (Zn(II), Mn(II), Co(II), Mg(II), and Ni(II)) results in complete loss of enzymatic activity [27]. These facts clearly show that oxygenase activity by the Fe(II) ion is completely different from those of other metal ions, especially Mn(II). This should be because *Mn(II) compounds are in favor of the "two-electron oxidation" reaction and readily oxidized to a Mn(IV) ion, whereas "one-electron reaction" for the Fe(II) species.* This is clearly exemplified in our recent results, i.e., Mn(II) ion of the Mn(II) complex, Mn(ntb) $ClCl_4$, which is very stable in air, is readily oxidized to a Mn(IV) ion in the presence of reducing agent, cycloyhexane-carboxyaldehyde in air (see Fig. 7) [28].

These are indicating that excess accumulation of manganese ion gives a serious damage to the function of the biological iron-containing hydroxylases such as tyrosine or triphtophan hydroxylases, which are necessary for the synthesis of transmitters such as dopamine and serotonin, and this will induce abnormalities in iron metabolism, leading to increasing labile plasma iron in brain (see section "Labile

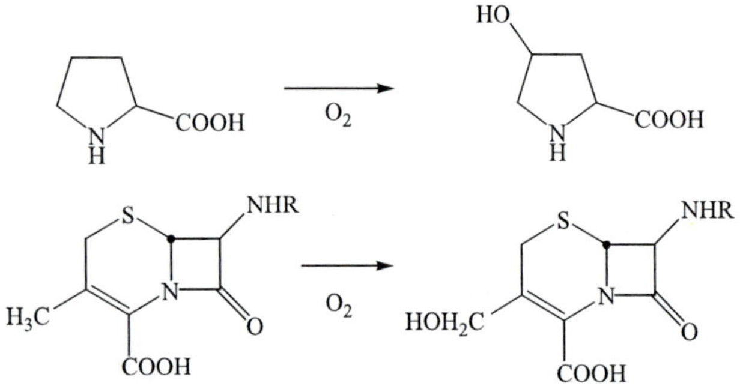

Fig. 5 Examples of α-keto acid-dependent oxygenases

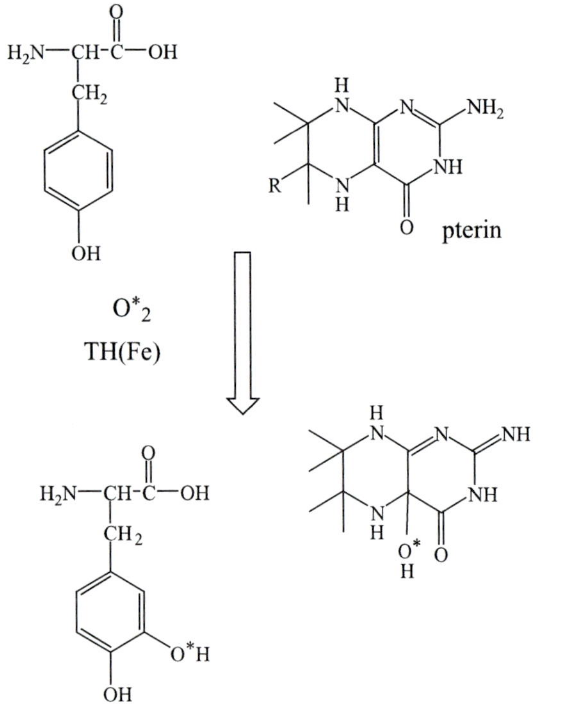

Fig. 6 Formation of dopa from tyrosine by tyrosine-hydroxylase (TH). Pterin acts as co-enzyme in this reaction

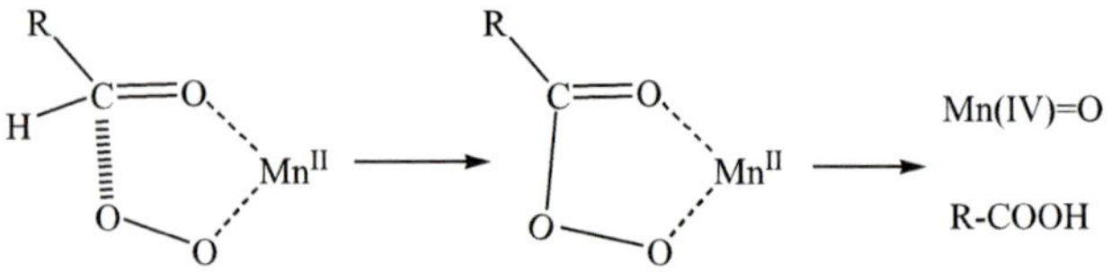

Fig. 7 Formation of Mn(IV) = O species in the presence of aldehyde

plasma iron: formation and toxicity"), which is consistent with the observed facts [12–14]. Similar abnormalities in iron metabolism were also induced by the excess accumulation of Al^{3+} ion in the brain [15, 16].

Labile plasma iron and iron-overload syndrome

Labile plasma iron: formation and toxicity

Human iron metabolism and absorption have been the subject of a recent review. Normal human males contain 3–5 g of iron (often less in females) and of this, two-thirds is in circulating red cells as hemoglobin and 15–25 % in storage as ferritin and hemosiderin [29]. The remaining iron is in muscle myoglobin (about 8 %) and in cytochromes and iron-containing enzymes. Plasma transferrin accounts for only 3 mg of Fe, but the daily exchange of iron through plasma transferrin is ten times this account. Transferrin, therefore, plays a central role in iron distribution. Iron delivery by transferrin to erythroid and many non-erythroid cells involves interaction of transferrin with specific receptors followed by endocyosis and recycling of apotransferrin and receptor. These receptors are present in low amounts on phagocytosis cells which receive their iron from degraded red cell hemoglobin. Genetic haematochromatosis is one of the most common genetic disorder in western populations, particularly among Celtic peoples. This disease is associated with greatly increased (sometimes 50-fold) deposits of storage iron, predominantly as haemosiderin, in the liver and other tissues due to abnormally high absorption from the gut, and the excess iron cannot be eliminated and the elevated body iron leads to increased iron in storage (called iron-overload syndrome).

Several data support the hypothesis that haemosiderin is a degradation product of ferritin, and this has been confirmed for phytosiderin, an insoluble iron-containing product from pea seed, which was found to contain a peptide derived from the ferritin subunit [29]. Haemosiderin is typically insoluble, as isolated, in contrast to the soluble ferritin, but it can be solubilized by the *several amino acids or small peptides to form a soluble iron(III) chelate;* these are called as labile plasma iron or non-transferrin-bound iron (NTBI) [13–17], which may exist as soluble dimeric (or polymeric) compounds in brain, similar to $Fe_2(HPTP)Cl_4^+$ (Fig. 8), or to Fe(III)-(nta) chelate; the latter complex is of a μ-oxo bridged dimeric structure, $[Fe_2O(nta)_2(CO_3)]^{2-}$, as illustrated in Fig. 9 [15, 30]. *It should be noted here that these binuclear iron(III) compounds promote the formation of an iron(III)-oxygen species in the presence of hydrogen peroxide or of both oxygen and reducing agents such as glutathione reductase, and the notable reactivity of an iron (III)-oxygen species, similar to singlet oxygen ($^1\Delta_g$), is induced through the interaction with substrate and the peripheral organic groups* [15, 17, 30].

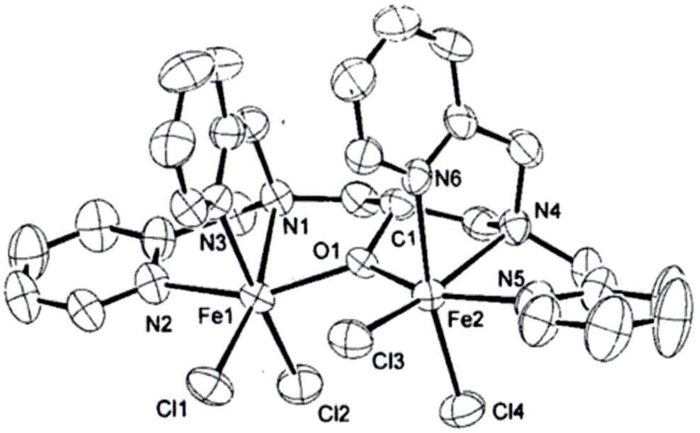

Fig. 8 Structure of $Fe_2(HPTP)Cl_4$ cation

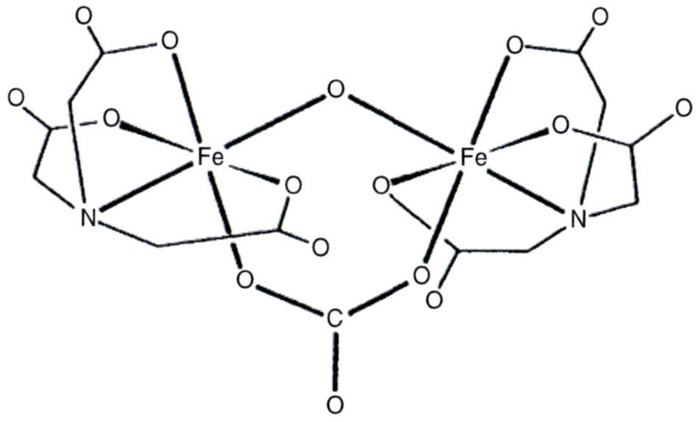

Fig. 9 Structure of Fe(III)-(nta) chelate

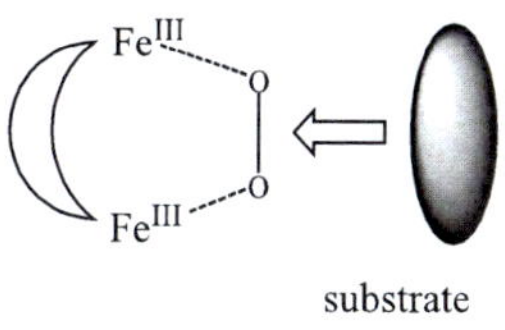

Fig. 10 Proposed intermediate

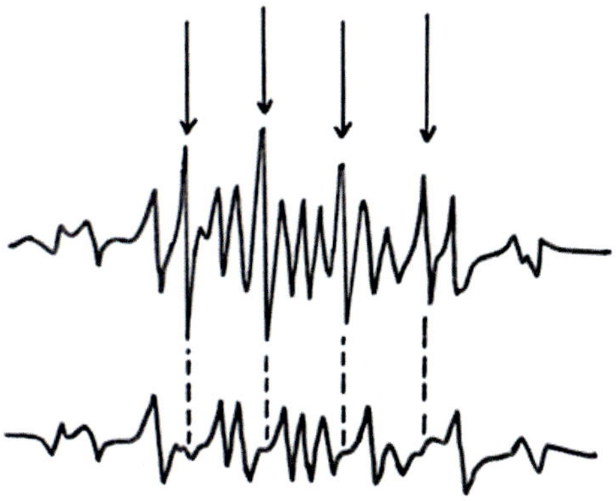

Fig. 11 ESR spectra of the solutions containing DMPO and A: $Fe_2(HPTP)(OH)(NO_3)_2^{2+}$ (*upper*) B: Fe(edta)Na (*lower*) [32]

Toxicity due to the binuclear iron(III) species: 1

We have found that the binuclear iron(III) complexes, such as $Fe_2(HPTP)Cl_4^+$ and $[Fe_2(HPTB)(OH)(NO_3)^2]^{2+}$, give much *TBARS* in the reaction with linolenic acid under aerobic conditions [15, 30]; here TBARS are malondialdehyde (which gives pink products with TBA (TBA = 2-thiobarbituric acid; see below); $\lambda_{max} = 532$ nm) or monoaldehyde derivatives (gives orange products), which are formed from the peroxidation of linolenic acid.

The above results have been elucidated on the assumption that oxygenation of linolenic acid by O_2 proceeds without the change of oxidation state of Fe(III) by forming an intermediate containing two iron(III) ions, O_2, and linolenic acid, as illustrated in Fig. 10; in this case, the substrate is linolenic acid. It should be noted that *the interaction between the two unpaired electrons of two iron(III) atoms and O_2 is necessary to activate the O_2*, which is promoted through interaction with the linolenic acid, and the reactivity of oxygen molecule in the intermediate is similar to that of singlet oxygen ($^1\Delta_g$) [15, 17, 30, 31].

The binuclear Al(III) complex, $Al_2(HPTP)(OH)Cl_2(ClO_4)_2$ was also isolated, and the activity for the oxygenation of linolenic acid was compared with that of the binuclear Fe(III) complex by measuring the quantities of the TBARS compounds. The result clearly shows that Al(III) complex exhibits no activity for the oxygenation of linolenic acid, and this can be attributed to the absence of unpaired d-electron in Al(III) complex, and this also supports the important role of unpaired d-electron of Fe(III) ion in the activation of O_2 in Fig. 10.

A similar fact was also observed when DMPO, which is one of the famous spin-trapping agents for OH· radical, was added to the solution containing binuclear iron(III) complex, $Fe_2(HPTP)(OH)(NO_3)_2^{2+}$; strong four signals which correspond to the formation of DMPO-OH have appeared as illustrated in Fig. 11 [15, 32], whereas no such signal was detected by the addition of DMPO to the solutions of Fe $(edta)^-$ and $Al_2(HPTP)(OH)Cl_2(ClO_4)_2$. This clearly indicates that the formation of DMPO-OH is not due to the presence of OH· in the solution. The above mysterious fact was elucidated in a similar way as described for the oxygenation of linolenic acid through the formation of the intermediate illustrated in Fig. 10 (in this case, substrate is DMPO); O_2 is activated to interact with DMPO, leading to DMPO-OH formation. These clearly demonstrate that the presence of DMPO induced the formation of an iron(III)-oxygen species to give a DMPO-OH in the reaction with DMPO, and thus the results reported hitherto on the formation of DMPO-OH *should be reinvestigated.*

Toxicity due to the binuclear iron(III) species: 2

We have observed that in the solution containing binuclear iron(III) compound $[Fe_2(HPTB)(OH)(NO_3)^2]^{2+}$

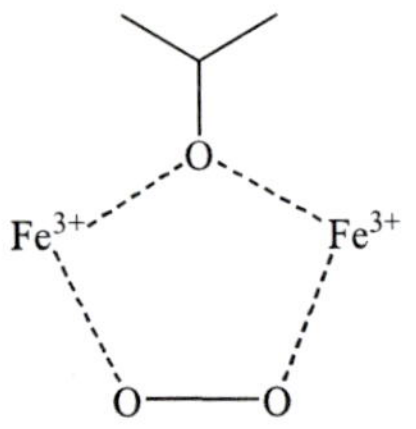

Fig. 12 Peroxide adduct of binuclear iron(III) compound

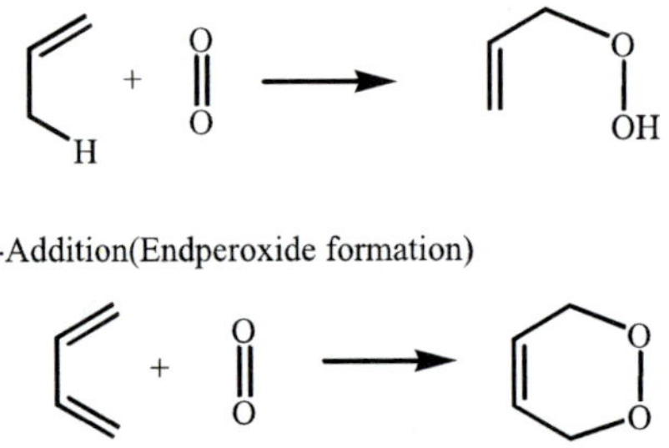

Fig. 13 Representative reactivity of singlet oxygen ($^1\Delta_g$)

and hydrogen peroxide, the peroxide adduct shown in Fig. 12 forms in a facile manner [15, 31].

It should be noted that the peroxide ion captured in the binuclear iron(III)-peroxide adduct exhibits high reactivity towards several organic compounds; its reactivity is similar to that of singlet oxygen ($^1\Delta_g$) [15, 30, 31], which is known to react with several organic compounds as illustrated in Fig. 13.

Thus, the labile plasma iron ions exhibit high toxicity towards human body when hydrogen peroxide is present in the plasma.

Hydrogen peroxide formation by binuclear iron (III) species

It is noteworthy that the oxidation of the TMPD, one of the famous one-electron donor, is greatly accelerated in the presence of the binuclear iron(III) complex, $Fe_2(HPTP)Cl_4^+$, or binuclear copper(II) complex [15, 32], to give *higher quantity of hydrogen peroxide* according to the similar reaction scheme shown in Fig. 10; this indicates that the *soluble dimeric* iron(III) species of labile plasma iron can produce hydrogen peroxide in the presence of reducing agent and O_2. These facts suggest that hydrogen peroxide may *readily form* in patients with iron overload, and this hydrogen peroxide should be one of serious origin in the pathogenesis of sporadic scrapie and BSE as described in the next section.

Copper(II)-hydroperoxide adduct in sporadic amyotophic lateral sclerosis and sporadic prion diseases

"Gain-of function" of copper(II)-hydroperoxide adduct in mutant SOD enzyme

ALS is a progressive paralytic disease characterized by selective degeneration of the upper and lower motor neurons [17, 33, 34]. Although ALS is predominantly a sporadic disease, ~10 % cases are inherited in an autosomal dominant manner (familial ALS (fALS)) and a subset of the fALS cases are caused by mutations in the SOD1 gene. The gene product of SOD1, cytoplasmic Cu, Zn-superoxide dismutase (SOD1), is a ubiquitously expressed enzyme that catalyzes the disproportionation reaction of superoxide radicals (see the Eqs. (1) and (2)) [35].

$$O_2^- + Cu(II) \rightarrow O_2 + Cu(I) \tag{1}$$

$$Cu(I) + O_2^- + 2H^+ \rightarrow O_2H_2 + Cu(II) \tag{2}$$

There are several lines of evidence suggesting that SOD1 mutations result in a gain, rather than loss of function that causes ALS. One hypothesis of the gain-of-function of SOD1 is that misfolding of the mutant alters the catalytic mechanism to allow the production of oxidants such as peroxynitrite and possibly hydrogen peroxide. Another major hypothesis is toxicity caused by intracellular aggregation of SOD1. SOD1 inclusion bodies, which also react with anti-ubiquitous antibodies, are a common pathological finding in motor neurons and neighboring astrocytes of ALS patients [33, 34]. Although SOD1 aggregates may be inherently toxic or cause motor neuron toxicity by sequestering chaperons and blocking proper functioning of the proteasome, origin of toxicity by SOD1 aggregates has not been elucidated.

We have studied the reactivity of a *copper(II)–OOH*, proposed as an important intermediate in the SOD reaction in order to obtain the comprehensive solution for the *correlation between the mutant SOD and pathogenesis of fALS*. For this purpose, we have synthesized many copper(II) compounds with the ligands containing N,N-*bis*(2-picolylmethyl)amine moiety, as illustrated in Fig. 14 [35]. The structural features of all the copper(II) compounds are essentially similar to each other (as an example, the crystal structure of $[Cu(bdpg)Cl]^+$ is illustrated in Fig. 15).

In the presence of hydrogen peroxide, a copper(II)-peroxide adduct formation as shown in Fig. 16 may occur. We have found that formation and reactivity of the peroxide adduct of the copper(II) compound is highly dependent on

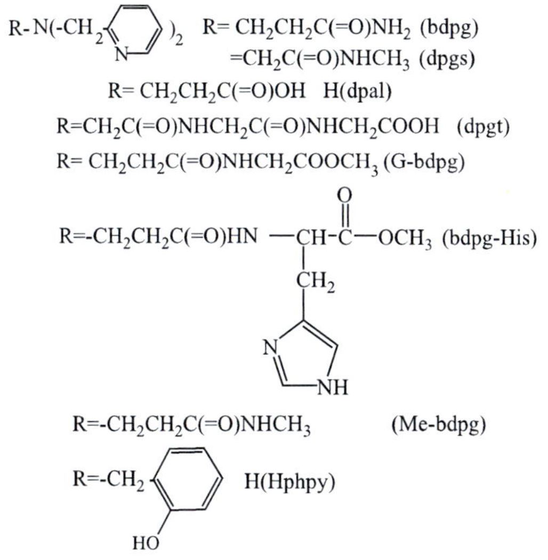

Fig. 14 Structures of the ligands used in this study

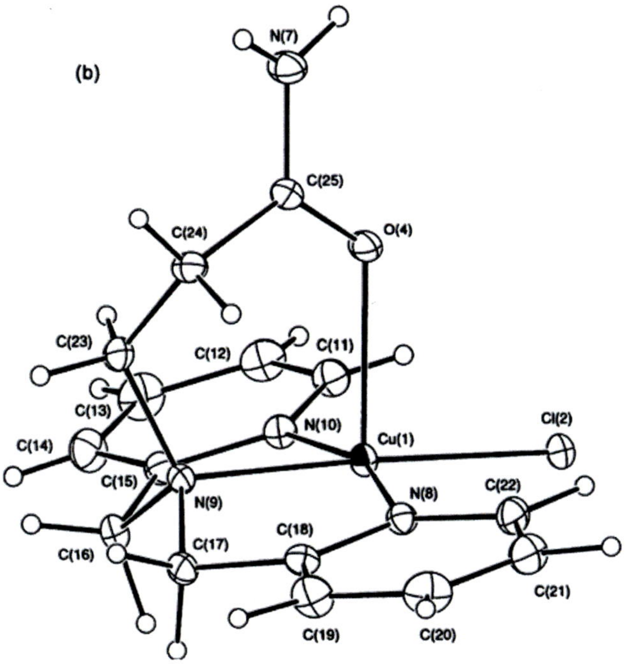

Fig. 15 Crystal structure of $[Cu(bdpg)Cl]^+$ cation

the R of the ligand system. This clearly demonstrates that formation and reactivity of the Cu(II)–OOH is controlled by the chemical interactions among copper(II)–OOH species, peripheral groups, and substrate (see Fig. 16) [15, 17, 30, 35], and in some cases *the hydrogen peroxide of copper (II)–OOH species is changed to exhibit high reactivity similar to that of the singlet oxygen* (${}^1\Delta_g$) [15, 30, 35], although hydrogen peroxide has been believed to be relatively inert and not toxic to cells.

Based on our experimental facts [15, 17, 35], we have concluded that the "*gain-of-function*" of the mutant SOD is

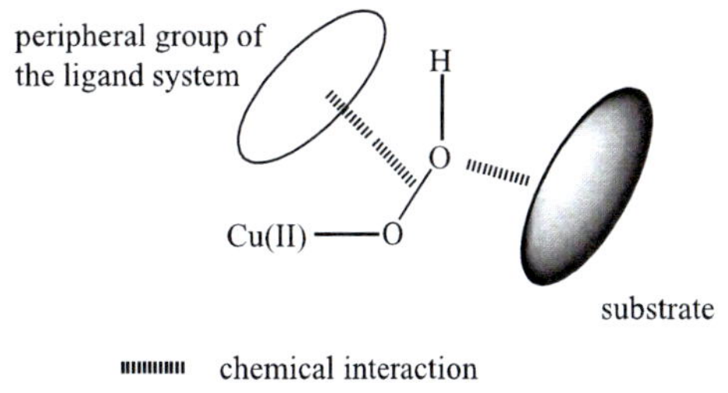

Fig. 16 Chemical interaction among Cu(II)–OOH, peripheral group, and substrate

due to the formation of a *long-lived highly reactive copper(-II)–OOH* as an intermediate in the process of SOD reaction. The chemical structures around the copper(II) in the mutant SOD are slightly changed, and this gives an unexpected effect on the reactivity of a copper(II)–OOH as observed in our papers. In the mutant *SOD*, the C–N bond cleavage by the Cu(II)–OOH is thought to cause great changes in the surface of SOD, leading to destabilization of the dimer contact of the SOD enzyme [36], and thus, *it is quite likely that formation and existence of a highly reactive Cu (II)–OOH species is an intrinsic origin for oxidative stress in the pathogenesis of fALS, which seems to be quite consistent with the* recent studies on the destabilization of the dimer contact of the SOD enzyme [17, 35].

Dissociation of dimeric SOD molecule into monomers

At present it has been widely recognized that protein aggregation is a common pathological feature of many neurological disorders, including Huntington's, Alzheimer's, and Parkinson's diseases and that SOD1 aggregates may be inherently toxic or cause motor neuron toxicity by sequestering chaperons and blocking proper functioning of the proteasome [17].

In 2004, Rakhit et al. reported that SOD1, normally a dimeric enzyme, dissociates to monomers prior to aggregation in both wild type and mutant proteins [36]. They used the "Dynamic Light Scattering (DLS)" method to detect the dissociation of dimeric SOD to monomers. Very recently we have reported that the capillary electrophoresis method (CE) is very suitable to investigate the conformational change of the proteins and aggregation states of the proteins in solution [17, 35].

We have observed that the drastic decrease of the peak strength due to the dimeric SOD molecule occurs when the copper(II)/ascorbic acid solution was added to the SOD molecule [37]; our experimental system was same as that reported by Rakhit et al. This clearly shows that the dissociation of the dimeric SOD molecule can be readily detected by the CE method. We also have found that the presence of

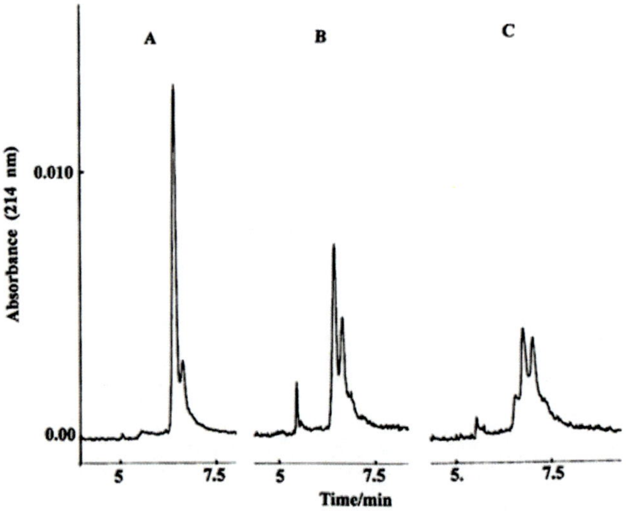

Fig. 17 CE profile of A: wild-type SOD solution, B; measured immediately after hydrogen peroxide was added to solution A, C; measured at 60 min after the addition of hydrogen peroxide [35, 38]

excess hydrogen peroxide induces the dissociation of dimeric structure of wild-type SOD molecule, because drastic decrease of the peak height due to the dimeric structure was observed (see Fig. 17) [17, 35, 38]. Thus, it seems quite likely that the oxidant in the system, the copper(II)/ascorbic acid solution, used by Kakhit et al. should be hydrogen peroxide, and that *sporadic ALS* should be closely related with the presence of excess hydrogen peroxide, and the same discussion may be applied to the elucidation of *sporadic prion diseases (see later).*

By using antibody methods to rapidly purified SOD1 and coupling this with mass spectrometry, Sato et al. have measured the relative accumulated levels of wild-type and mutant SOD1 in erythrocytes of 29 SOD1-mutated fALS patients [39]. They observed that the patients with undetectable SOD1 mutant had the shortest disease durations. Although age at disease onset was found to be uncorrelated with the amount of mutant SOD1, the evidence convincingly shows a strong inverse correlation between disease duration and mutant accumulation. In other words, an accelerated disease course is found for mutants that are less stable. This surprising discovery implies that it is the misfolded unstable forms of SOD1 mutants that contribute to toxicity underlying disease progression, and that despite its apparent importance for progression, SOD1 mutant stability is not correlated with disease onset. Thus dissociation of the dimeric SOD1 molecule to misfolded monomers should be an essential important process for ALS pathogenesis. As it has become apparent that hydrogen peroxide plays an important role in the formation of SOD1 monomers [17, 35, 37], we should pay attention to the formation of excess hydrogen peroxide in the human body, especially due to the reaction between a dimeric iron(III) species and glutathione cycle and other related systems as described in the section "Toxicity due to the binuclear iron(III) species: 2" [15, 30].

Copper(II)–OOH in sporadic bovine spongeform encephalopathies

PrP^{C} is a glycoprotein expressed on the surface of many cell types and its genetic code was identified only after the isolation of an abnormal isoform, PrP^{Sc}, from brains of mice that were infected with the disease scrapie. It is generally recognized that PrP^{C} is a copper-containing protein [at most 4 copper ions are present within the octarepeat region located in the unstructured N-terminus (Fig. 1 in section "Prion diseases and prion protein")]. Since 1996 there has been increasing evidence that PrP^{C} increases cellular resistance to oxidative stress. Cerebelle neurons and astrocytes from PrP^{C} knockout mice are more sensitive to superoxide toxicity, whereas cells with higher levels of PrP^{C} expression are more resistant to oxidative stress [8]. Analysis of recombinant mouse and chicken PrP^{C} has led to the *discovery of an important "gain-of-function"* following the formation of the PrP^{C} copper complex.

The copper at the synapse is released in vesicles, and the copper released in this way appears to be taken up rapidly by the neurons, and deployed within 30 min of this process. It is unknown in what form this copper is bound; however, it is probable that the copper is chelated to some peptides or amino acids because there is little free copper found in the body [40]. It has been pointed out that the copper(II) chelates compounds which across the membrane may originate from the cleavage of the PrP^{C} [8] and thus it seems quite likely that *these copper(II) chelates react with hydrogen peroxide to yield a Cu(II)–OOH species, giving serious effects towards the PrP^{C} such as oxygenation at methionine residue, conformational change, and degradation of protein, if hydrogen peroxide is present in the vicinity* (see Scheme 1).

As described in the section "Prion diseases and prion protein," the misfolded prions (PrP^{Sc}) ultimately kill neurons and leave the brain riddled with holes, like a sponge. In addition to PrP^{Sc}, another protease-resistant PrP of 27–30 kDa, which is called as PrP27-30, was extracted from affected brains. It should be noted here that PrP27-30 is derived from only PrP^{Sc} (not from PrP^{C}), and no difference in amino acid sequence between PrP^{C} and PrP^{Sc} has been identified. Based on these facts, we may assume that the chemical environment around the copper ion in the PrP^{Sc} should be different from those in the PrP^{C}; this situation is similar to the difference observed between the copper(II) ion in the wild-type and mutant SOD enzyme. Thus, it is most likely that the "gain-of-function" in the PrP^{Sc} due to a "*highly reactive*" *Cu(II)–OOH formation may appear* as

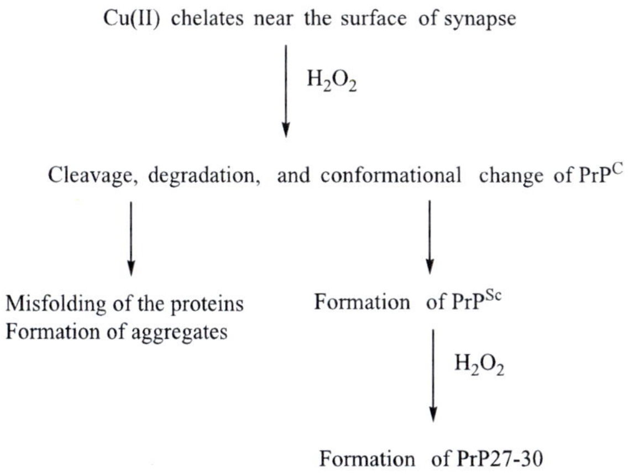

Scheme 1

described for the mutant SOD molecule, leading to the cleavage of the peptide bonds around the copper ion (near at about 90 site), giving PrP27-30, and also the conformational change of PrP^{C} to PrP^{Sc} (see Scheme 1).

We reported that some copper(II) complexes exhibit high catalytic activity to oxygenate the sulfur atom of methionine of amyloid beta-peptide in the presence of hydrogen peroxide [17, 35]. Oxidation of methionine residues in the prion protein by the hydrogen peroxide attracts recent interests; it has become apparent that Met 129, a residue located in a polymorphic position of human PrP and modulating risk of prion diseases, is easily oxidized as is Met 134 by hydrogen peroxide, and that H_2O_2-induced methionine oxidation leads to a modest increase in β-sheet structure. *Several experimental facts observed for the native prion proteins* [13, 41–46] *seem to be consistent with our results as described above,* and all these findings support our proposal [15, 17, 35] that *hydrogen peroxide should be the serious origin for the oxidative stress in sporadic prion diseases.*

New chelation therapy for prion diseases and several neurodegenerative disorders

Manganism in both the sporadic and infected prion diseases

Prion diseases may be divided into two groups: sporadic and infected types. We have demonstrated that pathogenesis of the sporadic prion diseases is closely related with the excess accumulation of manganese ions in the brain. Excess accumulation of manganese ion in the brain induces the abnormality in the synthesis of several neurotransmitters, and iron metabolism in the brain increases the quantity of labile plasma iron, and these iron ions cause a serious damage to the proteins, DNA, and other important substances.

Recent studies on mice experimentally infected with scrapie suggested that large increase in the levels of manganese ion occurs in blood and brain prior to the onset of symptoms of the prion disease; at the same time, elevated manganese ions in several central nervous systems and increase of labile plasma iron in certain regions of the brain were observed. The reason for the above facts observed in the infected type is not clear at present, but it is quite likely that much hydrogen peroxide is produced by the misfolded $Cu\text{-}PrP^{Sc}$ protein through its SOD-like function, and this hydrogen peroxide may induce the increasing of the labile manganese ion in the blood and brain as described in section "Manganism and labile manganese ions". Thus, it is reasonable to conclude that the prion diseases including both the sporadic and infected types should be elucidated by combined toxicity due to the labile plasma manganese and iron ions.

New chelation therapy for the neurodegeneration

The above discussions clearly demonstrate that the removal of the labile plasma iron and manganese ions should be one of the best ways to prevent the prion diseases of sporadic type and other neurodegenerative disorders. For this purpose, we have prepared new chelates to capture both the labile plasma iron and manganese ions effectively and to remove both the ions without toxicity in vitro. We hope that our new chelates should make notable contribution to the prevention and therapeutics for the prion disease, Parkinson's disease, amyotrophic lateral sclerosis (ALS), Alzheimer's disease, schizophrenia, and dementia, which is now in progress in Japan.

References

1. Chani AC, Ferguson NM, Donnell CA, Anderson RM (2000) Nature 406:583
2. Beale AJ (2001) J R Soc Med 94:207
3. Houston F, Foster JD, Chong A, Hunter N, Bostock CJ (2000) Lancet 356:955
4. Cohen FE, Prusiner SB (1998) Annu Rev Biochem 67:793
5. Collinge J (2001) Annu Rev Neurosci 24:519
6. Prusiner SB (1996) Trends Biochem Sci 21:482
7. Caughey B (2001) Trends Biochem Sci 25:235
8. Brown D (2001) Trends Neurosci 24:85
9. Wong BS, Chen SG, Colucci M, Xie Z, Pan T, Liu T, Li R, Gambetti P, Sy MS, Brown DR (2001) J Neurochem 78:1400
10. Dobson AW, Erikson KM, Aschner M (2004) Ann N Y Acad Sci 1012:115
11. Kaiser J (2003) Science 300:926
12. Hesketh S, Sassoon J, Knight R, Hopkins J, Brown DR (2007) J Anim Sci 85:1596
13. Fernaeus S, Reis K, Bedecs K, Land T (2005) Neurosci Lett 389:133

14. Fernaeus S, Halldin J, Bedecs K, Land T (2005) Mol Brain Res 133:266
15. Nishida Y (2004) Med Hypothesis Res 1:227–245
16. Nishida Y (2003) Z Naturforsch 58c:752
17. Nishida Y (2011) Monatsh Chem 142:375
18. Shiraki H, Yase Y (1991) In: Vinken PI, Bruyn GW, Klawans HL (eds) Handbook of clinical neurology, vol 15, pp 273–300
19. Gerlach M, Schachar DB, Riederer P, Youdim MBH (1994) J Neurochem 63:793
20. Youdim MBH, Riederer P (1997) Sci Am 1997:52
21. Heilig EA, Thonpson KJ, Molina RM, Ivanov AR, Brain JD, Resnick MW (2006) Am J Physiol Lung Cell Mol Physiol 290:L1247
22. Abe K, Chiba Y, Nishida Y (2008) Z Naturforsch 63c:154
23. Nishida Y, Ito Y, Satoh T (2007) Z Naturforsch 62c:608
24. Sutoh Y, Nishino S, Nishida Y (2005) Chem Lett 34:140
25. Abragam A, Bleaney B (1970) Electron paramagnetic resonance of transition ions. Clarendon, London
26. Okuno T, Nishida Y (1996) Polyhedron 15:1509–1515
27. Que L Jr, Ho RYN (1996) Chem Rev 96:2607
28. Sutoh Y, Nishida Y (2005) Synth React Inorg Metal-org Nano-metal Chem 35:575
29. Harrison PM, Arosio P (1996) Biochem Biophys Acta 1275:161
30. Nishida Y (2009) TCIMAIL 141:2. http://www.tciamerica.com/tcimail/backnumber/article/141drE.pdf
31. Nishida Y, Takeuchi M (1987) Z Naturforsch 42b:52
32. Nishida Y, Nasu M, Akamatu T (1992) J Chem Soc Chem Commun 1992:94
33. Yamanaka K, Cleveland DW (2005) Neurology 65:1859
34. Alessandra G, Hider RC (2005) Br J Pharm 146:1041
35. Nishida Y (2007) TCIMail 135:2. http://www.tciamerica.com/tcimail/backnumber/135drE.pdf
36. Rakhit R, Crow JP, Lepock JR, Kondejewski LH, Cashman NR, Chakrabartty A (2004) J Biol Chem 279:15499
37. Abe K, Nishida Y (2008) Z Naturforsch 63c:151
38. Chiba Y, Sutoh Y, Nishida Y (2006) Z Naturforsch 61c:273
39. Sato T, Nakanishi T, Yamamoto Y, Andersen PM, Ogawa Y, Fukada K, Zhou Z, Aoike F, Sugai F, Nagano S, Hirata S, Ogawa M, Nakano R, Ohi T, Kato T, Nakagawa M, Hamasaki T, Shimizu A, Sakoda S (2005) Neurology 65:1954
40. Rae TD, Schmidt PJ, Pufahl RA, Culotta VC, O'Halloran TV (1999) Science 284:805
41. MaMahon EHM, Mange A, Nishida N, Creminon C, Casanova D, Lehmann S (2001) J Biol Chem 276:2286
42. Requena JR, Groth D, Legname G, Sradtman ER, Prusiner SB, Revine RL (2001) Proc Natl Acad Sci USA 98:7170
43. Watt NT, Taylor DR, Gillott A, Thomas DA, Perera WS, Hooper NM (2005) J Biol Chem 280:35914
44. Tabler BJ, Turnbull S, Fullwood NJ, German M, Allsop D (2005) Biochem Soc Trans 33:548
45. Tabler BJ, Agnaf OMEA, Turnbull S, German MJ, Paleologou KE, Hayashi Y, Kooper LJ, Fullwood NJ, Allsop D (2005) J Biol Chem 280:35789
46. Watt NT, Hopper NM (2005) Biochem Soc Trans 33:1123

Metal ions and beta amyloid: conformational modifications and biological aspects

Alberto Granzotto and Paolo Zatta

Abstract

Many molecular modifications such as senile plaques and neurofibrillary tangles are known to be associated with Alzheimer's disease and other neurodegenerative diseases. In this connection, metal dyshomeostasis has aroused great interest and considerable support in recent years as relevant pathological cofactors of neurodegeneration. It has been largely demonstrated both in vivo and in vitro that aberrant metal ion metabolism can lead to the development and/or worsening of several neurological disorders. In this chapter, we will focus recent biophysical findings on β-amyloid structural modifications triggered by metal ions and we will provide insights into the biological consequences of these phenomena.

Keywords

Alzheimer • Amyloid • Neurological disorders • Metals and brain • Aluminium

Abbreviations

AD	Alzheimer's disease
Aβ	β-amyloid
BBB	Blood–brain barrier
ESI-MS	Electrospray ionisation mass spectrometry
β-APP	β-amyloid precursor protein
τ	τ (tau) protein

Introduction

Biological systems rely on a huge number of protein interactions as they undergo a wide diversity of physiological functions. This is the case, for instance, for the ionotropic glutamatergic receptor, where its full function relies on the correct subunit arrangement for the formation of highly ion-specific tetrameric structures.

Although the cell quality-control systems provide for the correct folding of proteins during cell life, proteins miss-folded can still aggregate, which leads to a series of pathologies known as the "*conformational diseases*" including Parkinson's disease, Huntington's disease, prion disease, amyotrophic lateral sclerosis and Alzheimer's disease (AD) [1]. In particular AD is characterised by the miss-folding, aggregation and deposition of two proteins: tau (τ), a microtuble-associated protein, and β-amyloid (Aβ), a proteolytic cleavage by-product of the Aβ precursor protein (β-APP) [2].

Over the last two decades, a large amount of data has been reported in the literature relating to Aβ production and aggregation, and to its interactions with other sub-cellular elements. Despite this, the trigger(s) that sets off Aβ production and accumulation is not entirely understood, as well as its interactions with the τ protein.

In this review, we provide an insight into the roles of metal ions in Aβ aggregation and into some of the biological and pathological aspects of this phenomenon.

P. Zatta (✉)
Metalloproteins Unit, Department of Biology, National Research Council (CNR), Institute for Biomedical Technologies, University of Padova, Via Colombo 3, 35121 Padova, Italy
e-mail: zatta@bio.unipd.it

W. Linert and H. Kozlowski (eds.), *Metal Ions in Neurological Systems*,
DOI 10.1007/978-3-7091-1001-0_7, © Springer-Verlag Wien 2012

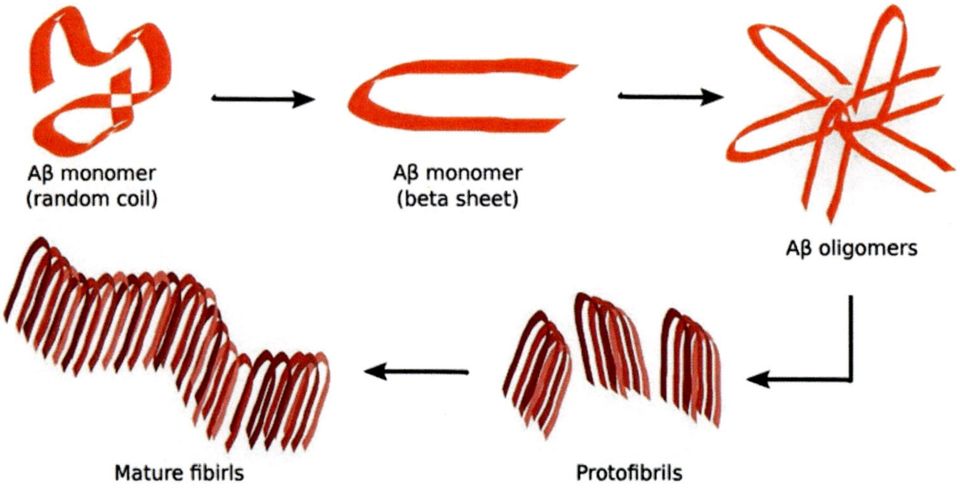

Fig. 1 The Aβ aggregation process follows a well-defined pathway. First, the Aβ monomers with a random-coil structure acquire a β-sheet conformation. Then, these monomers aggregate into oligomeric structures that contain a variable number of Aβ peptides (i.e. 2–50). These oligomers turn into higher ordered structures known as protofibrils, and then these protofibrils lead to the formation of mature fibrils that deposit as senile plaques

From β-amyloid to senile plaques

Aβ is a 39–43 amino-acid peptide and it is the main constituent of senile plaques. In patients with AD, the most common forms of Aβ have 40 or 42 amino acid residues, with the latter as the most amyloidogenic and prone to aggregation [3]. As mentioned earlier, Aβ derives from β-APP, which is itself a 695–770 amino-acid-residue transmembrane protein that has a physiological role that still remains largely unknown; the involvement of β-APP in metal homeostasis and cell trafficking and signalling has been proposed [4].

The metabolism of β-APP involves three enzymes: the α-, β- and γ-secretases. Only when β-APP is cleaved by the last two of these Aβ is formed. This is the so-called *amyloidogenic pathway* [5]. On the contrary, during the physiological non-amyloidogenic pathway, β-APP is metabolised by α-secretase and γ-secretase, which releases neither toxic nor pro-aggregation by-products [6]. Once it has been released, Aβ follows a well-known and well-defined process of aggregation (Fig. 1).

The in vitro aggregation kinetics of the 42-residue Aβ peptide can be summarised as follows (see Fig. 1): (1) random coiled Aβ monomers turn quickly into (2) β-sheet Aβ monomers. This shift leads to the formation of (3) soluble, low-molecular-weight oligomers, which become (4) paranuclei (higher ordered structures), and then (5) protofibrils, and eventually (6) larger fibrils, which represent the main constituent of senile plaques [7]. Together with insoluble deposits of the τ protein, which are known as neurofibrillary tangles, senile plaques are the macroscopic event and hallmark of AD.

However, senile plaques are the downstream event of a more complex process. Over these last two decades, research interest has shifted "upstream", with the report that prefibrillar Aβ species, and especially Aβ oligomers, are significantly more toxic than mature Aβ fibrils [8]. These findings are in agreement with clinical observations, where it has been reported that patients lacking senile plaques can show AD-like cognitive impairment.

While in vitro Aβ aggregation is performed in a highly controlled environment, the behaviour of Aβ in vivo can be influenced by a large number of variables. Within these, metal ions have aroused great interest for three main reasons: (1) patients with AD show pronounced metal dyshomeostasis in the brain; (2) high metal concentrations have been found in senile plaques (Table 1) and (3) it is well established that metals can influence and/or alter the Aβ aggregation pathway [9].

In the following sections, the roles of some of the metal ions (Al, Fe, Cu and Zn) that can influence the folding behaviour of Aβ will be considered and critically discussed, along with the AD "metal hypothesis". Furthermore, a brief insight into the biological aspects of AD and Aβ will be provided.

Aluminium

Aluminium [Al(III)] is the most abundant element in the Earth crust, although it remains to be demonstrated that it is involved in any specific vital biological processes. For this reason, the discovery of relative high Al(III) concentrations in senile plaques of *post-mortem* brains from patients with

Table 1 Metal ion concentrations detected in the cores of senile plaques from patients with AD

Metal ion	Concentration (μg/g senile plaque)
Al(III)	40[a]
Cu(II)	30[b,c]
Fe(III)	53[b,c]
Zn(II)	87[b,c]

[a][10]
[b][50]
[c][51]

AD aroused great interest, with the consequent suggestion of a possible role for Al(III) in the pathology of AD ([10]; for a recent review see Ref. [11]).

Along with all of the other electrically charged elements and molecules, Al(III) cannot be passively transported through the blood–brain barrier (BBB). Once it has been absorbed through the digestive system, Al(III) enters the bloodstream; here, it appears to be linked mainly to citrate and transferrin. However, when it reaches the brain vessels, Al(III) can indeed pass across the BBB via a transferrin-receptor-mediated endocytosis mechanism [12]. Once in the cerebrospinal fluid, Al(III) can influence the Aβ folding process, although its role in the pathology of AD is still debated and controversial [13]. Furthermore, it has been demonstrated that Al(III) can pass through the BBB already complexed with Aβ. In this case, the Aβ–Al(III) complex has more ready access to brain cells than Aβ alone [14].

In contrast to Cu(II) and Zn(II) (see below), the Al(III) complexes that are formed with Aβ have been studied to a lesser extent. Nevertheless, data reported from our and other laboratories have indicated that Al(III) can maintain Aβ in its oligomeric or pre-fibrillar state and can promote Aβ exposure of hydrophobic clusters [15, 16]. On the contrary, other studies have supported a role for Al(III) in the coordination of higher Aβ structures, such as fibrils, and in the promotion of their deposition.

Focusing on the chemical level, the presence of binding sites for Al(III) on Aβ has not been well established yet. Two different binding mechanism for Al(III) with Aβ were proposed several years ago by Fasman [17]), and more recently by our group [18]. In the former study, it was suggested that Al(III) can coordinate with four Aβ amino acids: Asp, Ser, Tyr and Glu, probably because of their high –OH-group content. In our study, we broadened the possible interaction sites to the 1–16 and 20–35 amino-acid sequences [18].

As more recently reported by Kawahara and Kato-Negishi[19]), the ability of Al(III) to coordinate Aβ and to modify its folding properties is due to two properties of Al (III): (1) it has a strong positive charge that is coupled to (2) a small ionic radius (50 pm), as compared to the other metal ions discussed here. These features mean that Al(III) can be considered as an effective protein cross-linker. Consequently, a role for Al(III) in τ folding needs to be investigated, because of its great number of phosphorylated sites; indeed, these R-OPO_3^{2-} sites are the targets of choice for Al(III)-like metals.

ESI-MS (electrospray ionisation mass spectrometry) data, recently reported by Bolognin et al. [15], showed that a bare Al(III) ion can bind to a single Aβ peptide, although Chen et al. [16] have hypothesised that two Al(III) ions can coordinate each Aβ peptide. In the latter study, the authors correctly reported a lack of data concerning the Al(III) concentration in their stoichiometric experiments, which thus questioned the results they obtained. This arose because Al(III) can form hydroxide complexes at neural pH [20]; however, the use of aluminium lactate can help to avoid, or at least delay, Al(III) hydroxide precipitation [18].

Collectively, even though several biophysical and immunological techniques have been used to demonstrate that Al (III) can "freeze" Aβ in oligomeric and highly hydrophobic structures [15], two key data appear to be missing: (1) the structure of the exact Aβ–Al oligomeric complex; and (2) the association constant (K_a) of this complex. In this connection, Bolognin et al. [15] hypothesised that Al(III) can form Aβ oligomers *tout court*, while Chen et al. [16] proposed the formation of Aβ–Al annular protofibril structures.

The present lack of studies does not allow us to provide data concerning the issue of a K_a for this Aβ–Al interaction; however, at the same time, it is possible to state with confidence that K_a (Aβ–Al) is greater than the K_a for deferoxamine mesylate ($K_a = 10^{-22}$ M), indeed, this iron/aluminium chelating agent can reverse the Al(III) influence on the Aβ oligomerisation process [18, 21].

Copper

Copper (Cu(II)) is an essential metal ion involved in several biological processes and analytically found in senile plaques at lower levels (400 μM) together with other metal ions, such as Zn(II) (1 mM) and Fe(III) (1 mM). A potential role for Cu (II) in AD has aroused interest for two main reasons: the influence of Cu(II) on Aβ conformational changes, and the reduction of Cu(II) to Cu(I). This latter is particularly relevant for Aβ-derived reactive oxygen species (ROS) [22], in that the Aβ–Cu(II) complex exerts its toxicity via ROS production (see Ref. [23]). The electron that is necessary to reduce Cu(II) to Cu(I) can be donated by both internal amino acids of Aβ or its external reductant molecules [24, 25].

Focusing on the structural level, a role for Cu(II) in Aβ aggregation has been widely studied, and a large body of evidence supports the idea that Cu(II) might be involved in the acceleration of Aβ deposition into amorphous aggregates [15, 26]. State-of-the-art coordination chemistry of Aβ/Cu (II) has shown that there are four putative residues that have been proposed as Cu(II) binding sites on Aβ: His6, His13, His14, Tyr10 [27–29]. Nevertheless, Cu(II) can bind other residues in the N-terminus of Aβ (e.g. Asp1, Glu11) [30, 31]. In agreement with Miller et al. [30, 31], this variability might be due to the different conditions under which the aggregation processes have been performed.

Conformational changes due to the Cu(II)/Aβ interaction appear to result in reduced exposure of the Aβ–Cu(II) hydrophobic clusters, as compared with Aβ alone or with its complexes with other metal ions, such as Zn(II) and Al(III) (see above) [32]. This event might lead to decreased interactions between the Aβ–Cu(II) complex and the hydrophobic cellular phospholipids [33], even though it has been proposed that in the presence of Cu(II), Aβ forms channel-like structures in cell membranes [34].

Recent findings supported by ESI-MS have reported that Aβ is metallated by a bare Cu(II) ion [15, 16]. This interaction appears to increase Aβ random coil content, which leads to the formation of non-fibrillar amorphous aggregates. Indeed, it has been shown that an elevated β-sheet content is required for fibril formation [35], while the random coil content leads to disordered aggregate deposition.

Recently, different Aβ–Cu(II) affinities have been proposed for the $A\beta_{1-40}$ peptide, which depend on the Aβ secondary structure: 0.14 μM^{-1} for $A\beta_{1-40}$ in a random-coil structure and 0.05 μM^{-1} for $A\beta_{1-40}$ in the beta-sheet stimulated conformation [36]. This scenario is further complicated by the variable molar ratios of Aβ and Cu(II) in the extracellular space. It has been reported that a sub-equimolar Aβ/Cu(II) ratio leads to amorphous and stable aggregates; vice versa, supra-equimolar ratios can lead to the formation of more toxic oligomeric structures [24, 25]. This hypothesis was recently confirmed by Pedersen et al. [37], where the discovery of distinct Cu(II)-concentration-dependent Aβ-aggregation pathways supports a key role for metal homeostasis in the folding of Aβ and, consequently, for its toxicity.

Together with our recent findings [15, 32], these data support the idea that Aβ–Cu(II) complex exerts its toxicity via ROS production.

Iron

As for Al(III), iron (Fe(III)) has also been studied to a lesser extent than some other metals, despite its key role in several biological functions (e.g. as a cofactor or an O_2 carrier, among other functions) and its redox properties. Here, we focus our attention on Fe(III) instead of the reduced Fe(II) form.

Data in the literature support the idea that Fe(III) can lead to the formation of a heterogeneous population of amorphous aggregates, thus shifting from oligomers to larger, high-molecular-weight structures. ESI-MS analyses has shown that Aβ can bind two Fe(III) ions [15, 16]. These Fe(III) ions appear to be coordinated via His13, His14 and Tyr10, as suggested by Alì-Torres et al. [38]. The same study also supported the idea that Aβ forms more stable complexes when it binds to Fe(III) rather than to Fe(II). It has been reported recently that Fe(III) increases the Aβ random coil content [39], which promotes the deposition of amorphous aggregates, as described for Cu(II). This conformational change is associated with decreased exposure of hydrophobic clusters [15, 16], which reduces the possible interactions between the Aβ–Fe(III) complexes and the lipid bilayers of the cell [33]. Again, as with Cu(II), Fe(III) interactions with Aβ can catalyse the generation of hydrogen peroxide (H_2O_2); consequently, a lack of detoxifying enzymes or an accumulation of Aβ–Fe species (both Fe(II) or Fe(III)) can trigger ROS formation via the Fenton reaction [9].

Very little is known about the Aβ affinity for Fe(III), as the lack of studies does not provide much data relating to this complex. Despite this, as for Aβ–Al(III), it is possible to assume that the Aβ affinity for Fe(III) is lower than that of the Fe(III)-chelating agents (e.g. deferoxamine mesylate); indeed, these compounds can revert the Aβ–Fe(III) aggregation process [21].

Collectively, the data reported here support the concept that Aβ–Fe(III) exerts its toxicity through two independent mechanisms. One mechanism involves ROS production, and the other involves the changes in Aβ conformation. This latter appears to be less convincing for two reasons: (1) Fe (III) only delays Aβ deposition in large amorphous aggregates, as the Aβ–Fe(III) oligomers are limited in time and tend to deposit into senile plaques; (2) data that support this hypothesis appear poor [39], because they do not discriminate between toxicity due to ROS production or to conformational changes in Aβ; moreover, the Aβ concentration used was largely higher than that of other studies reported in the literature (10 μM vs. 0.5 μM).

Zinc

The role of zinc (Zn(II)) in the pathophysiology of the central nervous system has been widely debated, and its involvement in neurodegenerative disorders appears to be well established. Its deregulation also appears to have an important role in AD [40]. At the same time, Zn(II) has key roles in synaptic functions, neurotransmission and cell signalling. Zn (II) in cells is usually maintained at low basal concentrations through three mechanisms: (1) Zn(II) transporters; (2) Zn importing proteins and (3) the buffering action of the metallothioneins. In addition to the metallothioneins, Zn(II) is stored at high concentrations (~1 mM) in presynaptic vesicles and co-released with glutamate during neurotransmission [41]. Once in the synaptic cleft, Zn(II) can bind to Aβ, promoting its conformational modifications.

Due to the importance of Zn(II) in cell physiology, its ability to modify the Aβ structure and the Aβ aggregation pathway has been largely characterised. As for the other metal ions, to assess Aβ–Zn(II) stoichiometry, ESI-MS has been carried out. The data reported in the literature support the idea that a single Zn(II) ion binds to Aβ [15, 16]. According to many nuclear magnetic resonance studies ([42, 43]; reviewed in Ref. [44]), the bare Zn(II) ion binds the N-terminal region of Aβ, which probably involves the same Cu(II)-binding residues, His6, His13 and His14, even though four other potential binding sites have been proposed: Asp1, Glu3, Asp7 and Glu11 [30, 31].

The morphology of Aβ–Zn(II) aggregates has been investigated by different groups, with the use of several biophysical and immunological techniques. The results reported appear comparable and have become largely accepted. Atomic force microscopy and transmission electron microscopy, together with dot-blotting, have shown that Zn (II) promotes Aβ deposition into amorphous aggregates that can coexist with heterogeneous oligomers [15, 16, 30, 31].

Despite the similarities between Aβ–Cu(II) and Aβ–Zn (II) (which can be attributed to their comparable ionic radii: 74 pm for Zn(II), and 73 pm for Cu(II)), Zn(II) is more effective in the promotion of Aβ exposure of the hydrophobic clusters [15, 16, 32]. This supports the idea that the Zn (II) binding sites are different from the Cu(II) ones (see above).

The apparent dissociation constants of Zn(II) from Aβ were reviewed by Faller and Hureau [44], where they hypothesised that the Aβ–Zn(II) K_d lies in the range of 1 μM–20 μM. This variability is mainly due to the different conditions that are used to assess K_d values (e.g. buffer, metal-ion concentration, protein concentration, metal/protein stoichiometry).

Brief insights into the biological aspects

A plethora of Aβ mechanisms of toxicity have been reported over the last decade, although only a fraction of these have addressed AD metal dyshomeostasis or the roles of these metal ions in Aβ miss-folding. The most studied mechanism through which Aβ-metal complexes exert their toxicity is the production of ROS. Aβ produces H_2O_2 in the presence of biological reducing agents [45]. Aβ ROS generation is promoted by the presence of the transition metals, such as Cu (II) and Fe(III), which can lead to the formation of free radical species through the Fenton reaction (for a detailed review, see Ref. [22]). The metal ions that are not involved in redox reactions (i.e. the aforementioned Al and Zn) appear to be involved indirectly in ROS production. As recently demonstrated by Duce et al. [46], Aβ–Zn can inhibit iron-export ferroxidase activity; this results in Fe(II) accumulation, which then leads to oxidative stress in cortical neurons. Also, Al(III) has a pro-oxidative role, as it promotes Fe(II)-induced lipid peroxidation [47]. Unfortunately, data concerning the role of Aβ–Al(III) in lipid peroxidation is missing; the shedding of light on this issue would be of great interest.

As well as ROS production, we have recently demonstrated a second way through which Aβ-metal complexes exert their toxic effects. In comparing Aβ-metal aggregation data with the effects of Aβ-metal complexes on membrane models and in toxicity essays, we found a strict correlation between Aβ-metal exposure of hydrophobic clusters and membrane damage. This effect, together with other data in the literature that we have reported here, might explain why our Aβ–Al(III) complex was the most effective in the reduction of cell viability in our cellular model. Indeed, Aβ–Al(III) has three characteristics that justify its toxicity under our experimental conditions: (1) it is the most effective Aβ-metal complex for the retaining of its oligomeric structure; (2) it is the most effective Aβ-metal complex for the exposure of hydrophobic clusters and (3) it can induce lipid peroxidation, as mentioned earlier. The other Aβ-metal complexes express no more than two of these features.

In addition to these two main mechanisms, others have been reported in the literature, and in particular, Aβ interactions with synaptic receptors, such as the metabotropic glutamate receptors and NMDA receptors. In these cases, direct binding of Aβ to the receptor might not occur. It appears more likely that Aβ can indirectly modulate synaptic receptors through its membrane association [48]; this hypothesis confirms the need for highly hydrophobic Aβ oligomers that can penetrate into lipid bilayers.

This scenario is further complicated by the difficulty of isolating well characterised and homogeneous Aβ or Aβ-metal aggregates. Furthermore, data reported in literature show large differences in data obtained with synthetic or with naturally occurring Aβ oligomers. Indeed, the latter require much lower concentrations to exert comparable toxic effects on cellular models [49], highlighting possible structural, as well as biochemical (e.g. glycosylation), differences between these in vitro and in vivo Aβ aggregates.

Conclusions

As generally reported in the literature and briefly summarised here, it is clear that metal ions have heterogeneous influences on Aβ miss-folding and deposition (Fig. 2). This variability modifies the pathways through which Aβ and its metal complexes exert their toxicity: from ROS production, to cell membrane damage [52].

The ability to change the impact of metal ions in Aβ aggregation pathway appears a possible and promising

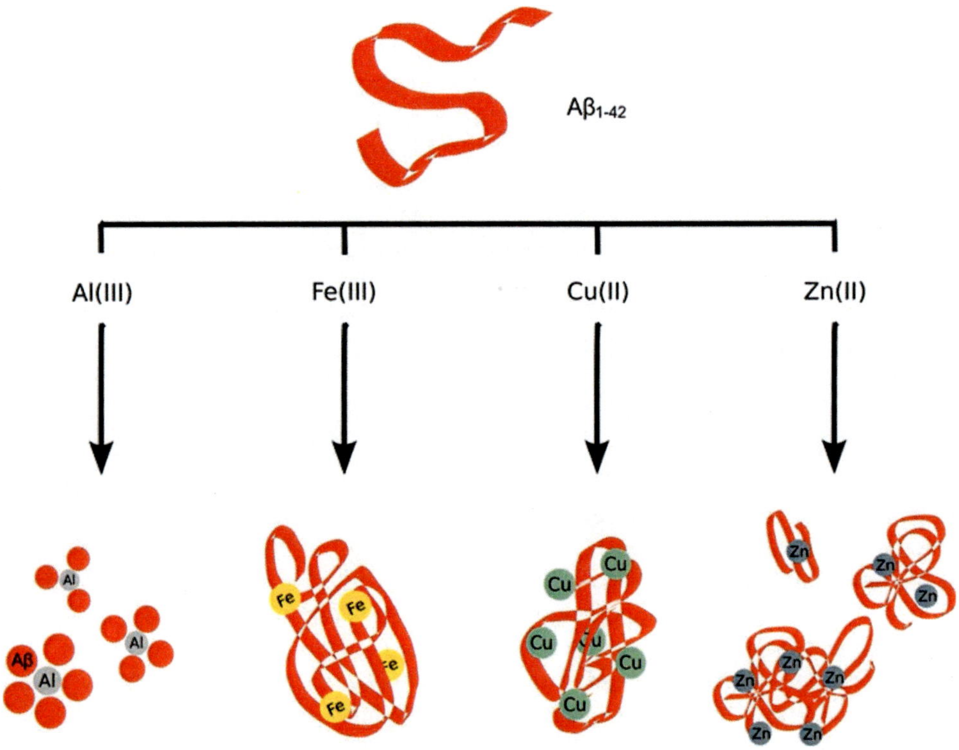

Fig. 2 Aluminium, iron, copper and zinc differentially alter Aβ aggregation. Al(III) promotes the formation of highly hydrophobic Aβ oligomers (i.e. from trimers to hexa-heptamers); Fe(III) promotes Aβ deposition into amorphous structures; Cu(II) leads to the formation of disordered/amorphous structures and Zn(II) triggers the formation of amorphous aggregates

therapeutic challenge. However, metal ions that arc involved in therapeutic strategies (e.g. chelation therapy) should be approached with caution. First of all, focusing on a single metal ion can lead to generation of a cascade of events that will involve the homeostasis of other life-essential metal ions, as seen by the so-called *domino effect* [9]. Accordingly, a new therapeutic approach has focused on molecules that can compensate for cellular metal-ion dysregulation, potentially by sequestering essential ions from senile plaques and "ferrying" these into the cell without compromising the homeostasis of other metal ions.

Moreover, a process to revert this Aβ-metal aggregation might cause more harm than good. This has been reported for toxic species of Aβ in solution that can be deposited in stable high-molecular-weight metal aggregates.

Despite these promising therapeutic strategies, multifactorial pathologies like AD should not be addressed by focusing on a single feature (e.g. Aβ accumulation, metal dyshomeostasis) without further considerations of others (e.g. τ-hyperphosphorylation, APOE4, synaptic failure, ROS production). Thus, approaching the treatment of patients with AD in this way might be reductionist and could ultimately be ineffective.

References

1. Kopito RR, Ron D (2000) Conformational disease. Nat Cell Biol 2: E207–E209
2. Querfurth HW, LaFerla FM (2010) Alzheimer's disease. N Engl J Med 362:329–344
3. Yin YI, Bassit B, Zhu L, Yang X, Wang C et al (2007) γ-Secretase substrate concentration modulates the $A\beta_{42}/A\beta_{40}$ ratio. J Biol Chem 282:23639–23644
4. Zhang Y, Thompson R, Zhang H, Xu H (2011) APP processing in Alzheimer's disease. Mol Brain 4:3
5. Hardy JA, Higgins GA (1992) Alzheimer's disease: the amyloid cascade hypothesis. Science 256:184–185
6. Wilquet V, Strooper BD (2004) Amyloid-β precursor protein processing in neurodegeneration. Curr Opin Neurobiol 14:582–588
7. Chiti F, Dobson CM (2006) Protein misfolding, functional amyloid, and human disease. Annu Rev Biochem 75:333–366
8. Sakono M, Zako T (2010) Amyloid oligomers: formation and toxicity of Aβ oligomers. FEBS J 277:1348–1358
9. Zatta P, Drago D, Bolognin S, Sensi SL (2009) Alzheimer's disease, metal ions and metal homeostatic therapy. Trends Pharmacol Sci 30:346–355
10. Yumoto S, Kakimi S, Ohsaki A, Ishikawa A (2009) Demonstration of aluminum in amyloid fibers in the cores of senile plaques in the brains of patients with Alzheimer's disease. J Inorg Biochem 103:1579–1584
11. Bolognin S, Zatta P (2011) In: Milardi D, Rizzarelli E (eds) Neurodegeneration, Royal Society of Chemistry, Cambridge
12. Yokel RA (2002) Brain uptake, retention, and efflux of aluminum and manganese. Environ Health Perspect 110:699–704

13. Tomljenovic L (2011) Aluminum and Alzheimer's disease: after a century of controversy, is there a plausible link? J Alzheimers Dis 23:567–598
14. Banks WA, Niehoff ML, Drago D, Zatta P (2006) Aluminum complexing enhances amyloid beta protein penetration of blood-brain barrier. Brain Res 116(1):215–221
15. Bolognin S, Messori L, Drago D, Gabbiani C, Cendron L et al (2011) Aluminum, copper, iron and zinc differentially alter amyloid-Aβ_{1-42} aggregation and toxicity. Int J Biochem Cell Biol 43:877–885
16. Chen W, Liao Y, Yu H, Cheng IH, Chen Y (2011) Distinct effects of Zn^{2+}, Cu^{2+}, Fe^{3+}, and Al^{3+} on amyloid-β stability, oligomerization, and aggregation. J Biol Chem 286:9646–9656
17. Fasman GD (1996) Aluminum and Alzheimer's disease: model studies. Coord Chem Rev 149:125–165
18. Ricchelli F, Drago D, Filippi B, Tognon G, Zatta P (2005) Aluminum-triggered structural modifications and aggregation of β-amyloids. Cell Mol Life Sci 62:1724–1733
19. Kawahara M, Kato-Negishi M (2011) Link between aluminum and the pathogenesis of Alzheimer's disease: the integration of the aluminum and amyloid cascade hypotheses. Int J Alzheimers Dis 2011:276393
20. Martin RB (1992) Aluminum speciation in biology. Ciba Found Symp 169:5–18, discussion 18–25
21. House E, Collingwood J, Khan A, Korchazkina O, Berthon G et al (2004) Aluminum, iron, zinc and copper influence the *in-vitro* formation of amyloid fibrils of Aβ_{42} in a manner which may have consequences for metal chelation therapy in Alzheimer's disease. J Alzheimers Dis 6:291–301
22. Hureau C, Faller P (2009) Aβ-mediated ROS production by Cu ions: structural insights, mechanisms and relevance to Alzheimer's disease. Biochimie 91:1212–1217
23. Hung Y, Bush A, Cherny R (2010) Copper in the brain and Alzheimer's disease. J Biol Inorg Chem 15:61–76
24. Smith DG, Cappai R, Barnham KJ (2007) The redox chemistry of the Alzheimer's disease amyloid β peptide. Biochim Biophys Acta 1768:1976–1990
25. Smith DP, Ciccotosto GD, Tew DJ, Fodero-Tavoletti MT, Johanssen T et al (2007) Concentration dependent Cu^{2+}-induced aggregation and dityrosine formation of the Alzheimer's disease amyloid-β peptide. Biochemistry 46:2881–2891
26. Tougu V, Karafin A, Zovo K, Chung RS, Howells C et al (2009) Zn(II)- and Cu(II)-induced non-fibrillar aggregates of amyloid-β_{1-42} peptide are transformed to amyloid fibrils, both spontaneously and under the influence of metal chelators. J Neurochem 110:1784–1795
27. Faller P (2009) Copper and zinc binding to amyloid-β: coordination, dynamics, aggregation, reactivity and metal-ion transfer. ChemBioChem 10:2837–2845
28. Hong L, Carducci TM, Bush WD, Dudzik CG, Millhauser GL et al (2010) Quantification of the binding properties of Cu^{2+} to the amyloid β peptide: coordination spheres for human and rat peptides and implication on Cu^{2+}-induced aggregation. J Phys Chem B 114:11261–11271
29. Eury H, Bijani C, Faller P, Hureau C (2011) Copper(II) coordination to amyloid β: murine *versus* human peptide. Angew Chem Int Ed Engl 50:901–905
30. Miller Y, Ma B, Nussinov R (2010) Polymorphism in Alzheimer Aβ amyloid organization reflects conformational selection in a rugged energy landscape. Chem Rev 110:4820–4838
31. Miller Y, Ma B, Nussinov R (2010) Zinc ions promote Alzheimer Aβ aggregation via population shift of polymorphic states. Proc Natl Acad Sci USA 107:9490–9495
32. Granzotto A, Bolognin S, Scancar J, Milacic R, Zatta P (2011) β-Amyloid toxicity increases with hydrophobicity in the presence of metal ions. Monats Chem 142:421–430
33. Suwalsky M, Bolognin S, Zatta P (2009) Interaction between Alzheimer's amyloid-β and amyloid-β-metal complexes with cell membranes. J Alzheimers Dis 17:81–90
34. Curtain CC, Ali F, Volitakis I, Cherny RA, Norton RS et al (2001) Alzheimer's disease amyloid-β binds copper and zinc to generate an allosterically ordered membrane-penetrating structure containing superoxide-dismutase-like subunits. J Biol Chem 276:20466–20473
35. Lin C, Huang H, Jiang Z (2010) Cu(II) interaction with amyloid-β peptide: a review of neuroactive mechanisms in AD brains. Brain Res Bull 82:235–242
36. Yang CA, Chen YH, Ke SC, Chen YR, Huang HB, Lin TH, Chen YC (2010) Correlation of copper interaction, copper-driven aggregation, and copper-driven H_2O_2 formation with Aβ_{40} conformation. Int J Alzheimers Dis 2011:607861
37. Pedersen JT, Ostergaard J, Rozlosnik N, Gammelgaard B, Heegaard NHH (2011) Cu(II) mediates kinetically distinct, non-amyloidogenic aggregation of amyloid-β peptides. J Biol Chem 286:26952–26963
38. Alì-Torres J, Rodriguez-Santiago L, Sodupe M, Rauk A (2011) Structures and stabilities of $Fe^{2+/3+}$ complexes relevant to Alzheimer's disease: An *ab-initio* study. J Phys Chem A 115 (45):12523–12530
39. Liu B, Moloney A, Meehan S, Morris K, Thomas SE et al (2011) Iron promotes the toxicity of amyloid β peptide by impeding its ordered aggregation. J Biol Chem 286:4248–4256
40. Sensi SL, Paoletti P, Bush AI, Sekler I (2009) Zinc in the physiology and pathology of the CNS. Nat Rev Neurosci 10:780–791
41. Corona C, Pensalfini A, Frazzini V, Sensi SL (2011) New therapeutic targets in Alzheimer's disease: brain deregulation of calcium and zinc. Cell Death Dis 2:e176
42. Nair NG, Perry G, Smith MA, Reddy VP (2010) NMR studies of zinc, copper, and iron binding to histidine, the principal metal-ion-complexing site of amyloid-β peptide. J Alzheimers Dis 20:57–66
43. Syme CD, Viles JH (2006) Solution ^{1}H-NMR investigation of Zn^{2+} and Cd^{2+} binding to amyloid-β peptide (Aβ) of Alzheimer's disease. Biochim Biophys Acta 1764:246–256
44. Faller P, Hureau C (2009) Bioinorganic chemistry of copper and zinc ions coordinated to amyloid-β peptide. Dalton Trans 7:1080–1094
45. Behl C (1994) Hydrogen peroxide mediates amyloid β protein toxicity. Cell 77:817–827
46. Duce JA, Tsatsanis A, Cater MA, James SA, Robb E et al (2010) Iron-export ferroxidase activity of β-amyloid precursor protein is inhibited by zinc in Alzheimer's disease. Cell 142:857–867
47. Zatta P, Kiss T, Suwalsky M, Berthon G (2002) Aluminum(III) as a promoter of cellular oxidation. Coord Chem Rev 228:271–284
48. Ittner LM, Gotz J (2011) Amyloid-β and tau – a toxic *pas de deux* in Alzheimer's disease. Nat Rev Neurosci 12:65–72
49. Li S, Jin M, Koeglsperger T, Shepardson NE, Shankar GM et al (2011) Soluble Aβ oligomers inhibit long-term potentiation through a mechanism involving excessive activation of extrasynaptic NR2B-containing NMDA receptors. J Neurosci 31:6627–6638
50. Lovell M (1998) Copper, iron and zinc in Alzheimer's disease senile plaques. J Neurol Sci 158:47–52
51. Frederickson CJ, Koh J, Bush AI (2005) The neurobiology of zinc in health and disease. Nat Rev Neurosci 6:449–462
52. Granzotto A, Zatta P (2011) Resveratrol acts not through anti-aggregative pathways, but mainly via its scavenging properties against Aβ and Aβ-metal complex toxicity. PLoS One 6:e21565

Beta-amyloid toxicity increases with hydrophobicity in the presence of metal ions

Alberto Granzotto, Silvia Bolognin, Janez Scancar, Radmila Milacic, and Paolo Zatta

Abstract

Alzheimer's disease is a multifactorial neurodegenerative disorder characterized by the pathological brain deposition of neurofibrillary tangles and senile plaques. The latter consist mainly of insoluble β-amyloid (Aβ) fibril deposition. Aβ aggregation and deposition can be increased by several factors, including metal ions. In this study we investigated the role played by metal ions in affecting Aβ oligomerization in the presence and in the absence of its hydrophobic fragment $A\beta_{17-28}$. This was done not as a physiological investigation, but as a paradigmatic study to confirm the key role of Aβ superficial hydrophobicity as a relevant aggravating factor that contributes to the toxicity of Aβ and Aβ–metal complexes. The structural conformations of Aβ–metal complexes were monitored through fluorescence and turbidity measurements as well as transmission electron microscopy. Results reported herein indicate that various metals differentially influence Aβ conformation, with aluminum being the only metal ion for which we are able to determine a dramatic enhancement of peptide oligomer formation with a consequent toxic effect. This scenario was further enhanced by the presence of $A\beta_{17-28}$, which resulted in a marked toxicity in a neuroblastoma cell culture as a consequence of the enhancement of the hydrophobicity of the amyloid and amyloid–metal complexes.

Keywords

Amyloid-beta • Aluminum • Alzheimer's disease • Metal ions • Hydrophobicity

Introduction

Even though more than 100 years have passed since Alzheimer's disease (AD) was first described, there are still no satisfactory explanations for the conditions that lead to the primum movens and the consequent development of the disease. Although the multifactorial character of the disease makes it difficult to identify the potential etiopathogenetic contributory factors, histologically AD is characterized by the loss of neuronal density in the cerebral cortex, the presence of intraneuronal neurofibrillary tangles (NFTs) consisting of hyperphosphorylated tau proteins (τ) [1], and senile plaques (SPs) formed by the extraneuronal deposition of β-amyloid protein (Aβ). Aβ is a byproduct of the metabolism of a transmembrane precursor (AβPP), and it is cleaved by two proteolytic enzymes called β- and γ-secretase. Recent studies have proposed that Aβ follows a complex process of oligomerization/aggregation. This starts with the formation of monomers, followed by soluble, low molecular weight (LMW) oligomeric structures that exhibit an increase in β-sheet content [2]. Oligomers associate rapidly to form higher-order insoluble aggregates called fibrils. Fibrils are the main constituent of SPs, although the presence of SPs is not necessarily an indication of AD. Recent evidence underlines a strong association between histological features of AD and dementia in patients in

P. Zatta (✉)
Department of Biology, Padua "Metalloproteins" Unit, CNR-Institute for Biomedical Technologies, University of Padua, Padua, Italy
e-mail: zatta@bio.unipd.it

W. Linert and H. Kozlowski (eds.), *Metal Ions in Neurological Systems*,
DOI 10.1007/978-3-7091 1001-0_8, © Springer-Verlag Wien 2012

their mid-1970s [3]. Nevertheless, many authors identify the important role played by soluble Aβ oligomers in the initial steps of AD, due to the oligomers' potent synaptotoxicity [4]. The progression of Aβ aggregation is influenced by many factors, including the presence of metal ions; many studies suggest that endogenous biometals such as copper, zinc and iron as well as exogenous biometals such as aluminum play a potential role in Aβ aggregation and bioavailability [5]. Accordingly, several reports have demonstrated a marked accumulation of metal ions in the SP and in the neurophils of AD patients [6, 7]; this evidence has been also quantitatively confirmed by Leskovjan et al. [8] for copper, iron, and zinc. The Aβ aggregation model is further complicated by other variables, such as Aβ_{1-42} hydrophobicity. Meanwhile, although the role of lipophilicity in promoting Aβ_{1-42} aggregation [9, 10] seems to be clear, the importance of Aβ_{1-42} surface hydrophobicity as a possible mechanism of toxicity has not been sufficiently investigated. With this experimental model, carried out using a nonphysiological Aβ_{1-42} truncated fragment as a hydrophobicity enhancer, we would like to stress the importance of lipophilicity in promoting Aβ_{1-42} toxicity, mainly in the presence of metal ions.

Results and discussion

Fluorescence

Thioflavin T (ThT) is known to rapidly bind the β-sheet-rich aggregated form of peptides. ThT fluorescence over time for samples of Aβ_{1-42}, Aβ_{1-42}–metal complexes and Aβ_{17-28} were examined.

A lag time prior to the formation of β-sheet structures which occurs after ~5 h for both Aβ_{17-28} and Aβ_{17-28} + Fe was observed. The aggregation was enhanced in the presence of the other metals tested. Al is the metal that most affects the process. The absence of a lag time in the sample Aβ_{17-28} + Al may suggest the possibility of an immediate conversion of monomeric Aβ_{17-28} into ThT-reactive species, such as oligomers [14] and protofibrils [15].

Then we used ThT to follow changes in β-sheet content in samples of Aβ_{1-42}/Aβ_{1-42}–metal complexes to which Aβ_{17-28} was initially added.

As shown in Fig. 1, the fluorescence of Aβ_{1-42} + Aβ_{17-28} increased in a sigmoidal manner, indicating the formation of aggregating structures. This curve is consistent with a nucleation-dependent model [16]. The nucleation phase was skipped and the elongation phase emerged rapidly as soon as the two peptides interacted with each other. This general trend is not observed for either Aβ_{1-42}–Fe + Aβ_{17-28} or for Aβ_{1-42}–Cu + Aβ_{17-28}. A marked increase in ThT fluorescence was observed in Aβ_{1-42}–Al + Aβ_{17-28}

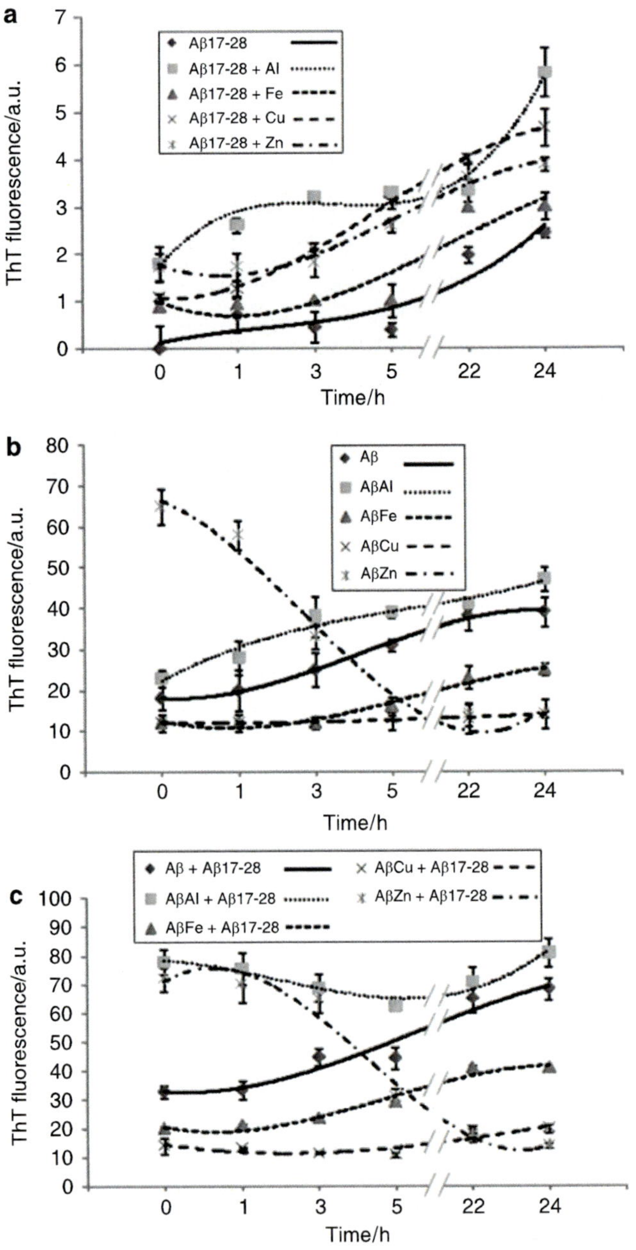

Fig. 1 Time dependence of the fluorescence emission intensity of ThT bound to Aβ_{17-28} both in the absence and in the presence of Al, Cu, Fe, and Zn at a concentration of 5 μM (**a**), to Aβ_{1-42}–metal complexes (**b**), and to Aβ_{17-28} in the presence of Aβ_{1-42}–metal complexes (**c**). The Aβ_{17-28}, Aβ_{1-42}, and Aβ_{1-42}–metal complex peptide concentrations were 5 μM. ThT (20 μM) fluorescence at 482 nm (λ_{exc} = 450 nm) was followed for 24 h. The emissions due to the free dye and buffer were subtracted. The data represented are the mean ± SD of three individual experiments

compared with Aβ_{1-42}–Al alone, suggesting that the interaction between the oligomers of the metal complex was involved in promoting nucleation assembly. The fact that Aβ_{1-42}–Zn precipitated indicates that the equilibrium of the protein species is shifted toward the formation of amorphous aggregates. The mechanism was not reversible, since mixing

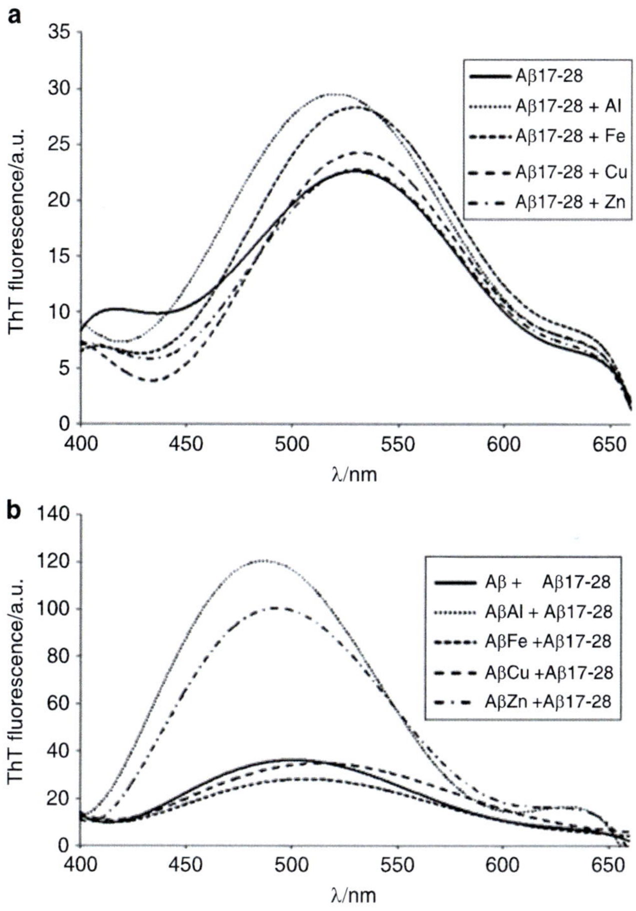

Fig. 2 Fluorescence emission spectra of ANS (25 μM) after interaction with $A\beta_{17-28}$ both in the absence and in the presence of Al, Cu, Fe, and Zn at a concentration of 5 μM (**a**), and with $A\beta_{17-28}$ in the presence of $A\beta_{1-42}$–metal complexes (**b**). Emission spectra were recorded from 400 to 700 nm with excitation at 360 nm. The $[A\beta_{17-28}]/[A\beta_{1-42}]$ ratio was equal to 1. The signals due to the free dye and buffer were subtracted. The peptide samples (5 μM) were left to incubate for 24 h at room temperature, and then the fluorescence was measured

the solution did not re-suspend the aggregate. All samples (except $A\beta_{1-42}$–Zn + $A\beta_{17-28}$) showed a sigmoidal curve characterized by a ~3/5 h lag time, followed by a ~19 h period where ThT fluorescence increased. After 24 h of incubation at room temperature a plateau was observed.

$A\beta_{1-42}$ + $A\beta_{17-28}$, $A\beta_{1-42}$–Al + $A\beta_{17-28}$, $A\beta_{1-42}$–Fe + $A\beta_{17-28}$, $A\beta_{1-42}$–Cu + $A\beta_{17-28}$, and $A\beta_{1-42}$–Zn + $A\beta_{17-28}$ were each tested for surface hydrophobicity by following the 8-anilino-1-naphthalene sulfonic acid (ANS) probe. According to Uversky et al. [17], changes in ANS fluorescence (an increase in intensity and a blue shift of the emission maximum) are characteristic of the interaction of this dye with the solvent-exposed hydrophobic clusters of partially folded peptides and proteins.

Figure 2 shows that $A\beta_{1-42}$–Al induced an increase in ANS fluorescence intensity and a blue shift of the emission maximum compared with the other $A\beta$–metal complexes. This implies that the peptide is converted into a more folded conformation with solvent-exposed hydrophobic clusters. This conversion was higher in the presence of $A\beta_{1-42}$–Al + $A\beta_{17-28}$.

Surprisingly, $A\beta_{1-42}$–Zn in the presence of $A\beta_{17-28}$ decreased its propensity to expose hydrophobic clusters. Simultaneously, we did not observe any significant difference in ANS fluorescence with $A\beta_{1-42}$–Fe, $A\beta_{1-42}$–Cu, and $A\beta_{1-42}$ alone in the presence of the $A\beta_{1-42}$ truncated fragment. In our opinion, these more hydrophobic complexes can be considered a good model for studying the effects of hydrophobicity in relation to $A\beta_{1-42}$–metal complex toxicity. In fact, lipophilicity could play a crucial role in increasing the deleterious effects of $A\beta_{1-42}$–metal complexes.

Turbidity

To clarify the effect of $A\beta_{17-28}$ on $A\beta_{1-42}$/$A\beta_{1-42}$–metal complex fibrillization, aggregation was also assayed by measuring the turbidity at 405 nm wavelength. A turbidity assay gives information on the quantity but not the quality of the aggregates.

First of all, the capacity of $A\beta_{17-28}$ to create complexes with metal ions was tested. When Al was present, there was a dramatic increase in the $A\beta_{17-28}$ aggregation rate which stabilized after 24 h of incubation. The other metals (Fe, Cu, and Zn) exerted only negligible effects. Stabilization of the aggregation process was observed after 24 h of incubation (see Fig. 3).

As for the role played by $A\beta_{17-28}$ in solution with $A\beta_{1-42}$–metal complexes, as suggested by ThT fluorescent assay, the presence of $A\beta_{17-28}$ stimulated aggregate formation in the $A\beta_{1-42}$–Al samples while the other conditions were not affected, especially when compared with $A\beta_{1-42}$–Al + $A\beta_{17-28}$. For $A\beta_{17-28}$, we observed a plateau in the aggregation process after 24 h, which was also seen for $A\beta_{1-42}$–metal complexes.

Transmission electron microscopy

To assess the morphology of aggregates formed in the presence of $A\beta_{17-28}$, the peptide fragment was incubated with $A\beta_{1-42}$ and $A\beta_{1-42}$–metal complexes. Aliquots were removed at time zero and after 24 h of incubation at room temperature. In the presence of $A\beta_{1-42}$–Al + $A\beta_{17-28}$, we observed small spherical oligomers resembling those previously reported in the presence of $A\beta_{1-42}$–Al alone by Drago et al. [18] (see Fig. 4).

After 24 h of incubation, the presence of $A\beta_{17-28}$ stimulated the formation of protofibrillar structures that could not be detected in the presence of $A\beta_{1-42}$–Al alone

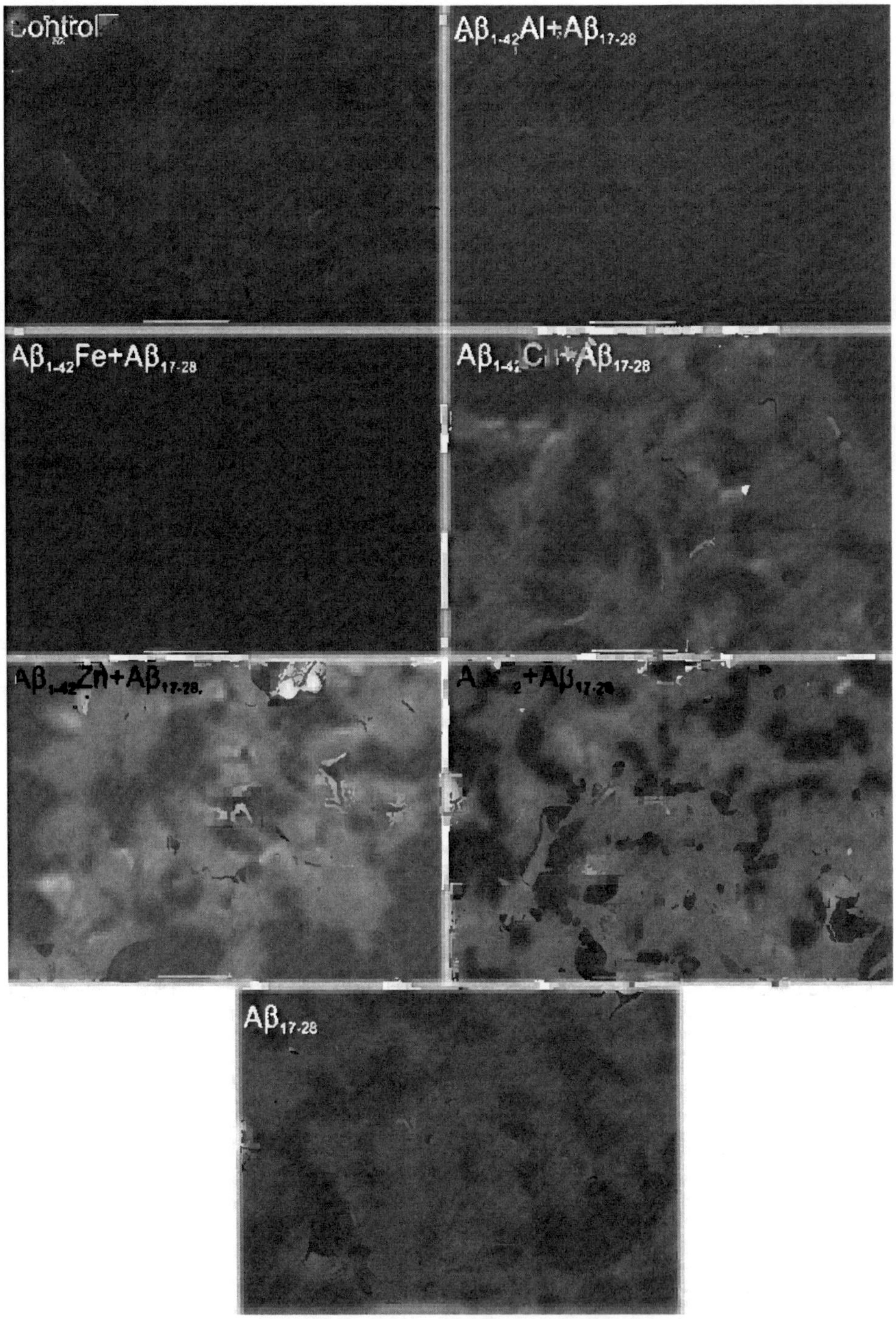

Fig. 3 SEM of neuroblastoma cells treated with the $A\beta_{1-42}$–metal complexes in the presence of $A\beta_{17-28}$ and with $A\beta_{17-28}$ alone after 24 h of incubation at room temperature

(data not shown). These spherical oligomers contained an extended β-sheet structure, as detected by ThT fluorescence. $A\beta_{1-42}$–Cu + $A\beta_{17-28}$ showed minimal aggregation at time zero but started to aggregate into well-structured fibrils after 24 h. $A\beta_{1-42}$–Fe + $A\beta_{17-28}$ appeared to form small fibrils together with amorphous aggregates at time zero; then, after 24 h, fibrils became the dominant species. These fibrils often formed large groups containing randomly oriented fibers. At time zero, the samples $A\beta_{1-42}$ + $A\beta_{17-28}$ showed minimal aggregation, consistent with the majority of $A\beta$ being in nonaggregated form. A significant increase in the rate of aggregation was observed after 24 h and this correlated with the increase in ThT fluorescence. $A\beta_{1-42}$–Zn + $A\beta_{17-28}$ showed no aggregation at time zero but a longer incubation period (24 h) resulted in the appearance of fibrils, which seemed to coexist with oligomers. Taken together, TEM images suggested that the presence of $A\beta_{17-28}$ enhanced $A\beta_{1-42}$ and $A\beta_{1-42}$–metal complex fibrillization, stimulating

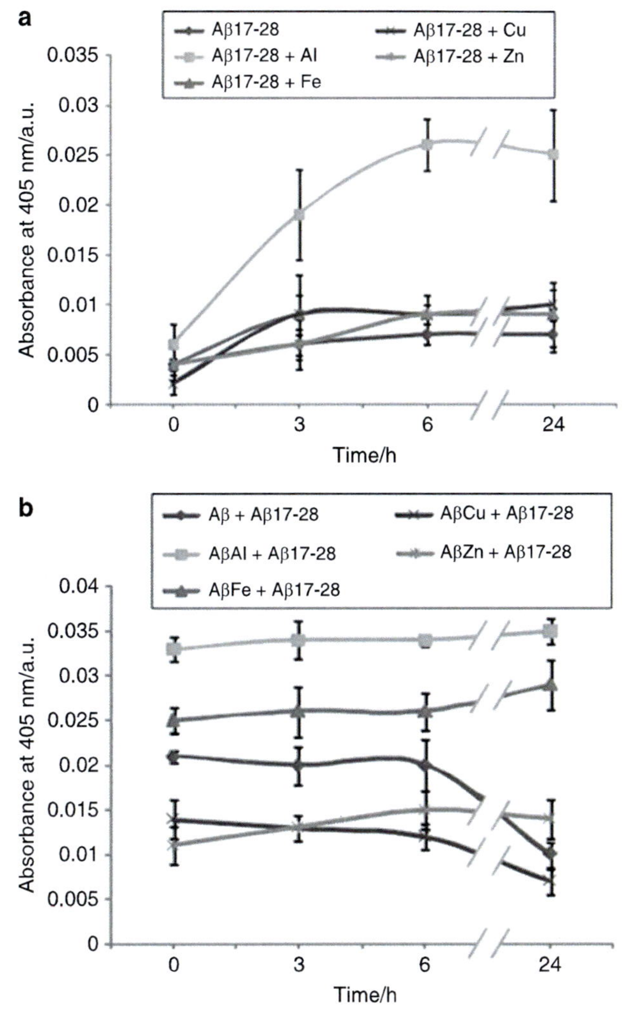

Fig. 4 Turbidity kinetic assay of $A\beta_{17-28}$ performed in both the absence and the presence of Al, Cu, Fe, and Zn at a concentration of 5 μM, and of $A\beta_{17-28}$ performed in the presence of $A\beta_{1-42}$–metal complexes. The $A\beta_{17-28}$, $A\beta_{1-42}$, and $A\beta_{1-42}$–metal complex peptide concentrations were 5 μM. Turbidity was measured at 405 nm. The data represented are the mean ± SD of three individual experiments. All readings were corrected for background absorbance

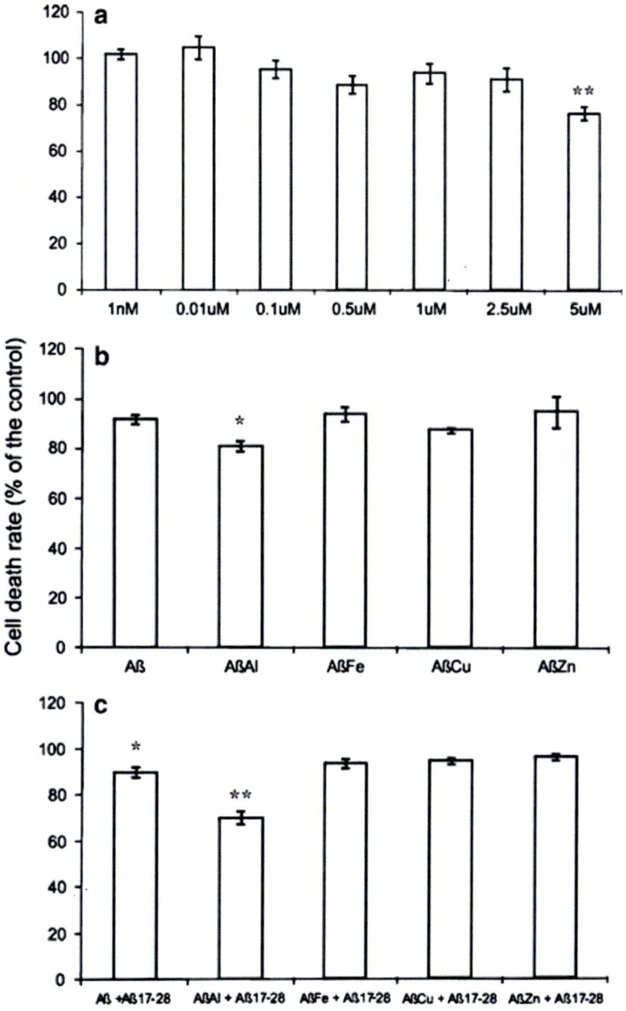

Fig. 5 Viability of human neuroblastoma cells measured by MTT assay. The dependence of neurotoxicity (% cell death as compared with control) on the concentration of $A\beta_{17-28}$ is shown (**a**). SH-SY5Y cells were incubated for 24 h with $A\beta_{1-42}$ alone, $A\beta_{1-42}$–metal complexes (peptide concentration = 0.5 μM), and with (**b**) or without (**c**) $A\beta_{17-28}$ (0.1 μM). The peptide mixture was allowed to incubate at room temperature for 24 h so that the peptides could interact before they were added to the cell medium. The data represented are the mean ± SD of three individual experiments, each done in triplicate. $^{**}P < 0.01$ versus control (chart B reprinted with permission from [23])

both fibril elongation and nucleation. When the peptide was present the type of aggregate formed seemed to be more regular, but in the case of $A\beta_{1-42}$–Al a lower rate of aggregation compared with the other treatments was still maintained after 24 h.

Toxicity to cell culture

We initially examined the concentration dependence of the toxicity of $A\beta_{17-28}$. SHSY5Y were treated with $A\beta_{17-28}$ in a range of concentrations between 0.01 and 5 μM (see Fig. 5).

The toxicity to SHSY5Y was evaluated by a standard MTT assay. We found that the fragment concentration that inhibits 50 % (IC_{50}) of the cell viability was higher than 5 μM. The other concentrations tested showed no or minimal toxicity. The first nontoxic concentration was 0.1 μM. $A\beta_{17-28}$–metal complexes showed a negligible effect. We also tested the toxicity of metal ions alone (at a concentration of 5 μM) to exclude the possibility that the effects observed were due to the metal itself rather than the complex; we did not observe any toxicity.

Then, to establish a structure–activity relationship, we also determined the effect of $A\beta_{17-28}$–metal complexes (prepared by incubating $A\beta_{17-28}$ with a 50 μM solution of metal ions at a concentration ratio of 1:1). We had previously demonstrated [18] that the $A\beta_{1-42}$–Al complex is more effective at decreasing cell viability than $A\beta_{1-42}$ alone or the other $A\beta_{1-42}$–metal complexes. Here we showed that an interaction between $A\beta_{1-42}$–Al and $A\beta_{17-28}$ produced a new toxic species which was significantly more toxic not only when compared with the control but also when compared with $A\beta_{1-42}$–Al. Simultaneously, we observed that all the other treatments with $A\beta_{1-42}$–metals + $A\beta_{17-28}$ showed a modest increase in cell toxicity compared with $A\beta_{1-42}$–metal complexes alone. The latter statement holds true except for $A\beta_{1-42}$–Cu + $A\beta_{17-28}$; its toxicity was significantly lower when compared with $A\beta_{1-42}$–Cu alone.

Consistent with the MTT data, morphological changes resulted in a reduction in the cellular body, neuritis, and neuronal cell number. We can exclude the idea that the toxicity was due merely to the activation of apoptotic pathways, because after 24 h of treatment, caspases 3 and 6 were not activated (data not shown). This implies that $A\beta_{1-42}$ and its metal complexes use a different mechanism of toxicity.

Scanning electron microscopy

Scanning electron microscopy (SEM) can give us a qualitative idea of toxicity of the treatment. We thus consider SEM images to be complementary data to the data gained through MTT cell viability assay.

In agreement with the MTT assay, the treatment with $A\beta_{17-28}$ did not show any differences in cell morphology compared with the control, thus confirming that $A\beta_{17-28}$ alone is not toxic.

As stated by several authors [19–21], the toxicity of $A\beta_{1-42}$ could originate from bilayer disruption and cell leakage. This statement is qualitatively confirmed by Fig. 6.

The cell membrane appears deeply damaged, but not homogeneously. Cell leakage is more evident in cells treated with $A\beta_{1-42}$–Al + $A\beta_{17-28}$ and $A\beta_{1-42}$–Cu + $A\beta_{17-28}$, while we do not observe significant variations in the presence of the other complexes, which matches with the MTT data.

Discussion

Alzheimer's disease, as well as other neurodegenerative diseases, could be classified as "conformational disorders" in that they are all characterized by a conformational rearrangement of specific proteins to form insoluble aggregates that deposit in the tissues [22].

It has been previously demonstrated by this laboratory that the propensity of $A\beta$ to aggregate can increase in the presence of some metal ions, and this aggregation led to an increase of in vitro toxicity to neuroblastoma cell cultures, especially when bound to Al^{3+}, forming the complex $A\beta$–Al [23]. To further understand the intriguing interrelationship

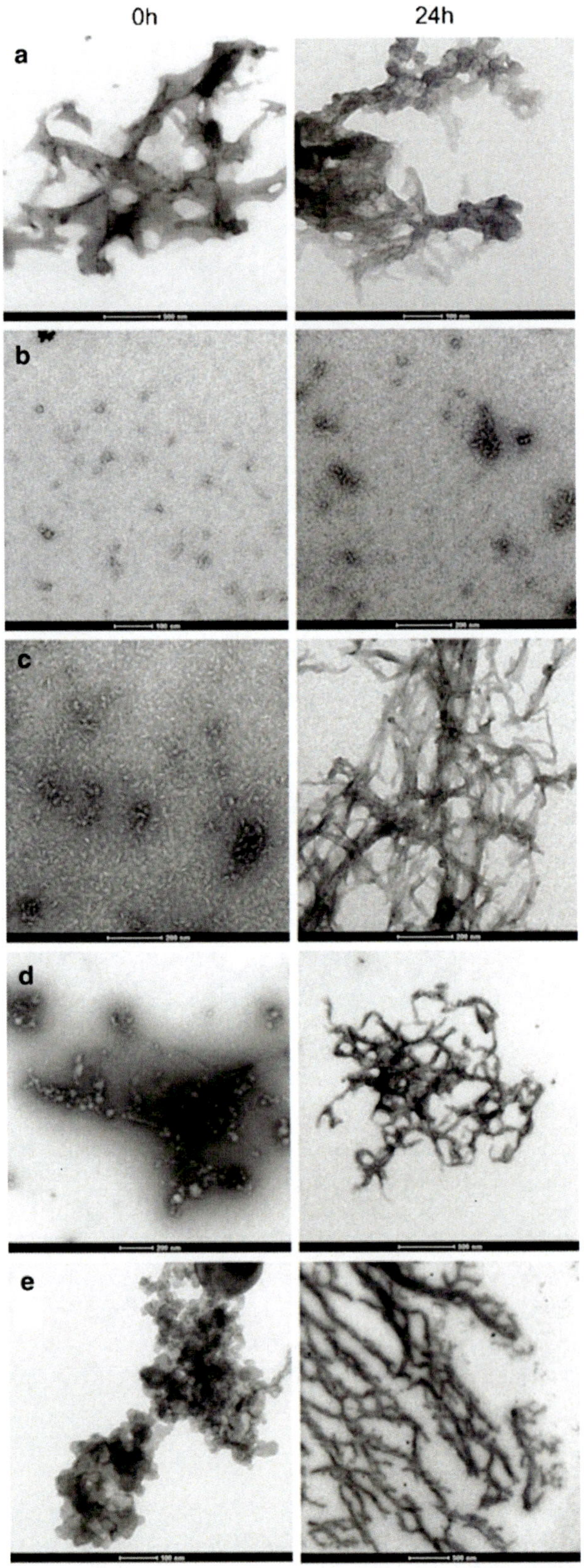

Fig. 6 TEM micrographs of $A\beta_{1-42}$ and $A\beta_{1-42}$–metal complexes in the presence of $A\beta_{17-28}$ at time 0 and after 24 h of incubation at room temperature. **(a)** $A\beta_{1-42}$ + $A\beta_{17-28}$, **(b)** $A\beta_{1-42}$–Al + $A\beta_{17-28}$, **(c)** $A\beta_{1-42}$–Cu + $A\beta_{17-28}$, **(d)** $A\beta_{1-42}$–Fe + $A\beta_{17-28}$, **e** $A\beta_{1-42}$–Zn + $A\beta_{17-28}$

between Aβ and metal ions, the interaction between $A\beta_{1-42}$ and the $A\beta_{17-28}$ truncated fragment was studied.

This $A\beta_{1-42}$ truncated fragment was chosen for several reasons: it contains one of the most lipophilic parts of the $A\beta_{1-42}$ amino acid sequence but is also not toxic at the concentrations used; it is not a physiological fragment that could be involved in biochemical pathways, and it has the ability to stimulate the exposure of hydrophobic clusters without significantly affecting $A\beta_{1-42}$ and $A\beta_{1-42}$–metal complex oligomerization and fibrillization processes (as demonstrated by ThT assay and TEM micrographs).

Since a direct correlation between Aβ toxicity and hydrophobicity has been demonstrated [25], we investigated the role of one of the Aβ's most hydrophobic sequences in influencing Aβ and Aβ–metal complex aggregational and toxic properties [24, 26, 27]. In spite of the fact that $A\beta_{17-28}$ is not physiologically relevant in AD, such a peptide is used as a case study for the specific purpose of emphasizing how hydrophobicity plays a crucial role in $A\beta_{1-42}$ toxicity; furthermore, this toxicity is increased when the peptide is complexed with metal ions, particularly Al^{3+}.

For both $A\beta_{1-42}$ [23] and $A\beta_{17-28}$, binding to Al resulted in an increase in superficial hydrophobicity and aggregation, as shown by ANS and ThT fluorescence measurements, respectively. In the presence of the $A\beta_{17-28}$–Al complex, the values of ThT fluorescence significantly increase with time (Fig. 1) compared with $A\beta_{17-28}$ alone and the other $A\beta_{17-28}$–metal complexes. In addition, it has been demonstrated that Al^{3+} has the ability to considerably enhance the exposure of the $A\beta_{17-28}$ hydrophobic cluster, as can be seen in Fig. 2. Also, the turbidity assay confirmed that Al greatly stimulates $A\beta_{17-28}$ aggregation; in fact, an increase in absorbance, as reported in Fig. 3, is due to the presence of an increased number of aggregates in solution.

The peculiar structural conformation of the $A\beta_{1-42}$–Al complex resulted in significant toxicity to the SHSY5Y cell culture, as shown in Fig. 5b. This result, according to early studies on ThT dye [28], may seem contradictory because the nonfibrillar species should be more toxic than the fibrillar ones [4]. Recent investigations, however, have reported that ThT binds to Aβ protofibrils [15] but—surprisingly—also to soluble oligomers [14]. For these reasons, we can hypothesize that the increased fluorescence of ThT was due to the presence of a large amount of oligomeric species in solution, as also confirmed by TEM micrographs (Fig. 4).

After preliminary analysis of $A\beta_{17-28}$ in the presence of metal ions, the interaction between the truncated fragment and $A\beta_{1-42}$–metal complexes was studied. Our data suggest that $A\beta_{17-28}$ enhances $A\beta_{1-42}$–Al oligomerization, as is apparent from ThT fluorescence (Fig. 1c) [14] and TEM images (Fig. 4). $A\beta_{1-42}$–Al + $A\beta_{17-28}$ oligomers are characterized by the exposure of hydrophobic residues, as revealed by the marked increased in ANS fluorescence compared with the other $A\beta_{1-42}$–metal complexes.

Although Al is an exogenous metal ion, it has been demonstrated to possess the capacity to strongly influence Aβ conformational changes, especially when compared with endogenous biometals such as Fe, Cu, and Zn.

It is worth noting that $A\beta_{17-28}$ stimulates $A\beta_{1-42}$–metal complex (with Fe, Cu, and Zn) fibrillization rather than oligomerization, in contrast to what was seen for $A\beta_{1-42}$–Al; this was primarily detected by electron microscopy. Accordingly, we should observe a dramatic increase in ThT fluorescence due to $A\beta_{1-42}$–metal ions + $A\beta_{17-28}$ aggregation. However, the increase of fluorescence was actually negligible. We can thus speculate that $A\beta_{1-42}$–metal complexes + $A\beta_{17-28}$ fibrils precipitated due to their high molecular weight and lipophilicity. In fact, we did not observe changes, even after mechanical stirring (data not shown).

Recent reports have highlighted the critical role played by Aβ soluble oligomers [2, 29, 30]. Coherently, we observed a high cellular toxicity upon $A\beta_{1-42}$–Al + $A\beta_{17-28}$ treatment, with this treatment being the only complex that formed oligomers in our experimental condition, while in the presence of the other $A\beta_{1-42}$–metal complexes + $A\beta_{17-28}$ we observed mainly insoluble fibrillary structures. The toxicity of $A\beta_{1-42}$–Al + $A\beta_{17-28}$ is significant not only when compared to that of the control, but even more so when compared with that of $A\beta_{1-42}$–Al alone. It should be remembered that we initially used a nontoxic $A\beta_{17-28}$ concentration. It is possible to establish that the toxicity of $A\beta_{1-42}$–Al + $A\beta_{17-28}$ was due to the increased hydrophobicity that characterizes these models.

It is still not yet completely clear how Aβ induces toxicity [31]. Nevertheless, several mechanisms of neurotoxicity have been proposed, such as a capacity to induce mitochondrial apoptotic pathways [32], pro-oxidant effects [33], and the ability to induce apoptosis through the activation of caspases [34]. Besides these complex apoptotic mechanisms, it has been proposed that $A\beta_{1-42}$ and its metal complexes (especially $A\beta_{1-42}$–Al) could interact significantly with cell lipid bilayers [20], perturbing membrane fluidity and thus their physiological properties, leading to a generalized increase in bilayer permeability which could induce cell death [35, 36]. Accordingly, it is not surprising that $A\beta_{1-42}$–Al produces the most relevant alteration; this is probably due to its elevated superficial hydrophobicity (as confirmed by ANS fluorescence) and its stable oligomeric structure (as confirmed by TEM). From these observations we can deduce that the enhanced toxicity induced by $A\beta_{1-42}$–Al + $A\beta_{17-28}$ is due to its increased superficial hydrophobicity, which could act as "shrapnel." In fact, our results highlight how the neurotoxicity is directly related to the $A\beta_{1-42}$–metal + $A\beta_{17-28}$ complex superficial hydrophobicity (except for $A\beta_{1-42}$–Zn + $A\beta_{17-28}$, which has a propensity to precipitate). In our opinion, the surface hydrophobicity is key to understanding the toxicity of $A\beta_{1-42}$ for two reasons: first, because several reports highlight the role played by $A\beta$ in the disruption of membrane fluidity [20, 37]; second, we can assume that the exposure of hydrophobic clusters by $A\beta_{1-42}$ and by its metal complexes facilitates the interaction with the lipophilic cellular bilayer, bringing $A\beta_{1-42}$ oligomers from a higher to a lower protein energy state [38].

In conclusion, data reported herein underline the key role played by superficial hydrophobicity in modulating $A\beta_{1-42}$–metal complex toxicity. In particular, the MTT assay indicated that cellular toxicity can be enhanced in the presence of several metal ions, but the metals play different roles. In fact, different morphological structures can be observed with different $A\beta$–metal ion complexes, depending on the amino acid residues that coordinate the different metal ions [38]. Meanwhile, Al increases the formation of low-dimensional and highly hydrophobic aggregates (such as oligomers, as shown by TEM); the other metal ions favor the formation of large, amorphous aggregates. This paper has not focused only on the $A\beta_{1-42}$ aggregation pathway in the presence of metal ions; the main focus has been on the importance of superficial hydrophobicity as a crucial feature to discern whether different $A\beta_{1-42}$ species have a greater or lesser ability to cause neuronal toxicity. In our opinion, $A\beta_{1-42}$–metal + $A\beta_{17-28}$ neurotoxicity is not merely due to oxidative stress mechanisms mediated by the presence of metal ions; indeed, the toxicity of $A\beta_{1-42}$–Al + $A\beta_{17-28}$ is greater than those of both the control and $A\beta_{1-42}$–Al alone.

Materials and methods

Materials

Human β-amyloid 1–42 was purchased from Invitrogen. β-amyloid truncated fragment 17–28, thioflavin T (ThT), ANS,L-lactic acid aluminum salt, $FeCl_3$, $CuCl_2$, $ZnCl_2$, and 3-(4,5-dimethylthiazol-2-yl)-2,5-diphenyltetrazolium bromide (MTT) were purchased from Sigma-Aldrich (St. Louis, MO, USA). Experiments with $CuCl_2$ were carried out in PBS pH 7.4 buffer, while all other experiments were developed in 0.1 M Tris/HCl pH 7.4 buffer plus 0.15 M NaCl (standard medium).

Preparation of $A\beta$–metal complexes

Human $A\beta_{1-42}$ was dissolved in hexafluoroisopropanol (HFIP) for 40 min at room temperature and then separated into aliquots. HFIP was removed under vacuum in a Speed Vac (Sc110 Savant Instruments). This treatment was repeated three times (modified protocol from Dahlgren et al. [11]). $A\beta$ fragment 17–28 (1 mg) was dissolved in 2 cm^3 of HFIP for 3 h at room temperature and then separated into aliquots. The solvent was removed under vacuum as done for human $A\beta_{1-42}$; this treatment was repeated twice more with a latency period in HFIP of 40 min. The $A\beta_{1-42}$–metal complexes were prepared by 24 h of dialysis against 10 mM metal solutions ([$CH_3CH(OH)COO]_3Al$, $FeCl_3$, $CuCl_2$, $ZnCl_2$) at $T = 4$ °C using Spectra/Por® Float-A-Lyser® tubes (Spectrum Labs) with a molecular weight cut-off (MWCO) of 1000. Then, $A\beta_{1-42}$ metal complexes were dialyzed against distilled water (three water changes) for 24 h to remove the excess of metals. The same treatment was also performed with $A\beta$ alone [12]. Aliquots of $A\beta_{1-42}$, $A\beta_{1-42}$–metal complexes, and $A\beta_{17-28}$ were stored at −20 °C until used.

Fluorescence measurements

Fluorescence measurements were performed with a PerkinElmer LS 50B spectrofluorimeter equipped with a thermostatic cell holder and magnetic stirring. The experiments were carried out at 25 °C. Fluorescence tests with ThT (12 μM) were developed on solutions containing 5 μM $A\beta$ fragment 17–28 and the metal ions Al^{3+}, Fe^{3+}, Cu^{2+}, and Zn^{2+} (1:1 ratio), and on solutions containing 5 μM human $A\beta_{1-42}$ alone or complexed with Al^{3+}, Fe^{3+}, Cu^{2+}, and Zn^{2+} along with 5 μM $A\beta$ fragment 17–28. Development was followed for 24 h by monitoring the increase in the fluorescence intensity at 482 nm with excitation at 450 nm.

Turbidity measurements

Turbidity assays were performed in 300 mm^3 96-well plates (Falcon). The absorbance of all samples was measured at 405 nm using a Microplate SPECTRAmax®. The solutions were stirred for 25 s before reading to suspend the aggregates. The absorbance due to the buffer alone was subtracted from that of the buffer plus metal ions. Turbidity measurements were carried out to quantify the presence of aggregates due to the interaction between human $A\beta_{1-42}$ and its truncated fragment, and between $A\beta_{17-28}$ and metal ions (Al^{3+}, Fe^{3+}, Cu^{2+}, and Zn^{2+}).

Transmission electron microscopy

All samples at a protein concentration of 10 μM, and after an incubation period of 24 h, were absorbed onto glow-discharged carbon-coated Butwar films on 400-mesh copper grids. The grids were negatively stained with 1 % uranyl acetate and observed at 40,000× by transmission electron microscopy (Tecnai G2, FEI). The samples observed contained Aβ_{1-42} and its metal complexes with Aβ_{17-28} (1:1 ratio) or Aβ_{17-28} in solution with Al^{3+}, Fe^{3+}, Cu^{2+}, and Zn^{2+} (concentration ratio 1:1).

Neuroblastoma cells

SH-SY5Y human neuroblastoma cells were purchased from the ECACC (European Collection of Cell Culture, Salisbury, UK). The medium in which they were cultured contained DMEM/F12 (Gibco, Carlsbad, CA, USA) with 15 % (v/v) fetal bovine serum (FBS, Sigma-Aldrich, St. Louis, MO, USA), 100 units/cm^3 penicillin and 100 μg/cm^3 streptomycin (Gibco, Carlsbad, CA, USA) and 1 % (v/v) MEM nonessential amino acid (NEAA) (Sigma-Aldrich, St. Louis, MO, USA). Cells were stored at 37 °C with 5 % CO_2 in a humidified atmosphere (90 % humidity). Cells were used until passage 35. The culture medium was replaced every 2 days.

Cell viability assay

Cell viability was determined through an MTT reduction assay. SH-SY5Y cells were seeded into 24-well plates at a density of 7×10^4 cells per well in 1 cm^3 culture medium. An FBS culture medium (2 %) containing (1) Aβ_{1-42}, (2) Aβ_{1-42}–metal complexes with or without Aβ_{17-28}, (3) Aβ_{17-28} with or without metals, or (4) metal ions alone (Al^{3+}, Fe^{3+}, Cu^{2+}, and Zn^{2+}) was added to the cells for 24 h. MTT (100 mm^3, 5 mg/cm^3) was added to each well and incubated in the dark at 37 °C for 3 h. After that, the cells were lysed with 1 cm^3 of acidic isopropanol (0.04 M HCl in absolute isopropanol) [13]. Color intensity was measured with a 96-well ELISA plate reader at 550 nm (Microplate SPECTRAmax®). All MTT assays were performed three times in triplicate. Viability was defined as the relative absorbance of the treated versus the untreated, expressed as a percentage.

Scanning electron microscopy of human neuroblastoma cells

SH-SY5Y cells were seeded onto glass cover slips and treated with Aβ_{1-42} and Aβ_{1-42}–metal complexes with or without the interaction with Aβ_{17-28}. After 24 h of incubation, the cells on glass cover slips were fixed with formaldehyde pH 7.4 and dehydrated in a graded ethanol series. Then the samples were critical point dried with CO_2 in a HCP- 2 Hitachi 2 Critical Point Dryer and gold-coated for examination under a JEDL JSM-6490 scanning electron microscope. The working pressure was 4.2–4.3 bar and the temperature was 5 °C. Untreated cells (control) were also examined for comparison.

Statistical analysis

MTT, turbidity, and ThT fluorescence assays were statistically analyzed by ANOVA followed by the Student–Newman–Keuls *t* test as a post hoc test. Results were reported to be highly statistically significant if $P < 0.01$ and statistically significant if $P < 0.05$. Results are presented as mean ± standard deviation.

Acknowledgment This work was supported by PRIN 2007.

References

1. Iqbal K, Liu F, Gong CX, Alonso AD, Grundke-Iqbal I (2009) Acta Neuropathol 118:53
2. Ono K, Condron MM, Teplow DB (2009) Proc Natl Acad Sci USA 106:14745
3. Savva GM, Wharton SB, Ince PG, Forster G, Matthews FE, Brayne C (2009) N Engl J Med 360:2302
4. Nygaard HB, Strittmatter SM (2009) AMA Arch Neurol 66:1325
5. Zatta P, Drago D, Bolognin S, Sensi SL (2009) Trends Pharmacol Sci 30:346
6. Lovell MA, Robertson JD, Teesdale WJ, Campbell JL, Markesbery WR (1998) J Neurol Sci 158:47
7. Miller LM, Wang Q, Telivala TP, Smith RJ, Lanzirotti A, Miklossy J (2006) J Struct Biol 155:30
8. Leskovjan AC, Lanzirotti A, Miller LM (2009) Neuroimage 47:1215
9. Zbilut JP, Webber CL, Colosimo A, Giuliani A (2000) Protein Eng 13:99
10. Kim W, Hecht MH (2006) Proc Natl Acad Sci USA 103:15824
11. Dahlgren KN, Manelli AM, Stine WB, Baker LK, Krafft GA, LaDu MJ (2002) J Biol Chem 277:32046
12. Drago D, Folin M, Baiguera S, Tognon G, Ricchelli F, Zatta P (2007) J Alzheimers Dis 11:33
13. Shearman MS, Hawtin SR, Tailor VJ (1995) J Neurochem 65:218
14. Maezawa I, Hong HS, Liu R, Wu CY, Cheng RH, Kung MP, Kung HF, Lam KS, Oddo S, LaFerla FM, Jin LW (2008) J Neurochem 104:457
15. Ferreira ST, Vieira MNN, De Felice FG (2007) IUBMB Life 59:332
16. Naiki H, Gejyo F, Nakakuki K (1997) Biochemistry 36:6243
17. Uversky VN, Winter S, Lober G (1996) Biophys Chem 60:79
18. Drago D, Bolognin S, Zatta P (2008) Curr Alzheimer Res 5:500
19. Friedman R, Pellarin R, Caflisch A (2009) J Mol Biol 387:407
20. Suwalsky M, Bolognin S, Zatta P (2009) J Alzheimers Dis 17:81
21. Yankner BA, Lu T (2009) J Biol Chem 284:4754
22. Kopito RR, Ron D (2000) Nat Cell Biol 2:E207
23. Drago D, Bettella M, Bolognin S, Cendron L, Scancar J, Milacic R, Ricchelli F, Casini A, Messori L, Tognon G, Zatta P (2008) Int J Biochem Cell B 40:731
24. Millucci L, Ghezzi L, Bernardini G, Santucci A (2010) Curr Protein Pep Sc 11:457

25. Atamna H (2009) J Bioenerg Biomembr 41:457
26. Tjernberg LO, Callaway DJE, Tjernberg A, Hahne S, Lilliehook C, Terenius L, Thyberg J, Nordstedt C (1999) J Biol Chem 274:12619
27. Melquiond A, Dong X, Mousseau N, Derreumaux P (2008) Curr Alzheimer Res 5:244
28. Levine H (1993) Protein Sci 2:404
29. Xue WF, Hellewell AL, Gosal WS, Homans SW, Hewitt EW, Radford SE (2009) J Biol Chem 284:34272
30. Zhang A, Qi W, Good TA, Fernandez EJ (2009) Biophys J 96:1091
31. Shah SB, Nolan R, Davis E, Stokin GB, Niesman I, Canto I, Glabe C, Goldstein LSB (2009) Neurobiol Dis 36:11
32. Deshpande A, Mina E, Glabe C, Busciglio J (2006) J Neurosci 26:6011
33. Behl C, Davis JB, Lesley R, Schubert D (1994) Cell 77:817
34. Nakagawa T, Zhu H, Morishima N, Li E, Xu J, Yankner BA, Yuan JY (2000) Nature 403:98
35. Kremer JJ, Pallitto MM, Sklansky DJ, Murphy RM (2000) Biochemistry 39:10309
36. Lashuel HA, Hartley D, Petre BM, Walz T, Lansbury PT (2002) Nature 418:291
37. Eckert GP, Wood WG, Muller WE (2005) Subcell Biochem 38:319
38. Miller Y, Ma B, Nussinov R (2010) Chem Rev 110:4820

Aluminium in the human brain

Christopher Exley and Emily R. House

Abstract

An inevitable consequence of humans living in the Aluminium Age is the presence of aluminium in the brain. This non-essential, neurotoxic metal gains entry to the brain throughout all stages of human development, from the foetus through to old age. Human exposure to myriad forms of this ubiquitous and omnipresent metal makes its presence in the brain inevitable, while the structure and physiology of the brain makes it particularly susceptible to the accumulation of aluminium with age. In spite of aluminium's complete lack of biological essentiality, it actually participates avidly in brain biochemistry and substitutes for essential metals in critical biochemical processes. The degree to which such substitutions are disruptive and are manifested as biological effects will depend upon the biological availability of aluminium in any particular physical or chemical compartment, and will under all circumstances be exerting an energy load on the brain. In short, the brain must expend energy in its 'unconscious' response to an exposure to biologically available aluminium. There are many examples where 'biological effect' has resulted in aluminium-induced neurotoxicity and most potently in conditions that have resulted in an aluminium-associated encephalopathy. However, since aluminium is non-essential and not required by the brain, its biological availability will only rarely achieve such levels of acuity, and it is more pertinent to consider and investigate the brain's response to much lower though sustained levels of biologically reactive aluminium. This is the level of exposure that defines the putative role of aluminium in chronic neurodegenerative disease and, though thoroughly investigated in numerous animal models, the chronic toxicity of aluminium has yet to be addressed experimentally in humans. A feasible test of the 'aluminium hypothesis', whereby aluminium in the human brain is implicated in chronic neurodegenerative disease, would be to reduce the brain's aluminium load to the lowest possible level by non-invasive means. The simplest way that this aim can be fulfilled in a significant and relevant population is by facilitating the urinary excretion of aluminium through the regular drinking of a silicic acid-rich mineral water over an extended time period. This will lower the body and brain burden of aluminium, and by doing so will test whether brain aluminium contributes significantly to chronic neurodegenerative diseases such as Alzheimer's and Parkinson's.

Keywords

Metal • Neurodegenerative disease • Neuropathology • Neurotoxicity • Alzheimer's disease • Parkinson's disease

C. Exley (✉)
The Birchall Centre, Lennard-Jones Laboratories, Keele University,
ST5 5BG Staffordshire, UK
e-mail: c.exley@chem.keele.ac.uk

W. Linert and H. Kozlowski (eds.), *Metal Ions in Neurological Systems*,
DOI 10.1007/978-3-7091-1001-0_9,

Aluminium is present in the human brain

While aluminium is present throughout human brain tissue, it is a point of debate as to what level of brain aluminium constitutes a 'normal' amount. Since aluminium is non-essential and is not known to be beneficial [1], then its presence in brain tissue, at any level, could be construed as abnormal. There are no 'normal' levels of brain aluminium, only levels that equate with an individual's age, their state of health and, perhaps, the presence of certain diseases. Quantitative values of brain aluminium in tissue homogenates have been determined using a range of analytical techniques and have been expressed numerically in a number of different ways [2–29]. It is not always straightforward to decide which of such a range of measurements should be interpreted as absolute values and which would be better understood as relative determinations. Of paramount importance to the confidence that is placed in absolute measurements are their provenance, a complete record of their natural history from sampling through to their analysis and statistical significance. The data must be supported by quality assurance programmes that have taken account of issues with tissue sampling and their contamination from extraneous sources and that also demonstrate the precision and sensitivity of the chosen analytical method. Finally, the analysis and interpretation of high quality data of tissue aluminium levels must take into account that aluminium is unlikely to be distributed homogeneously in brain tissue, and so computations of mean or median values either should be avoided or should be based upon as many replicates as is practical. Despite the problems associated with accurate and reproducible measurements of brain aluminium in tissue homogenates, the scientific literature has over several decades consistently recorded values for 'normal' brains as being in the range 0.1–4.5 μg Al/g tissue (dry wt) with the higher values (>2.0) being measured in brains taken from the non-demented elderly. There is a clear trend for an increase in brain aluminium content with age [23]. Brain aluminium is also increased in a number of disease states including: Alzheimer's disease (up to 11.5 μg/g dry wt); dialysis encephalopathy (up to 14.1 μg/g dry wt); congophilic amyloid angiopathy (up to 23.0 μg/g dry wt); and various aluminium-related encephalopathies (up to 47.4 μg/g dry wt). Semi-quantitative evidence has also demonstrated elevated levels of aluminium in: neurones and neurofibrillary tangles in Alzheimer's disease [30–33] and dementia pugilistica [34]; Lewy bodies in Parkinson's disease [35]; lipofuscin in Alzheimer's disease and aged brains [36]; senile plaques in Alzheimer's disease [37]; pathological lesions in Fahr's and Behcet's disease [38, 39]; myelin in progressive leukoencephalopathy [40]; and neurones, glia and the choroid epithelium in aluminium-induced encephalopathies [41, 42]. Aluminium can also be identified in brain tissue using histochemistry and light microscopy (Fig. 1).

Where is aluminium in the brain?

The quantitative data of the presence and content of aluminium in human brain tissue are complemented by detailed studies of its multifarious locations. There are probably five or six major sinks or compartments for aluminium, and their aluminium loads are likely to be in some form of equilibrium (fast, slow, etc.) with one another. For example, the major compartments and their likely constituent sinks/sources for aluminium would include: (1) the blood–brain barrier including endothelia, choroid epithelia, cerebrospinal fluid, pericytes and the basal laminas; (2) the brain interstitial fluid including, proteins (transferrin, albumin); neurotransmitters (glutamate, gamma amino butyric acid, acetylcholine, dopamine); nucleotides (ATP, ADP, AMP); amino acids (aspartate, serine, tyrosine); small organic anions (citrate, lactate); (3) non-neuronal cells (astrocytes, oligodendrocytes, microglia, mononuclear migratory cells); (4) neurones; and (5) pathological features (senile plaques, neurofibrillary tangles, Lewy bodies, lipofuscin). Subcellular compartments for aluminium would include nuclei, mitochondria, liposomes, ferritin, cytosolic pools of citrate and ATP, and neuronal stores of neurotransmitters such as glutamate (Fig. 2). The experimental evidence for the distribution of aluminium in the brain supports a higher content of aluminium in gray (e.g. 0.40 μg/g dry wt) as opposed to white (e.g. 0.34 μg/g dry wt) matter [4, 16, 23], and especially so in brains affected by an aluminium-induced encephalopathy (e.g. 8.72 and 0.75 μg/g dry wt for gray and white matter, respectively [24]). The preferential accumulation of aluminium in gray matter is supported by many studies that have demonstrated high levels of aluminium in neuronal bodies and often specifically neuronal nuclei [9, 12, 13, 19, 20, 24, 30, 32, 33, 42]. High levels of aluminium have also been identified in glia, again specifically in liposomes and in nuclei [20, 24, 35, 42]. The choroid epithelia and adjacent supporting structures of the blood–brain barrier as well as the oligodendrocytes that constitute the myelin sheath of axons are all sites of significant focal accumulations of aluminium [20, 24, 38, 42]. The strong association of aluminium with the blood–brain barrier (up to ca. 50 μg/g dry wt) most probably reflects this structure's role as aluminium's main point of both entry into and exit out of the brain, whereas that with myelin is indicative of a significant chemical affinity between it and aluminium [43]. There are also good chemical reasons for the co-localisation of aluminium with the neuropathological features, senile plaques, neurofibrillary tangles, Lewy bodies and lipofuscin

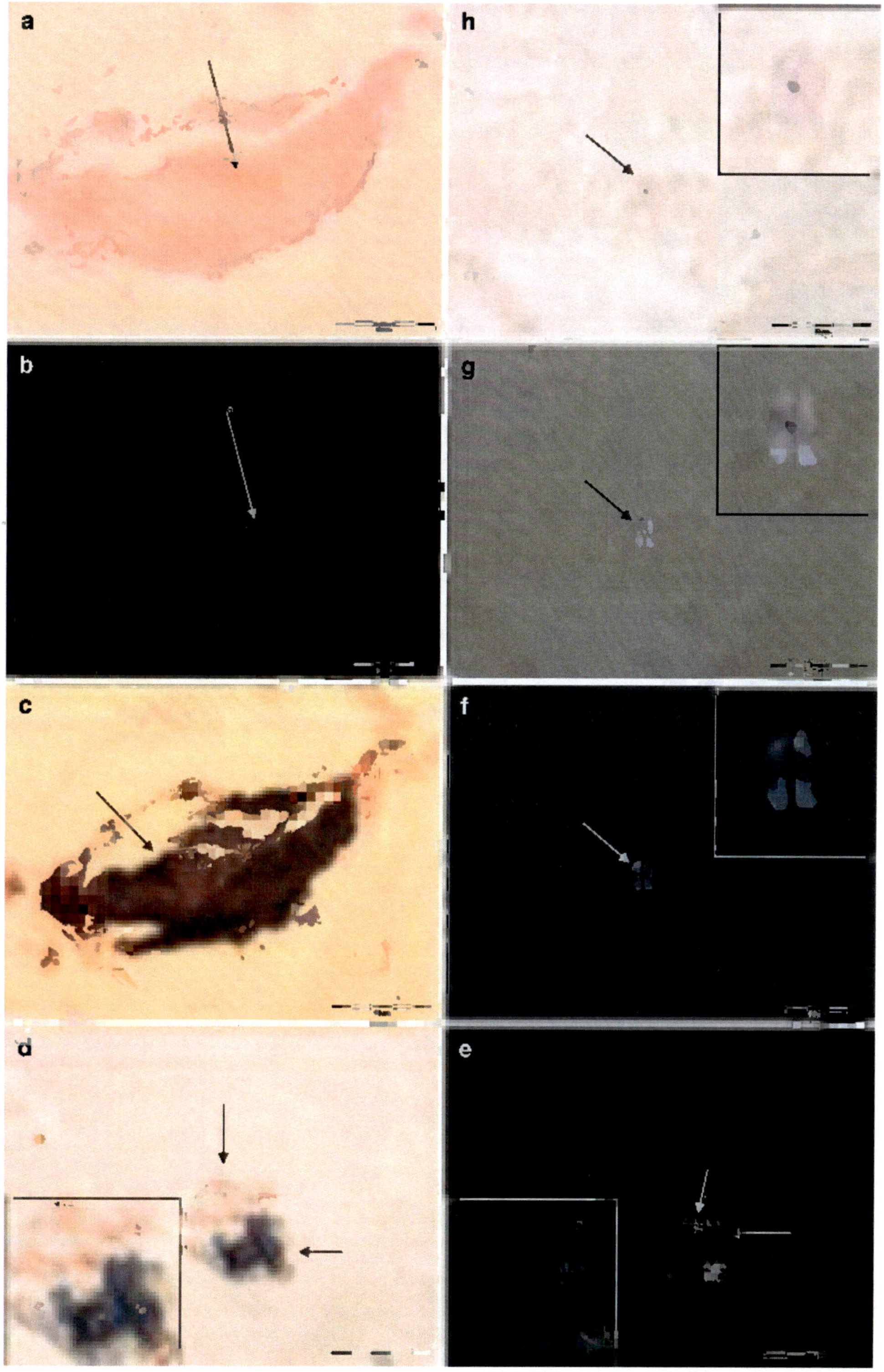

Fig. 1 The identification of aluminium in human brain tissue using histochemical methods and light microscopy. **(a)** Section of temporal lobe: *arrow* shows a senile plaque-like structure stained for amyloid using Congo red. **(b)** The same section under polarised light showing (*arrow*) apple-green birefringence characteristic of β-sheets of amyloid. **(c)** The same section stained for aluminium (*arrow*) using modified haematoxylin. The *scale bar* for **(a–c)** is 50 µm. **(d)** Section of temporal lobe showing a senile plaque-like structure that has been stained for both amyloid and aluminium (indicated by *arrows* and *inset*). **(e)** The same section under polarised light showing apple-green birefringence and spherulites (*arrows* and *inset*) [69]. The *scale bar* for **(d, e)** is 100 µm. **(h)** Section of temporal lobe stained using modified haematoxylin and showing an amyloid spherulite positively stained for aluminium (*arrow* and *inset*). The same section under partial **(g)** and full **(f)** crossed polarisers shows (*arrows* and *insets*) the spherulitic signature of a Maltese cross pattern of light extinction and that the core of the spherulite stains positively for aluminium. The *scale bar* for **(f, g)** is 50 µm

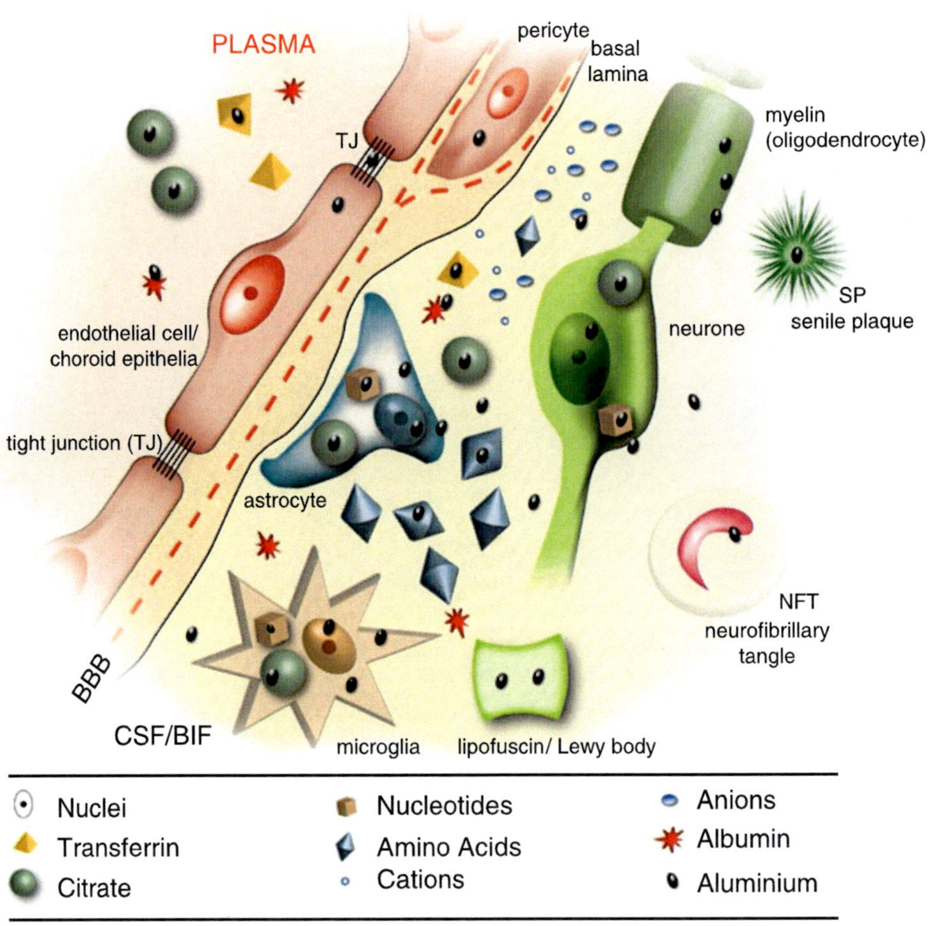

Fig. 2 A schematic of the possible distribution of aluminium in plasma, the blood–brain barrier (BBB), the cerebrospinal fluid (CSF), the brain interstitial fluid (BIF), and the cellular and pathological compartments of the human brain

as each of these have significant component parts (beta amyloid [44], tau [45], alpha synclein [46] and lipofuscin [36], respectively) with strong affinities for binding aluminium. It remains to be determined whether the presence of aluminium in these structures is also indicative of a role in their formation, as has been suggested recently for both neurofibrillary tangles [47] and senile plaques [48]. The cerebrospinal fluid and brain interstitial fluid will act as reservoirs of aluminium that are in continuous exchange with all other compartments. The 'normal' concentration of aluminium in the brain interstitial fluid is probably less than 5.0 $\mu g/dm^3$, though it has been measured at levels as high as 190 $\mu g/dm^3$ in acute aluminium intoxication [24, 42]. The major ligand for aluminium in brain interstitial fluid is probably citrate at ca. 250 μM [49], whereas other competitive ligands would include transferrin (ca. 1 μM) [49], glutamate (ca. 10 μM), pyroglutamate (ca. 180 μM) [50], and the nucleotides ATP, ADP and AMP (ca. 5 μM) [51]. All of these ligands along with insoluble phases involving hydroxide, phosphate and possibly silicic acid will participate in competitive equilibria that under kinetic control will drive and determine the fate of the brain interstitial fluid aluminium load. The neuronal microenvironment and the various cell types it supports are together potentially awash with aluminium that has gained entry to the brain either across the blood–brain barrier or via the olfactory route [52]. The known persistence of aluminium within the brain probably reflects the longevity of neurones, which are significant sinks for aluminium, while the increased permeability with age of epithelial barriers such as those of the gut, lung, olfactory system and the blood–brain barrier must contribute towards its accumulation over lifetimes. However, neither presence nor location of aluminium necessarily infers biological effect or neurotoxicity.

The neurotoxicity of aluminium

Since aluminium is not known to participate in any essential brain biochemistry, its neurotoxicity might simply be defined in terms of its biological availability. Since biological availability infers a biological response, then aluminium could be considered neurotoxic whenever its presence results in an aluminium-induced change in brain biochemistry. Quite simply, all biologically available aluminium in the brain is neurotoxic, and it is only the degree to which the toxicity is manifested as an unwanted biochemical change that defines the acuity and 'phenotype' of its neurotoxicity. Most, if not all, of the toxicity of aluminium is through the biochemical reactivity of $Al_{(aq)}^{3+}$, which is bound avidly by oxygen (and fluoride)-based ligands and functional groups [1]. It is often the rate of delivery of $Al_{(aq)}^{3+}$ to target ligands that determines the extent of its toxicity in any given system. Indeed, once aluminium has entered the brain its rate of delivery of $Al_{(aq)}^{3+}$ to target groups is the only limit upon its neurotoxicity, and so the identification of sinks/sources of biologically available aluminium is critically important in predicting its neurotoxicity. Numerous animal models have demonstrated the neurotoxicity of aluminium when the delivery of $Al_{(aq)}^{3+}$ is potentiated [53]. Similar degrees of potency are also observed in fatal aluminium-induced encephalopathies in humans, though such are probably the only examples of confirmed aluminium neurotoxicity in humans. Aluminium-induced encephalopathies are acute events in which an accelerated neuronal loss is accompanied by a miscellany of additional aberrant processes including alterations in the processing of tau protein [15] and the deposition of aluminium in neurones, glia and the choroid epithelium [12, 17, 19, 20, 24, 27, 41, 42]. In these acute episodes in which high concentrations of toxin are delivered to the brain from the blood, the disturbance of the selectivity of the blood–brain barrier [54] is probably an early event in a rapidly progressing cascade of deleterious events that culminate in neuronal necrosis. In these acute events the constancy of the neuronal microenvironment is lost and brain biochemistry is overwhelmed both directly and indirectly by the aluminium challenge. While aluminium-induced encephalopathies serve to demonstrate the acute neurotoxicity of aluminium, they are unlikely models of lower levels of exposure to brain aluminium. Under conditions of chronic exposure to aluminium, the selectivity of the blood–brain barrier may not be seriously compromised, and aluminium will gain entry to the brain by piggy-backing upon normal transport mechanisms as well as through more indirect processes such as residual leakiness and fluid phase endocytosis. Immediately upon entry $Al_{(aq)}^{3+}$ will be continuously shuttled between chemical and physical compartments, leading to both its removal from the brain back across the blood–brain barrier and its retention in the brain through associations with extracellular and intracellular sinks. It is important to emphasise that the fate of brain aluminium is not under any form of homeostatic control; the analogy is with a game of bagatelle or pinball with aluminium accumulating slowly where its persistence will be supported by the local chemistry [55]. Similarly it is this same chemistry that will also dictate how a sink might also act as a source of biologically available $Al_{(aq)}^{3+}$. For example, extracellular and intracellular pools of citrate will accommodate significant concentrations of $Al_{(aq)}^{3+}$, as soluble complexes, and they will also act as sources of biologically reactive $Al_{(aq)}^{3+}$, promoting, for example, the pro-oxidant activity of aluminium through formation of the putative aluminium superoxide semi-reduced radical ion, $AlO_{2(aq)}^{\cdot 2+}$ [56]. The remarkable propensity for aluminium to promote oxidative events makes such highly likely throughout the brain, with membrane lipids, nucleic acids and free radical-mediated signalling as prime targets for oxidative damage. However, the high likelihood of aluminium-induced oxidative damage must, in the main, be countered by the brain's sophisticated mechanisms of antioxidant protection. These have evolved alongside oxidative metabolism, and though they might be expected to protect against aluminium-induced oxidative damage, their upregulation will add to cellular energy requirements and hence, eventually, deficits. In addition, the protection afforded by antioxidants may be focally as opposed to universally distributed throughout the brain and so free radical damage initiated by, for example, the co-deposition of iron, aluminium and amyloid in senile plaques [57] may not be so easily countered. In summary, under conditions of chronic (everyday) exposure to aluminium, we would expect the brain to be subject to aluminium-induced oxidative stress, though probably without sustaining any short-term damage.

Extracellular and intracellular citrate pools are likely sources of $Al_{(aq)}^{3+}$ to many biochemical targets in the brain. The establishment of micromolar concentrations of aluminium within such pools will allow $Al_{(aq)}^{3+}$ to compete effectively with millimolar concentrations of competitive cations such as Mg^{2+} and Ca^{2+} and displace such essential metals from coordination sites in enzymes, signalling molecules, receptors, transporters, channels, nucleic acids and many other biochemical ligands [1]. While each of such target systems would be influenced by aluminium under the finite conditions of in vitro preparations, the inherently robust and flexible nature of in vivo physiology must dictate that only when the number of displacements reaches a particular threshold will the functioning of the system be disrupted and neurotoxicity manifested. These 'thresholds' may be reached over decades of chronic exposure, though in the interim each displacement of an essential metal by aluminium will use up some of the energy currency of the brain.

While the majority of the neurotoxicity of aluminium is predicted to emanate from the action of $Al_{(aq)}^{3+}$ at a target site, the accumulation of aluminium within the cytosolic pool of ATP might also result in neurotoxicity following the secretion of ATP into the brain interstitial fluid and the action of Al-ATP at extracellular purinergic receptors [58]. ATP is arguably the single most important extracellular signalling molecule in the brain [59], and upon being bound by specific P2X and P2Y receptors acts like a gain control on many other receptor-based signalling systems, such as the NMDA receptor complex. When Al-ATP substitutes for Mg-ATP at ATP receptors, the likely result is that the dissociation of the agonist-receptor complex will be delayed due to the enhanced stability of the Al-nucleotide complex. This will have the effect of extending the overall stimulus and hence signalling event beyond its normal lifetime. In the case of the NMDA receptor complex, this could result in a transient elevation in cytosolic [Ca^{2+}] for which the cell has to expend additional energy to bind and/or remove it from the cell cytosol [58]. The net effect of such a persistent affect of biologically available aluminium would be an underlying excitotoxicity that in time would deplete neuronal energy reserves and instigate neuronal cell death by apoptosis.

The consistent observation of aluminium in the brain associated with neuronal and glial nuclei [9, 19, 24, 30–32] must highlight nucleic acids as additional sinks for $Al_{(aq)}^{3+}$, not only in cell nuclei but also in the cytosol and in mitochondria. The phosphate-rich nuclear compartment is an obvious site for aluminium to bind and accumulate over time. Indeed, the persistent occurrence of aluminium in all cell nuclei has recently prompted speculation that aluminium through its compaction of chromatin may actually serve a biological purpose in silencing the expression of genetic information [60]. Aluminium is a powerful cross-linking agent, a property that is used in many industrial processes, including leather tanning, and its potential longevity in neuronal nuclei is envisaged to prevent or at least slow down the unravelling of DNA. Whether such a process could be under any sort of homeostatic control remains to be investigated. Certainly biologically reactive aluminium is bound by nucleic acids [61], and through such interactions could influence and modify many aspects of a neurone's genetic machinery.

The majority of the potential neurotoxicity of aluminium outlined thus far is assumed to result from the binding of $Al_{(aq)}^{3+}$ by myriad oxygen (and possibly fluoride)-based ligands. However, there is a burgeoning interest in the putative neurotoxicity of nanoparticulates, many of which are aluminium-based [62]. There is evidence that nanoparticulates of aluminium are found in the brain and that they either form in situ or that they enter the brain across the blood–brain barrier or via the olfactory system. There is currently very little understanding about the mechanisms of toxicity of aluminium-based nanoparticulates and almost no understanding of their putative neurotoxicity. However, there are well-defined precedents for the toxicity of such forms of aluminium, and in particular from data that purport to describe the activity, one could say immunotoxicity, of aluminium-based adjuvants [63]. Unfortunately there is, as yet, no consensus on whether the immunobiology of aluminium-based adjuvants is mediated through their particulate or dissolved forms. Certainly aluminium either directly as a particulate or indirectly following the dissolution of nanoparticulates could induce an inflammatory action in the human brain, and this has been demonstrated in animal models [64]. The immunopotency of aluminium-based adjuvants outside their role as adjuvants in vaccine and allergy therapies seems to have been largely ignored as a potential mechanism of aluminium toxicity throughout the body [65] and especially in the nervous system [66]. The consistent observation of significant accumulations of aluminium in the brain should at least be a warning of the potential for such to participate in neuroinflammatory toxicity.

The brain is an obvious target for aluminium toxicity. Neurotoxicity is evident under acute conditions such as encephalopathies, and it is predicted but not necessarily recognised as such under chronic or everyday exposures to environmental aluminium. The mechanisms of neurotoxicity are potentially myriad, while their manifestations as biochemical changes are probably quite subtle for all but the most vulnerable groups. While the latter must include the foetus and neonate, there are few indications as to the identities of others who are susceptible to the neurotoxicity of aluminium. Since the advent of the Hall–Héroult process (and thereafter Bayer process) towards the end of the nineteenth century and our ability to extract aluminium from its inert ores on an industrial scale, we have all been living in the Aluminium Age [67]. Now, in the twenty-first century, we can no longer completely avoid environmental exposure to aluminium. Since there is as yet no proven requirement for aluminium in any living organism, never mind in humans, it would be prudent to reduce our everyday exposure to avoid aluminium entering the body and persisting in the human brain [68]. We have begun to show that this can be achieved by using nature's own way of avoiding biologically available aluminium. We have shown that regular consumption of silicon-rich mineral waters both reduce our gastrointestinal uptake of aluminium and, importantly, facilitate our urinary excretion of systemic aluminium [48]. Life on Earth evolved in spite of a crust of aluminosilicate [1]. However, the Hall–Héroult process and the subsequent arrival of an Aluminium Age have let the aluminium genie out of the bottle. Our final wish should be that the unique inorganic chemistry of aluminium and silicic acid will help to put the genie back where it can be used effectively but, most importantly, safely.

Acknowledgments Andrew Lawrence (Keele, KUDIS) is thanked for help in preparing Fig. 2. G Forster and Professor PG Ince (Royal Hallamshire Hospital, Sheffield) are thanked for help in providing brain tissues from MRC CFAS.

References

1. Exley C (2009) Trends Biochem Sci 34:589
2. Crapper DR, Krisnan SS, Dalton AJ (1973) Science 180:511
3. Legendre GR, Alfrey AC (1976) Clin Chem 22:53
4. Freundlich M, Zilleruelo G, Abitbol C, Strauss J, Faugere M-C, Malluche HH (1985) Lancet 7:527
5. Bishop NJ, Robinson MJ, Lendon M, Hewitt CD, Day JP, O'Hara M (1989) Arch Dis Child 64:1316
6. Bozynski MEA, Sedman AB, Naglie RA, Wright EJ (1989) J Parenter Enteral Nutr 13:428

7. Van Ginkel MF, van der Voet GB, de Wolff FA (1990) Clin Chem 36:658
8. Yasui M, Yase Y, Ota K, Mukoyama M, Adachi K (1991) Neurotoxicology 12:277
9. Lukiw WJ, Krishnan B, Wong L, Kruck PA, Bergeron C, Crapper McLachlan DR (1992) Neurobiol Aging 13:115
10. Good PF, Perl DP, Bierer LM, Schmeidler J (1992) Ann Neurol 31:286
11. Xu N, Majidi V, Markesbery WR, Ehmann WD (1992) Neurotoxicology 13:735
12. Candy JM, McArthur FK, Oakley AE, Taylor GA, Chen CPL-H, Mountfort SA, Thompson JE, Chalker PR, Bishop HE, Beyreuther K, Perry G, Ward MK, Martyn CN, Edwardson JA (1992) J Neurolog Sci 107:210
13. Lovell MA, Ehmann WD, Markesbery WR (1993) Ann Neurol 33:36
14. Moreno A, Dominguez C, Ballbriga A (1994) Acta Paediatr 83:25
15. Harrington CR, Wischik CM, McArthur FK, Taylor GA, Edwardson JA, Candy JM (1994) Lancet 343:993
16. Bush VJ, Moyer TP, Batts KP, Parisi JE (1995) Clin Chem 41:284
17. Hantson P, Mahieu P, Gersdorff M, Sindic C, Lauwerys R (1995) Clin Toxicol 33:645
18. Andrási E, Farkas E, Scheibler H, Réffy A, Bezúr L (1995) Arch Gerontol Geriatr 21:89
19. Galassi G, Cappelli G, Crisi G, Botticelli AR, Lursvarghi E, Winkelmann MD, Lovell MA, Ehmann WD, Markesbery WR (1995) Trace Elem Electrolytes 12:68
20. Reusche E, Koch V, Friedrich H-J, Nünninghoff D, Stein P, Rob P-M (1996) Clin Neuropathol 15:342
21. Deibel MA, Ehmann WD, Candy JM, Ince PG, Shaw PJ, Markesbery WR (1997) Trace Elem Electrolytes 14:51
22. Beauchemin D, Kisilevsky R (1998) Anal Chem 70:1026
23. Roider G, Drasch G (1999) Trace Elem Electrolytes 16:77
24. Reusche E, Pilz P, Oberascher G, Lindner B, Egensperger R, Gloeckner K, Trinka E, Iglseder B (2001) Hum Pathol 32:1136
25. Meshitsuka S, Koeda T, Hara T, Takeshita K (2001) Dev Med Child Neurol 43:286
26. De Wolff FA, Berend K, van der Voet GB (2002) Forensic Sci Int 128:41
27. Zatta P, Zambenedetti P, Reusche E, Stellmacher F, Cester A, Albanese P, Meneghel G, Nordio M (2004) Nephrol Dial Transplant 19:2929
28. Andrási E, Páli N, Molnár Z, Kösel S (2005) J Alzheimers Dis 7:273
29. Exley C, Esiri MM (2006) J Neurol Neurosurg Psychiatry 77:877
30. Perl DP, Brody AR (1980) Science 208:297
31. Walton JR (2006) Neurotoxicology 27:385
32. Yumoto S, Horino Y, Mokuno Y, Kakimi S, Fujii K (1996) Nucl Instrum Methods Phys Res B 109/110:362
33. Solomon B, Koppel R, Jossiphov J (2001) Brain Res Bull 55:253
34. Bouras C, Giannakopoulos P, Good PF, Hsu A, Hof PR, Perl DP (1997) Eur Neurol 38:53
35. Hirsch EC, Brandel J-P, Galle P, Javoy-Agid F, Agid Y (1991) J Neurochem 56:446
36. Tokutake S, Oyanagi S (1995) Gerontology 52:131
37. Yumoto S, Kakimi S, Ohsaki A, Ishikawa A (2009) J Inorg Biochem 103:1579
38. Bouras C, Giannakopoulos P, Good PF, Hsu A, Hof PR, Perl DP (1996) Acta Neuropathol 92:351
39. Aranyosiova M, Kopani M, Rychly B, Jakubovsky J, Velic D (2008) Appl Surf Sci 255:1123
40. Itoh M, Suzuki Y, Sugai K, Ozuka N, Ohsawa M, Otsuki T, Goto Y (2008) J Child Neurol 23:938
41. Reusche E, Seydel U (1993) Acta Neuropathol 86:249
42. Shirabe T, Irie K, Uchida M (2002) Neuropathology 22:206
43. Exley C, Mamutse G, Korchazhkina O, Pye E, Strekopytov S, Polwart A, Hawkins C (2006) Mult Scler 12:533
44. Exley C, Price NC, Kelly SM, Birchall JD (1993) FEBS Lett 324:293
45. Scott CW, Fieles A, Sygowski LA, Caputo CB (1993) Brain Res 628:77
46. Uversky VN, Li J, Fink AL (2001) J Biol Chem 276:44284
47. Walton JR (2009) Neurotoxicology 30:1059
48. Exley C, Korchazhkina O, Job D, Strekopytov S, Polwart A, Crome P (2006) J Alzheimers Dis 10:17
49. Van Landeghem GF, Dhaese PC, Lamberts LV, Barata JD, DeBroe ME (1997) Nephrol Dial Transplant 12:1692
50. Eckstein JA, Ammerman GM, Reveles JM, Ackermann BL (2008) J Neurosci Methods 171:190
51. Czarnecka J, Cieslak J, Michal K (2005) J Chromatogr B 822:85
52. Perl DP, Good PF (1987) Lancet 1:1028
53. Kumar V, Gill KD (2009) Arch Toxicol 83:965
54. Banks WA, Kastin AJ (1983) Lancet 2:1227
55. Beardmore J, Exley C (2009) J Inorg Biochem 103:205
56. Exley C (2004) Free Radic Biol Med 36:380
57. Khan A, Dobson J, Exley C (2006) Free Radic Biol Med 40:557
58. Exley C (1999) J Inorg Biochem 76:133
59. Abbracchio MP, Burnstock G, Verkhratsky A, Zimmerman H (2009) Trends Neurosci 32:19
60. Lukiw WJ (2010) J Inorg Biochem 104:1010
61. Karlik SJ, Eichhorn GL, McLachlan DRC (1980) Neurotoxicology 1:83
62. Chen L, Yokel RA, Hennig B, Toborek M (2008) J Neuroimmune Pharmacol 3:286
63. Exley C, Siesjö P, Eriksson H (2010) Trends Immunol 31:103
64. Becaria A, Lahiri DK, Bondy SC, Chen DM, Hamadeh A, Li H, Taylor R, Campbell A (2006) J Neuroimmunol 176:16
65. Perl DP, Fogarty U, Harpaz N, Sachar DB (2004) Inflamm Bowel Dis 10:881
66. Campbell A, Bondy SC (2000) Cell Mol Biol 46:721
67. Exley C (2003) J Inorg Biochem 97:1
68. Exley C (2009) Aluminium and medicine. In: Merce ALR, Felcman J, Recio MAL (eds) Molecular and supramolecular bioinorganic chemistry: applications in medical sciences. Nova Science Pub Inc, New York, p 45
69. Exley C, House E, Collingwood JF, Davidson M, Cannon D, Donald AM (2010) J Alzheimers Dis 20:1159

Aluminium-dependent human diseases and chelating properties of aluminium chelators for biomedical applications

Valeria Marina Nurchi, Guido Crisponi, Valerio Bertolasi, Gavino Faa, and Maurizio Remelli

Abstract

The aim of this chapter is to give a general view on the current status of the role of aluminium in human health and disease. The main aspects of aluminium metabolism in humans are covered, summarizing the state of knowledge on the absorption, transport, tissue distribution, and excretion of aluminium, giving particular emphasis to the main metabolic pathways of this metal ion in the bones and in the brain. Some features concerning the solution chemistry of aluminium are considered, giving special care to the speciation of aluminium hydroxides in municipal water and in water for dialysis. The function of different chelators utilized in clinical practice in the therapy of aluminium-depending diseases is discussed, providing some insight on the chelators recently proposed. As a last point the crystal structures of selected Al(III) complexes are presented and thoroughly discussed.

Keywords

Aluminium • Aluminium overload • Chelating agent • Complex formation constant • Aluminium-related diseases

Introduction

Aluminium is the most abundant metal of the Earth's crust. Nevertheless, its existence was ascertained only in 1808 by Davy who gave it the actual name. In 1825, Oersted was successful in separating some impure metallic aluminium, by reacting its chloride salt with potassium amalgam and then distilling the mercury off. Sainte-Claire Deville introduced the first commercial process for aluminium production; the dynamo invention enabled Heroult and Hall in 1886 to patent separately in France and USA the electrolytic process in use today. Aluminium production, already 8,000 tons in 1900, has been continuously increasing: in 2006 global aluminium production was about 50 million tons, 16 million of which constituted by recycled metal.

Aluminium shows a unique combination of properties—lightweight, strong, flexible, recyclable—making it suitable for a wide range of applications, spanning from automotive and aeronautic transport, to packaging and drinks, construction and spatial industries [1, 2]. Aluminium compounds are widely used in paper and dye production, in textile and food industry, as anti-caking agents, for leaving baked goods, emulsifying cheeses, binding meats, thickening prepared sauces, colouring desserts and for buffering, stabilizing, curing and texturing foods [3, 4], as components of both cosmetic preparations [5] and drugs in human and veterinary medicine [6]; among them, buffered aspirin containing aluminium glycinate has been commonly using as analgesic from many years. Moreover, some aluminium salts are used in water purification, as well as in brewing and sugar refining [7].

Until some decades ago, the content of aluminium in natural waters was mostly insignificant, with only some exceptions, like volcanic regions. In fact, native Al(III) released by minerals is normally immobilized to the soil as

G. Crisponi (✉)
Dipartimento di Scienze Chimiche, Università di Cagliari, Cittadella Universitaria, 09042 Monserrato, Italy
e-mail: crisponi@unica.it

W. Linert and H. Kozlowski (eds.), *Metal Ions in Neurological Systems*,
DOI 10.1007/978-3-7091-1001-0_10, © Springer-Verlag Wien 2012

insoluble hydroxide. Through human activity, biota are experiencing a growing exposure to biologically reactive aluminium [8, 9]; atmospheric acidification and, especially, acid rains [10] are causing a progressive export of aluminium from rocks to surface waters where absorbable cationic aluminium species come in contact with plants and animals (and humans) [11]. Martin underlines that Al(III) and other metal ions can "damage fish and humans more than does the increased acidity" [12].

Aluminium commodities permeated all human activities until the second half of the twentieth century, when this metal was recognized as the main cause of serious diseases, like dialysis, osteodystrophy and dementia. As a consequence, the chemical and biomedical research on this metal ion dramatically increased: many reviews and some books are now available in literature on aluminium and its biological role, including transport, distribution, speciation and toxicity [13–27].

Aluminium metabolism

Aluminium is not essential to humans [2]; one reason can be its inertness from the biochemical point of view. Nevertheless, a recent theory suggests that its exclusion from successful biochemical pathways is mainly due to its very low natural availability [8, 9]. Other theories are focused on the role of silicon: the formation of hydroxyl-alumino silicates would reduce the biological availability of aluminium itself [28, 29].

Aluminium, given its high affinity for phosphate, is still being used in clinical practice with limitations, although the use of aluminium hydroxide is not recommended anymore in dialysis units [30]. The risk for aluminium overload is not restricted to subjects affected by chronic kidney diseases undergoing dialysis: general populations may be exposed to aluminium toxicity, when aluminium sulphate is used as a sedimentation agent for treating city water (see below).

Food is considered to provide majority of aluminium absorbed by humans, daily exposure to aluminium from food products being estimated between 3 and 10 mg [31]. Tea, and recently also coffee, infusions are recognized as a major source of dietary aluminium exposure to humans [32]. Some associations may interfere with the percentage of the absorbed trace metal: in particular, adding milk to tea infusions has been shown to significantly decrease the bioavailability of aluminium [33]. On the other hand, the concomitant administration of aluminium-rich compounds and citrate, such as lemon and tea, has been shown to markedly increase aluminium absorption [34].

The oral bioavailability of aluminium is low, less than 1 % in laboratory animals and human volunteers. Oral aluminium bioavailability from water appears to be about 0.3 % [35]. High phosphate concentrations in the gut lumen have been shown to decrease aluminium absorption, whereas citrate increases it. In the presence of fluoride in drinking water, the intestinal absorption of aluminium in food is markedly increased [36]. Food is the primary source for aluminium under physiological conditions, but Al bioavailability from food has not been adequately determined. The concentration of aluminium in food is extremely variable. Plants such as tea accumulate Al(III) by storing a chelated version in older leaves , which have been shown to contain as much as 3 % aluminium. Different methods of beverage preparation may change significantly their aluminium content. In particular, the difference between coffee beverage prepared in aluminium and stainless steel moka pots was proved significant, due to aluminium leaching to the coffee infusion [37]. Most plant food contains little aluminium, normally less than 5 μg/g of wet weight. At the extremes of a spectrum, we find tomatoes showing very low aluminium levels, ranging from 0.2 up to 1.1 μg/g, and marjoram and thyme, which contain 500–1,000 μg/g of the metal. The concentration of natural aluminium in animal-derived food is low, normally below 1 μg/g, with higher values (19 μg/g) reported in Swiss cheese.

The risk assessment of aluminium in food should always take into account not only the amount of metal ingested, but even the possibility of marked differences in aluminium absorption due to differences in composition of the foods containing aluminium [38]. Contamination of food during processing, cooking and storage has been shown to result in high aluminium intakes among consumers. In particular, when acidic foods are cooked in aluminium-made pans, leaching of relevant amounts of the metal has been demonstrated, aluminium levels reaching concentrations as high as 50 mg/L. An international panel of the European Food Safety Agency (EFSA), on request of the European Commission, has recently given a scientific opinion on safety of aluminium from dietary intake. In this study, EFSA determined the tolerable weekly intake (TWI) for aluminium, corresponding to 1 mg Al/kg of body weight per week. Dietary surveys in people from different countries estimated the daily intake of aluminium from overall diets including additives to range from 14 up to 280 mg/week. These levels clearly exceed the tolerable weekly intake recommended by the European Community and show that in many countries a considerable number of people, especially children, are at risk of aluminium intoxication [39].

Human exposure to aluminium takes myriad forms, including the use of illicit drugs, being aluminium biologically accumulated in the plant-based products or as a processing contaminant of such products. Elevated concentrations of aluminium have been reported in users of heroin [40], indicating an increase in the body burden of the metal. Different clinical symptoms related to aluminium toxicity have been reported in subjects undergoing inhalation of cocaine vaporized on aluminium foils, in tobacco and cannabis smokers, after ingestion of oral methadone solutions, and after intravenous injection of boiled methadone [41].

Excess of aluminium in the diet may have relevant consequences on bioavailability of other trace elements and, in particular, on iron and calcium metabolism. Aluminium has been shown to exert an inhibitory effect on calcium absorption in the small intestine, probably by blocking calcium ions entry across the brush border of the enterocytes [42].

Some common widely used pharmaceutical products, such as antiacids and antidiarrhoeic drugs which are non-prescription medications, contain high quantities of aluminium compounds, and have been administered for many years for the treatment of peptic disorders [43]. Among the categories at risk of aluminium intoxication are patients affected by Down's syndrome, a genetic disease caused by trisomy of chromosome 21, who show a higher frequency of Alzheimer's disease as compared to normal population. In Down's syndrome subjects, gastrointestinal absorption of aluminium has been shown to be markedly increased [44], suggesting a relationship between aluminium burden and human dementia.

The absorption of Al may be markedly increased by industrial and medicinal exposure, and perhaps by antiperspirant use, which can significantly increase the metal absorption through unusual routes. Aluminium may be absorbed through the underarm skin when contained in antiperspirant, or may be inhaled with dust in polluted environment. Aluminium chlorohydrate present in aerosol antiperspirants may enter our body by the transdermal absorption, particularly in women when utilized by spraying onto underarm abraded skin [45].

The uptake of aluminium by the inhalation route, in normal circumstances, can be ignored, the estimated daily intake by inhalation being 4.4 μg. Aluminium industry is not considered a source of occupational hazard [46], even though concentrations of the metal in blood and urine, higher in welders and workers producing aluminium than in non-exposed referents, were measured [47]. Occupational exposure to aluminium by inhalation was also reported for bauxite miners [48]. It is interesting to report that, at the beginning of the last century, Canadian miners were deliberately exposed to inhalation of the so-called "McIntyre powder", containing ultra-fine aluminium particles, in order to prevent lung silicosis [49]. However, this "preventive" treatment was later demonstrated to cause aluminium intoxication and brain damage.

Another route for aluminium uptake is its injection into our body as a vaccine adjuvant. The two adjuvants most commonly used are aluminium hydroxyphosphate and aluminium oxyhydroxide, both insoluble salts. Aluminium intake by vaccination should be considered as an important source of body burdens, especially in children: it has been calculated that 20 vaccinations, each containing 0.5 mg of aluminium, could release into our body 10 mg of aluminium, i.e., the intake reached from normal diet in 4,000 days. Recently, long-term persistence of vaccine-derived aluminium hydroxide has been demonstrated to be associated with chronic cognitive dysfunction [50].

Intestinal aluminium absorption

The intimate mechanism by which aluminium ions are transported across the brush border of the intestinal cells is not yet well known. The chemical speciation of aluminium contained in vegetables or in animal foodstuffs probably represents a key factor in the metal absorption; in fact, other than that naturally present at low concentrations in staple foods, aluminium is also used as a food additive.

Intestinal absorption of aluminium contained in food is subject to many systemic and local factors, including gastric acidity, variability of compounds with which Al is complexed in the gut lumen, and food quality. The influence of food on Al absorption is demonstrated by the observation that it is higher in the post-prandial state. Moreover, it has been shown that the chemical species of aluminium influences its uptake by Caco-2 cells, an in vitro model of gastrointestinal absorption [51]. Complexing agents present in food, such as dietary citrate, fluoride, maltolate, lactate and silicate have been demonstrated to affect aluminium absorption [52]. Recently, an elegant study from a Swedish group at the University of Uppsala [38] clearly showed that fatty acids common in food, such as oleic acid and DHA, may increase the paracellular absorption of aluminium, probably acting on the tight junctions between the Caco-2 cells. The authors evidenced at the ultrastructural level the ability of oleic acid to change the morphology of tight junctions, giving a morphomolecular basis to the hypothesis that fatty acids could enhance aluminium absorption through the paracellular route. The intimate mechanism at the basis could be the ability of fatty acids to produce an increased number of functional pores, remodelling tight junctions between enterocytes, and increasing their permeability to aluminium as well as to other trace elements.

Intestinal pH is one of the most important physiological factors affecting aluminium absorption, since an acid environment is essential for freeing aluminium ions from complexes formed in food and in intestinal mucosal secretions, setting the conditions for their absorption. The presence in the diet of chelating agents as citric acid, present in large amounts in fruits, has been shown to enhance aluminium absorption.

Aluminium transport in blood

At physiological pH, aluminium ions normally undergo hydrolysis (see below). However, the presence in blood of complexing anions and of other binding species can completely inhibit hydrolysis. In blood and tissue fluids, aluminium ions are found complexed with organic molecules [53]. Aluminium ions, entering portal blood, are immediately bound by the iron transporter transferrin, the only binding aluminium protein in human serum [54]. Transferrin has been shown to bind about 90 % of circulating aluminium, ranging from 80 up to 94 % [55]. Aluminium binds to the iron-binding sites on the transferrin molecule but, under physiological conditions, it does not replace iron ions, since the strength of the aluminium bond is much weaker than that of iron [56]. The binding affinity of iron to transferrin is about 10^2 times greater than that of aluminium to this molecule. Given the low binding affinity of aluminium for blood transferrin, it has been suggested that aluminium probably binds to the molecule only because, under physiological conditions, plasma transferrin is about 50 μM deficient in iron, i.e., unsaturated [57, 58]. A competition between aluminium and iron for transferrin has been reported in patients undergoing dialysis affected by the so-called "renal" anaemia caused by iron deficiency. In an in vitro system, iron deficiency has been found to increase binding of aluminium to transferrin, developing a real competition between the two metals which could explain the molecular basis for aluminium deposition in many dialysis patients [59].

About 10 % of serum aluminium is bound to low molecular mass species: this fraction is composed of Al-citrate, Al-phosphate and ternary Al-citrate-phosphate complexes, whose percentage is individually variable and depending on the health status of individuals [55]. Citrate is the second aluminium binding molecule in human plasma, the percentage of aluminium species associated with citrate being evaluated about 7–8 % [35]. A competition between aluminium and calcium ions for citrate has been reported: the binding affinity of aluminium is much higher and it easily replaces calcium in such complexes [53]. This competition with calcium in serum transport could, at least in part, explain a disruption in calcium metabolism, frequently cited among the proposed mechanisms of aluminium toxicity. Minor percentages, below 1 %, of circulating aluminium species are also bound to free hydroxide and to phosphate. About 10 % of blood aluminium is bound to erythrocytes. Aluminium binding to red blood cells appears much stronger than to plasma proteins, half retention time in erythrocytes ranging from 20 up to 40 days [53]. Since aluminium concentrations in erythrocytes appear to be much more stable than plasma levels, in clinical practice measurement of red blood cell aluminium levels is considered a useful tool as an indicator of aluminium exposure and intake in studies on possible aluminium toxicity.

Tissue distribution

In comparison with many other polyvalent cations, including gallium, lanthanides and actinides, aluminium binds rather weakly to plasma components, maximizing its potential for transfer to binding sites within tissues [53]. It has been estimated that 45–75 % of intravenously injected Al^{26} is excreted in the first 24 h after injection [60]. Since only 0.5 % of injected aluminium remains in human plasma 1 day after injection, it follows that aluminium transfer to different tissues is very rapid. Aluminium has been shown to accumulate in all tissues of mammals, preferentially in kidneys, liver, heart, bones and brain [61]. The molecular mechanism through which tissues uptake aluminium probably differs from tissue to tissue and depends on the type of compound. Uptake of aluminium ions via transferrin receptor-mediated exocytosis might have toxic consequences, as demonstrated in bone marrow cells [62]. The uptake of aluminium-transferrin (Al-Tf) complexes results in metabolic effects different from those observed after uptake of aluminium-citrate (Al-citr) complexes, suggesting that Al-Tf might sequester aluminium in intracellular compartments different from Al-citr. Therefore, different mechanisms and different molecular pathways probably underlay aluminium uptake in different tissues: Tf-mediated endocytosis, uptake via ion channels, active transport mediated by unknown carriers. Data on the aluminium content in different human organs and tissues have been collected by the International Commission on Radiological Protection (ICRP) on Reference Man [63]. This study demonstrated that skeleton concentrates the major part of tissue aluminium, with a percentage around 54 %, followed by soft tissues (muscles 14 %, adipose tissue 5 %, connective tissue 3 %) reaching all together 22 %, skin 13 %, liver 3 %, gut 2 % and brain 1 %. The high values of skin aluminium content are questionable, given the possibility of skin contamination by unabsorbed surface deposits of environmental aluminium [53]. In pathological conditions, it has been shown that aluminium concentration in different organs may change dramatically. In patients affected by chronic renal insufficiency undergoing dialysis, the major site of aluminium accumulation was the skeleton (2.7 g) followed by liver (1 g), muscles (400 mg), and brain (100 mg) [64]. This altered aluminium distribution in dialysis patients, characterized by a marked preference to bone and liver, could be due to iron deficiency. Important consequences of iron depletion on aluminium metabolism are: the increased expression of transferrin receptor in several tissues; the increased availability of binding sites free for aluminium on the transferrin molecule; the increased affinity between transferrin and aluminium ions at low Fe-Tf saturations [59]. Taken together, these molecular changes may explain the preferential aluminium deposition in tissues strongly expressing transferrin receptor, including liver (hepatocytes), bone (osteoblasts) and parathyroid glands [65] frequently observed in clinical practice in dialysis patients. The increase in aluminium content in tissues has been reported to cause oxidative stress and a decrease, at the intracellular level, of reduced glutathione [66].

Aluminium stored in different tissues may be revealed by many histochemical methods. The main staining methods are aluminon method, Morin stain, nephtochrome green B method, solochrome blue-black R method, solochrome azurine method and acid solochrome cyanine method. Recently, a novel histochemical method that stains aluminium in the brain of patient with Alzheimer's disease was introduced by Walton [67].

The major deposition site for many divalent and polyvalent metals, including aluminium, is the skeleton, and for this reason these metals are often referred to as bone-seekers [53]. Roughly half of the aluminium absorbed in the intestine in young rats goes to the skeleton within 2 h. Skeletal deposits represent the reservoir for the long-term retention of aluminium within the human body [53], typical half times being 10–20 years. The cycle of aluminium in bones starts with the transfer of the metal ions from transferrin and citrate in the blood stream to bone surfaces, where aluminium becomes attached. Within the skeleton, aluminium ions are first deposited on bone surfaces, including internal endosteal and external periosteal surfaces, trabecular surfaces and the surface of the vascular channels that permeate compact bones. Subsequently, aluminium ions may be transferred to osteoclasts, large multinucleated bone resorbing cells, or incorporated in the bone matrix. Osteoclasts may release aluminium to macrophages in the bone marrow. Macrophages could act as a temporary store of aluminium ions, before releasing the metal to citrate and/or transferring, allowing aluminium re-entering the blood stream [53].

Aluminium deposits are present at the mineralized bone front on both growing and resting bones. In recent years, the association between increased aluminium bone stores in dialysed patients and the development of osteomalacia, previously known as "renal osteodystrophy" [68] has been well established. When sufficient quantities of aluminium accumulate, the process of bone formation is disrupted and an osteodystrophy, subsequently better defined as "adynamic bone disease" (ABD) [69] or "aluminium-induced bone disease" (AIBD), develops ending with spontaneous fractures. Aluminium does not deposit in osteoid, the unmineralized matrix which precedes bone formation, but rapidly migrates through it to deposit at the mineralizing front, commonly referred as "lamina limitans". The presence of aluminium ions at this site inhibits mineralization of the osteoid matrix, leading to osteomalacia. Aluminium delivered by transferrin to osteoblasts, thanks to the high expression of transferrin receptor on their surface, exerts an antiproliferative effect on osteoblasts, followed by a low bone turnover and osteomalacia [70]. In man, ABD has been reported in patients who underwent chronic parenteral nutrition with high aluminium contaminations and in dialysed patients [71].

Aluminium-transferrin complexes are also taken up by parathyroid cells, resulting in reduced parathyroid hormone secretion and hypoparathyroidism which plays a relevant role in the development of ABD [65]. Aluminium uptake via transferrin-mediated endocytosis might be linked to the increasing prevalence of ABD in the current dialysis population, by inducing a state of hypoparathyroidism. Iron status may play a role in aluminium toxicity. In fact, with the introduction of erythropoietin, the dialysis population may have a greater risk of iron deficiency [72], giving an advantage to aluminium in its competition with iron for transferrin [59]. The intimate mechanism by which aluminium interferes with parathyroid hormone secretion is not yet clear. By in vitro studies, it has been demonstrated that cytoskeleton-associated and soluble neurofilaments subunits show a marked susceptibility to aluminium exposure [73]. Moreover, aluminium has been shown to bind to a number of secretory granule proteins, interfering with the process of exocytosis and, in particular, blocking the release of parathyroid hormone-containing secretory granules [65], causing hypoparathyroidism and adinamic bone disease.

Brain uptake and efflux

If it were not for the widely accepted link between aluminium and dialysis-related dementia, aluminium deposits in the brain would be of little interest, given their small size and the little impact of the brain upon aluminium metabolism. In recent years, research interest has been stimulated by reports that some cognition/neurobehavioural deficits could be associated with increased aluminium body burdens [74].

Aluminium may enter the brain through multiple routes: from blood, either through choroid plexuses or the blood–brain barrier (BBB); from the nasal cavity into olfactory nerves followed by direct distribution into the brain [24, 25]. The large size (about 77,000 Da) and low lipophilicity of transferrin preclude its ability to diffuse through the pericellular pathway or endothelial cell membrane [75]. The transport of essential and of non-essential metal ions across tight membrane barriers such as the BBB is mediated by specific transport mechanisms, which regulate the brain levels of different metals [76]. These transporters often mediate the transport of multiple metal ions. Data on tissue aluminium content obtained from patients with chronic renal insufficiency treated for long periods with dialysis show high levels of aluminium in the brain, clearly demonstrating the ability of the metal to penetrate the BBB. The intimate mechanisms by which aluminium ions may cross the BBB are not well known. About 90 % of aluminium in brain extracellular fluid is predicted to be Al-citrate and only 4 % Al-transferrin. Since transport of many trace elements, including copper as L-histidine complex, iron by transferrin receptor-mediated endocytosis, mercury as a cysteine complex, has been shown to be carrier-mediated, therefore BBB permeation of aluminium cannot solely be attributed to diffusion and is probably carrier-mediated as Al-citrate [24, 25]. Different candidates that may mediate Al-citrate transport across the BBB have been proposed: among them, the best candidates are the monocarboxylate transporter (MCT) and an organic anion-transporting polypeptide (oatp) [24, 25]. MCT1 has been found in rat brain microvessels, on the luminal and abluminal endothelial cell plasma membrane [77]. In rats, the half-life of brain aluminium has been estimated to be about 150 days: in animals receiving Deferoxamine injections, it decreased to 55 days. These data clearly indicate that aluminium accumulated in brain can be mobilized by an iron chelator such as Deferoxamine, and suggest the existence of a carrier-mediated mechanism to protect the brain from aluminium, by effluxing it across the BBB into blood [24, 25].

Aluminium excretion

The primary route of systemic aluminium elimination is renal clearance [78]. Half of aluminium absorbed at intestinal level in rats is excreted in the urine, most of it within 48 h [79]. Al elimination in man has been shown to be primarily renal, with less than 2 % excreted in bile [80]. Different mechanisms of renal excretion of aluminium have been reported: glomerular filtration and secretion in the distal nephron are generally considered the most important mechanisms of aluminium excretion [81]. Recently, it has been shown that intake of a single dose of analgesics containing aluminium is able to up-regulate the expression of the renin gene, increasing the synthesis of renin, eventually resulting in possible induction of essential hypertension [82]. Reported values for aluminium concentration in urine range from 0.68 up to 8 μg/L. The amount of aluminium excreted per day is extremely variable, according to the variations in the volume of urine excreted by different individuals: values ranging from 1.8 up to 12 μg per day have been reported in different studies. In studies on human volunteers utilizing ^{26}Al, the majority of injected aluminium has been demonstrated to be excreted within the first few days in urine [83].

Solution chemistry of aluminium

Al(III) (ionic radius: 0.54 Å) is a hard metal ion, like high spin iron(III) (ionic radius: 0.65 Å) and gallium(III) (ionic radius: 0.62 Å) [84]. In Al(III) minerals the metal ion is normally surrounded by six oxygen atoms; at the same time, most of its biological binding sites are negatively charged oxygen donors such as carboxylates, phenolates, catecholates and phosphates. It is worth noting that a very strong ligand for Al(III) is also the fluoride ion [85].

The hydrolytic chemistry of aluminium greatly affects its solubility and bioavailability in biological environments. At room temperature and intermediate ionic strength values, the exa-aquo ion, $[Al(H_2O)_6]^{3+}$, is the main species below pH 4.5; at higher pH, coordinated water molecules gradually deprotonate, giving rise to the soluble mononuclear species $[Al(OH)]^{2+}$, $[Al(OH)_2]^+$ and $[Al(OH)_3]$ (coordinated water molecules are not reported, for the sake of simplicity). Some polynuclear species have been also reported ($[Al_2(OH)_2]^{4+}$ and $[Al_3(OH)_4]^{5+}$), whose amount mainly depends on total aluminium concentration and equilibration time [22]. If the total aluminium concentration is high enough, at neutral pH a precipitate of $Al(OH)_3$ is observed; at alkaline pH, a further proton is released by aluminium hydroxide, thus leading to the formation of the soluble species $[Al(OH)_4]^-$. It is worth of note that the above hydrolytic equilibria can be strongly affected by the presence of competing ligands, like F^-, Cl^-, SO_4^{2-} or SiO_4^{2-}.

A detailed investigation on this topic has been recently described, where two particular cases were considered [26, 27]:

- Water from municipal treatment plants.
- Water for dialysis, purified by reverse osmosis treatment.

In the treatment of municipal drinking waters, aluminium salts are widely used as coagulants to decolorize, reduce organic matter, turbidity and microorganism levels. This is obtained through the addition of a suitable aluminium salt (usually sulphate) at appropriate pH, followed by flocculation, sedimentation and filtration. The slow precipitation of gelatinous $Al(OH)_3$, surrounding any solid particle of whatsoever origin, like dust particles and bacteria, purifies and clarifies water. The resulting purified water is thus a saturated solution of aluminium hydroxide. In Fig. 1 the distribution plots at the ionic strength 0.16 M and total Al (III) amount of 2 mM (reproducing the conditions usually employed in water-treatment plants), are reported; they are calculated by using literature constant values [57, 58]. Al (III) hydroxide precipitation starts at pH a bit higher than 4 and it is the unique species at neutral pH; $Al(OH)_3$ starts dissolving above pH 8, with the formation of the negatively charged species $[Al(OH)_4]^-$. More in details: (1) since the purified water is in contact with the $Al(OH)_3$ solid-phase, the soluble-species concentrations do not depend on the amount of added flocculants; (2) Fig. 1 shows that polynuclear species do not significantly contribute the total amount of soluble aluminium species. The same calculations were repeated [26, 27] considering the presence of about 5 mg/L of fluoride (often present in high amount in natural waters), again taking the corresponding equilibrium constants from

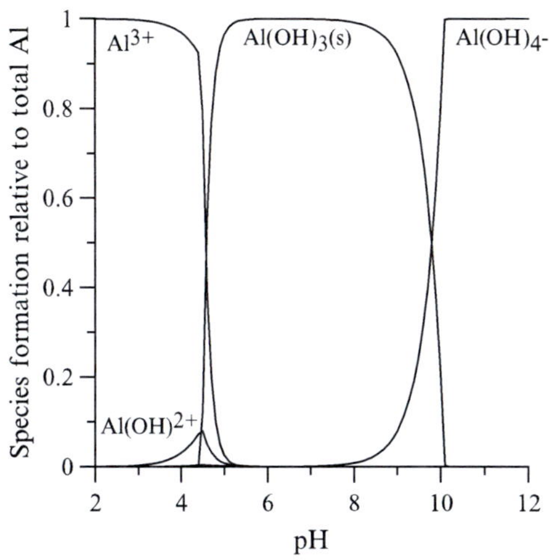

Fig. 1 Aluminium hydroxo-species distribution plots at 0.16 M ionic strength, at room temperature and total aluminium concentration = 0.002 M

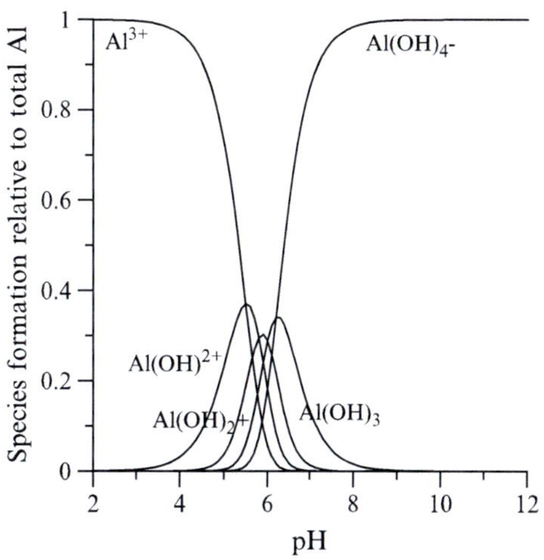

Fig. 2 Aluminium hydroxo-species distribution plots at 0.16 M ionic strength at room temperature and total aluminium concentration of 5.0×10^{-7} M

literature [18, 85, 86]. It is interesting to note that, under these conditions, the contribute of simple hydroxo species is almost unchanged (essentially depending only on pH), but the formation of fluoride complexes $[AlF_2]^+$, $[AlF_3]$ and $[AlF_4]^-$ strongly enhances the aluminium solubility, especially at pH 6. These results are in good agreement with those very recently reported by Frankowski et al. who described a new chromatographic method for speciation analysis of aluminium fluoride complexes in soil samples [87]. The complexing behaviour of Al(III) towards F^- and OH^- anions has been recently inspected through the density functional theory and the polarized continuum model [88].

Deionized water is nowadays available in each dialysis centre; it is characterized by an aluminium content lower than 6×10^{-8} M. The speciation of aluminium hydroxides at this concentration level is reported in Fig. 2. It can be observed that no solid precipitate is formed: all aluminium is in the form of soluble species at pH 6 prevailing $[Al(OH)_2]^+$ and $[Al(OH)_3]$; whereas at pH 8, $[Al(OH)_4]^-$ becomes again the principal species.

Chelating agents

The discovery of the potential aluminium toxicity gave the cue to design and study aluminium chelators, able to mobilize aluminium deposits and reduce the body burden of the metal, with special concern to dialysis-encephalopathy and osteomalacia. Management of aluminium overload first involves discontinuation of aluminium exposure. The introduction of water-treatment systems in dialysis centres greatly contributed to drastic decrease in aluminium-overload diseases [30]. In fact, the use of double reverse-osmosis devices for haemodialysis, producing a purified water with very low aluminium concentrations, not only avoids the passage of aluminium to plasma, but allows the metal removal from plasma itself, the extent of removal depending on the gradient between the free diffusible plasma and the dialysate aluminium concentration [89].

When the interruption of exposure to aluminium does not lead to any clinical improvement, a chelation therapy is requested [13]. Deferoxamine (Scheme 1) was the first aluminium chelator to be introduced in clinical practice for the treatment of aluminium-related osteomalacia [90, 91]: it was administered with good efficacy either intramuscularly or intraperitoneally [92, 93]. Treatment of aluminium bone toxicity with Deferoxamine requires a prolonged therapeutic protocol, lasting for not less than 6 months [94]; a side effect of this long therapy might be the parallel reduction of brain aluminium burden [95]. Deferoxamine treatment was reported to be successful even in cases of acute encephalopathy due to severe aluminium intoxication following aluminium bladder irrigation [96]. Therapy with Deferoxamine is not without some risk: an intensive treatment can give rise to acute neurological toxicity [13]; therefore, Deferoxamine treatment is indicated only for patients showing serum aluminium levels higher than 200 µg/L or whose aluminium bone concentration is ten times greater than normal value [97].

In recent years, other aluminium chelators have been developed and progressively introduced in clinical setting; among them ascorbate and Feralex-G (Scheme 1) used either alone or in combination with Deferoxamine [98]. A "metal-targeted strategy" has been also proposed for the treatment of Alzheimer's disease, pointing to the brain metal redistribution rather than its scavenging and removal [99].

Different new ligands for aluminium have been synthesized in the last 10 years; for some of them the complex formation equilibria were thoroughly investigated and biological activity evaluated. A particular attention has been devoted to 3-hydroxy-4-pyridinones (3,4-HPs), mono-anionic *N*-heterocyclic bidentate *{O,O}*-chelators with high affinity for *hard* metal ions (the most important molecule of this class, Deferiprone, is shown in Scheme 1). They can be easily extra-functionalized to modify their chelating properties and/or to improve their bioavailability. Hider et al., in their studies for the treatment of iron overload, introduced several bidentate or polydentate 3,4-HP chelators [100, 101]; some of them were later studied for the removal of aluminium [23, 102] or plutonium [103].

The strategy of functionalizing a bidentate mono (3,4-HP) unit with carbohydrate moieties was suggested to improve the cell transport of the ligand: Kruck and Burrow [104] described the synthesis of the above-mentioned Feralex-G, a glucopyranose derivative of Deferiprone (2-Deoxy-2-(*N*-arbamoylmethyl-[*N*9-29-methyl-39-hy-droxypyrid-49-one])-D-glucopyranose); Chaves et al. [105] proposed two *N*-glycosyl mono-(3,4-HP) derivatives.

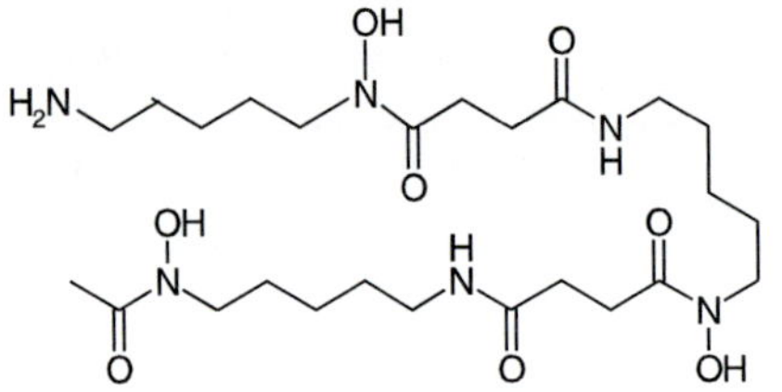

Deferoxamine

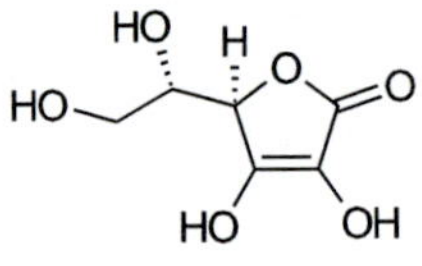

Ascorbic acid

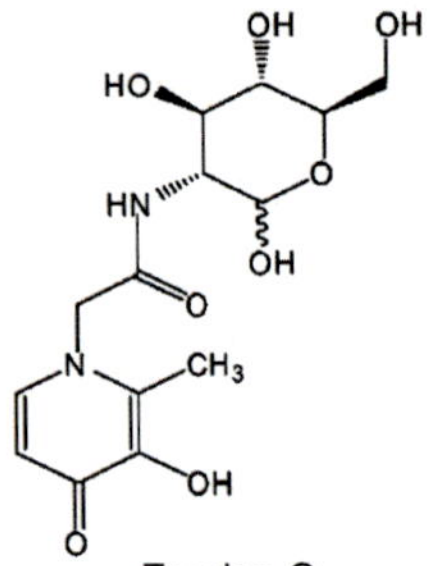

Feralex-G

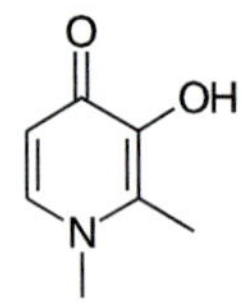

Deferiprone
(1,2-Dimethyl-3-hydroxypyridin-4-one)

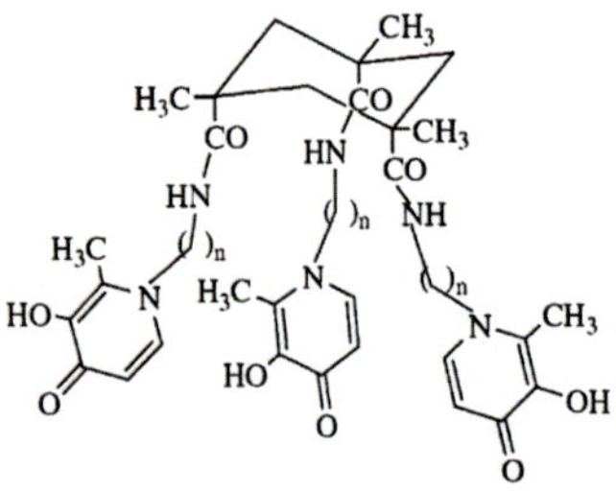

KEMPPr(3,4-HP)$_3$ (*n*=3) or KEMPBu(3,4-HP)$_3$ (*n*=4)

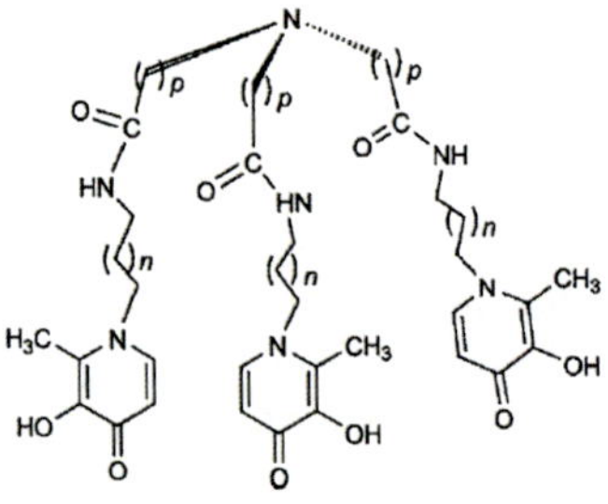

NTA(PrHP)$_3$ *p*=2, *n*=1
NTA(BuHP)$_3$ *p*=1, *n*=2

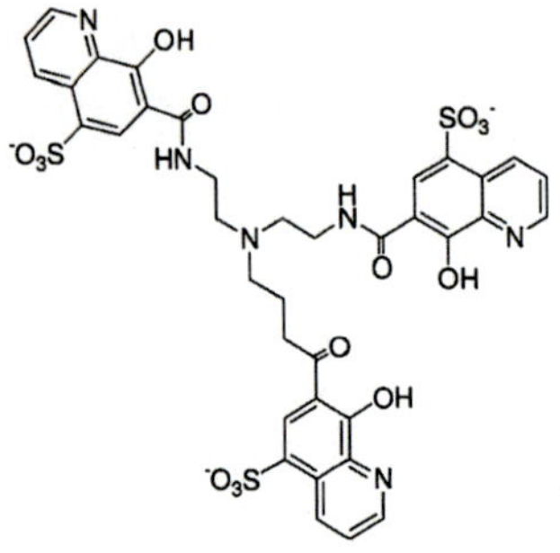

O-TRENSOX

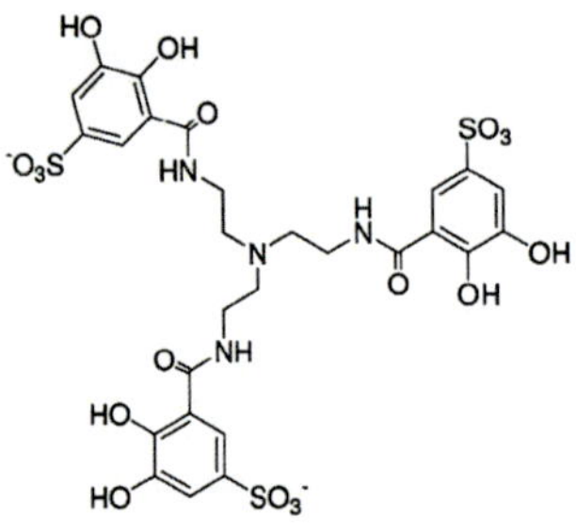

TRENCAMS

Scheme 1 continued

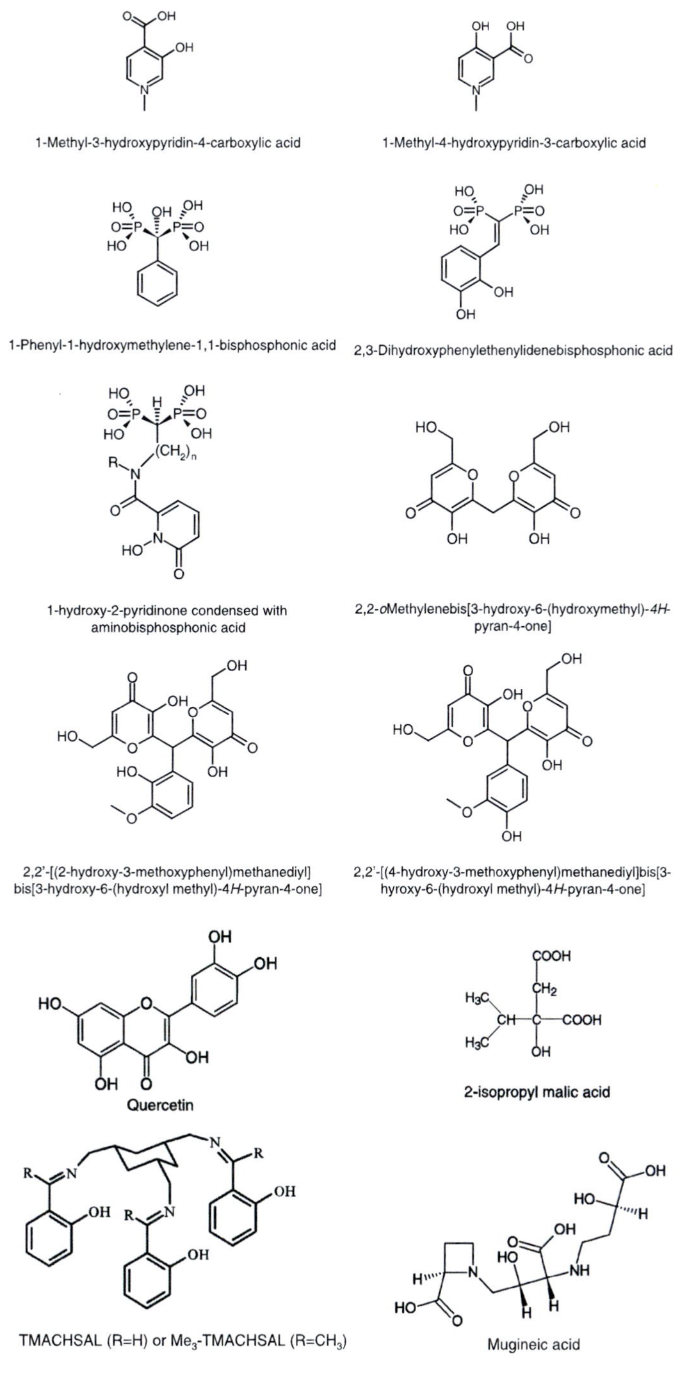

Scheme 1

Different polydentate 3,4-HP chelators have been proposed by the group of Santos:

- A tetradentate iminodiacetic acid bis-(3,4-HP) ligand, and two *N*-glycosyl mono-(3,4-HP) derivatives [105];
- A ternary system where an arylpiperazine-containing bis-hydroxypyridone was mixed with a biomimetic mono-hydroxypyridinone, ornithine derivative in the presence of Fe(III), Ga(III) or Al(III) ions [106];
- Two exadentate ligands, the tris–hydroxypyridinone-based compounds KEMPPr(3,4-HP)$_3$ and KEMPBu(3,4-HP)$_3$ (Scheme 1), derived from the KEMP acid scaffold to which three 3-hydroxy-4-pyridinone chelating moieties are attached via two different size spacers [107];
- Two new tripodal tris(3-hydroxy-4-pyridinone) hexadentate chelators NTA(BuHP)$_3$ and NTP(PrHP)$_3$ (NTA = nitrilotriacetic acid, NTP = nitrilotripropionic acid, HP = hydroxypyridone, Scheme 1) [108].

These exadentate ligands proved to be strong sequestering agents for the group III metal ions, with potential pharmacological applications in metal-chelation therapy; in particular, $NTA(BuHP)_3$ and $NTP(PrHP)_3$, characterized by a mild hydrophilic character, revealed a very strong chelating affinity for Fe(III) and Al(III) (pFe = 27.9 and pAl = 22.0 for $NTA(BuHP)_3$; pFe = 29.4 and pAl = 22.4 for $NTP(PrHP)_3$).

Other families of chelating agents have been taken into consideration as aluminium chelators. Biaso et al. [109] studied two tripodal molecules, O-TRENSOX, containing three 8-hydroxy-5-sulfonate-quinoline linked to tris(2-amino ethyl)amine (TREN), and the analogous triscatechol TRENCAMS (Scheme 1). The corresponding, excellent, pAl values are 20.0 and 26.2 for O-TRENSOX and TRENCAMS, respectively.

A number of hydroxypyridinecarboxylic acids (HPs) have been synthesized and studied in solution as aluminium ligands [110–117]. These ligands have the two coordinating –OH and –COOH groups in different positions (2,3; 3,2; 3,4 and 4,3), and can be variously methylated; some of them are shown in Scheme 1. HPs proved to be good chelators for Al (III), even though they are generally not strong and selective enough to compete with Deferoxamine or Deferiprone.

Bisphosphonate ligands (Scheme 1) are very efficient chelating agents for aluminium, with pAl values higher than that of Deferiprone and comparable to that of Deferoxamine [118]. In order to improve bisphosphonate chelating properties, catechol–bisphosphonate conjugates and mixed bisphosphonates–hydroxypyridinonate compounds were synthesized by Ding et al. [119] and by Bailly et al. [120], respectively. Solution-equilibria studies, in the presence of Fe (III) or Al(III) ions, showed that the shortness of the linker hinders the simultaneous tetradentate coordination for both the ligands [121, 122].

An interesting aluminium chelator, 2,2-*o*-Methylenebis[3-hydroxy-6-(hydroxymethyl)-4 H-pyran-4-one] (Scheme 1), constituted by two kojic acid units linked together by a methylene bridge, was proved by Fox and Taylor an efficient ligand for the in vitro mobilization of ferritin-bound iron [123]. Our recent solution studies showed that it forms very stable iron and aluminium dinuclear complexes [124]; the same is true for its vanillin or *o*-vanillin derivatives (Scheme 1) [125].

Four curcuminoid analogues and the corresponding Al (III) complexes were recently synthesized, spectroscopically characterized and examined for their possible anticancer activity. All the compounds were found to be cytotoxic towards Ehrlich ascites carcinoma cells and cultured L929 cells; the activity of aluminium chelates was proved to be higher than that of free curcuminoid analogues [126].

Two novel tripodal imine-phenol ligands, TMACHSAL (cis,cis-1,3,5-tris{(2-hydroxybenzilidene)aminomethyl} cyclohexane) and Me_3-TMACHSAL (cis,cis-1,3,5-tris{[(2-hydroxyphenyl)ethylidene]aminomethyl}cyclohexane) (Scheme 1) have been synthesized and their complexing ability towards Al(III) and Ga(III) ions has been investigated by potentiometric and spectrophotometric methods [127]. At neutral pH, both ligands form tris(iminophenolate) type complexes with Al(III) and Ga(III), with high stability and good selectivity.

Chelating properties of some natural ligands towards aluminium were also investigated. Cornard and Merlin [128] examined the interaction of quercetin (3,3′,4′,5,7-pentahydroxyflavone, Scheme 1), one of the most common natural flavanols, with Al(III): they reported the formation of two different complexes with 1:2 and 2:1 metal/ligand stoichiometries. Kobayashi et al. [129] recognized an aluminium-binding substance (ABS) secreted by *Saccharomyces cerevisiae*, able to solubilize Al(III) at neutral pH. ABS was identified as 2-isopropylmalic acid (Scheme 1) and its ability to complex aluminium was investigated by different spectroscopic techniques, thus showing the formation of at least four Al(III) species. Yoshimura et al. studied the phytosiderophores, mugineic acid (Scheme 1) and epi-hydroxymugineic acid, for their ability to bind Al(III) [130]; they bind aluminium under physiological conditions even though much less effectively than Fe(III). The related compound nicotianamine was investigated as well, but its complexing ability towards Al(III) proved negligible.

Crystal structures of selected Al(III) complexes

This section reports a survey of the crystal structures, retrieved from the CSD (Cambridge Structural Database) [131], of Al(III) complexes with organic chelators as potential drugs for the treatment of aluminium overload. Both the ligands and the aluminium complexes should posses a unique combination of properties: water solubility, hydrolytic stability and lipophilicity. The different types of chelating agents have been divided following their various coordination modes towards the Al(III) central cation.

Al-(6O) complexes

Bidentate (2O) chelators

Several bidentate monobasic ligands such as 3-hydroxy-2-methylpyridin-4(1*H*)-one, 3-hydroxypyridin-2(1*H*)-one and 3-hydroxy-4*H*-pyran-4-one derivatives have been employed as new binding groups for the Al(III) cation. The related octahedral complexes with Al(III) cation, whose crystal structures have been determined, are shown in Scheme 2. The different substituents R have been utilized to change the ligand lipophilicity but maintaining similar thermodynamic properties.

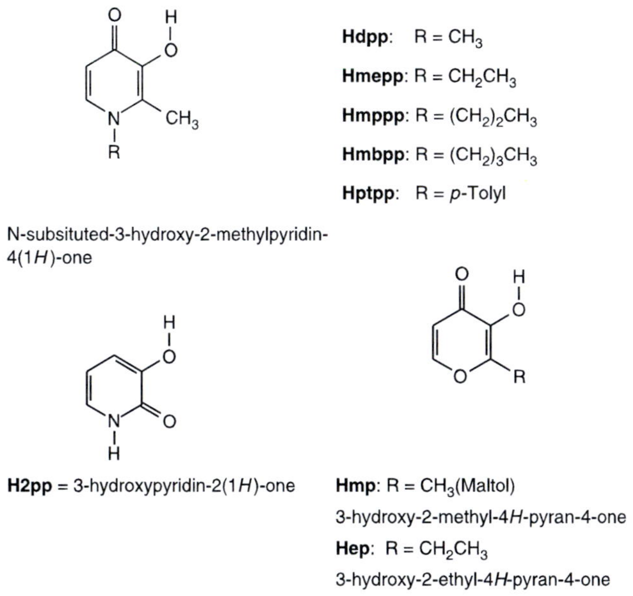

Scheme 2

Table 1 List of the crystal structures of Al-(6O) complexes with bidentate (2O) chelators

Compound	CSD code	Configuration	References
$Al(dpp)_3 \cdot 12H_2O$	FIMLOC10	*fac*	[132]
$Al(mepp)_3 \cdot 12H_2O$	JASMIZ	*fac*	[133]
$Al(mepp)_3 \cdot 3H_2O$	YAJGIZ	*fac*	[134]
$Al(mppp)_3 \cdot 3H_2O$	TANJAT	*fac*	[135]
$Al(mbpp)_3 \cdot 3H_2O$	TANHUL	*fac*	[135]
$Al(ptpp)_3 \cdot 5.5H_2O$	VIPZUP	*fac*	[136]
$Al(2pp)_3$	CAQDON	*fac*	[110, 111]
$Al(mp)_3$	DUGJEU	*mer*	[137]
$Al(mp)_3 \cdot 0.5H_2O$	MOBJOC	*mer*	[138]
$Al(ep)_3$	MACFEC	*mer*	[139]

Table 1 reports the crystal data for Al(III) complexes with these bidentate chelators. All complexes of N-Substituted-3-hydroxy-2-methylpyridin-4(1*H*)-one are hydrates and exhibit distorted octahedral coordination, in *fac* configuration, that is, the longer Al-O_{ket} bonds are *cis* to each other and the shorter Al-O_{enol} ones are also *cis*. The higher complexation strength arises from the great stabilization of a positive charge on the pyridine nitrogen, due to the inductive effect of the methyl group in position 2, which allows a higher negative charge on the chelating oxygen. Figure 3 reports the structure of the neutral Al(III) complex *N*-ethyl-3-hydroxy-2-methylpyridin-4(1*H*)-one (YAJGIZ).

The 3-hydroxypyridin-2(1*H*)-one derivatives are other potential promising Al/Fe chelators but are not used to the same extent as the 3-hydroxy-2-methylpyridin-4(1*H*)-ones. Figure 4 reports the crystal structure of $Al(2pp)_3$ neutral complex (CAQDON) which displays a *fac* configuration.

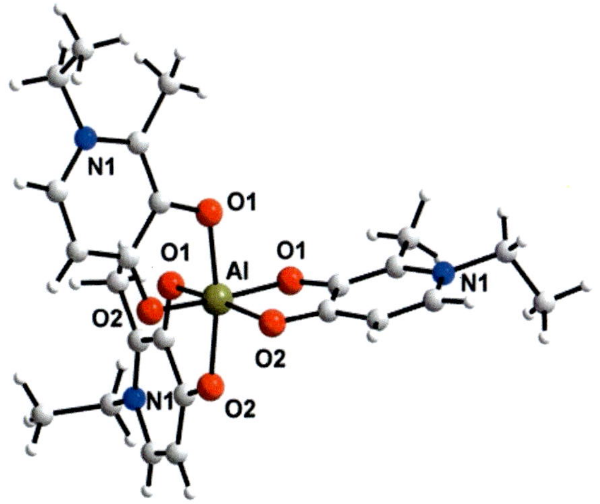

Fig. 3 Crystal structure of Al(III) complex with *N*-ethyl-3-hydroxy-2-methylpyridin-4(1*H*)-one ligands (YAJGIZ)

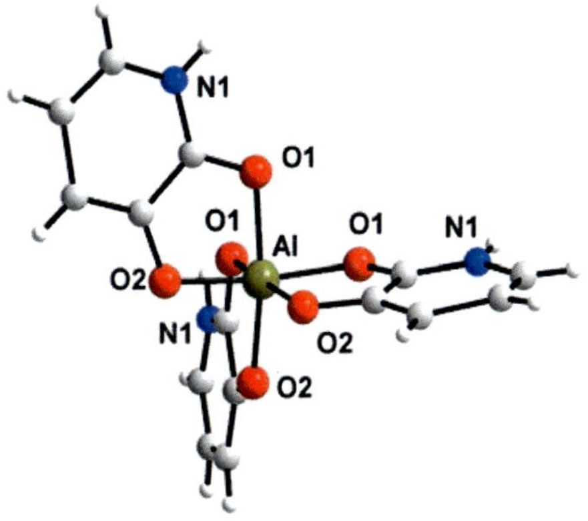

Fig. 4 Crystal structure of the $Al(2pp)_3$ complex (CAQDON)

The 3-hydroxy-4*H*-pyran-4-one chelators are similar to N-substituted-3-hydroxy-2-methylpyridin-4(1*H*)-one ones and have excellent chelator properties. The structure of Al (III) complex with 3-hydroxy-2-methyl-4H-pyran-4-one ligands, displaying *mer* configuration, is reported in Fig. 5.

Tridentate (3O) chelators

Tridentate ligands which are sterically constrained to a facial coordination ("tripod ligands") such as 1,3,5-triamino-1,3,5-trideoxy-*cis*-inositol (taci) and 1,3,5-tris-dimethylamino-1,3,5-trideoxy-*cis*-inositol (dtci) (Scheme 3) have received attention because they compete barely with hydrolysis of Al (III) in aqueous solution. Al(III) forms mononuclear 1:2 complexes with taci and dtci ligands in octahedral geometry. The crystal structures showed that the proton of the hydroxyl groups are transferred to the amine groups of taci and tdci during the coordination. The sequestering ability of taci is lower than that of tdci because the stability of $[Al(taci)_2]^{3+}$

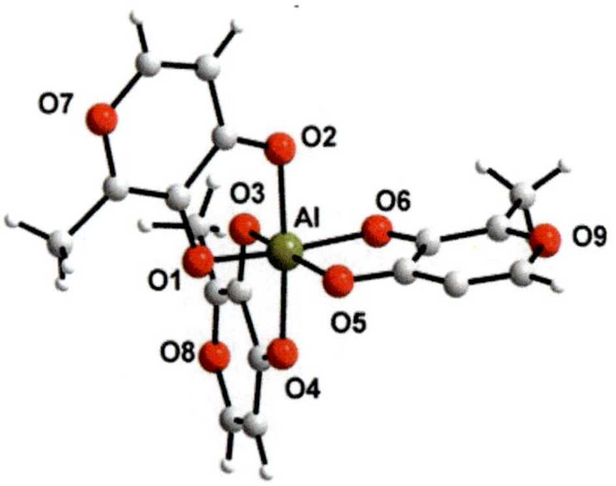

Fig. 5 Crystal structure of Al(III) complex with 3-hydroxy-2-methyl-4*H*-pyran-4-one chelator, $Al(mp)_3$ (DUGJEU)

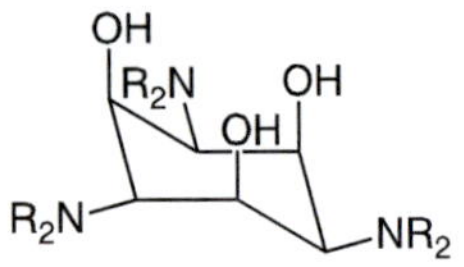

taci: R = H
1,3,5-Triamino-1,3,5-trideoxy-*cis*-inositol
dtci: R = CH_3
1,3,5-Tris-dimethylamino-1,3,5-trideoxy-*cis*-inositol

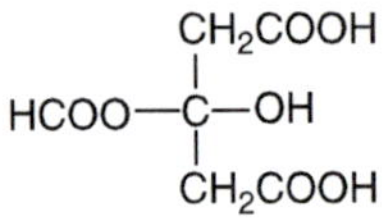

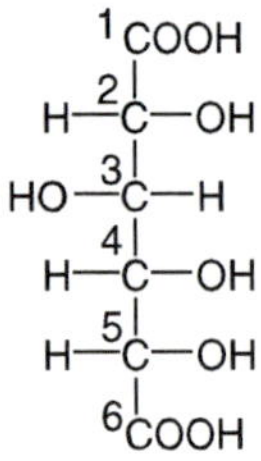

H_4citr: Citric Acid **H_4Dsacc:** D-saccharic acid

Scheme 3

(log K = 18.8) is significantly reduced with respect to that of $[Al(dtci)_2]^{3+}$, (log K = 26.4). In Table 2 the crystal data for these complexes are reported; Figure 6 shows the structure of the cationic complex $[Al(taci)_2]^{3+}$.

Al(III), as a typical hard metal, prefers the coordination of negatively charged oxygen donors, such as carboxylates, phenolates, alcoholates and phosphates. Among them the tridentate ligands such as citric acid, tartaric acid and sugar carboxylates (Scheme 3) are able to hinder the hydrolysis of the Al(III) under physiological conditions. In the polydentate coordination of these ligands, the alcoholate and carboxylate oxygen atoms behave as donors. Table 2 reports the crystal data of Al(III) complexes with citrate and D-saccharate anions.

Citric acid, a tricarboxylic acid, is the most prevalent ion binder in plasma. The presence of citrate anions is also related to the elevate absorption and toxic effect of aluminium in biological tissues. At about pH = 8 citrate is fully deprotonated and two citrate ligands are linked to Al(III).

Table 2 List of crystal structures of Al-(6O) complexes with tridentate (3O) chelators

Compound	CSD code	Reference
$Al(taci)_2Br_3{\cdot}7H_2O$	PEJTIH	[140]
$Al(tdci)_2Cl_3{\cdot}15H_2O$	ZAWMUF	[141]
$(NH_4)_5Al(citr)_2{\cdot}2H_2O$	HIXBEV01	[142]
$(NH_4)_4Al(citr)(Hcitr){\cdot}3H_2O$	QICFEN	[142]
$K_4Al(citr)(Hcitr){\cdot}4H_2O$	QICJIV	[142]
$(NH_4)_4Al_2(citr)(Hcitr)_2{\cdot}4H_2O$	QADXEZ	[143]
$2[Al_3(citr)_2(OH)_2(H_2O)_4]^-{\cdot}NO_3^-{\cdot}Al(H_2O)_6^{3+}$	IQECEM	[144]
$K_2[Al_2(Dsacc)_2(H_2O)_2]{\cdot}H_2O$	AYEGOA	[145]

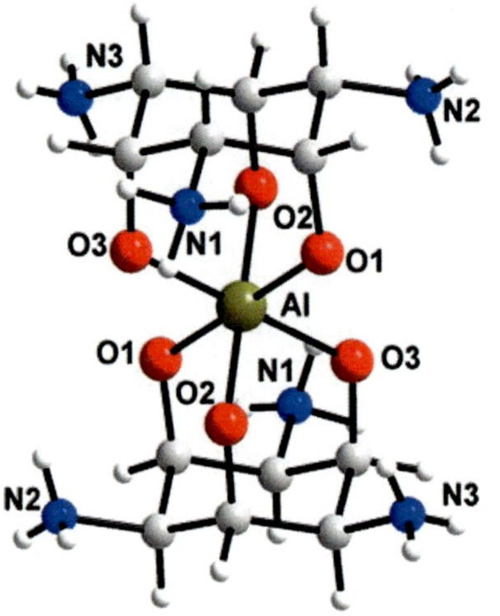

Fig. 6 Crystal structure of the cationic complex $Al(taci)_2^{3+}$ (ZAWMUF)

Each citrate binds Al(III) through the alcoholate and carboxylate groups in the equatorial plane and through a terminal carboxylate in the axial position. The other terminal carboxylate group, also deprotonated, remains free. The structure of the complex anion $[Al(citr)_2]^{5-}$ (HIXBEV01) is reported in Fig. 7a. At pH of about 4.5 the structure of the complex anion displays an octahedral geometry with two citrate ligands coordinated to Al(III), one of the citrate is fully deprotonated whereas the other only triply deprotonated. The hydrogen on the latter citrate lies on the terminal carboxylate group which remains unbound. The structure of the anion $[Al(citr)(Hcitr)]^{4-}$(QICFEN) is reported in Fig. 7b. The first dinuclear citrate Al(III) complex was isolated at low pH $\approx$ 3.5. The crystal structure shows a dinuclear assembly of two aluminium ions, octahedrally coordinated to three citrate ligands of different protonation states. The structure of the complex anion $[Al_2(citr)(Hcitr)_2]^{4-}$(QADXEZ) is displayed in Fig. 7c. A trinuclear aluminium citrate complex was observed at very low pH (1.66). The anion contains two quadruply deprotonated citrate ligands. All three of the Al(III) atoms are coordinated in distorted octahedral geometry. Al1 is

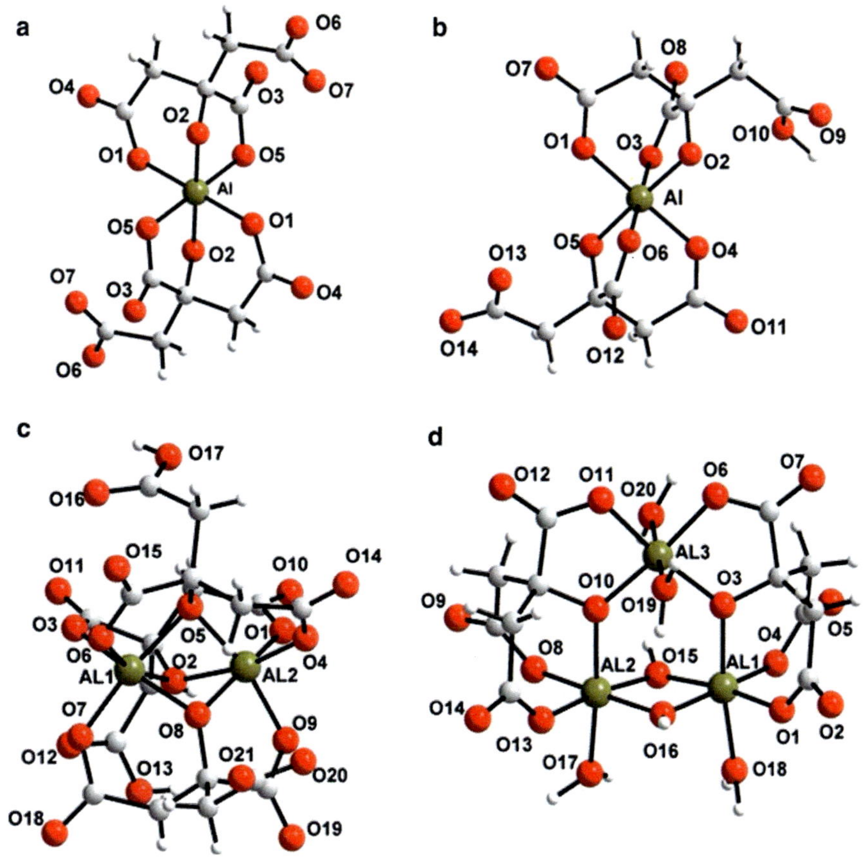

Fig. 7 Crystal structures of the complex anions: (**a**) $[Al(citr)_2]^{5-}$ (HIXBEV01); (**b**) $[Al(citr)(Hcitr)]^{4-}$(QICFEN); (**c**) $[Al_2(citr)(Hcitr)_2]^{4-}$(QADXEZ); (**d**) $[Al_3(citr)_2(OH)_2(H_2O)_4]^{-}$ (IQECEM)

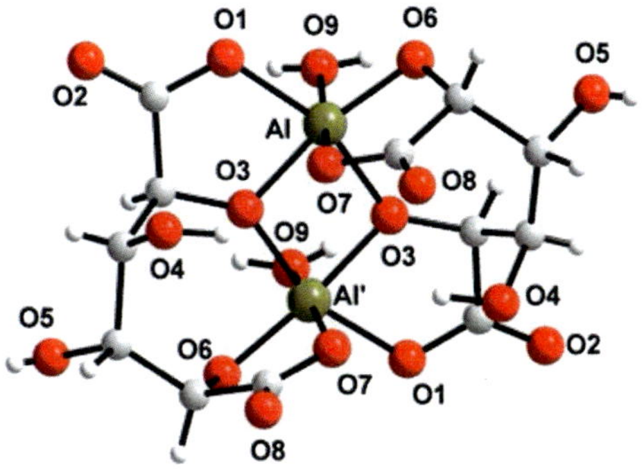

Fig. 8 Crystal structure of the anionic complex $[Al_2(Dsacc)_2\ (H_2O)_2]^{2-}$ (AYEGOA)

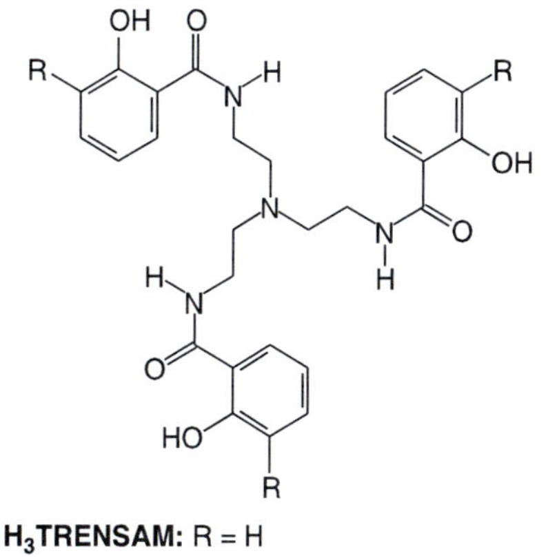

H_3TRENSAM: R = H

tris[(2'-hydroxybenzoyl)-2-aminoethyl]-amine

H_3TREN(3M)SAM: R = OCH_3

tris[(2'-hydroxy-3-methoxybenzoyl)-2-aminoethyl]-amine

Scheme 4

linked to Al2 via two μ_2-hydroxide bridges. Al3 completes its octahedral coordination by means of a carboxylate oxygen from two different citrate ligands and two coordinated waters. The crystal structure of the anion $[Al_3(citr)_2(OH)_2(H_2O)_4]^{-}$(IQECEM) is shown in Fig. 7d.

D-saccharic acid produces a dinuclear anionic complex $[Al_2(Dsacc)_2(H_2O)_2]^{2-}$ between pH 5.5 and 7.5. The crystal structure shows a distorted octahedral geometry around each Al(III) ion. Each aluminium atom is coordinated to two alcoholate and one carboxylate group of the same molecular unit, to one alcoholate and one carboxylate group of the other molecular unit and to one water molecule. The anionic complex $[Al_2(Dsacc)_2(H_2O)_2]^{2-}$ (AYEGOA) is given in Fig. 8.

Table 3 List of crystal structures of Al-(6O) complexes with hexadentate (6O) chelators

Compound	CSD code	Reference
Al(TRENSAM)·HCl·CH_3OH	HAKNIQ	[146]
Al[TREN(3M)SAM]·HCl	HAKQIT	[146]

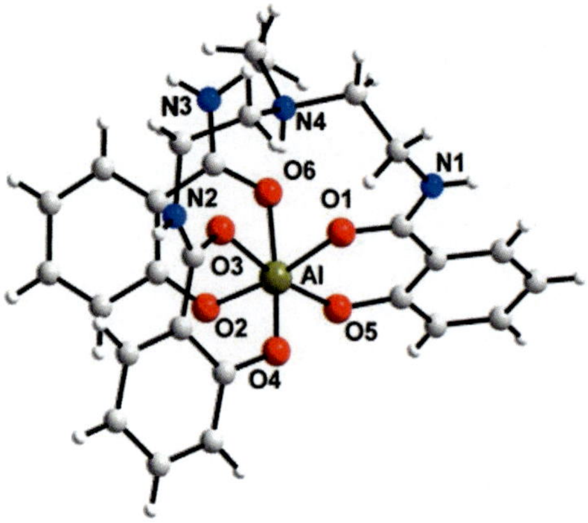

Fig. 9 Crystal structure of cationic complex $[Al(HTRENSAM)]^+$ (HAKNIQ)

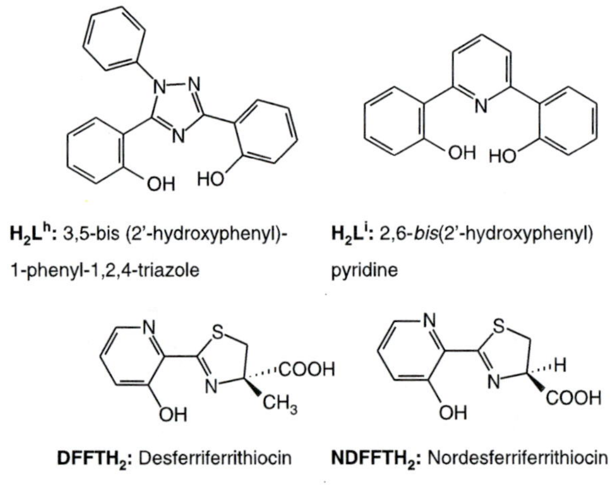

Scheme 5

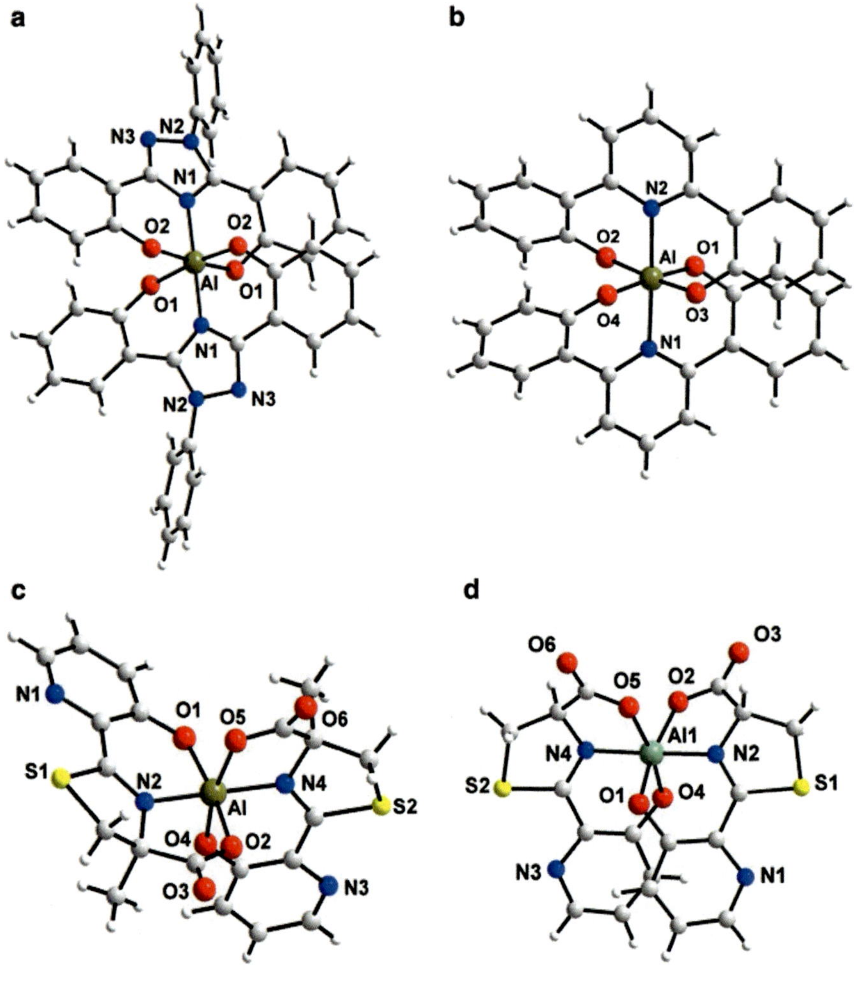

Fig. 10 Crystal structure of the complex anions: (**a**) $[Al(L^h)_2]^-$ (SAJFUF); (**b**) $[Al(L^i)_2]^-$(HALXEY); (**c**) $[Al(DFFT)_2]^-$ (TOVJET) and (**d**) $[Al(NDFFT)_2]^-$ (TOVJIX)

Hexadentate (6O) chelators

The coordination chemistry of the new model ligands TRENSAM and TREN(3M)SAM evidences the previously proposed tris-salicylate mode of binding for ferric enterobactin. Accordingly, they form 1:1 metal–ligand complexes with Fe(III) and Al(III) through tris-salycilato coordination (i.e., the chelate is a six-membered metallocycle with the phenolic and amide oxygens binding the metal centre). The Fe(III) and Al(III) complexes are isostructural. The central TREN nitrogens are protonated and the complexes crystallize as hydrochloric salts. The structures of the chelators are shown in Scheme 4 and the crystal data of Al complexes are given in Table 3. The crystal structure of the cationic complex $[Al(HTRENSAM)]^+$ (HAKNIQ) is reported in Fig. 9.

Al-(4O,2N) complexes

Tridentate (2O,N) chelators

For steric considerations the triazole derivative H_2L^h (Scheme 5), which form six-membered chelate rings, should display a preference for small metal cations such as Fe(III) and Al(III). For the similar pyridine base ligand H_2L^i (Scheme 5) the pyridine nitrogen donor is more nucleophilic than the triazole one, and this would further improve the complex stability. The Al(III) and Fe(III) complexes are isotypic. Both the chelators are candidate for the detoxification of Al(III). Figure 10a, b report the crystal structures of the anionic Al(III) complexes $[Al(L^h)_2]^-$ (SAJFUF) and $[Al(L^i)_2]^-$(HALXEY), respectively.

DFTTH2 and NDFFTH2 ligands (Scheme 5) show typical activity as possible drugs for the treatment of iron and aluminium overload. While DFFTH2 is significantly toxic, NDFFTH2 exhibits a lower toxicity. Both molecules are tridentate ligands coordinating via the phenolate oxygen atom, the N atom in the thiazoline ring and the carboxylate oxygen atom. The molecule of DFTTH2 is chiral with S absolute configuration at the quaternary carbon atom. Also the Al(III) and Fe(III) complexes are chiral and their stability constant are very high, $\log K = 23.6$ and 31.0 for Al(III) and Fe(III), respectively. For NDFFH2 the chiral carbon has a configuration R and the stability constants $\log K$ are 22.0 and 29.1 for Al(III) and Fe(III) complexes, respectively. The ligand structures are shown in Scheme 5 and the crystal structures of the complex anions $[Al(DFFT)_2]^-$ (TOVJET) and $[Al(NDFFT)_2]^-$ (TOVJIX) are given in Fig. 10c, d, respectively (Table 4).

Table 4 List of crystal structures of Al-(4O,2N) complexes with tridentate (2O,N) chelators

Compound	CSD code	References
$Na[Al(L^h)_2]·H_2O·C_2H_5OH$	SAJFUF	[147, 148]
$Na[Al(L^i)_2]·2C_2H_5OH$	HALXEY	[147, 148]
$K[Al(DFFT)_2]·2C_2H_5OH$	TOVJET	[149]
$K[Al(NDFFT)_2]·1.166H_2O·C_2H_5OH$	TOVJIX	[149]

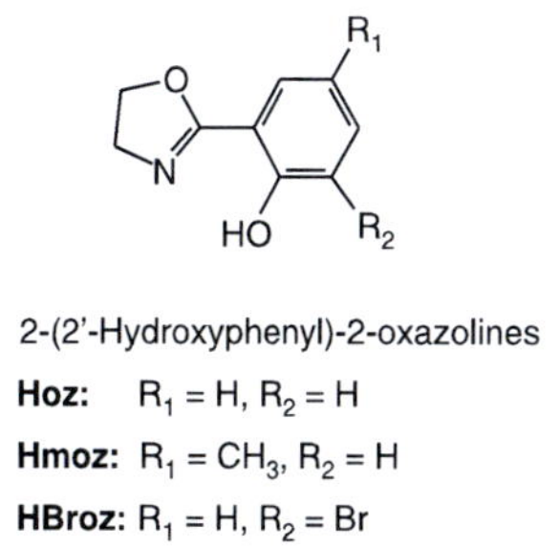

Scheme 6

Table 5 List of crystal structures of Al-(3O,3N) complexes with bidentate (O,N) chelators

Compound	CSD code	Reference
$Al(oz)_3·CH_3OH$	WADJUG	[150]
$Al(moz)_3$	WADKIV	[150]
$Al(Broz)_3·CH_3OH$	WIJYAP	[151]

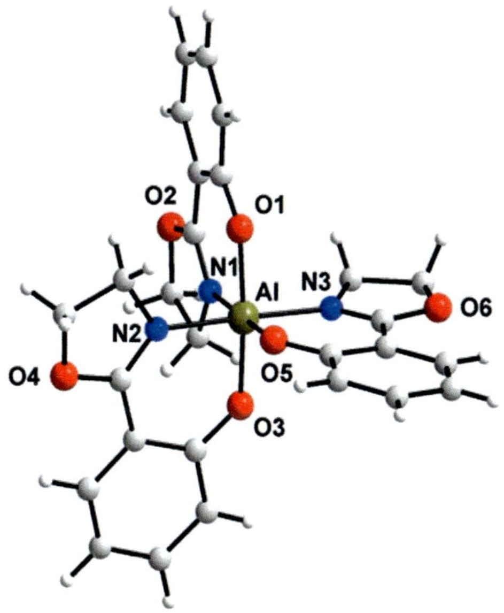

Fig. 11 Crystal structure of the neutral complex $Al(oz)_3$ (WADJUG)

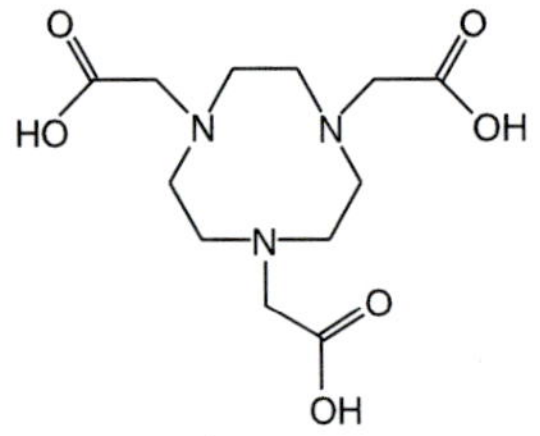

H_3TCTA: 1,4,7-Tris(aceticacid)-1,4,7-triazacyclononanone

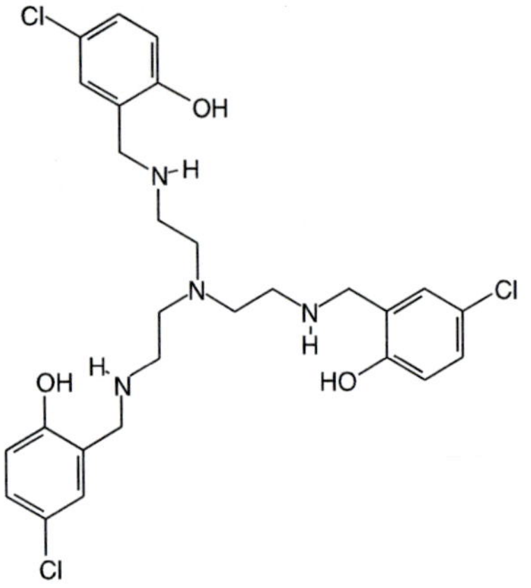

H_3L2: Tris(5'-chloro-2'-hydroxybenzylaminoethyl)amine

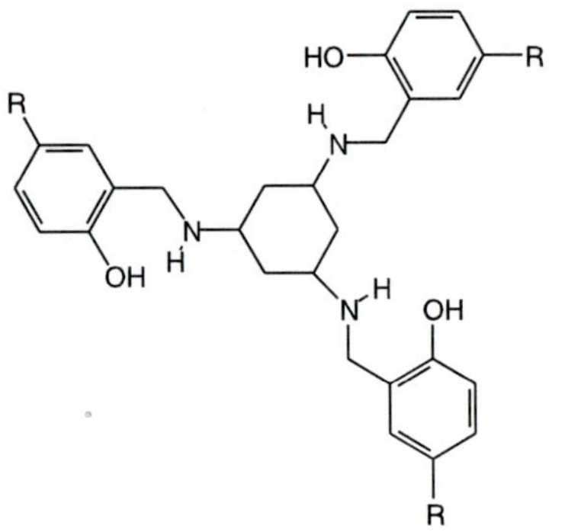

(salH2)$_3$tachH$_3$: R = H
1,3,5-*cis*-Tris((2'-hydroxybenzyl) amino)cyclohexane
(NO$_2$salH2)$_3$tachH$_3$: R = NO2
1,3,5-*cis*-Tris((2'-hydroxy-5'-nitrobenzyl) amino)cyclohexane

Scheme 7

Al-(3O,3N) complexes

Bidentate (O,N) chelators

The (2′-hydroxyphenyl)-2-oxazolines (Scheme 6) act as bidentate ligands via the phenolate oxygen and the oxazoline ring nitrogen to form neutral complexes of Al(III). The 2-(2′-Hydroxyphenyl)-2-oxazoline is natural occurring and is found in certain classes of siderophores. The X-ray crystallographic structures reveal that Al(III) complexes are *pseudo* octahedral in *mer* configuration. Table 5 reports the crystal data of complexes with Al(III) cation and Fig. 11 reports the crystal structure of the neutral octahedral [Al(oz)$_3$] complex (WADJUG).

Hexadentate (3O,3N) chelators

Polydentate ligands with trianionic pendant arm forming three-dimensional cavities are of interest because of the high stability of their trivalent metal complexes and the high selectivity in their binding. Scheme 7 reports the structures of some polydentate chelators and Table 6 gives the crystal data of their Al(III) complexes.

Table 6 List of crystal structures of Al-(3O,3N) complexes with hexadentate (3O,3N) chelators

Compound	CSD code	References
Al(TCTA)	VIPCEC01	[152]
[Al(HL2)]ClO$_4$·2CH$_3$OH	WACYOO	[153]
Al(salH2)tach·3.5H$_2$O	YOZWEP	[154]
Al(NO$_2$salH2)tach·C$_2$H$_5$OH	HECWUG10	[154]

[Al(TCTA)] is the neutral complex between Al(III) and the anionic chelator 1,4,7-tris(acetate)-1,4,7-triazacyclononanone. The Al cation is coordinated to the three nitrogens of the macrocycle and three oxygens of the pendant acetate groups. The complex has slightly distorted octahedral geometry. The [Al(TCTA)] crystal structure (VIPCEC01) is shown in Fig. 12a. [Al(TCTA)] stability constant is very high, with log $K > 30$, and the complex is very stable in aqueous solution over a wide range of pH (0–13) for many days at room temperature.

In the cationic complex [Al(HL2)]$^+$ the Al(III) is coordinated to the tripodal chelator by three aminic nitrogens and the three O^- negative oxygens of the phenolate groups,

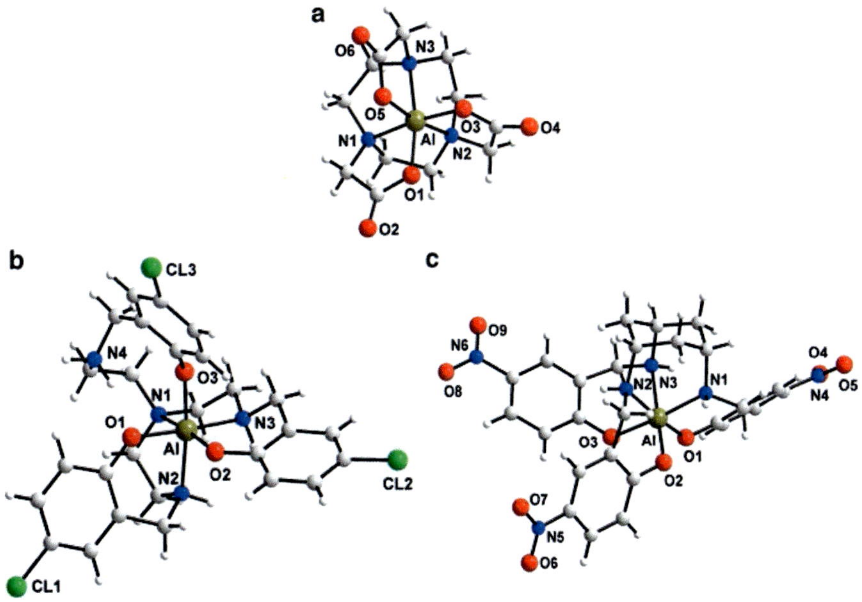

Fig. 12 Crystal structures of the complexes (**a**) Al(TCTA) (VIPCEC01); (**b**) [Al (HL2)]$^+$(WACYOO) and (**c**) Al (NO_2salH2)tach (HECWUG10)

leaving the protonated N4H$^+$ nitrogen uncoordinated. The coordination geometry is distorted octahedrally. The crystal structure of the cation complex [Al(HL2)]$^+$ is shown in Fig. 12b.

Another set of important hexadentate ligands is based on a cyclohexanetriamine backbone as, for instance, a ligand system that has three phenolic groups linked to the *cis* form of 1,3,5-triaminocyclohexane. The *cis* stereochemistry of the cyclohexane moiety can facilitate the complexation. In the structures of [Al(salH2)$_3$tach] and [Al(NO_2salH2)$_3$tach], the aluminium in an octahedral environment and the ligands act as hexadentate. The structures have an approximately C3 symmetry. The crystal structure of the neutral [Al(NO_2salH2)$_3$tach] (HECWUG10) complex is shown in Fig. 12c.

References

1. Macdonald TL, Martin RB (1988) Aluminum ion in biological-systems. Trends Biochem Sci 13:15–19
2. Exley C (2003) A biogeochemical cycle for aluminium? J Inorg Biochem 97:1–7
3. Humphreys S, Bolger M (1997) In: Zatta PF, Alfrey AC (eds) A public health analysis of dietary aluminium. World Scientific, Singapore
4. Yokel RA, Hicks CL, Florence RL (2008) Aluminum bioavailability from basic sodium aluminum phosphate, an approved food additive emulsifying agent, incorporated in cheese. Food Chem Toxicol 46:2261–2266
5. Hem SL (2002) Elimination of aluminum adjuvants. Vaccine 20: S40–S43
6. Greger JL, Sutherland JE (1997) Aluminum exposure and metabolism. Crit Rev Clin Lab Sci 34:439–474
7. Martin RB (1986) The chemistry of aluminum as related to biology and medicine. Clin Chem 32:1797–1806
8. Exley C (2009) Aluminium and medicine. In: Mercè ALR, Felcman J, Cecio J (eds) Molecular and supramolecular bioinorganic chemistry, Applications in medical sciences. Nova Biomedical Books, Hauppauge, NY
9. Exley C (2009) Darwin, natural selection and the biological essentiality of aluminium and silicon. Trends Biochem Sci 34:589–593
10. Jorgensen SE, Jensen A (1984) Met Ions Biol Syst 18:61–74
11. Kopacek J, Hejzlar J, Kana J, Norton SA, Porcal P, Turek J (2009) Trends in aluminium export from a mountainous area to surface waters, from deglaciation to the recent: effects of vegetation and soil development, atmospheric acidification, and nitrogen-saturation. J Inorg Biochem 103:1439–1448
12. Martin RB (1994) Aluminum – a neurotoxic product of acid-rain. Acc Chem Res 27:204–210
13. Nicolini M, Zatta PF, Corain B (Eds.) (1991) Aluminum in Chemistry, Biology and Medicine, Cortina International-Raven Press, Verona, New York
14. Aramini JM, Saponja JA, Vogel HJ (1996) Spectroscopic studies of the interaction of aluminum(III) with transferrins. Coord Chem Rev 149:193–229
15. Berthon G (1996) Chemical speciation studies in relation to aluminium metabolism and toxicity. Coord Chem Rev 149:241–280
16. Harris WR (1996) Binding and transport of aluminum by serum proteins. Coord Chem Rev 149:347–365
17. Kiss T, Zatta P, Corain B (1996) Interaction of aluminium (III) with phosphate-binding sites: biological aspects and implications. Coord Chem Rev 149:329–346
18. Martell AE, Hancock RD, Smith RM, Motekaitis RJ (1996) Coordination of Al(III) in the environment and in biological systems. Coord Chem Rev 149:311–328
19. Berthon G (2002) Aluminium speciation in relation to aluminium bioavailability, metabolism and toxicity. Coord Chem Rev 228:319–341
20. Bodor A, Banyai I, Toth I (2002) Slow dynamics of aluminium-citrate complexes studied by 1H- and 13C-NMR spectroscopy. Coord Chem Rev 228:163–173
21. Crichton RR, Florence A, Ward RJ (2002) Aluminium and iron in the brain-prospects for chelation. Coord Chem Rev 228:365–371
22. Rubini P, Lakatos A, Champmartin D, Kiss T (2002) Speciation and structural aspects of interactions of Al(III) with small biomolecules. Coord Chem Rev 228:137–152
23. Santos MA (2002) Hydroxypyridinone complexes with aluminium. In vitro/vivo studies and perspectives. Coord Chem Rev 228:187–203

24. Yokel RA (2002) Aluminum chelation principles and recent advances. Coord Chem Rev 228:97–113
25. Yokel RA (2002) Brain uptake, retention, and efflux of aluminum and manganese. Environ Heal Perspect 110:699–704
26. Crisponi G, Nurchi VM, Faa G, Remelli M (2011) Human diseases related to aluminium overload. Mon Chem 142:331–340
27. Crisponi G, Nurchi VM, Bertolasi V, Remelli M, Faa G (2012) Chelating agents for human diseases related to aluminium overload. Coord Chem Rev 256:89–104
28. Exley C, Pinnegar JK, Taylor H (1997) Hydroxyaluminosilicates and acute aluminium toxicity in fish. J Theor Biol 189:133–139
29. Exley C, Schneider C, Doucet FJ (2002) The reaction of aluminium with silicic acid in acidic solution: an important mechanism in controlling the biological availability of aluminium? Coord Chem Rev 228:127–135
30. Arenas MD, Malek T, Gil MT, Moledous A, Nunez C, Alvarez-Ude F (2008) Use of the aluminium phosphate-binders in hemodialysis in the ultrapure water era. Nefrologia 28:168–173
31. Yokel RA, Florence RL (2008) Aluminum bioavailability from tea infusion. Food Chem Toxicol 46:3659–3663
32. Malik J, Szakova J, Drabek O, Balik J, Kokoska L (2008) Determination of certain micro and macroelements in plant stimulants and their infusions. Food Chem 111:520–525
33. Milacic R (2005) In: Cornelis R, Caruso J, Crews H, Heumann K (ed) Handbook of elemental speciation II – species in environment, food, medicine and occupational health. Wiley, New York
34. Froment DP, Molitoris BA, Buddington B, Miller N, Alfrey AC (1989) Site and mechanism of enhanced gastrointestinal absorption of aluminum by citrate. Kidney Int 36:978–984
35. Yokel RA, McNamara PJ (2001) Aluminium toxicokinetics: an updated MiniReview. Pharmacol Toxicol 88:159–167
36. Varner JA, Jensen KF, Horvath W, Isaacson RL (1998) Chronic administration of aluminum-fluoride or sodium-fluoride to rats in drinking water: alterations in neuronal and cerebrovascular integrity. Brain Res 784:284–298
37. Frankova A, Drabek O, Havlik J, Szakova J, Vanek A (2009) The effect of beverage preparation method on aluminium content in coffee infusions. J Inorg Biochem 103:1480–1485
38. Aspenstrom-Fagerlund B, Sundstrom B, Tallkvist J, Ilback NG, Glynn AW (2009) Fatty acids increase paracellular absorption of aluminium across Caco-2 cell monolayers. Chem Biol Interact 181:272–278
39. Krewski D, Yokel RA, Nieboer E, Borchelt D, Cohen J, Harry J, Kacew S, Lindsay J, Mahfouz AM, Rondeau V (2007) Human health risk assessment for aluminium, aluminium oxide, and aluminium hydroxide. J Toxicol Environ Health B Crit Rev 10:1–269
40. Exley C, Ahmed U, Polwart A, Bloor RN (2007) Elevated urinary aluminium in current and past users of illicit heroin. Addict Biol 12:197–199
41. Yong RL, Holmes DT, Sreenivasan GM (2006) Aluminum toxicity due to intravenous injection of boiled methadone. New Engl J Med 354:1210–1211
42. Orihuela D (2009) Inhibitory effect of aluminium on calcium absorption in small intestine of rats with different thyroid hormone status. J Inorg Biochem 103:1542–1547
43. Domingo JL (1995) Reproductive and developmental toxicity of aluminum – a review. Neurotoxicol Teratol 17:515–521
44. Moore PB, Edwardson JA, Ferrier IN, Taylor GA, Lett D, Tyrer SP, Day JP, King SJ, Lilley JS (1997) Gastrointestinal absorption of aluminum is increased in Down's syndrome. Biol Psychiatry 41:488–492
45. Flarend R, Bin T, Elmore D, Hem SL (2001) A preliminary study of the dermal absorption of aluminium from antiperspirants using aluminium-26. Food Chem Toxicol 39:163–168
46. Bertholf RL, M. R. Wills MR, Savory J (1988) In: H. G. Seiler and H. Sigel (ed) Handbook on toxicity of inorganic compounds. Marcel Dekker, New York
47. Sjogren B, Lundberg I, Lidums V (1983) Aluminum in the blood and urine of industrially exposed workers. Br J Ind Med 40:301–304
48. de Kom JFM, Dissels HMH, van der Voet GB, de Wolff FA (1997) Serum aluminium levels of workers in the bauxite mines. J Toxicol Clin Toxicol 35:645–651
49. McDonald B, Haszard R, Spence A, Osbourne K (1996) 492A mortality study of Alzheimer's disease and aluminium exposure through inhalation of McIntyre powder in Cornish Tin Miners. Neurobiol Aging 17:S122–S123
50. Couette M, Boisse MF, Maison P, Brugieres P, Cesaro P, Chevalier X, Gherardi RK, Bachoud-Levi AC, Authier FJ (2009) Long-term persistence of vaccine-derived aluminum hydroxide is associated with chronic cognitive dysfunction. J Inorg Biochem 103:1571–1578
51. Zhou YZ, Yokel RA (2005) The chemical species of aluminum influences its paracellular flux across and uptake into Caco-2 cells, a model of gastrointestinal absorption. Toxicol Sci 87:15–26
52. Zhou Y, Harris WR, Yokel RA (2008) The influence of citrate, maltolate and fluoride on the gastrointestinal absorption of aluminum at a drinking water-relevant concentration: A Al-26 and C-14 study. J Inorg Biochem 102:798–808
53. Priest ND (2004) The biological behaviour and bioavailability of aluminium in man, with special reference to studies employing aluminium-26 as a tracer: review and study update. J Environ Monit 6:375–403
54. Wrobel K, Gonzalez EB, Sanzmedel A (1995) Aluminum and silicon speciation in human serum by ion-exchange high-performance liquid-chromatography electrothermal atomic-absorption spectrometry and gel-electrophoresis. Analyst 120:809–815
55. Milacic R, Murko S, Scancar J (2009) Problems and progresses in speciation of Al in human serum: an overview. J Inorg Biochem 103:1504–1513
56. Garcia Alonso JI, Lopez Garcia A, Perez Parajon J, Blanco Gonzalez E, Sanz Medel A, Cannata JB (1990) High-performance liquid-chromatography methods for studying protein-binding of aluminum in human serum in the absence and in the presence of desferrioxamine. Clin Chim Acta 189:69–80
57. Martin R (1991) Aluminium in biological systems. In: Aluminium in chemistry, biology and medicine. Raven Press, New York
58. Martin RB (1991) Fe^{3+} and Al^{3+} hydrolysis equilibria – cooperativity in Al^{3+} hydrolysis reactions. J Inorg Biochem 44:141–147
59. VanLandeghem GF, D'Haese PC, Lamberts LV, DeBroe ME (1997) Competition of iron and aluminum for transferrin: the molecular basis for aluminum deposition in iron-overloaded dialysis patients? Exp Nephrol 5:239–245
60. Talbot RJ, Newton D, Priest ND, Austin JG, Day JP (1995) Intersubject variability in the metabolism of aluminum following intravenous-injection as citrate. Hum Exp Toxicol 14:595–599
61. Gonzalez MA, Bernal CA, Mahieu S, Carrillo MC (2009) The interactions between the chronic exposure to aluminum and liver regeneration on bile flow and organic anion transport in rats. Biol Trace Elem Res 127:164–176
62. Garbossa G, Gutnisky A, Nesse A (1994) The inhibitory-action of aluminum on mouse bone-marrow cell-growth – evidence for an erythropoietin-mediated and transferrin-mediated mechanism. Miner Electrolyte Metab 20:141–146
63. ICRP (1975) Report of Task Group on Reference Man, ICRU Publication 23. Pergamon Press, Oxford, England
64. Flendrig JA, Kruis H, Das HA (1976) Aluminum and dialysis dementia. Lancet 307:1235–1235

65. Smans KA, D'Haese PC, Van Landeghem GF, Andries LJ, Lamberts LV, Hendy GN, De Broe ME (2000) Transferrin-mediated uptake of aluminium by human parathyroid cells results in reduced parathyroid hormone secretion. Nephrol Dial Transplant 15:1328–1336
66. Gonzalez MA, Alvarez MdL, Pisani GB, Bernal CA, Roma MG, Carrill MC (2007) Involvement of oxidative stress in the impairment in biliary secretory function induced by intraperitoneal administration of aluminum to rats. Biol Trace Elem Res 116:329–348
67. Walton JR (2006) Aluminum in hippocampal neurons from humans with Alzheimer's disease. Neurotoxicology 27:385–394
68. Nebeker HG, Coburn JW (1986) Aluminum and renal osteodystrophy. Annu Rev Med 37:79–95
69. Malluche HH, Monierfaugere MC (1992) Risk of adynamic bone-disease in dialyzed patients. Kidney Int 42:S62–S67
70. Kasai K, Hori MT, Goodman WG (1991) Transferrin enhances the antiproliferative effect of aluminum on osteoblast-like cells. Am J Physiol 260:E537–E543
71. Torres A, Lorenzo V, Hernandez D, Rodriguez JC, Concepcion MT, Rodriguez AP, Hernandez A, Debonis E, Darias E, Gonzalezposada JM, Losada M, Rufino M, Felsenfeld AJ, Rodriguez M (1995) Bone-disease in predialysis, hemodialysis, and capd patients – evidence of a better bone response to PTH. Kidney Int 47:1434–1442
72. Drueke TB, Barany P, Cazzola M, Eschbach JW, Grutzmacher P, Kaltwasser JP, Macdougall IC, Pippard MJ, Shaldon S, vanWyck D (1997) Management of iron deficiency in renal anemia: guidelines for the optimal therapeutic approach in erythropoietin-treated patients. Clin Nephrol 48:1–8
73. Shea TB, Beermann ML, Wang FS (1995) Relative susceptibility of cytoskeleton-associated and soluble neurofilament subunits to aluminum exposure in intact-cells – a possible mechanism for reduction of neurofilament axonal-transport during aluminum neurotoxicity. Mol Chem Neuropathol 24:203–219
74. Altmann P, Cunningham J, Dhanesha U, Ballard M, Thompson J, Marsh F (1999) Disturbance of cerebral function in people exposed to drinking water contaminated with aluminium sulphate: retrospective study of the Camelford water incident. BMJ 319:807–811
75. Yokel RA, Allen DD, Ackley DC (1999) The distribution of aluminum into and out of the brain. J Inorg Biochem 76:127–132
76. Bressler JP, Olivi L, Cheong JH, Kim Y, Maerten A, Bannon D (2007) Metal transporters in intestine and brain: their involvement in metal-associated neurotoxicities. Hum Exp Toxicol 26:221–229
77. Gerhart DZ, Enerson BE, Zhdankina OY, Leino RL, Drewes LR (1997) Expression of monocarboxylate transporter MCT1 by brain endothelium and glia in adult and suckling rats. Am J Physiol Endocrinol Metab 273:E207–E213
78. Stoehr G, Luebbers K, Wilhelm M, Hoelzer J, Ohmann C (2006) Aluminum load in ICU patients during stress ulcer prophylaxis. Eur J Int Med 17:561–566
79. Drueke TB (2002) Intestinal absorption of aluminium in renal failure. Nephrol Dial Transplant 17:13–16
80. Williams JW, Vera SR, Peters TG, Luther RW, Bhattacharya S, Spears H, Graham A, Pitcock JA, Crawford AJ (1986) Biliary-excretion of aluminum in aluminum osteodystrophy with liver-disease. Ann Int Med 104:782–785
81. Monteagudo FSE, Isaacson LC, Wilson G, Hickman R, Folb PI (1988) Aluminum excretion by the distal tubule of the pig-kidney. Nephron 49:245–250
82. Ezomo OF, Matsushima F, Meshitsuka S (2009) Up-regulation in the expression of renin gene by the influence of aluminium. J Inorg Biochem 103:1563–1570
83. Priest ND, Newton D, Day JP, Talbot RJ, Warner AJ (1995) Human metabolism of Al-26 and Ga-67 injected as citrates. Hum Exp Toxicol 14:287–293
84. Shannon RD (1976) Revised effective ionic-radii and systematic studies of interatomic distances in halides and chalcogenides. Acta Crystallogr Sect A 32:751–767
85. Martin RB (1996) Ternary complexes of Al^{3+} and F- with a third ligand. Coord Chem Rev 149:23–32
86. Agarwal RP, Moreno EC (1971) Stability constants of aluminium fluoride complexes. Talanta 18:873–880
87. Frankowski M, Ziola-Frankowska A (2010) Speciation analysis of aluminium and aluminium fluoride complexes by HPIC-UVVIS. Talanta 82:1763–1769
88. Bogatko S, Cauet E, Geerlingst P (2011) Influence of F- coordination on Al^{3+} hydrolysis reactions from density functional theory calculations. J Phys Chem C 115:6910–6921
89. Graf H, Stummvoll HK, Meisinger V, Kovarik J, Wolf A, Pinggera WF (1981) Aluminum removal by hemodialysis. Kidney Int 19:587–592
90. Brown DJ, Ham KN, Dawborn JK, Xipell JM (1982) Treatment of dialysis osteomalacia with desferrioxamine. Lancet 2:343–345
91. Nebeker HG, Milliner DS, Ott SA, Sherrard DJ, Alfrey AC, Abuelo JG, Wasserstein A, Coburn JW (1984) Aluminum-related osteomalacia – clinical-response to desferrioxamine. Kidney Int 25:173–173
92. Ciancioni C, Poignet JL, Mauras Y, Panthier G, Delons S, Allain P, Man NK (1984) Plasma aluminum and iron kinetics in hemodialyzed patients after iv infusion of desferrioxamine. Trans Am Soc Artif Intern Organs 30:479–482
93. Molitoris BA, Alfrey PS, Miller NL, Hasbargen JA, Kaehney WD, Alfrey AC, Smith BJ (1987) Efficacy of intramuscular and intraperitoneal deferoxamine for aluminum chelation. Kidney Int 31:986–991
94. Yokel RA, Ackrill P, Burgess E, Day JP, Domingo JL, Flaten TP, Savory J (1996) Prevention and treatment of aluminum toxicity including chelation therapy: status and research needs. J Toxicol Environ Health 48:667–683
95. Yokel RA, Rhineheimer SS, Sharma P, Elmore D, McNamara PJ (2001) Entry, half-life, and desferrioxamine-accelerated clearance of brain aluminum after a single Al-26 exposure. Toxicol Sci 64:77–82
96. Nakamura H, Rose PG, Blumer JL, Reed MD (2000) Acute encephalopathy due to aluminum toxicity successfully treated by combined intravenous deferoxamine and hemodialysis. J Clin Pharmacol 40:296–300
97. Wang G, Zhu P, Wang S (1996) [The causes, diagnosis and treatment of aluminum toxicity in patients with chronic renal failure undergoing dialysis]. Zhonghua nei ke za zhi [Chinese journal of internal medicine] 35:36–40
98. Kruck TP, Cui JG, Percy ME, Lukiw WJ (2004) Molecular shuttle chelation: the use of ascorbate, desferrioxamine and Feralex-G in combination to remove nuclear bound aluminum. Cell Mol Neurobiol 24:443–459
99. Hegde ML, Bharathi P, Suram A, Venugopal C, Jagannathan R, Poddar P, Srinivas P, Sambamurti K, Rao KJ, Scancar J, Messori L, Zecca L, Zatta P (2009) Challenges associated with metal chelation therapy in Alzheimer's disease. J Alzheimers Dis 17:457–468
100. Liu ZD, Hider RC (2002) Design of clinically useful iron(III)-selective chelators. Med Res Rev 22:26–64
101. Hider RC, Liu ZD (2003) Emerging understanding of the advantage of small molecules such as hydroxypyridinones in the treatment of iron overload. Curr Med Chem 10:1051–1064

102. Santos MA, Gil M, Gano L, Chaves S (2005) Bifunctional 3-hydroxy-4-pyridinone derivatives as potential pharmaceuticals: synthesis, complexation with Fe(III), Al(III) and Ga(III) and in vivo evaluation with (67)Ga. J Biol Inorg Chem 10:564–580
103. Fukuda S (2005) Chelating agents used for plutonium and uranium removal in radiation emergency medicine. Curr Med Chem 12:2765–2770
104. Kruck TPA, Burrow TE (2002) Synthesis of feralex a novel aluminum/iron chelating compound. J Inorg Biochem 88:19–24
105. Chaves S, Dron PI, Danalache FA, Sacoto D, Gano L, Santos MA (2009) Combined chelation based on glycosyl-mono- and bis-hydroxypyridinones for aluminium mobilization: solution and biodistribution studies. J Inorg Biochem 103:1521–1529
106. Gama S, Gil M, Gano L, Farkas E, Santos MA (2009) Combined chelation of bi-functional bis-hydroxypiridinone and mono-hydroxypiridinone: Synthesis, solution and in vivo evaluation. J Inorg Biochem 103:288–298
107. Grazina R, Gano L, Sebestik J, Santos MA (2009) New tripodal hydroxypyridinone based chelating agents for Fe(III), Al(III) and Ga(III): synthesis, physico-chemical properties and bioevaluation. J Inorg Biochem 103:262–273
108. Chaves S, Marques SM, Matos AMF, Nunes A, Gano L, Tuccinardi T, Martinelli A, Santos MA (2010) New tris (hydroxypyridinones) as iron and aluminium sequestering agents: synthesis, complexation and in vivo studies. Chemistry 16:10535–10545
109. Biaso F, Baret P, Pierre J-L, Serratrice G (2002) Comparative studies on the iron chelators O-TRENSOX and TRENCAMS: selectivity of the complexation towards other biologically relevant metal ions and Al^{3+}. J Inorg Biochem 89:123–130
110. Di Marco VB, Bombi GG, Tapparo A, Powell AK, Anson CE (1999) Complexation of aluminium(III) with 3-hydroxy-2(1H)-pyridinone. Solution state study and crystal structure of tris(3-hydroxy-2(1H)-pyridinonato)aluminium(III). J Chem Soc Dalton Trans 2427–2432
111. Di Marco VB, Tapparo A, Bertani R, Bombi GG (1999) Complex formation between aluminium(III) and two pyridine derivatives: 2-hydroxynicotinic and 3-hydroxypicolinic acid. Ann Chim 89:535–546
112. Di Marco VB, Yokel RA, Ferlin MG, Tapparo A, Bombi GG (2002) Evaluation of 3,4-hydroxypyridinecarboxylic acids as possible bidentate chelating agents for aluminium(III): synthesis and metal-ligand solution chemistry. Eur J Inorg Chem 2648–2655
113. Di Marco VB, Dean A, Ferlin MG, Yokel RA, Li HT, Venzo A, Bombi GG (2006) Methyl-hydroxypyridinecarboxylic acids as possible bidentate chelating agents for aluminium(III): synthesis and metal-ligand solution chemistry. Eur J Inorg Chem 1284–1293
114. Bombi GG, Di Marco VB, Marton D, Moro S, Reheman A, Tapparo A, Viero L (2007) 2-Hydroxy-3-carboxy-dlhydrocinnamic acid: complexation properties towards aluminium(III) and iron(III). Polyhedron 26:3419–3427
115. Dean A, Ferlin MG, Brun P, Castagliuolo I, Badocco D, Pastore P, Venzo A, Bombi GG, Di Marco VB (2008) Evaluation of 2-methyl-3-hydroxy-4-pyridinecarboxylic acid as a possible chelating agent for iron and aluminium. Dalton Trans 1689–1697
116. Dean A, Ferlin MG, Brun P, Castagliuolo I, Yokel RA, Badocco D, Pastore P, Venzo A, Bombi GG, Di Marco VB (2009) 1,6-Dimethyl-4-hydroxy-3-pyridinecarboxylic acid and 4-hydroxy-2-methyl-3-pyridinecarboxylic acid as new possible chelating agents for iron and aluminium. Dalton Trans 1815–1824
117. Dean A, Ferlin MG, Brun P, Castagliuolo I, Yokel RA, Venzo A, Bombi GG, Di Marco VB (2011) Evaluation of 4-hydroxy-6-methyl-3-pyridinecarboxylic acid and 2,6-dimethyl-4-hydroxy-3-pyridinecarboxylic acid as chelating agents for iron and aluminium. Inorg Chim Acta 373:179–186
118. Gumienna-Kontecka E, Silvagni R, Lipinski R, Lecouvey M, Marincola FC, Crisponi G, Nurchi VM, Leroux Y, Kozlowski H (2002) Bisphosphonate chelating agents: complexation of Fe(III) and Al(III) by 1-phenyl-1-hydroxymethylene bisphosphonate and its analogues. Inorg Chim Acta 339:111–118
119. Ding H, Xu G, Wang J, Zhang Y, Wu X, Xie Y (2004) Catechol—bisphosphonate conjugates: new types of chelators for metal intoxication therapy. Heteroat Chem 15:549–555
120. Bailly T, Burgada R, Prangé T, Lecouvey M (2003) Synthesis of tetradentate mixed bisphosphonates–new hydroxypyridinonate ligands for metal chelation therapy. Tetrahedron Lett 44:189–192
121. Crisponi G, Nurchi VM, Pivetta T (2008) Potentiometric and spectrophotometric equilibrium study on Fe(III) and new catechol-bisphosphonate conjugates. J Inorg Biochem 102:209–215
122. Crisponi G, Nurchi VM, Pivetta T, Galezowska J, Gumienna-Kontecka E, Bailly T, Burgada R, Kozlowski H (2008) Towards a new attenuating compound: a potentiometric, spectrophotometric and NMR equilibrium study on Fe(III), Al(III) and a new tetradentate mixed bisphosphonate-hydroxypyridinonate ligand. J Inorg Biochem 102:1486–1494
123. Fox RC, Taylor PD (1998) Bis(5-hydroxy-2-hydroxymethyl-pyran-4-one-6-yl)methane: a novel ligand for the intracellular mobilisation of ferritin-bound iron. Bioorg Med Chem Lett 8:443–446
124. Nurchi VM, Crisponi G, Lachowicz JI, Murgia S, Pivetta T, Remelli M, Rescigno A, Niclos-Gutierrez J, Gonzalez-Perez JM, Dominguez-Martin A, Castineiras A, Szewczuk Z (2010) Iron(III) and aluminum(III) complexes with hydroxypyrone ligands aimed to design kojic acid derivatives with new perspectives. J Inorg Biochem 104:560–569
125. Nurchi VM, Lachowicz JI, Crisponi G, Murgia S, Arca M, Pintus A, Gans P, Niclos-Gutierrez J, Dominguez-Martin A, Castineiras A, Remelli M, Szewczuk Z, Lis T (2011) Kojic acid derivatives as powerful chelators for iron(III) and aluminium(III). Dalton Trans 40:5984–5998
126. John VD, Krishnankutty K (2010) Antitumour studies of aluminium complexes of synthetic curcuminoids main group. Met Chem 33:157–166
127. Sahoo SK, Baral M, Bera RK, Kanungo BK (2011) Spectrophotometric and potentiometric studies on the binding abilities of two novel tripodal imine-phenol ligands towards Al(III) and Ga(III). J Solut Chem 40:1187–1199
128. Cornard JP, Merlin JC (2002) Spectroscopic and structural study of complexes of quercetin with Al(III). J Inorg Biochem 92:19–27
129. Kobayashi A, Edo I, Furihata K, Yoshimura E (2005) Secretion of an aluminum chelator, 2-isopropylmalic acid, by the budding yeast, *Saccharomyces cerevisiae*. J Inorg Biochem 99:1260–1263
130. Yoshimura E, Kohdr H, Mori S, Hider RC (2011) The binding of aluminum to mugineic acid and related compounds as studied by potentiometric titration. Biometals 24:723–727
131. Allen FH, Bellard S, Brice MD, Cartwright BA, Doubleday A, Higgs H, Hummelink T, Hummelinkpeters BG, Kennard O, Motherwell WDS, Rodgers JR, Watson DG (1979) Cambridge crystallographic data center – computer-based search, retrieval, analysis and display of information. Acta Crystallogr B 35:2331–2339
132. Nelson WO, Karpishin TB, Rettig SJ, Orvig C (1988) Aluminum and gallium compounds of 3-hydroxy-4-pyridinones – synthesis, characterization, and crystallography of biologically-active complexes with unusual hydrogen-bonding. Inorg Chem 27:1045–1051
133. Nelson WO, Rettig SJ, Orvig C (1989) Aluminum and gallium complexes of 1-ethyl-3-hydroxy-2-methyl-4-pyridinone – a new exoclathrate matrix. Inorg Chem 28:3153–3157

134. Xiao G, van der Helm D, Hider RC, Dobbin PS (1992) Structure-stability relationships of 3-hydroxypyridin-4-one complexes. J Chem Soc Dalton Trans 3265–3271
135. Simpson L, Rettig SJ, Trotter J, Orvig C (1991) 1-Normal-propyl-3-hydroxy-2-methyl-4-pyridinone and 1-normal-butyl-3-hydroxy-2-methyl-4-pyridinone complexes of Group 13 (IIIA) metal-ions. Can J Chem 69:893–900
136. Zhang Z, Rettig SJ, Orvig C (1991) Lipophilic coordination compounds: aluminum, gallium, and indium complexes of 1-aryl-3-hydroxy-2-methyl-4-pyridinones. Inorg Chem 30:509–515
137. Finnegan MM, Rettig SJ, Orvig C (1986) A neutral water-soluble aluminum complex of neurological interest. J Am Chem Soc 108:5033–5035
138. Yu P, Phillips BL, Olmstead MM, Casey WH (2002) Rates of solvent exchange in aqueous aluminium(III)-maltolate complexes. J Chem Soc Dalton Trans 2119–2125
139. Odoko M, Yamamoto K, Hosen M, Okabe N (2003) fac-Tris(2-ethyl-4-oxo-4H-pyran-3-olato-kappa O-2(3), O-4)iron(III) and its aluminium(III) analog. Acta Crystallogr C 59:M121–M123
140. Hegetschweiler K, Ghisletta M, Fassler TF, Nesper R, Schmalle HW, Rihs G (1993) 1,3,5-Triamino-1,3,5-trideoxy-cis-inositol, a new ligand with a remarkable versatility for metal-ions. 3. Preparation and characterization of the aluminum(III), gallium(III), and thallium(III) complexes. Inorg Chem 32:2032–2041
141. Hegetschweiler K, Kradolfer T, Gramlich V, Hancock RD (1995) The design of selective chelating-agents – 1,3,5-trideoxy-1,3,5-tris-(dimethylamino)-cis-inositol, a powerful ligand for hard and highly-charged metal-ions. Chem Eur J 1:74–88
142. Matzapetakis M, Kourgiantakis M, Dakanali M, Raptopoulou CP, Terzis A, Lakatos A, Kiss T, Banyai I, Iordanidis L, Mavromoustakos T, Salifoglou A (2001) Synthesis, pH-dependent structural characterization, and solution behavior of aqueous aluminum and gallium citrate complexes. Inorg Chem 40:1734–1744
143. Dakanali M, Raptopoulou CP, Terzis A, Lakatos A, Banyai I, Kiss T, Salifoglou A (2003) A novel dinuclear species in the aqueous distribution of aluminum in the presence of citrate. Inorg Chem 42:252–254
144. Malone SA, Cooper P, Heath SL (2003) Synthesis and structure of a new aluminium citrate trimer from aqueous solution at very low pH. Dalton Trans 4572–4573
145. Lakatos A, Bertani R, Kiss T, Venzo A, Casarin M, Benetollo F, Ganis P, Favretto D (2004) Al-III ion complexes of saccharic acid and mucic acid: a solution and solid-state study. Chem Eur J 10:1281–1290
146. Cohen SM, Meyer M, Raymond KN (1998) Enterobactin protonation and iron release: Hexadentate tris-salicylate ligands as models for triprotonated ferric enterobactin. J Am Chem Soc 120:6277–6286
147. Steinhauser S, Heinz U, Bartholoma M, Weyhermuller T, Nick H, Hegetschweiler K (2004) Complex formation of ICL670 and related ligands with Fe-III and Fe-II. Eur J Inorg Chem 4177–4192
148. Steinhauser S, Heinz U, Sander J, Hegetschweiler K (2004) Complex formation of 2,6-bis-(2′-hydroxyphenyl)pyridine with Al-III, Fe-III and Cu-II. Zeitschrift Fur Anorganische Und Allgemeine Chemie 630:1829–1838
149. Langemann K, Heineke D, Rupprecht S, Raymond KN (1996) Nordesferriferrithiocin. Comparative coordination chemistry of a prospective therapeutic iron chelating agent. Inorg Chem 35:5663–5673
150. Hoveyda HR, Karunaratne V, Rettig SJ, Orvig C (1992) Coordination chemistry of 2-(2′-hydroxyphenyl)-2-oxazolines with aluminum, gallium, and indium – 1st tris(ligand)metal(III) complexes of this naturally-occurring binding group. Inorg Chem 31:5408–5416
151. Hoveyda HR, Orvig C, Rettig SJ (1994) mer-Tris[2-(5-bromo-2-hydroxyphenyl)-2-oxazolinato]aluminium(III) methanol solvate. Acta Crystallogr C 50:1906–1909
152. Bossek U, Hanke D, Wieghardt K, Nuber B (1993) Pendent arm macrocyclic complexes – crystal-structures of Al(TCTA) and In (TS-TACN). Polyhedron 12:1–5
153. Liu S, Rettig SJ, Orvig C (1992) Polydentate ligand chemistry of Group 13 metals: effects of the size and donor selectivity of metal ions on the structures and properties of aluminum, gallium, and indium complexes with potentially heptadentate (N4O3) amine phenol ligands. Inorg Chem 31:5400–5407
154. Bollinger JE, Mague JT, Banks WA, Kastin AJ, Roundhill DM (1995) Lipophilic hexadentate aluminum complexes of new phenolate-derivatized cyclohexanetriamine ligands and their effect on the peptide-transport system (PTS-1). Inorg Chem 34:2143–2152

Aluminum increases toxic effects of amyloid β-peptides on the human erythrocyte membrane and molecular models

Mario Suwalsky, Pedro L. Hernandez, and Carlos P. Sotomayor

Abstract

The amyloid β-peptide (Aβ) and aluminum have been found, among other components, in the senile plaques from Alzheimer's disease patients. Aggregated Aβ and aluminum are toxic to neurons but the mechanism of accumulation and toxicity remains poorly understood. It has been proposed that Aβ and aluminum toxicity results from Aβ– and aluminum–membrane interactions. For this reason it was thought of interest to study the effect that Aβ and aluminum could have on cell membranes. With this aim, Aβ(1–40), Aβ(1–42), and Al(III) were incubated with intact human erythrocytes, isolated unsealed human erythrocyte membranes (IUM), and molecular models of the erythrocyte membrane. These consisted in bilayers of dimyristoylphosphatidylcholine (DMPC) and dimyristoylphosphatidylethanolamine (DMPE), phospholipids classes located in the outer and inner monolayers of the erythrocyte membrane, respectively. Their capacity to perturb the bilayer structures of DMPC and DMPE was assessed by X-ray diffraction, IUM were studied by fluorescence spectroscopy, and intact human erythrocytes were observed by scanning electron microscopy (SEM). It was found that Aβ(1–40) and Aβ(1–42) in the presence of Al(III) altered the erythrocyte morphology, in IUM induced an ordering effect at the bilayer hydrophobic region, and the structure of DMPC bilayers was perturbed, effects that were different and stronger of those induced by each Aβ and Al(III) separately.

Keywords

Alzheimer's disease • Amyloid β-peptide • Aluminum • Erythrocyte membrane • Lipid bilayer

Abbreviations

Aβ	β-amyloid peptides
DMPC	Dimyristoylphosphatidylcholine
DMPE	Dimyristoylphosphatidylethanolamine
DPH	1,6-Diphenyl-1,3,5-hexatriene
GP	Generalized polarization
IUM	Isolated unsealed human erythrocyte membranes
laurdan	6-Dodecanoyl-2-dimethylaminonaphtalene
r	Anisotropy
SEM	Scanning electron microscopy

M. Suwalsky (✉)
Faculty of Chemical Sciences, University of Concepcion, Concepcion, Chile
e-mail: msuwalsk@udec.cl

Introduction

Alzheimer's disease (AD) is an age-related disorder characterized by progressive cognitive decline and neurodegeneration. One of the key features of this disease is the presence of amyloid plaques associated with neuritic degeneration [1]. Amyloid plaques are composed predominantly of 40–42 residue peptides, the amyloid β-peptide

W. Linert and H. Kozlowski (eds.), *Metal Ions in Neurological Systems*,
DOI 10.1007/978-3-7091-1001-0_11, © Springer-Verlag Wien 2012

(Aβ), being Aβ(1–42) the most toxic [2]. The Aβ peptides are proteolitically cleaved from the transmembrane amyloid precursor protein (APP) by the β- and γ-secretases with the γ-cleavage site within the membrane bilayer [3]. Soluble Aβ peptides interact with cell membranes and have been proposed to affect membrane integrity leading to apoptosis [4]. The molecular mechanisms of these interactions are still unclear. It has been suggested that membrane alterations may proceed to cell death by either an oxidative stress mechanism caused by the Aβ peptide and synergized by transition metal ions, or through formation of ion channels by peptide interfacial self-aggregation [5]. It has also been suggested that Aβ peptides modulate membrane functions by a non-receptor-mediated mechanism, potentially as a result of altering the physicochemical properties of membrane lipids and proteins [6]. Shin et al. [7] found that the Aβ42 is essential for the early development of AD but not sufficient to promote the formation of plaques unless succeeded by Aβ40 deposition. Studies performed in rat synaptic plasma membrane indicated that soluble Aβ40 was located in its hydrophobic core whereas the aggregated form was associated with the phospholipids head group area; however, both soluble and aggregated Aβ42 interacted only with the membrane lipid bilayer hydrocarbon core [6]. Furthermore, it has been reported that both Aβ40 and Aβ42 decrease the fluidity of the acyl chains and head groups of human brain plasma, endosomal and lysosome cell membranes, whereas that of the Golgi bilayer fluidity increased [8]. On the other hand, studies indicated that the ability of both Aβ40 and Aβ42 to disrupt and/or aggregate phospholipids vesicles was mediated through electrostatic interactions with the phospholipids head groups [9]. However, Müller et al. [10] reported that Aβ specifically disturb the acyl-chain layer of cell membranes, while by contrast membrane properties at the level of the polar head groups are much less affected. Given that Aβ are generated in a membrane environment, that its pathological behavior may be due to interactions with membranes, and the unclear nature of these interactions, their understanding is important for deciphering the biological role of Aβ.

Aluminum, an environmental abundant nonredox cation has long been implicated as a potential risk factor in the AD [11–15]. The involvement of Al as a risk factor for AD is suggested by its presence in a high focal concentration in senile plaques and hyperphosphorylated neurofibrillary tangles; however, others attributed the presence of the metal to contamination of the tissue samples by Al from fixatives and staining reagents, point of view refuted by the use of more precise techniques (reviewed by Zatta et al. [15], Exley and Korchazhkina [16], and Gupta et al. [17].

With the aim to better understand the molecular mechanisms of the interaction of Aβ and aluminum with cell membranes, we have utilized paradigmatic well-established models. They regard intact human erythrocytes and molecular models of the erythrocyte membrane. Erythrocytes were chosen because, although less specialized than many other cell membranes they carry on enough functions in common with them such as active and passive transport, and the production of ionic and electric gradients to be considered representative of the plasma membrane in general. On the other hand, circulating blood cells are exposed to soluble Aβ, with binding detected at physiological levels of 5 ng/ml [18]. It has been also reported that erythrocytes of subjects with AD show increased levels of lipid peroxidation, perturbations in the physical state of membrane proteins, and irregular shape distortions [19–22]. Intact human erythrocytes incubated with Aβ, aluminum, and their mixtures were observed by scanning electron microscopy (SEM), and their isolated unsealed membranes (IUM) were analyzed by fluorescence spectroscopy. The molecular model consisted of dimyristoylphosphatidylcholine (DMPC) and dimyristoylphosphatidylethanolamine (DMPE) bilayers, representative of phospholipid classes located in the outer and inner monolayers of the human erythrocyte membrane, respectively [23, 24]. The capacity of Aβ, aluminum, and of their mixtures to perturb the bilayer structures of DMPC and DMPE was determined by X-ray diffraction. These systems and techniques have been used in our laboratories to determine the interaction and the membrane-perturbing effects of other metal ions such as Pb^{2+} [25], Cd^{2+} [26], Au^{3+} [27], Ti^{4+} [28], Fe^{2+} and Fe^{3+} [29], Cr^{3+} and Cr^{6+} [30], and Zn^{2+} [31].

Materials and methods

Scanning electron microscopy studies on human erythrocytes

0.05 cm^3 of blood were obtained from healthy male volunteers by aspiration into tuberculine syringes containing 1 cm^3 heparinized saline/buffered solution (154 mM NaCl, 10 mM Tris–HCl pH 7.4; 50 UI/ml heparine); this mixture was centrifuged at 1,000 rpm for 10 min, the supernatant extracted and the pellet washed three times with 1 cm^3 of saline. The following specimens were prepared from the pellet by mixing with 0.1 cm^3 of saline: (a) control, (b) $AlCl_3$, and (c) Aβ samples by adding 0.1 cm^3 each in a range of concentrations. Samples were incubated at 37 °C for 1 h and then fixed overnight at 5 °C by adding one drop of each sample to plastic tubes containing 1 cm^3 of 2.5 % glutaraldehyde in saline, reaching a final fixation concentration of about 2.4 %; samples were centrifuged at 1,000 rpm for 10 min, and the fixed samples were directly placed on Al stubs, air dried at 37 °C for 30 min to 1 h, and gold-coated for 3 min at 10^{-1} Torr in a sputter device (Edwards S150,

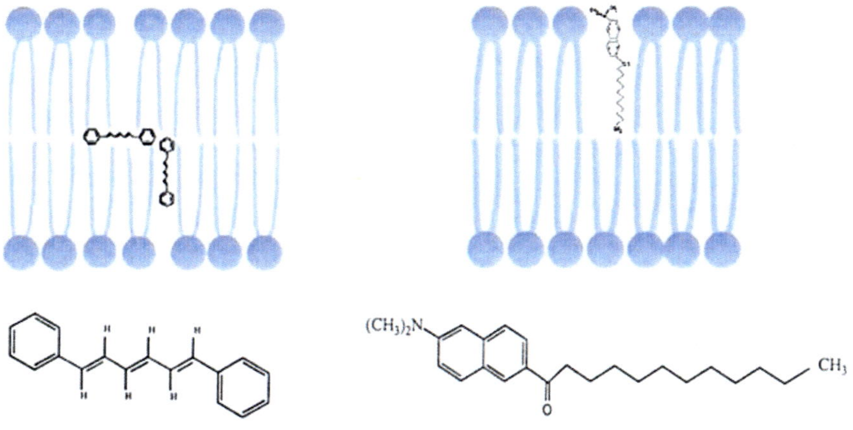

1,6-diphenyl-1,3,5-hexatriene (DPH)

6-dodecanoyl-2-dimethylaminonaphthalene (laurdan)

Fig. 1 Schematic formula of fluorophores DPH, laurdan, and their location in the lipid bilayer. As DPH locates at the hydrophobic acyl chains of the bilayer, laurdan is placed in the hydrophilic/hydrophobic interface

Sussex, England). Resulting specimens were examined in an Etec Autoscan SEM (Etec Corp., Hayword, CA, USA).

X-ray diffraction studies of phospholipid multibilayers

The capacity of Aβ and Al(III) to perturb the structures of DMPC and DMPE multilayers was determined by X-ray diffraction. Synthetic DMPC (lot 80H-8371 A grade MW 677.9), DMPE (lot 68F-8350 A grade MW 635.9) from Sigma, Aβ(1–40) (lot Q9640C, MW 4331) and Aβ(1–42) (lot Q9644, MW 4515) from Biosource International Inc. (CA, USA) and $AlCl_3$ (Merck) were used without further purification. About 2 mg of each phospholipid mixed with 0.2 cm^3 of (a) distilled water, aqueous solutions of (b) $AlCl_3$, (c) Aβ, and (d) mixtures of $AlCl_3$ and Aβ in a range of concentrations were incubated for 1 h at 37 °C (DMPE was previously incubated at 58 °C for 20 min). 0.1 cm^3 of each sample were transferred to 1.5 mm dia special glass capillaries (Glas Technik and Konstruktion, Berlin, Germany) and centrifuged at 2,000 rpm for 10 min. Specimens were immediately subjected to X-ray diffraction in flat plate cameras provided with rotating devices. Specimen-to-film distances were 8 cm and 14 cm, standardized by sprinkling calcite powder on the capillary surface. Ni-filtered Cu $K\alpha$ radiation from a Philips PW 1140 X-ray generator (The Netherlands) was used. The reflection intensities on films versus observed spacing were obtained using a Bio-Rad GS-700 (CA, USA) microdensitometer and Bio-Rad Molecular Analyst/PC image software. The experiments were performed at 17.2 °C, which is below the main phase transition temperature of both DMPC and DMPE. Each experiment was repeated three times and in case of doubts additional experiments were carried out.

Fluorescence measurements of isolated unsealed human erythrocyte membranes

The influence of $AlCl_3$, Aβ(1–40), and Aβ(1–42) on the physical properties of isolated unsealed human erythrocyte membranes (IUM) was examined by fluorescence spectroscopy using DPH and laurdan (Molecular Probe, Eugene, OR, USA) fluorescent probes. DPH is widely used as a probe for the hydrophobic regions of the phospholipid bilayers because of its favorable spectral properties (Fig. 1). Their steady-state fluorescence anisotropy measurements were used to investigate the structural properties as it provides a measure of the hindered rotational diffusion of the fluorophor, restricted within a certain region such as the cone due to the lipid acyl chain packing order. With the fluorescent moiety within a shallow position in the bilayer, laurdan fluorescence spectral shifts provide information on the polarity and/or molecular dynamic properties at the phospholipid polar head groups level (Fig. 1). Quantification of the laurdan fluorescence shift was effected using the generalized polarization (GP) concept [32].

IUM were prepared from blood obtained from healthy male volunteers according to the method by Dodge et al. [33]. DPH and laurdan were incorporated into IUM suspensions by addition of small aliquots of concentrated solutions of the probes in dimethylsulfoxide and ethanol, respectively, in order to obtain a proportion of the probe to lipids of ca. 0.1 mol.%. After probe addition, samples were incubated at 37 °C for 1 h. Fluorescence measurements were performed on a K2 multifrequency phase shift and modulation spectrofluorometer (ISS, Inc., Champaign, IL, USA) interfaced to computers. Software from ISS was used for both data collection and analysis. IUM suspensions were measured at 37 °C using 5 mm path-length square quartz cuvettes. Sample temperature was controlled by an external bath circulator (Cole-Parmer, Chicago, IL, USA) and monitored before and after each measurement using an Omega digital thermometer (Omega Engineering, Inc., Stanford, CT, USA). Anisotropy measurements were performed in the L configuration using Glan Thompson prism polarizers (ISS) in both exciting and emitting beams. Both probes were excited at 360 nm. DPH emission was measured by means of a WG-420 Schott high-pass filter (Schott WG-420, Mainz, Germany) with negligible fluorescence. DPH fluorescence anisotropy (r) was calculated according to the definition: $r = (I_{||} - I_{\perp})/(I_{||} + 2I_{\perp})$, where $I_{||}$ and $I_{\perp}$ are the corresponding vertical and horizontal emission fluorescence intensities with respect to the vertically polarized excitation light [34]. Laurdan fluorescence spectral shifts were quantitatively evaluated using the GP concept (see above) which is defined by the expression $\mathrm{GP} = (I_b - I_r)/(I_b + I_r)$, where I_b and I_r are the emission intensities at the blue and red edges of the emission spectrum, respectively. These intensities have been measured at the emission wavelengths of 440 and 490 nm, which correspond to the emission maxima of laurdan in the gel and liquid crystalline phases, respectively [35]. $AlCl_3$, Aβ(1–40), and Aβ(1–42) were incorporated in IUM suspensions by addition of adequate aliquots of concentrated aqueous solution in order to obtain the different concentrations used in this work. Samples thus prepared were then incubated at 37 °C for ca. 1 h. Blank subtraction was performed in all measurements through unlabeled samples without probes. Data represent mean values and standard error of 10 measurements in two independent samples. Unpaired Student's t-test was used for statistical calculations.

Results

Scanning electron microscopy studies on human erythrocytes

Figure 2 shows the SEM image of intact human erythrocytes, and Fig. 3 the results observed after incubating the red cells with different concentrations of Al(III). As it can be observed, 1 μM and 10 μM Al(III) ions did not induce significant morphological changes to erythrocytes, whereas 25 μM and 100 μM Al(III) ions induced cell adhesion. Figures 4 and 5 show SEM images of human red cells incubated with Aβ(1–40) and Aβ(1–42), respectively. Both Aβ in the maximum concentration assayed (20 μM) did not produced any significant morphological change to the red cells. Erythrocytes incubated with the mixtures 20 μM Aβ (1–40) and 100 μM Al(III) (Fig. 6), and 20 μM Aβ(1–42) and 100 μM $AlCl_3$ (Fig. 7) show equinocytosis, a shape alteration characterized by the development of blebs and/or protuberances on the red cell surfaces.

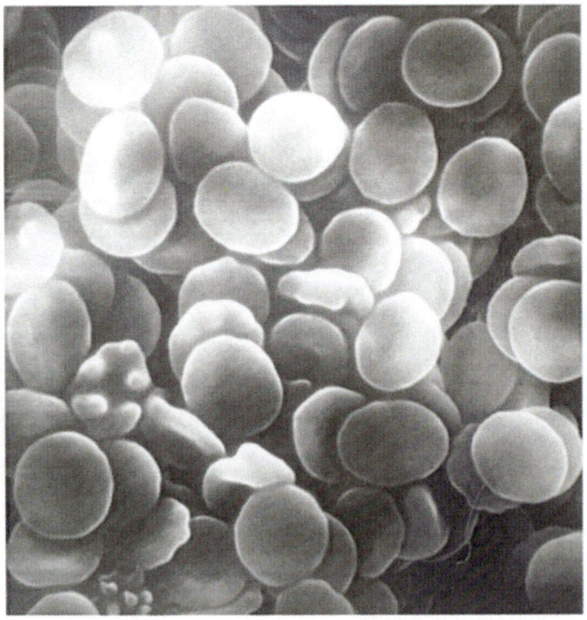

Fig. 2 Scanning electron microscopy (SEM) image (×2,500) of intact human erythrocytes

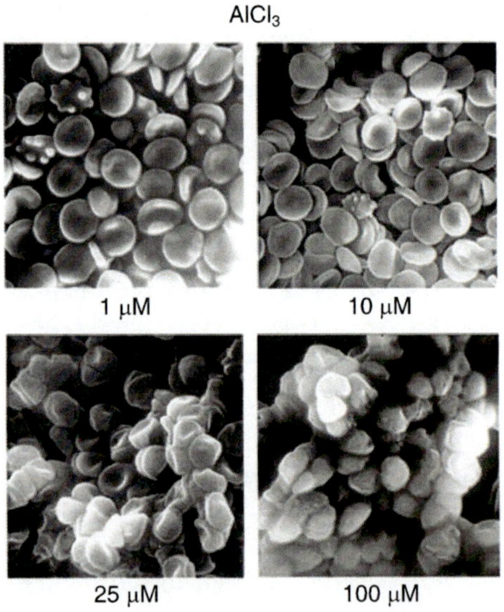

Fig. 3 Scanning electron microscopy (SEM) images (×2,500) of human erythrocytes incubated with $AlCl_3$ in a range of concentrations

X-ray diffraction studies

Interaction of each Al(III) and Aβ(1–40) alone and of their mixtures with DMPC bilayers

Figure 8a shows the results obtained after DMPC bilayers were mixed and had interacted with water and aqueous solutions of Al(III). As expected, pure water altered the structure of DMPC. In fact, its bilayer repeat (bilayer width plus the width of water layer between bilayers) expanded

Fig. 4 Scanning electron microscopy (SEM) images (×2,500) of human erythrocytes incubated with Aβ(1–40) in a range of concentrations

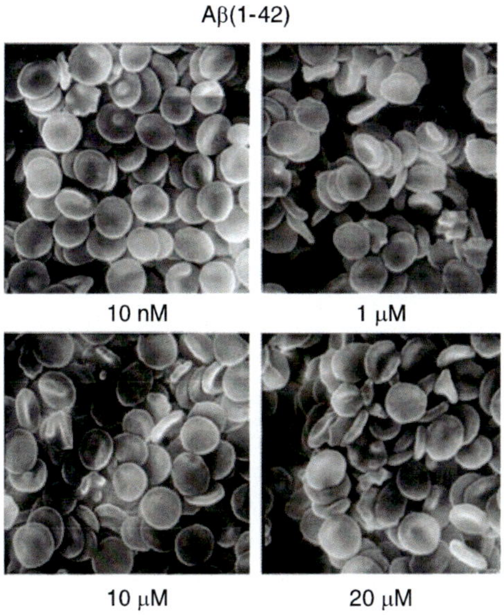

Fig. 5 Scanning electron microscopy (SEM) images (×2,500) of human erythrocytes incubated with Aβ(1–42) in a range of concentrations

from about 55 Å when dry [36] to 64.5 Å when immersed in water and the reflections were reduced to only the first two orders of the bilayer repeat. On the other hand, the structure became more fluid as indicated by the presence of a new and strong 4.2 Å reflection, which corresponds to the average separation of the fully extended acyl chains rotationally disordered in a hexagonal arrangement. It also discloses that DMPC after exposure to 1 and 10 μM Al(III) showed a considerable weakening of the lipid reflection intensities. From these results it can be concluded that Al(III) produced a significant structural perturbation of DMPC bilayers. Figure 9a shows the results obtained after DMPC was mixed

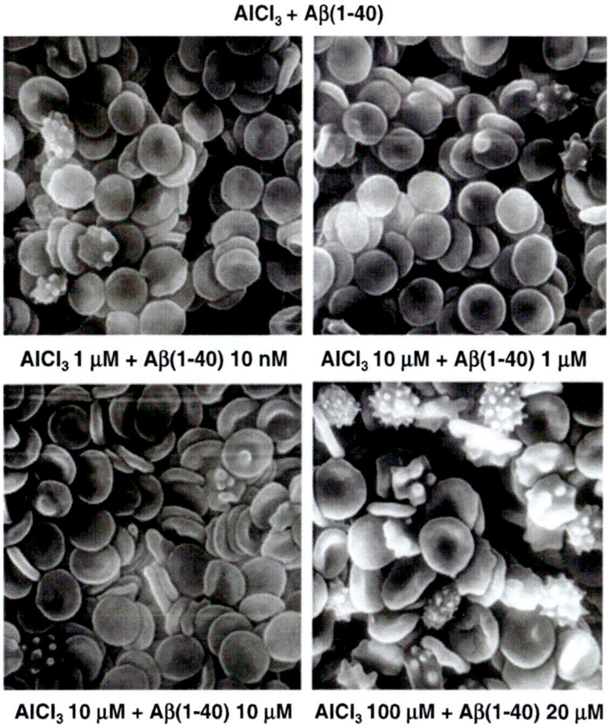

Fig. 6 Scanning electron microscopy (SEM) images (×2,500) of human erythrocytes incubated with $AlCl_3$ and Aβ(1–40) in a range of concentrations

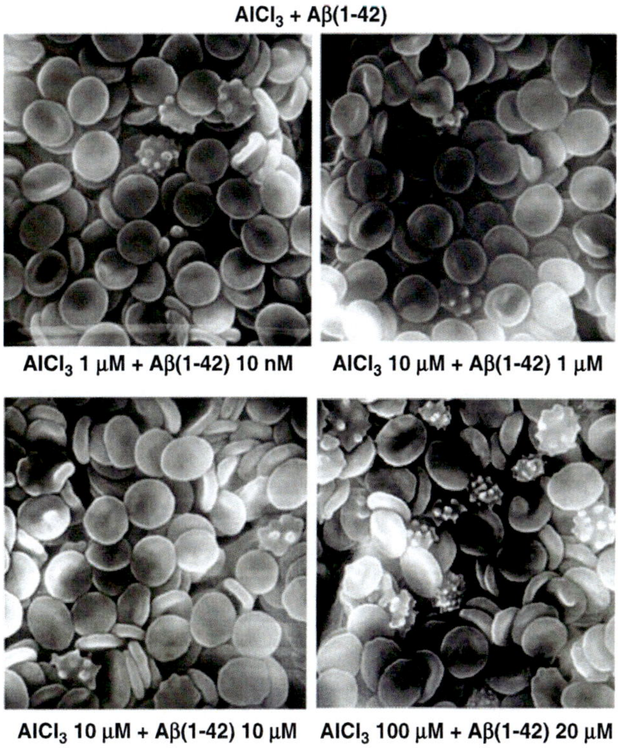

Fig. 7 Scanning electron microscopy (SEM) images (×2,500) of human erythrocytes incubated with $AlCl_3$ and Aβ(1–42) in a range of concentrations

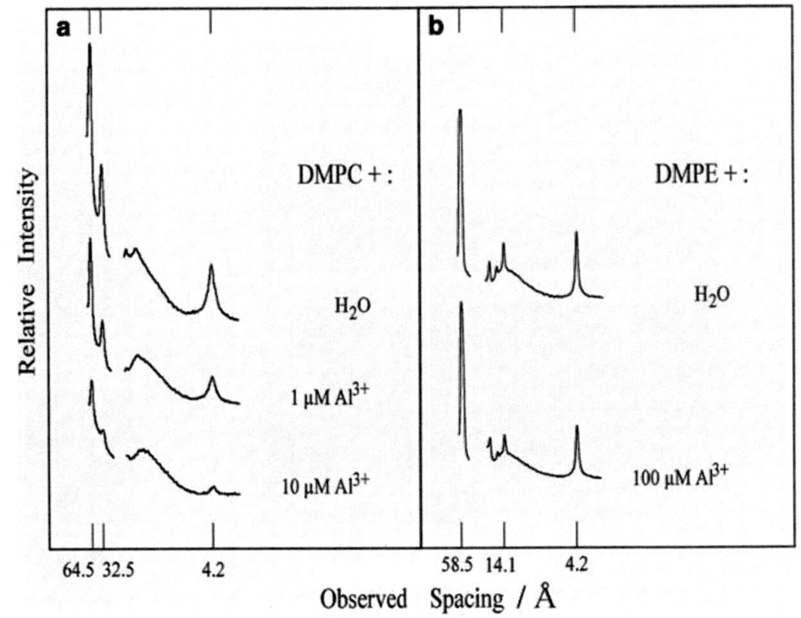

Fig. 8 Microdensitograms from X-ray diffraction of DMPC (**a**) and DMPE (**b**) incubated with aqueous solutions of $AlCl_3$

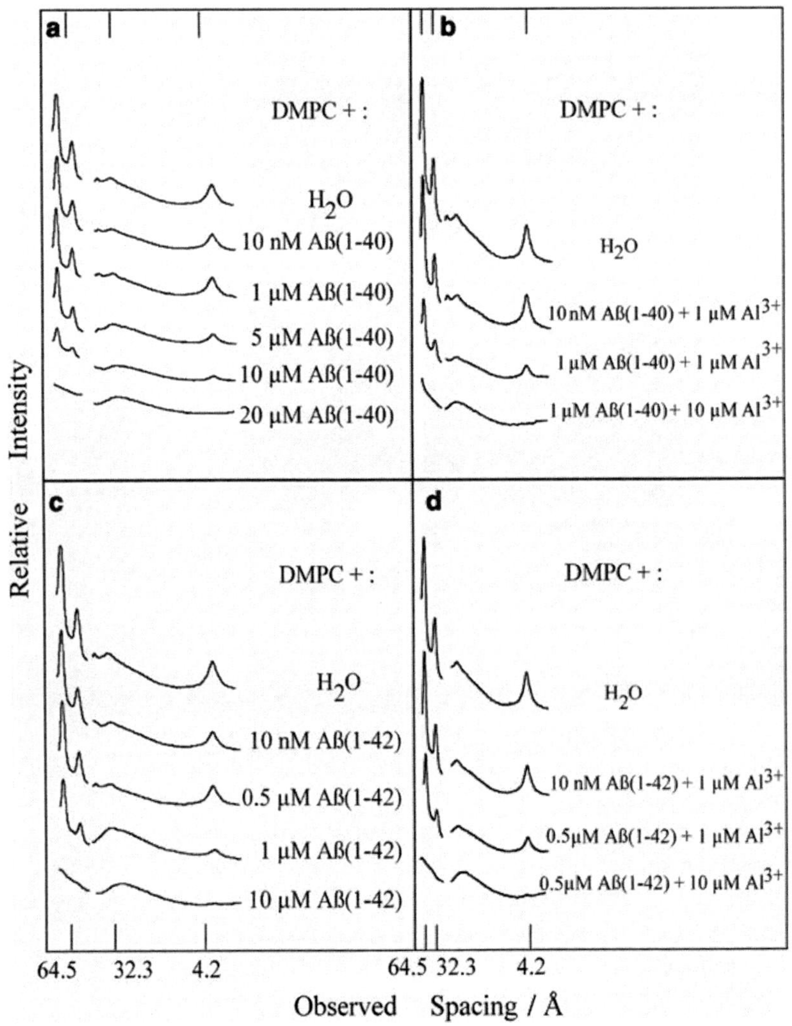

Fig. 9 Microdensitograms from X-ray diffraction of DMPC incubated with Aβ(1–40) (**b**); Aβ(1–40) + $AlCl_3$ (**b**); Aβ(1–42) (**c**); Aβ(1–42) + $AlCl_3$ (**d**)

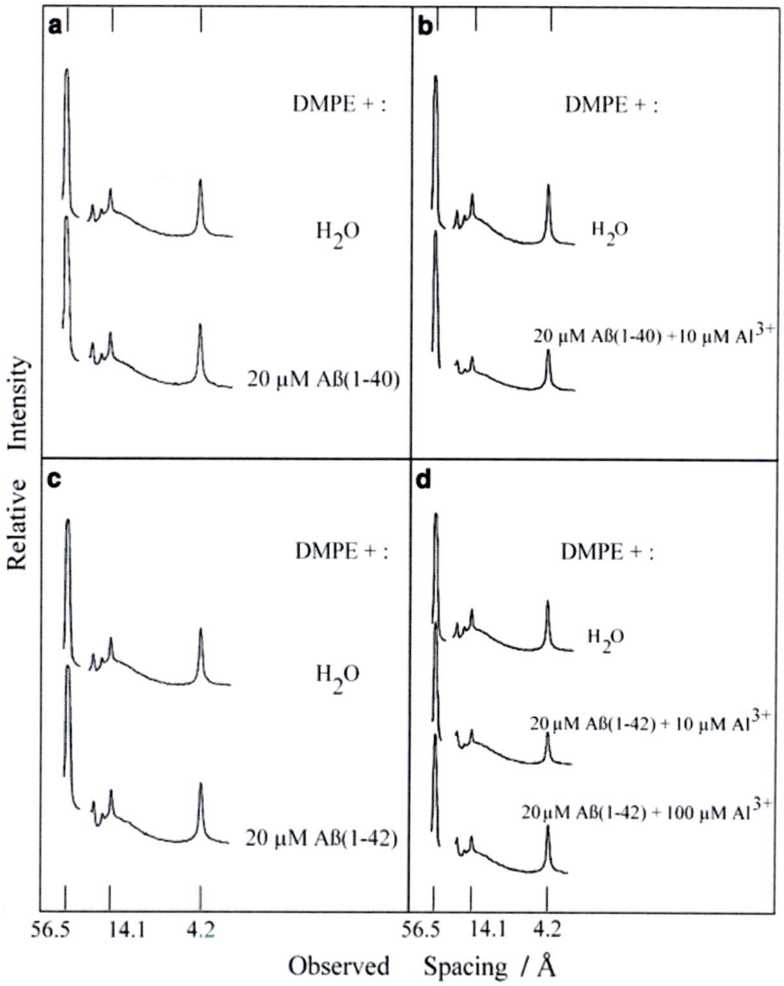

Fig. 10 Microdensitograms from X-ray diffraction of DMPE incubated with Aβ(1–40) (**a**); Aβ(1–40) + $AlCl_3$ (**b**); Aβ(1–42) (**c**); Aβ(1–42) + $AlCl_3$ (**d**)

and had interacted with aqueous solutions of Aβ(1–40). The figure discloses that increasing concentrations of Aβ(1–40) from 10 nM up to 10 μM gradually induced a decrease of all the reflection intensities, which means that the polar head and acyl chain groups were perturbed by the peptide. The fact that the reflections completely disappeared at a 20 μM Aβ(1–40) concentration indicated a total perturbation of the organized lipid structure. The combined effect of Al(III) and Aβ(1–40) is presented in Fig. 9b. As it can be appreciated, DMPC structure was completely perturbed when Aβ(1–40) and $AlCl_3$ concentrations were 1 μM and 10 μM, respectively. This effect was achieved with Aβ(1–40) and $AlCl_3$ concentrations lower than that induced by each compound interacting separately with DMPC.

Interaction of Aβ(1–42) alone and in the presence of Al(III) with DMPC bilayers

Figure 9c shows the effects of Aβ(1–42) on DMPC. As it can be appreciated, this peptide induced approximately the same results observed with Aβ(1–40). However, in this case the complete perturbation of the lipid structure was achieved when the peptide concentration was 0.5 μM, the half of that of Aβ(1–40). This finding implies that the interaction of Aβ (1–42) with DMPC is stronger than with Aβ(1–40). When this experiment was repeated in the presence of Al(III) (Fig. 9d) it was again observed that the complete perturbation of DMPC bilayer arrangement was achieved at lower concentrations than those separately induced by Aβ(1–42) and Al(III).

Interaction of each Al(III) and Aβ(1–40) alone and of their mixtures with DMPE bilayers

Figure 8b shows the results of incubation Al(III) with DMPE. As reported elsewhere, water did not significantly affect the bilayer structure of DMPE [36]. Figure 8b also shows that a Al(III) concentration as high as 100 μM did not affect the bilayer structure of DMPE. Results from similar experiments performed on DMPE incubated with Aβ(1–40) are exhibited in Figure 9a. As it can be noticed, a concentration of the peptides as high as 20 μM, that destroyed DMPC structure, did not affect that one of DMPE. Figure 10b shows the combined effect of highest assayed concentrations of

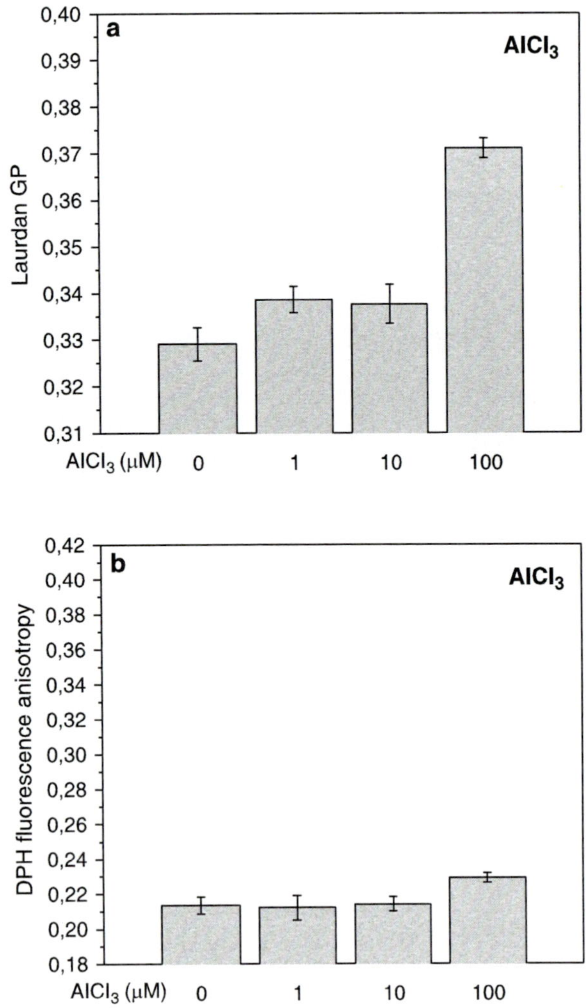

Fig. 11 Effects of $AlCl_3$ on the general polarization (GP) of laurdan (**a**) and on the anisotropy (*r*) of DPH (**b**) embedded in isolated unsealed erythrocyte membranes (IUM) at 37 °C

Aβ(1–40) and $AlCl_3$ (20 μM and 100 μM, respectively) upon DMPE bilayers. As it can be appreciated, DMPE structure was considerably less affected than that one of DMPC incubated with 1 μM Aβ(1–40) and 10 μM $AlCl_3$.

Interaction of Aβ(1–42) alone and in the presence of Al(III) with DMPE bilayers

Results from similar experiments performed on DMPE incubated with Aβ(1–42) are exhibited in Fig. 10c. It can be observed that the highest assayed concentration of this peptide (20 μM) did not induce any perturbation to the lipid structure. The addition of 10 and 100 μM $AlCl_3$ (Fig. 10d) did not produce a significant alteration of DMPE bilayer structure. This result can be compared with that one induced to DMPC by much lower concentrations of both Aβ(1–42) and $AlCl_3$ (0.5 μM and 10 μM, respectively).

Fluorescence measurements of isolated unsealed human erythrocyte membranes

$AlCl_3$, in the 1–100 μM range, induced a gradual increase of the laurdan GP that was more pronounced with $AlCl_3$ 100 μM (Fig. 11a). $AlCl_3$, studied in the same range of concentrations, caused only a slight increase in the DPH anisotropy at 100 μM concentration (Fig. 11b). On the other hand, Aβ(1–40) and Aβ(1–42), studied independently in the range 1–20 μM, did not induce changes in laurdan GP (Fig. 12a, b). However, both peptides in the same range of concentrations induced a gradual and significant increase in the DPH anisotropy, with the biggest increase upon the addition of 10 μM of both peptides (Fig. 11c, d).

The mixture of $AlCl_3$ with Aβ(1–40) (Fig. 13a) produced significant increases in laurdan GP, being markedly higher than the corresponding increase induced by $AlCl_3$ alone at 1 and 10 μM (Fig. 11a). Similar results were observed with the mixture $AlCl_3$ with Aβ(1–42), although higher effects were observed with the highest assayed concentrations (Fig. 13b). The addition of Aβ(1–42) together with $AlCl_3$ produced DPH anisotropy increases at all concentrations (Fig. 13d) that were higher than those observed when the peptide and the salt were assayed alone. In the case of Aβ(1–40) and $AlCl_3$, this condition occurred at 1 μM of both additives (Fig. 13c).

Discussion

The results of the X-ray diffraction analysis of both DMPC and DMPE with Aβ in the presence of Al(III) allow to conclude that (a) DMPC bilayer structure is more susceptible than that one of DMPE to be altered by both Aβ(1–40) and Aβ(1–42); (b) Aβ(1–42) induces higher structural perturbations than Aβ(1–40), and (c) Al(III) increases the perturbing capacity of both Aβ.

SEM results indicate that none of the assayed amyloid peptides significantly affected the shape of the human erythrocytes. On the other hand, $AlCl_3$ induced an effect like cellular adhesion to these cells. However, the incubation of erythrocytes with Aβ(1–40) and Aβ(1–42) in the presence of $AlCl_3$ resulted in relevant morphological alterations. In fact, SEM observations of specimens showed that the erythrocytes changed their normal biconcave shape to an echinocytic form, characterized by the formation of blebs and/or protuberances over the cell surface. According to the bilayer couple hypothesis [37], shape changes induced in erythrocytes by foreign molecules are due to a differential expansion of the two monolayers of the membrane. Thus, the spiculated shape (echinocyte) arises when the added compound locates in the outer monolayer, whereas a cup shape

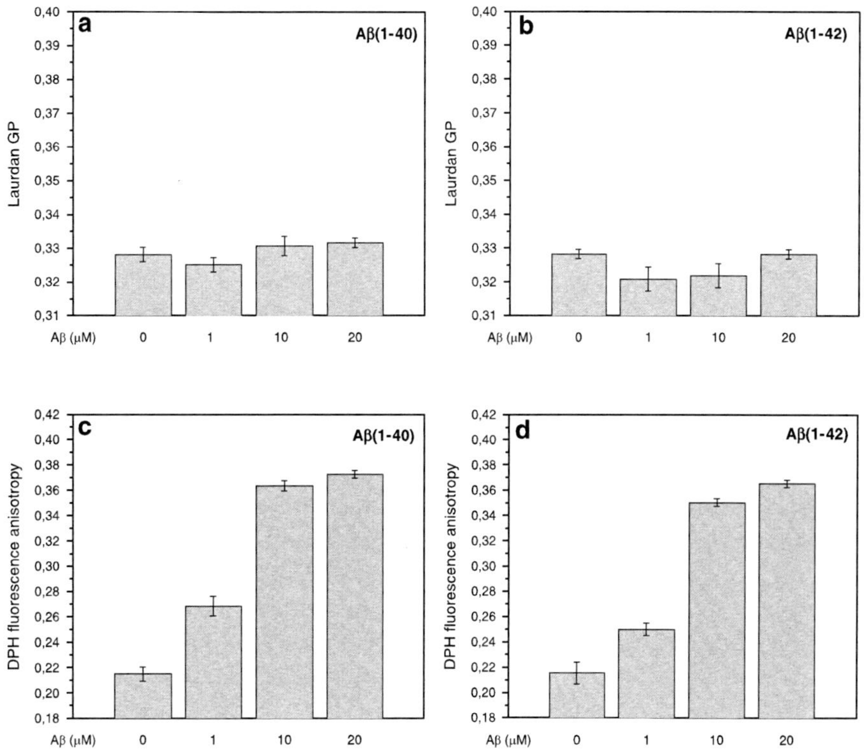

Fig. 12 Effects of Aβ(1–40) (**a**) and Aβ(1–42) (**b**) on the general polarization (GP) of laurdan and of Aβ(1–40) (**c**) and Aβ(1–42) (**d**) on the anisotropy (*r*) of DPH embedded in isolated unsealed erythrocyte membranes (IUM) at 37 °C

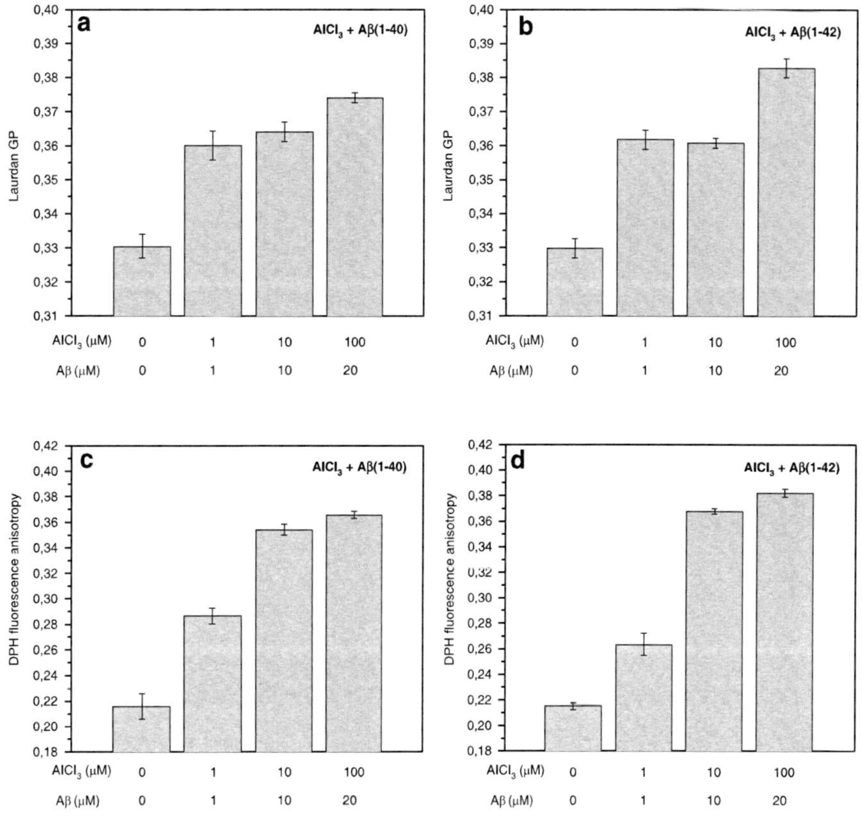

Fig. 13 Effects of $AlCl_3$ + Aβ (1–40) (**a**) and $AlCl_3$ + Aβ(1–42) (**b**) on the general polarization (GP) of laurdan, and of $AlCl_3$ + Aβ(1–40) (**c**) and $AlCl_3$ + Aβ(1–42) (**d**) on the anisotropy (*r*) of DPH embedded in isolated unsealed erythrocyte membranes (IUM) at 37 °C

(stomatocyte) is induced when the compound is inserted in the inner monolayer. The fact that echinocytes were observed is an indication that the peptides and Al(III) ions accumulated in the outer moiety of the red cell membrane.

This conclusion is supported by X-ray diffraction experiments performed in bilayers built-up of both DMPC and DMPE. They represent phospholipid classes located in the outer and inner monolayers of the human erythrocyte and many other cell membranes, respectively [23, 24]. Chemically the two lipids only differ in their terminal amino groups, these being $^{+}NH_3$ in DMPE and $^{+}N(CH_3)_3$ in DMPC. Moreover, both molecular conformations are very similar in their dry crystalline phases; their acyl chains are mostly parallel and extended with the polar groups lying perpendicularly to them; however, DMPE molecules pack tighter than those of DMPC. This effect, due to the DMPE smaller polar group and higher effective charge, makes for a very stable multilayer arrangement which is neither significantly perturbed by the presence of water [33] nor by several metal ions [25–31]. On the other hand, the gradual hydration of DMPC bilayers leads to water filling the highly polar interbilayer spaces. Consequently, there is an increase in its bilayer repeat from 54.5 Å when dry up to 64 Å when fully hydrated at a temperature below that of its main transition. This condition promoted the incorporation of the Al (III) ions and peptides into DMPC highly polar interbilayer space and the ensuing molecular perturbation of the phospholipid bilayer structure.

Results from the fluorescence spectroscopy studies in IUM showed that the mixture of $AlCl_3$ both with Aβ(1–40) as with Aβ(1–42) induced a gradual and significant increase in laurdan GP as well as in the DPH anisotropy. These effects were slightly greater with 20 μM Aβ(1–42). This can be interpreted as an ordering effect at the bilayer hydrophobic region of the lipids acyl chains, together with a decrease in water molecular dynamic and/or in the water content at the level of the lipids glycerol backbone. When comparing these results with those obtained with $AlCl_3$, Aβ(1–40), and Aβ (1–42) alone it can be concluded that the effect produced by the $AlCl_3$ and Aβ mixture in the laurdan GP is dominated by the $AlCl_3$ effect, being the net effect significantly greater than any of the effects of the components separately. The effect produced by $AlCl_3$ and Aβ mixture in the DPH anisotropy is dominated by the Aβ effect, being the net effect greater than any of the effects of the components separately. We hypothesize that these greater effects could be due to the interaction of Al^{3+} ions in interfacial positions with the amyloid peptides inserted in the membrane.

The set of these results would allow postulating a simplified mechanism of action of peptides Aβ in the presence of aluminum at the erythrocyte membrane level. The incorporation of Al(III) ions into the outer monolayer and the ensuing interaction with phosphatidylcholine phosphate groups [38] disrupts their arrangement resulting in a more fluid condition. On the other hand, Aβ interactions with Al (III), with at least four aluminum ions bound to each peptide molecule [39], stabilize certain conformational state of the peptides; this would confer to them a greater degree of penetration in the bilayer with which the effect of alteration of the molecular order is much greater. This interleave of the peptide includes the hydrocarbon chain region of the bilayer producing a net disorder in both bilayer zones (polar heads and the hydrophobic chains). The amphoteric and amphiphilic characteristics of the Aβ peptides endow these molecules with a capacity to interact with lipids [18]. The higher extent of Aβ(1–42) interactions with DMPC compared to Aβ(1–40) might be explained by the two additional hydrophobic amino acids on the C-terminus of Aβ(1–42) [40], which increased its insertion and destabilization of the lipid bilayer. As indicated by Curtain et al. [41] the interaction of Aβ with the lipid matrix of neuronal cell membranes plays an important role in the pathogenesis of AD. Based on these results, the controversy relative to the aluminum participation in the pathology of the AD is partly clarified: aluminum enhances the injurious effects of peptides Aβ in cell membranes [42].

Acknowledgements The authors thank Fernando Neira for his valuable technical assistance. This work was supported by a grant from FONDECYT (1090041).

References

1. Eckert GP, Wood WG, Muller WE (2001) Effects of aging and beta-amyloid on the properties of brain synaptic and mitochondrial membranes. J Neural Transm 108:1051–1064
2. Eckert GP, Wood WG, Muller WE (2005) Membrane disordering effects of beta-amyloid peptides. Subcell Biochem 38:319–337
3. Lau TL, Ambroggio EE, Tew DJ, Cappai R, Masters CL, Fidelio GD, Barnham KJ, Separovic F (2005) Amyloid-beta peptide disruption of lipid membranes and the effect of metal ions. J Mol Biol 356:759–770
4. Demeester N, Baier G, Enzinger C, Goethals M, Vandekerckhove J, Rosseneu M, Labeur C (2000) Apoptosis induced in neuronal cells by C-terminal amyloid beta-fragments is correlated with their aggregation properties in phospholipid membranes. Mol Membr Biol 17:219–228
5. Ambroggio EE, Kim DH, Separovic F, Barrow CJ, Barnham KJ, Bagatolli LA, Fidelio GD (2005) Surface behavior and lipid interaction of Alzheimer beta-amyloid peptide 1-42: a membrane-disrupting peptide. Biophys J 88:2706–2713
6. Mason RP, Jacob RF, Walter MF, Mason PE, Avdulov NA, Chochina SV, Wood WG (1999) Distribution and fluidizing action of soluble and aggregated amyloid beta-peptide in rat synaptic plasma membranes. J Biol Chem 274:18801–18807
7. Shin RW, Lee VM, Trojanowski JQ (1994) Aluminum modifies the properties of Alzheimer's disease PHF tau proteins in vivo and in vitro. J Neurosci 14(Pt 2):7221–7233
8. Waschuk SA, Elton EA, Darabie AA, Fraser PE, McLaurin JA (2001) Cellular membrane composition defines A beta-lipid interactions. J Biol Chem 276:33561–33568

9. McLaurin J, Chakrabartty A (1997) Characterization of the interactions of Alzheimer beta-amyloid peptides with phospholipid membranes. Eur J Biochem 245:355–363
10. Muller WE, Kirsch C, Eckert GP (2001) Membrane-disordering effects of beta-amyloid peptides. Biochem Soc Trans 29 (Pt 4):617–623
11. Yokel RA (2000) The toxicology of aluminum in the brain: a review. Neurotoxicology 21:813–828
12. Jansson ET (2001) Aluminum exposure and Alzheimer's disease. J Alzheimers Dis 3:541–549
13. Rondeau V (2002) A review of epidemiologic studies on aluminum and silica in relation to Alzheimer's disease and associated disorders. Rev Environ Health 17:107–121
14. Zatta P, Kiss T, Suwalsky M, Bretón M (2002) Aluminum(III) as promoter of cellular oxidation. Coord Chem Rev 228:271–284
15. Zatta P, Lucchini R, van Rensburg SJ, Taylor A (2003) The role of metals in neurodegenerative processes: aluminum, manganese, and zinc. Brain Res Bull 62:15–28
16. Exley C, Korchazhkina O (2001) The association of aluminium and β amyloid in Alzheimer's disease. In: Exley C (ed) Aluminium and Alzheimer's disease. Elsevier, Amsterdam, p 421
17. Gupta VB, Anitha S, Hegde ML, Zecca L, Garruto RM, Ravid R, Jagannatha Rao KS (2005) Aluminium in Alzheimer's disease: are we still at a crossroad? Cell Mol Life Sci 62:143–158
18. Kuo YM, Kokjohn TA, Kalback W, Luehrs D, Galasko DR, Chevallier N, Roher AE (2000) Amyloid-beta peptides interact with plasma proteins and erythrocytes: implications for their quantitation in plasma. Biochem Biophys Res Commun 268:750–756
19. Jayakumar R, Kusiak JW, Chrest FJ, Demehin AA, Murali J, Wersto RP, Rifkind JM (2003) Red cell perturbations by amyloid beta-protein. Biochim Biophys Acta 1622:20–28
20. Mattson MP, Begley JG, Mark RJ, Furukawa K (1997) Abeta25-35 induces rapid lysis of red blood cells: contrast with Abeta1-42 and examination of underlying mechanisms. Brain Res 771: 147–153
21. Mark RJ, Hensley K, Butterfield DA, Mattson MP (1995) Amyloid beta-peptide impairs ion-motive ATPase activities: evidence for a role in loss of neuronal Ca^{2+} homeostasis and cell death. J Neurosci 15:6239–6249
22. Goodall HB, Reid AH, Findlay DJ, Hind C, Kay J, Coghill G (1994) Irregular distortion of the erythrocytes (acanthocytes, spur cells) in senile dementia. Dis Markers 12:23–41
23. Devaux PF, Zachowsky A (1994) Distribution of phospholipids in erythrocyte membranes. Chem Phys Lipids 73:107–120
24. Boon JM, Smith BD (2002) Chemical control of phospholipid distribution across bilayer membranes. Med Res Rev 22:251–281
25. Suwalsky M, Villena F, Norris B, Cuevas YF, Sotomayor CP, Zatta P (2003) Effects of lead on the human erythrocyte membrane and molecular models. J Inorg Biochem 97:308–313
26. Suwalsky M, Villena F, Norris B, Cuevas F, Sotomayor CP (2004) Cadmium-induced changes in the membrane of human erythrocytes and molecular models. J Inorg Biochem 98:1061–1066
27. Suwalsky M, Zambenedetti P, Carpene E, Ibnlkayat M, Wittkowski W, Messori L, Zatta P (2004) Effects of chronic treatment with sodium tetrachloroaurate(III) in mice and membrane models. J Inorg Biochem 98:2080–2086
28. Suwalsky M, Villena F, Norris B, Soto MA, Sotomayor CP, Messori L, Zatta P (2005) Structural effects of titanium citrate on the human erythrocyte membrane. J Inorg Biochem 99:764–770
29. Suwalsky M, Martinez F, Cardenas H, Grzyb J, Strzalka K (2005) Iron affects the structure of cell membrane molecular models. Chem Phys Lipids 134:69–77
30. Suwalsky M, Castro R, Villena F, Sotomayor CP (2008) Cr(III) exerts stronger structural effects than Cr(VI) on the human erythrocyte membrane and molecular models. J Inorg Biochem 102:842–849
31. Suwalsky M, Novoa V, Villena F, Sotomayor CP, Aguilar LF, Ronowska A, Szutowicz A (2009) Structural effects of Zn(2+) on cell membranes and molecular models. J Inorg Biochem 103:797–804
32. Parasassi T, Gratton E (1995) Membrane lipid domains and dynamics as detected by laurdan fluorescence, J. Fluoresc. 5: 59–69.
33. Dodge JT, Mitchell C, Hanahan DJ (1963) The preparation and chemical characteristics of hemoglobin-free ghosts of human erythrocytes. Arch Biochem Biophys 100:119–30
34. Lakowicz JR (1999) Principles of Fluorescence Spectroscopy, Plenum
35. Parasassi T, De Stasio G, D'Ubaldo A, Gratton E (1990) Phase fluctuation in phospholipid membranes revealed by laurdan fluorescence, Biophys. J. 57:1179–1186.
36. Suwalsky M (1996) Phospholipids bilayers. In: Salamone JC (ed) Polymeric materials encyclopedia, vol 7. CRC, Boca Raton, FL, p 5073
37. Sheetz MP, Singer SJ (1974) Biological membranes as bilayer couples. A molecular mechanism of drug-erythrocyte interactions. Proc Natl Acad Sci USA 71:4457–4461
38. MacKinnon N, Ridgway J, Crowell KJ, Macdonald PM (2006) Aluminum binding to phosphatidylcholine lipid bilayer membranes: aluminum exchange lifetimes from 31P NMR spectroscopy. Chem Phys Lipids 139:85–95
39. Vyas SB, Duffy LK (1995) Interaction of synthetic Alzheimer beta-protein-derived analogs with aqueous aluminum: a low-field 27Al NMR investigation. J Protein Chem 14:633–644
40. Pillot T, Goethals M, Vanloo B, Talussot C, Brasseur R, Vandekerckhove J, Lins L (1996) Fusogenic properties of the C-terminal domain of the Alzheimer beta-amyloid peptide. J Biol Chem 271:28757–28765
41. Curtain CC, Ali FE, Smith DG, Bush AI, Masters CL, Barnham KJ (2003) Metal ions, pH, and cholesterol regulate the interactions of Alzheimer's disease amyloid-beta peptide with membrane lipid. J Biol Chem 278:2977–2982
42. Drago D, Folin M, Baiguera S, Tognon G, Ricchelli F, Zatta P (2007) Comparative effects of Abeta(1-42)-Al complex from rat and human amyloid on rat endothelial cell cultures. J Alzheimers Dis 11:33–44

Oxidative stress in neurodegeneration: targeting mitochondria as a therapeutic aid

Elena Gaggelli and Gianni Valensin

Abstract

Increasing evidence links oxidative stress caused by radical overproduction and/or impaired antioxidant defenses with the onset of neurodegenerative diseases. Mitochondria have been shown to play a major role due to their specialized ability in producing free radicals joined with their sensitivity to the toxic effects of free radicals. The present article provides a short review of the sources of free radicals, the sites of their generation, the mitochondrial detoxifying systems and the consequences of oxidative stress. As main examples, the roles played by mitochondria in Alzheimer's and Parkinson's diseases are briefly considered. Finally, drugs targeting antioxidant moieties to mitochondria are shortly described and summarized

Keywords

Drugs targeting • Mitochondria • Neurodegeneration • Oxidative stress • ROS • RNS

Introduction

Neurodegeneration is the umbrella comprising neurological disorders of cognitive abnormalities yielding progressive loss of structure or function of neurons in selected regions of the brain. Many similarities appear which relate these diseases to one another on a subcellular level. Discovering these similarities offers hope for therapeutic advances that could ameliorate many diseases simultaneously. All these disorders, in fact, share many common features among them and also with other disorders, such as type II diabetes, affecting organs other than the brain:

- There is a conspicuous common feature—aggregation and deposition of abnormal protein; this, in turn, determines the occurrence of excess misfolded proteins causing neuronal damage in any pathological case (Table 1) (Adapted from Ref. [1]).

  These aberrant proteins have a typical tendency to aggregate and form solid deposits as diverse as the plaques of Alzheimer's (AD) and prion (PrPD) diseases, the Lewy bodies of Parkinson's disease (PD), the nuclear and cytoplasmic inclusions of Huntington disease (HD), the Bunina bodies of familial amyotrophic lateral sclerosis (ALS), and many others. These proteins are normally soluble but they convert into insoluble aggregates that can form intractable and frequently toxic deposits not only in the brain but also in skeletal and muscular tissue and in the heart and the liver.

  Despite all control mechanisms, the failure of proteins to achieve their functional states may result in the onset of pathological and often fatal disorders. Newly translated misfolded proteins can still be managed by degradation pathways: (1) **ubiquitin–proteasome:** ubiquitin and other enzymes are the key for degradation of many proteins that cause proteinopathies including polyQ expansions (*vide infra*) and α-synucleins. Decreased proteasome activity is consistent with models in which intracellular protein aggregates form. It is still unknown whether or not these aggregates are a cause or a result of neurodegeneration; (2) **autophagy–lysosome pathways:** this form of programmed cell death (PCD) becomes the favorable route when a protein is aggregate-prone meaning it is a poor proteasome substrate. This can be split into two

G. Valensin (✉)
Department of Chemistry, University of Siena, Via Aldo Moro 2, Siena 53100, Italy
e-mail: gianni.valensin@unisi.it

W. Linert and H. Kozlowski (eds.), *Metal Ions in Neurological Systems*,
DOI 10.1007/978-3-7091-1001-0_12, © Springer-Verlag Wien 2012

Table 1 Aggregation and deposition of abnormal proteins in selected neurodegenerative disorders

Disease	Protein deposits	Toxic protein
Alzheimer's disease	Extracellular plaques	Aβ
	Intracellular tangles	Tau
Parkinson's disease	Lewy bodies	α-synuclein
Prion disease	Prion plaque	PrP^{sc}
Polyglutamine disease	Nuclear and cytoplasmic inclusions	Polyglutamine containing proteins
Familial amyotrophic lateral sclerosis	Bunina bodies	SOD1

forms of autophagy: **macroautophagy** and **chaperone-mediated autophagy (CMA)**. **Macroautophagy** is involved with nutrient recycling of macromolecules under conditions of starvation, certain apoptotic pathways, and if absent, leads to the formation of ubiquinated inclusions. Experiments in mice with neuronally confined macroautophagy-gene knockouts develop intraneuronal aggregates leading to neurodegeneration. **CMA** defects may also lead to neurodegeneration. Research has shown that mutant proteins bind to the CMA-pathway receptors on lysosomal membrane and in doing so block their own degradation as well as the degradation of other substrates. In some cases, however, proteins with a high propensity to misfold escape all protective mechanisms and assemble into large aggregates that can be highly organized and extremely stable. Amyloidoses, in which proteins assemble into amyloid fibrils, belong to such a family of disorders associated with polypeptide aggregation. The list of diseases associated with amyloid deposition keeps growing. More than 20 different human amyloid disorders are known so far, and they also include type II diabetes. Aging cells and several age-related pathologies are also characterized by accumulation of oxidatively damaged proteins. Oxidative modification of proteins and interaction with metal ions are known to potentially modify the proteins (*Vide infra*) and favor formation of abnormally folded intermediates and/or prefibrillar aggregates.

- All neurodegenerative diseases may be of familial or sporadic origin; familial diseases typically (1) make a few percent of diagnosed cases, (2) are determined by mutations in well-defined genes encoding for the amyloidogenic proteins found in brain deposits, and (3) manifest themselves at early ages.

  Genetic mutations are mostly located in completely unrelated genes. In many of the different diseases, the mutated gene has a common feature: a repeat of the CAG nucleotide triplet, encoding for the amino acid glutamine. A repeat of CAG results in a polyglutamine (polyQ) tract. Diseases showing this are known as polyglutamine diseases. Extra glutamine residues can acquire toxic properties through a variety of ways, including irregular protein folding and degradation pathways, altered subcellular localization, and abnormal interactions with other cellular proteins.

  Examples of inherited neurodegenerative diseases caused by the expansion of the CAG trinucleotide and polyQ tract are Huntington's disease and spinocerebellar ataxias. While polyglutamine-repeat diseases encompass many different neurodegenerative disorders, there are many more it does not apply to. The genetics behind each disease are different and often unknown.
- A close relationship has been emerging in these last years between the onset of neurological disorders and impaired homeostasis of metal ions such as copper, iron, manganese, and zinc. As a matter of fact the sequences of many amyloidogenic proteins encompass metal binding sites and metal binding has been shown to possibly directly or indirectly mediate oxidative modifications leading to aberrant conformations. In fact, Fenton chemistry (*vide infra*) verifies the ability of metal ions to produce reactive oxygen (ROS) and nitrogen (RNS) species and protein-bound redox-active metal ions are efficient catalysts of protein oxidation by external agents.
- Proteomic and other analyses performed on brains or brain extracts taken after death from patients affected by many disorders, such as AD, PD, and ALS, have demonstrated extensive signs of oxidative stress leading to oxidatively modified proteins, nucleic acids, and cell membrane lipids.
- Neurons face unique challenges in intracellular trafficking of signaling proteins. In response to both physiologic and pathologic stimuli, signals need to traverse dendrites or axon to affect transcriptional responses in the nucleus. Specifically, bidirectional trafficking of proteins such as transcription factors between the cytoplasm and the nucleus is a crucial aspect of signal transduction essential for the proper regulation of transcription. Disruption of efficient nucleocytoplasmic transport can significantly impair neuronal function and lead to neurodegeneration. Emerging evidence has been collected that altered nuclear transport is a possible mechanism involved in the pathogenesis of neurodegenerative disorders. In turn, altered nuclear transport arises in response to oxidative stress, which is thought to play a significant role in the pathogenesis of many neurodegenerative diseases (*vide infra*) [2].

In 1956, Harman proposed the "free radical theory of" aging and associated neurodegenerative diseases [3], according to which "the reaction of active free radicals, normally produced in the organisms, with cellular constituents initiates the changes associated with aging and yields activation of cytosolic stress signaling pathways". Ultimately, both life span and life quality are severely negatively affected. While some controversy regarding the generality of the Harman's theory continues, experimental evidence for the link between ROS overproduction and the biology of aging is steadily solidifying, as already stated.

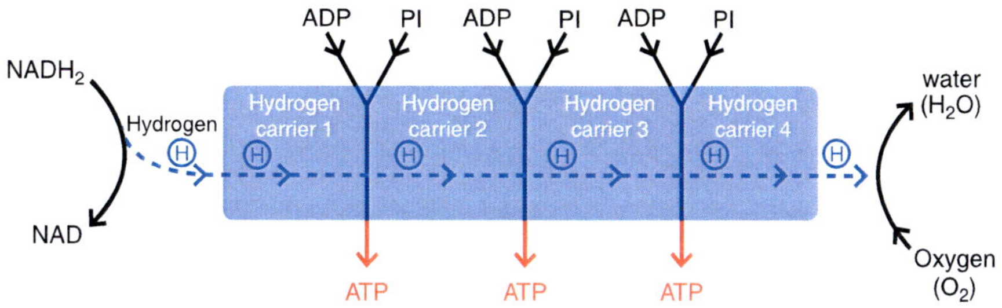

Fig. 1 Oxidative phosphorylation uses the hydrogen attached to the coenzymes from Krebs cycle and oxygen to produce water and 34 ATP molecules, releasing hydrogen-free form of the coenzymes in the process. Occurring in the inner mitochondrial membrane mediated by cytochromes, the process is also called an electron transport chain (Reprinted with permission from the web site http://www.biology-online.org/9/5_protein_activity.htm)

The central role of mitochondria

In recent years, increasing and convincing evidence has been collected that mitochondrial dysfunction and oxidative stress play a crucial role in the majority of neurodegenerative diseases. Mitochondria are the major source of intracellular ROS and are particularly vulnerable to oxidative stress. Oxidative damage to mitochondria has been shown to impair mitochondrial function and lead to cell death via apoptosis and necrosis. Because dysfunctional mitochondria are known to produce more ROS, a feed-forward loop is set up whereby ROS-mediated oxidative damage favors more ROS generation, resulting in a vicious cycle.

Just to furnish some examples taken from the huge amount of reports in the literature, the evidence for the major role directly played by mitochondria in drawing the onset of neurodegenerative diseases can be summarized as follows [4]:

- Decreased mitochondrial complex I activity has been reported in the substantia nigra of postmortem brain samples from patients affected by PD;
- Inhibitors of complex I of the mitochondrial electron transport chain (ETC) produce PD in humans and animals, and knockouts of parkin or DJ-1, proteins associated with familial PD, show defective mitochondrial function;
- Alzheimer's brains harbor somatic mitochondrial DNA (mtDNA) mutations that suppress mitochondrial transcription and replication and there is evidence that β-amyloid is targeted to mitochondria, where it has been shown to bind ABAD (β-Amyloid Binding Alcohol Dehydrogenase) and to inhibit cytochrome *c* oxidase;
- Mice overexpressing a mutant SOD1, a model for ALS, show compromised mitochondrial respiration and ATP production;
- Impaired mitochondrial function has been shown in patients with HD, as well as in animal models of HD.

Mitochondrial dysfunction leads to overproduction of ROS (oxidative stress): as a matter of fact, postmortem analyses have revealed overall elevated levels of oxidatively damaged proteins, lipids, and DNA in AD and PD brains. Oxidative modifications to proteins, such as α-synuclein in PD, β-amyloid in AD, and SOD1 in ALS, are likely to drive protein misfolding and aggregate formation.

Neurons contain several hundred mitochondria that form cable-like structures along neuronal projections to help the neuron meet its large energy demands. Neurons require energy to transport organelles and cargo along microtubules or actin fibers (motor molecules, such as dyneins, kinesins, and myosin, mediate this process) and to maintain ion gradients and the membrane potential with ATP-dependent Ca^{2+} and Na^+/K^+ pumps and ion channels. Additionally, neurotransmitter-vesicle loading at presynaptic terminals and Ca^{2+}-mediated neurotransmitter release into the synaptic cleft are also ATP-dependent events. Glutamate transporters mediate glutamate re-uptake from the synaptic cleft and, at the postsynaptic membrane, glutamate binding to NMDA (*N*-methyl-D-aspartate) receptors (NMDARs) evokes Ca^{2+} influx, which in turn can activate nitric oxide synthase (NOS) and stimulate the generation of nitric oxide (NO). Both NO and Ca^{2+} can directly modulate mitochondrial function by altering the levels of reactive oxygen species (hydrogen peroxide and superoxide) and ATP production. Mitochondria consume ca. 85–90 % of a cell's oxygen to support oxidative phosphorylation (OxPh) by harnessing oxidized fuel to the synthesis of ATP [Fig. 1]; this also implies intracellular production of ROS and RNS, which damage cellular components and accelerate aging of the organism.

The OxPh is accomplished by protein complexes transferring high-energy electrons from NADH and FADH2 to molecular oxygen through the electron transport chain (Fig. 2) [5]. The energy released by the flow of electrons through the ECT is used to pump protons out of the mitochondrial inner membrane (IMB) through complexes I, III, and IV and to harvest potential energy from the proton gradient across the IMB (MMP = mitochondrial membrane potential, Fig. 3) [6]. The MMP is then used to drive a membrane ATP synthase (complex V) to generate ATP. Complex IV (cytochrome oxidase) is normally responsible for donation of electrons to O_2; however, nonenzymatic reactions may prematurely occur with generation of ROS and RNS.

Mitochondrial metabolism and the OxPh cascade are key factors in the generation of ROS and RNS associated with a large number of disease states, including atherosclerosis, Alzheimer's disease, Parkinson's, neuronal death including ischemic and hemorrhagic stroke, acute and chronic degenerative cardiac myocyte death, and cancer [7].

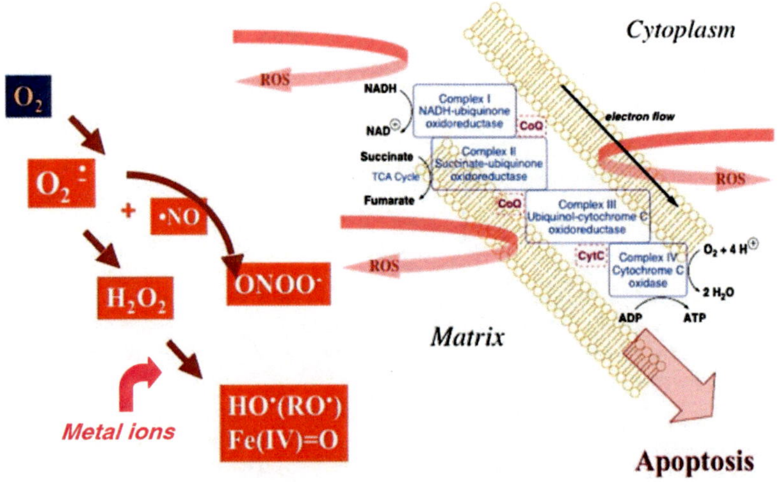

Fig. 2 Schematic model for generation of reactive oxygen species (ROS) and reactive nitrogen species (RNS) during oxidative phosphorylation in the mitochondrial membrane and matrix. The formation of superoxide radical anion initiates a cascade process that can induce programmed cell death (apoptosis) (Slightly modified from Ref. [5])

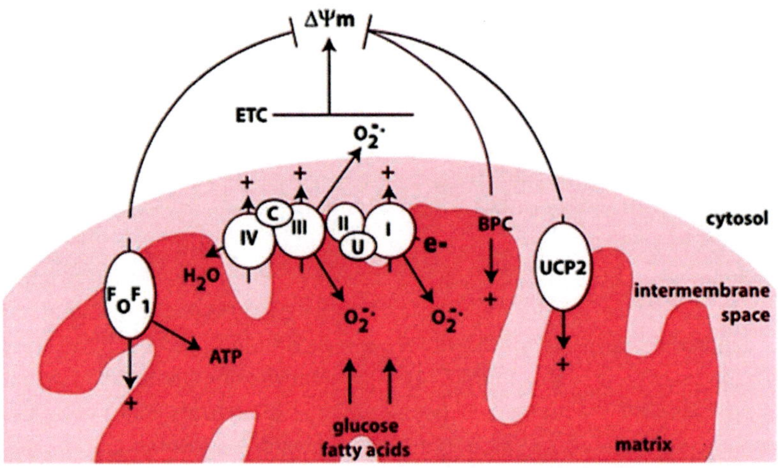

Fig. 3 Major determinants of potential ($\Delta\psi_m$)—$\Delta\psi_m$ is the result of proton translocation from mitochondrial matrix to intermembrane space by the ETC that includes complexes from I to IV and mobile electron carrier proteins ubiquinone (I) and cytochrome C. ETC is supplied by reducing equivalents primarily formed during combustion of glucose and fatty acids. $\Delta\psi_m$ is consumed through OxPh mediated by ATP synthase (F_oF_1) and through proton leak mediated by carrier proteins (UCP2 shown here) and by basal proton conductance. Production rate of superoxide at various sites correlates with ψ_m explaining why processes that consume the proton gradient may alleviate mitochondrial oxidative stress (Reprinted with permission from Ref. [6])

Sources of ROS and RNS

As shown in Scheme 1 [8, 9], ROS cover several redox states of oxygen, ranging from the relatively unreactive superoxide ($O_2^{\cdot-}$) and H_2O_2 to the highly reactive oxygen species (hROS), which may exist as free hydroxyl radicals ($HO^{\cdot}$), as bound ("crypto") radicals or as Fe(IV)-oxo (ferryl) species.

ROS play an important role in cell signaling, a process termed redox signaling [10, 11], with specific mechanisms of homeostatic control. Thus, to maintain proper cellular homeostasis, a finely tuned balance must be assured between ROS production and consumption. Excessive ROS or free radicals need to be either quenched by converting them into metabolically nondestructive molecules or be scavenged/neutralized right after their formation. ROS-induced damage is often irreversible and creates a positive feed-forward process, such that mitochondrial functions are impaired by ROS with favored further ROS generation. This vicious relationship finally yields oxidative stress and accelerated aging. Oxidative stress may be therefore defined as an

$$O_2 \xrightarrow{e^-} O_2^{\cdot-} \xrightarrow{e^-+H^+} H_2O_2 \xrightarrow{e^-+H^+} OH^{\cdot} \xrightarrow{e^-+H^+} OH^-$$

$$E°(mV)) \quad -160 \quad +940 \quad +380 \quad +2330$$

Scheme 1 (Adapted from Ref. [8])

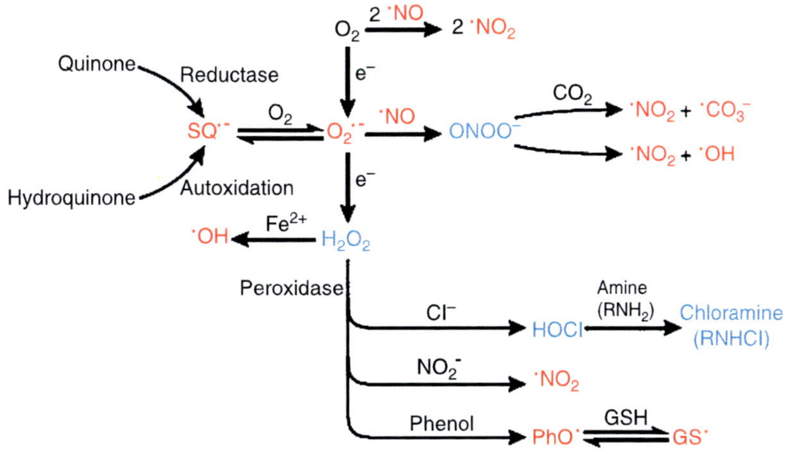

Scheme 2 The scheme covers most reactive species and selected routes of generation or interconversion. Not included are lipid or protein peroxyl radicals, singlet oxygen or carbon-centered radicals. Phenoxyl radicals (PhO·) are produced from tyrosine and other phenolic metabolites and xenobiotics. Aromatic amines and indoles are oxidized to radicals with similar properties. Semiquinone-like radicals ($SQ^{·-}$) are generated by autoxidation of a range of compounds including adrenaline and DOPA, or by enzymatic reduction of quinones such as ubiquinone or menadione. Flavonoids and other polyphenols can generate both semiquinone and phenoxyl radicals. Radicals equivalent to the glutathionyl radical (GS·) are generated from other thiols such as dihydrolipoic acid or cysteine residues. Different peroxidases vary in their substrate specificity, and only myeloperoxidase is capable of generating hypochlorous acid (HOCl). With bromide instead of chloride, equivalent reactive bromine species can be generated. One-electron oxidants (radicals) are shown in *red*; two-electron oxidants are shown in *blue*. Reprinted with permission from Ref. [17]

imbalance between the production of ROS and the ability of a biological system to detoxify the reactive intermediates and/or easily repair the resulting damage.

Many biochemical reactions are effective in forming H_2O_2 and superoxide but hROS are mainly generated in nonenzymatic processes, such as the breakdown of xenobiotics and reactions mediated by transition metal cations or by radiation.

In its simplest form the Fenton reaction, in which ferrous iron (or another redox active metal ion) catalyzes the formation of hydroxyl radicals from hydrogen peroxide can be written as a two-step process [8, 12]:

$$\mathrm{Fe(II)} + H_2O_2 \rightarrow \mathrm{Fe(III)} + HO^{\cdot} + OH^-$$
$$\mathrm{Fe(III)} + H_2O_2 \rightarrow \mathrm{Fe(II)} + HOO^{\cdot} + H^+$$

where $HOO^{\bullet}$ is the protonated form of superoxide ($pK_a = 4.8$), also known as hydroperoxyl or perhydroxyl radical.

In principle, the product of Fenton-like processes might alternatively be (i) the free hydroxyl radical, (ii) a bound "crypto" radical, or (iii) a Fe(IV)-oxo species [8, 12–16]:

$$\text{(i) } Fe^{2+} + H_2O_2 \rightarrow Fe^{3+} + HO^{\cdot} + OH^-$$
$$\text{(ii) } Fe^{2+} + H_2O_2 \rightarrow Fe^{3+} - HO^{\cdot} + OH^-$$
$$\text{(iii) } Fe^{2+} + H_2O_2 \rightarrow \mathrm{Fe\ (IV)}\ O_2^+ + H_2O$$

In case (i) the hydroxyl radical is free to leave the metal coordination sphere and react independently from it. In case (ii) the "crypto" radical oxidizes its substrate within the first or second coordination sphere of the metal. In both cases, the electron gap is placed on the oxygen atom. In contrast, in case (iii) the Fe(IV) species reacts primarily by the charged iron atom, which can lead to different reaction intermediates in comparison to a bound or free hydroxyl radical. The most reactive ROS and RNS species are summarized in Scheme 2 together with selected routes of generation or interconversion [17].

As already stated, mitochondria are an important source of ROS within most mammalian cells [18–23], and ROS production contributes to mitochondrial damage. Besides being implicated in aging by the accumulation of the ROS and mutations in the mitochondrial DNA (mtDNA), oxidative stress is associated with various pathological conditions. It plays a central role in chronic lung disease which is usually found in preterm infants subjected to high oxygen concentration [24]. Evidence has been collected that oxidative stress is a key player in ischemia due to oxygen reperfusion injury following hypoxia, leading to both cardiovascular diseases and strokes [25]. Diabetes initiated by hyperglycemia has a wide spectrum of disease manifestations as it leads to increased susceptibility of oxidative stress ultimately damaging various organs [26]. Upon administration of antioxidants like SOD and catalase, endothelial cell function is improved indicating oxidative stress as an important secondary factor in diabetes [27]. However, the main concern has been to definitely establish a strong and effective relation of oxidative stress with neurological disorders and to ascertain the redox signaling from the organelle to the rest of the cell (Fig. 4) [28].

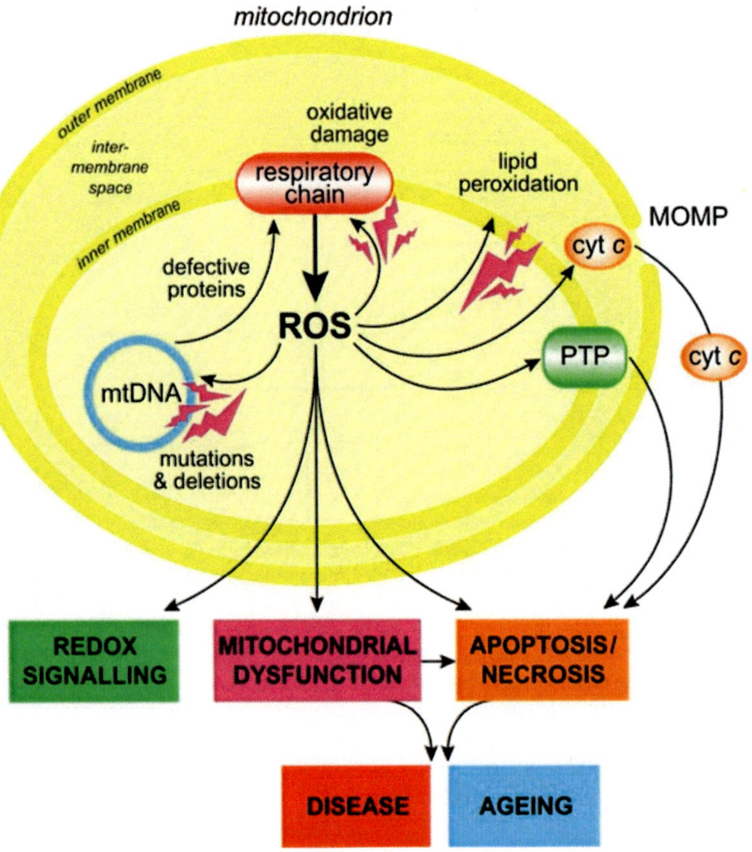

Fig. 4 ROS production by mitochondria can lead to oxidative damage to mitochondrial proteins, membranes, and DNA, impairing the ability of mitochondria to synthesize ATP and to carry out their wide range of metabolic functions, including the tricarboxylic acid cycle, fatty acid oxidation, the urea cycle, amino acid metabolism, hem synthesis, and FeS center assembly that are central to the normal operation of most cells. Mitochondrial oxidative damage can also increase the tendency of mitochondria to release intermembrane space proteins such as cytochrome *c* (cyt *c*) to the cytosol by mitochondrial outer membrane permeabilization (MOMP) and thereby activate the cell's apoptotic machinery. In addition, mitochondrial ROS production leads to induction of the mitochondrial permeability transition pore (PTP), which renders the inner membrane permeable to small molecules in situations such as ischemia/reperfusion injury. Consequently, it is unsurprising that mitochondrial oxidative damage contributes to a wide range of pathologies. In addition, mitochondrial ROS may act as a modulatable redox signal, reversibly affecting the activity of a range of functions in the mitochondria, cytosol, and nucleus (Reprinted with permission from Ref. [28])

The first report that the respiratory chain produced ROS came in 1966 [29]. It was then shown that isolated mitochondria produce H_2O_2 [30]. This H_2O_2 was later confirmed to arise from the dismutation of superoxide generated within mitochondria [31, 32]. The biological significance of superoxide production by mitochondria was verified by the discovery that mitochondria contain their own SOD, MnSOD [33]. Since then sources and consequences of mitochondrial ROS production have been widely and comprehensively investigated (Fig. 5) [34].

The location of superoxide within mitochondria is important because it does not diffuse across membranes. Complex I releases superoxide into the matrix [35], while complex III can release it both into the matrix as well as into the intermembrane space. Superoxide can be converted to H_2O_2 by mitochondrial matrix enzyme MnSOD or by CuZnSOD in the intermembrane space. H_2O_2 is more stable than superoxide and can diffuse out of the mitochondrion and into the cytosol (Fig. 6) [36].

In the presence of relatively high iron or copper concentrations, H_2O_2 can form the highly reactive hydroxyl radical via the Fenton reaction. The superoxide can also react with nitric oxide to form the highly reactive peroxynitrite. All the events leading to generation of ROS and RNS within mitochondria are summarized in Fig. 7.

Peroxynitrite ($pK_a = 6.8$) is a biologically generated reactive species as the product of the coupling reaction of two free radicals via the diffusion-controlled process:

$$NO + O_2^{\cdot -} \rightarrow ONOO^-$$

Peroxynitrite is continuously formed under basal metabolic conditions, but its biological actions become particularly effective under enhanced cell/tissue rates of superoxide

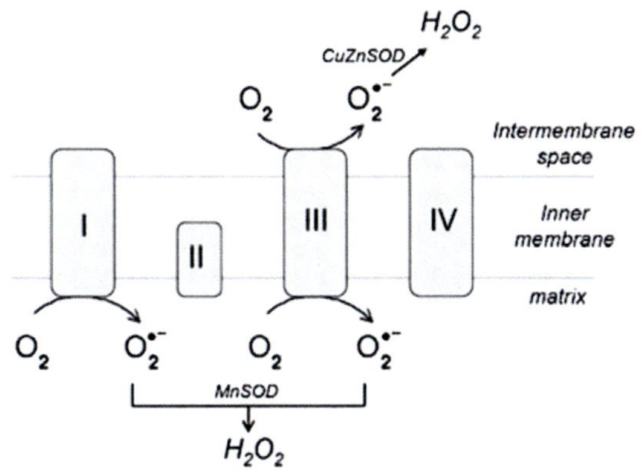

Fig. 5 Mitochondrial superoxide production and disposition (Reprinted with permission from Ref. [34])

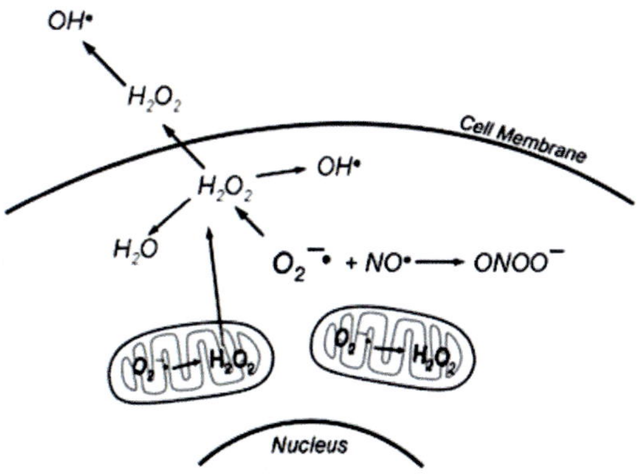

Fig. 6 Formation of intracellular ROS and RNS (Reprinted with permission from Ref. [36])

and/or nitric oxide generation as a result of its unique reactivity towards biomolecules and its tendency to initiate free radical processes. At physiological pH, $ONOO^-$ rapidly protonates to peroxynitrous acid, ONOOH, also a nonradical but nevertheless a very reactive agent, able to directly oxidize and nitrate proteins, lipids, and DNA [37].

Peroxynitrous acid can cause additional damage by undergoing homolytic fission to hydroxyl radical:

$$ONOOH \rightarrow NO_2^{\cdot} + OH^{\cdot}$$

Peroxynitrite also reacts with CO_2 [38]:

$$ONOO^- + CO_2 \rightarrow ONOOCOO^-$$
$$ONOOCOO^- \rightarrow NO_2^{\cdot} + CO_3^{\cdot-}$$

The chemical properties of peroxynitrite make it an important biological oxidant and, through this chemistry, a central pathogenic mediator in a variety of diseases, including cardiovascular, inflammatory, and neurodegenerative disorders.

Intracellular generation of ROS is an inevitable (and sometimes physiologically important) process [39]. In order to deal with it, several defense systems there occur in mitochondria, and cells in general. As a matter of fact, the true source of oxidative stress is the spatiotemporal imbalance of ROS production and detoxification, rather than ROS generation itself.

Sites of ROS generation

Given the redox potential of the $O_2^{\cdot-}/O_2$ couple ($E_{1/2} = -160$ mV) [40], the one-electron reduction of oxygen is thermodynamically affordable for several mitochondrial oxidoreductases [19]. Moreover, the superoxide is effectively removed from the reaction (*vide infra*) and a highly reduced state is possible for many redox carriers (*vide infra*), such that the reaction becomes virtually irreversible. Therefore the actual site of ROS production is under kinetic control. So far, a measurable ROS production has been reported for at least nine of the mammalian mitochondrial enzymes (labeled by purple stars in Fig. 8), being the capacity of producing ROS strongly variable among tissues and species. Among these sites, the following seven sources were shown to produce ROS in experiments with either isolated enzymes or mitochondria, but their contribution to mitochondrial ROS production under physiological conditions is not known.

- ***Cytochrome b5 reductase***, located in the OMM, oxidizes cytoplasmic NAD(P)H and reduces cytochrome *b*5 in the outer membrane. It may also reduce ascorbyl free radical and, therefore, be involved in regeneration of ascorbate in mammalian liver [41]. The enzyme is upregulated in the patients suffering from schizophrenia, thus implying a role in the etiology of the disease [42]. Cytochrome *b*5 reductase may produce superoxide with a high rate of ~300 nmol/min per mg protein [42].
- ***Monoamine oxidases*** (MAO-A and MAO-B), also located in the OMM, are ubiquitously expressed in various mammalian tissues and catalyze oxidation of biogenic amines accompanied by release of H_2O_2. MAOs of brain mitochondria play a central role in the turnover of monoamine neurotransmitters. Their potential for H_2O_2 generation exceeds that of other mitochondrial sources [43]. MAOs may be a major source of H_2O_2 in tissues in ischemia [44], and aging [45]. It has also been suggested that upregulation of MAO and the resulting elevated H_2O_2 production might be responsible for the mitochondrial damage in Parkinson's disease [46].
- ***Dihydroorotate dehydrogenase*** (DHOH), located at the outer surface of the IMM, is ubiquitously distributed in mammalian tissues. It catalyzes conversion of dihydroorotate to orotate. In the absence of its natural

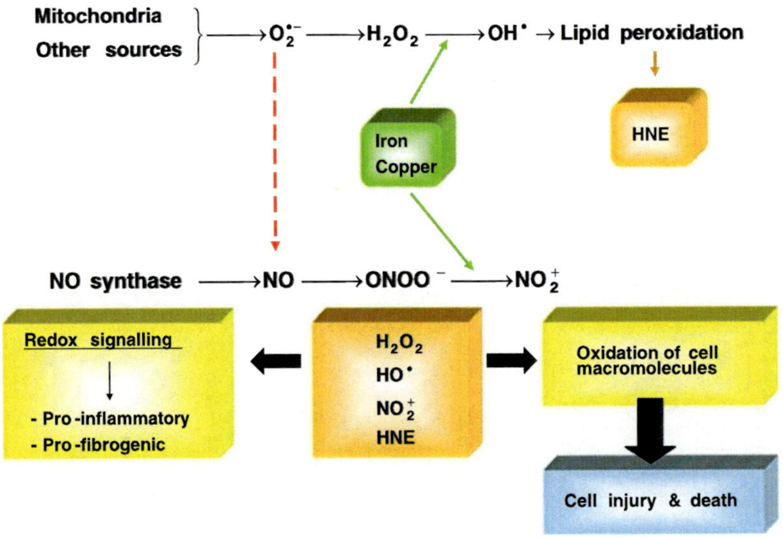

Fig. 7 Summary of generation and effects of ROS and RNS

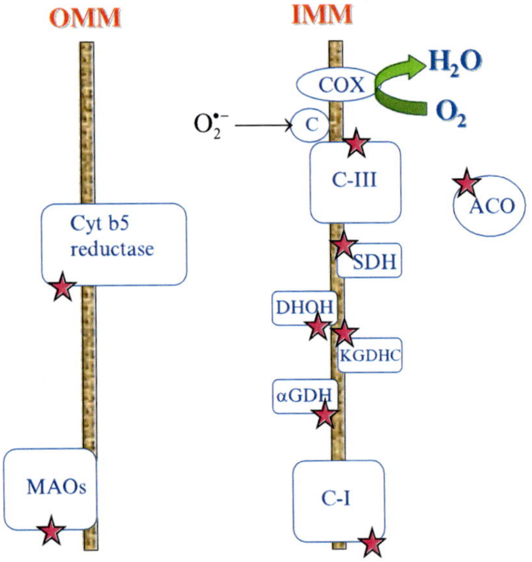

Fig. 8 Selected ROS-producing enzymes shown in the context of their location within mitochondria {*COX* cytochrome *c* oxidase, *c* cytochrome *c*, *C-III* complex III, *SDH* succinate dehydrogenase, *ACO* aconitase, *DHOH* dihydroorotate dehydrogenase, *KGDHC* α-ketoglutarate dehydrogenase complex, *αGDH* α-glycerophosphate dehydrogenase, *PDHC* pyruvate dehydrogenase complex, *C-I* complex I, *Cyt-b5 reductase* cytochrome *b*5 reductase, *MAOs* monoamine oxidases A and B, *OMM* outer mitochondrial membrane, *IMM* inner mitochondrial membrane} *Purple stars* indicate sources of ROS. Modified from Ref. [18]

electron acceptor, coenzyme Q of the IMM, reduced DHOH can produce H_2O_2 in vitro [47].

- ***Dehydrogenase of α-glycerophosphate*** (αGDH), also located at the outer surface of the IMM, is ubiquitously expressed in mouse tissues with brown fat, muscle, and brain possessing the highest activity [48]. It is a FAD-containing enzyme catalyzing oxidation of glycerol-3-phosphate to dihydroxyacetone phosphate and utilizing mitochondrial coenzyme Q as electron acceptor. Mitochondria oxidizing sn-glycerol-3-phosphate have been reported to produce H_2O_2 [49].
- ***Succinate dehydrogenase*** (SDH) is a flavoprotein located at the inner surface of the IMM. The enzyme oxidizes succinate to fumarate using coenzyme Q as electron acceptor. Isolated SDH incorporated in liposomes can produce ROS, most likely via its FAD, reduced in the absence of electron acceptor [50].
- ***Aconitase*** is localized in the mitochondrial matrix. It catalyzes conversion of citrate to isocitrate as part of the tricarboxylic acid cycle. The enzyme is inactivated upon oxidation of its iron–sulfur cluster by superoxide [51]. Upon inactivation, isolated aconitase induces production of hydroxyl radical, most likely mediated by released Fe^{2+} [52].
- ***α-Ketoglutarate dehydrogenase complex*** (KGDHC) is tightly associated with the matrix side of the IMM. It catalyzes oxidation of α-ketoglutarate to succinyl-CoA using NAD^+ as electron acceptor. KGDHC is composed of multiple copies of three enzymes: (e1) α-ketoglutarate dehydrogenase, (e2) dihydrolipoamide succinyl-transferase, and (e3) lipoamide dehydrogenase. The e3 component is a flavin-containing enzyme identical to the e3 component of another mitochondrial enzyme, *pyruvate dehydrogenase* (PDHC). Both PDHC and KGDHC have been demonstrated to generate superoxide and hydrogen peroxide [53, 54].

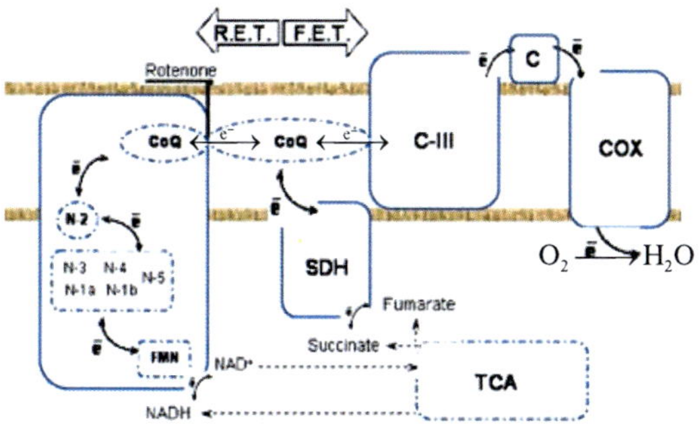

Fig. 9 Forward (F.E.T.) and reverse (R.E.T.) electron transfer in the respiratory chain {TCA, tricarboxylic acid cycle; SDH, succinate dehydrogenase; C-III, complex III; *c*, cytochrome *c*; COX, cytochrome *c* oxidase; FMN, flavin mononucleotide; N-1a, N-1b, N-2, N-3, N-4, N-5, iron–sulfur centers of complex I; CoQ, coenzyme Q} Modified from Ref. [18]

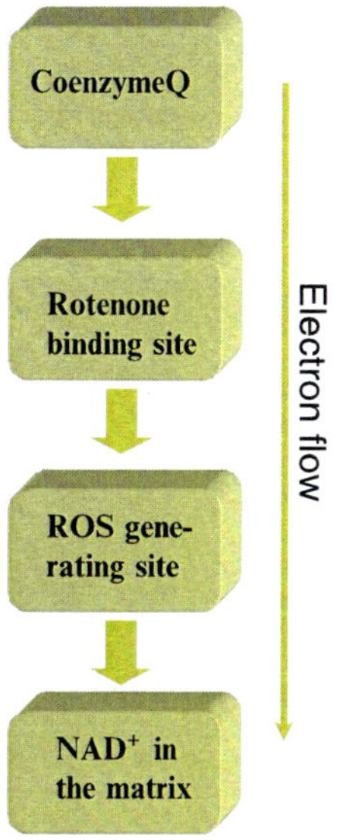

Fig. 10 Sequence of main electron transfer reactions in complex I

ROS production at complex I

Complex I (NADH-ubiquinone oxidoreductase) is an integral inner membrane multi-protein complex exposed to both matrix and intermembrane space. It oxidizes NADH using coenzyme Q as electron acceptor in a reversible reaction coupled with proton pump generating transmembrane potential [55]. This represents one of the two major entry points into the respiratory chain for reducing equivalents derived from tricarboxylic acid cycle (TCA) substrates (the other being SDH). Isolated complex I has long been known to generate superoxide in the presence of NADH in an ubiquinone-dependent way [56]. This suggested a similarity with the mechanism of ROS production observed in complex III.

Studies with both isolated complex I and submitochondrial particles demonstrated that a ROS producing site is located between flavin and the rotenone-binding site (Fig. 9) [57], and that there may be more than one superoxide producing site in that region [58].

For the sake of completeness, a less agreed suggestion has been made that the ROS producing site in complex I might be the flavin [59], or a complex of half-reduced NAD• radical bound to flavin [60].

One of the first reported mitochondrial reactions supporting ROS production was the reverse electron transfer (RET), initially discovered in experiments with submitochondrial particles [61]. RET is a set of reactions in the respiratory chain that allow electrons to be transferred against the gradient of redox potentials of electron carriers, from reduced coenzyme Q to NAD^+ instead of oxygen. To proceed, this thermodynamically unfavorable process has to be coupled to utilization of the energy of membrane potential. The reduction of coenzyme Q for this reaction requires FADH2-linked oxidizable substrate (e.g., succinate or α-glycerophosphate) and is accompanied by massive production of H_2O_2 [61]. Both NAD^+ reduction and ROS production require high membrane potential (provided by ATP hydrolysis). The ROS production is prevented by complex I inhibitors acting at the rotenone-binding site, indicating the sequence of electron transfer reactions shown in Fig. 10.

NAD^+ is not required but substantially enhances the ROS production [60]. RET supports very high rates of ROS production, ranging from 0.5 to 3 nmol H_2O_2/min per mg mitochondrial protein [61]. Consistent with high energy requirement for RET, this ROS production is sharply regulated by the amplitude of transmembrane potential [62], so that a 10 % decrease in the membrane potential inhibits 90 % of ROS production. Therefore, it is inhibited by any energy-utilizing process, e.g., ATP synthesis or Ca^{2+} uptake. RET-supported ROS production in intact mitochondria is inhibited by rotenone because it blocks the flow of electrons from coenzyme Q to complex I.

In the absence of complex I inhibitors, ROS production supported by NAD^+-linked substrates is stimulated by high membrane potential [62]. These conditions also favor high degree of reduction of redox carriers proximal to the proton pump of complex I. The dependence of ROS production rate on the amplitude of membrane potential is not as steep as in the case of RET, consistent with a more thermodynamically favorable process [45]. In any case ROS generation requires a highly reduced status of the involved redox carriers/centers, at least, more electronegative than the standard redox potential of the $NADH/NAD^+$ couple.

Although many unresolved issues are still to be clarified, complex I of the mitochondrial electron transport chain is widely accepted as a major site of mitochondrial ROS production [63, 64].

Q-cycle and the mechanism of ROS production at complex III

Complex III (*bc*1 complex, ubiquinone:cytochrome *c* reductase) of the mitochondrial respiratory chain is an enzyme complex oxidizing coenzyme Q (QH2) using cytochrome *c* as electron acceptor. The oxidation of coenzyme Q proceeds in a set of reactions known as the "Q-cycle" coupled to a vectorial translocation of protons, thereby generating transmembrane potential [65]. Complex III is capable of robust production of superoxide [30, 66, 67], which then rapidly dismutates to form H_2O_2 [39]. The consensus view of the underlying mechanism is schematized in Fig. 11.

The most relevant information on the site and the source of superoxide production was gained by using specific complex III inhibitors, namely (Fig. 12):

(a) Myxothiazol preventing the binding of QH_2 at the Q_o site
(b) Stigmatellin preventing the transfer of the first electron to ISP
(c) Antimycin A interrupting the transfer of the second electron to the Q_i site.

These specific inhibitors of complex III affect the production of superoxide in a very good agreement with their supposed effect on the formation of the putative semiquinone at the site Q_o.

According to the classical Q-cycle hypothesis inhibitors acting at the Q_i site, e.g., antimycin A, prevent the transfer of the second electron to the Q_i site thereby causing accumulation of unstable semiquinone at Q_o site; whereas inhibitors of the Q_o site inhibit semiquinone formation either by displacing quinol QH_2 at site Q_o (myxothiazol) or by specifically blocking the electron transfer reaction from quinol to ISP (stigmatellin) [68–70].

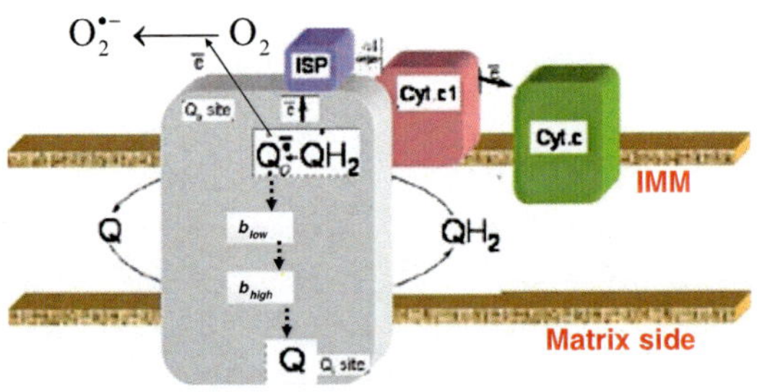

Fig. 11 Q-cycle model of coenzyme Q oxidation: the mechanism of superoxide formation in complex III. The reaction starts from the oxidation of the CoQ quinol (QH2) at the Qo site of complex III. The first electron is transferred to a high reduction potential chain consisting of the iron–sulfur protein (ISP, Rieske protein), cytochrome *c*1 (Cyt *c*1), and cytochrome *c* (Cyt *c*) and further to cytochrome *c* oxidase. The instable remaining semiquinone donates the second electron to the low reduction potential chain consisting of two cytochromes *b*, cyt *b*low and cyt *b*high, which serve as a pathway conducting electrons to the Qi site. There, these electrons reduce another CoQ molecule. To provide two electrons required for the complete reduction of CoQ quinone at the Qi site, the Qo site oxidizes two QH2 molecules in two successive turnovers. The first electron at the Qi site generates a stable semiquinone that is reduced to a quinol (QH2) by the second electron [65, 68, 69] (Modified from Ref. [18])

Mitochondrial ROS detoxifying systems

Mammalian mitochondria possess a complex multi-leveled ROS defense network of enzymes and nonenzymatic antioxidants. A main selection includes the following enzymatic subsystems.

- ***Membrane lipid peroxide removal systems.*** The peripheral layer of ROS defenses is formed by systems, mainly α-tocopherol (αTP), protecting membrane lipids from peroxidation. αTP is a ubiquitous lipid-soluble free radical scavenging antioxidant present in mitochondrial membranes. It reduces lipid radicals and can be regenerated by reduced coenzyme Q within the membrane or by water-soluble ascorbic acid at the water/membrane interface [71, 72].
- ***Phospholipid hydroperoxide glutathione peroxidase (PHGPx)*** is a selenoenzyme that utilizes glutathione as source of reducing equivalents. Broad selectivity allows it to reduce phospholipid hydroperoxides, H_2O_2, and other peroxides. It is the only enzyme known to reduce peroxidized phospholipids within membranes and it is thought to play an important role in cellular ROS defense system [73].
- ***MnSOD.*** The second layer of ROS defenses is formed by enzymes dealing with the primary ROS generated in mitochondria, i.e., superoxide radical and H_2O_2. The former is a substrate for mitochondrial manganese-containing superoxide dismutase (MnSOD). This enzyme is located exclusively inside the mitochondrial matrix; its only known function is to facilitate dismutation of superoxide radical to H_2O_2, thereby protecting mitochondrial iron–sulfur cluster containing enzymes from superoxide attack [74]. MnSOD does not require any cofactors so its efficiency is determined only by the amount of enzyme present. The MnSOD activity is high in liver, kidneys, and brain.
- ***Cytochrome c.*** The intermembrane space of mitochondria contains *ca.* 0.7 mM cytochrome *c* [75] that is capable of superoxide removal. Cytochrome *c* can be alternatively reduced by the respiratory chain or superoxide [76]. The reduced cytochrome *c* is regenerated by its natural electron acceptor, cytochrome *c* oxidase. Such re-oxidation generates proton-motive force that mitochondria can use to produce ATP [77]. This ability of a ROS -defense system to generate useful metabolic energy while detoxifying potentially harmful superoxide has earned it the title "the ideal antioxidant" [78].

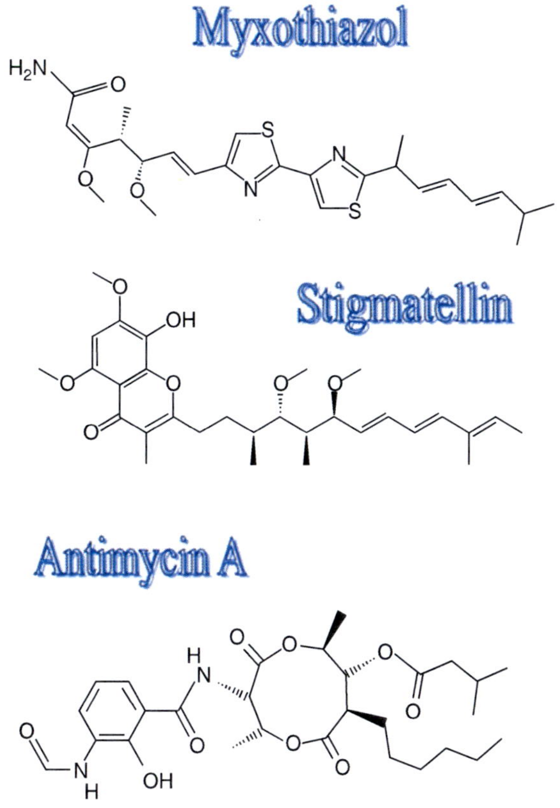

Fig. 12 Chemical structures of selected inhibitors of complex III of the mitochondrial respiratory chain

- ***Catalase.*** Superoxide dismutation leads to formation of another ROS, H_2O_2 that can be quite toxic and has to be detoxified by other enzymes. Catalase, one such enzyme, converts H_2O_2 into O_2 and H_2O and is commonly found in nearly all living organisms that are exposed to oxygen [79]. Catalase has one of the highest turnover numbers of all enzymes: one catalase molecule can convert 4×10^7 molecules of hydrogen peroxide to water and oxygen each second. Catalase is a tetramer of four polypeptide chains, each containing a porphyrin heme group that allows the enzyme to react with the hydrogen peroxide.
- ***Glutathione (GSH).*** Mitochondria contain ca. 10–12 % of total GSH amount in a cell, but due to their relatively small matrix volume the concentration of GSH in mitochondrial matrix is somewhat higher than that in the cytoplasm [80]. Mitochondria lack enzymes needed for GSH biosynthesis; the intramitochondrial pool of GSH is restored by rapid net uptake of GSH from the cytoplasm [81]. The concentration of GSH within mitochondria is in the range from 2 to 14 mM [81–83]; about ~90 % of GSH is in its reduced form. Actual concentrations of total (reduced + oxidized) glutathione in mitochondria vary depending on the metabolic state, age, and tissue. However, given the low micromolar estimates for steady-state levels of H_2O_2 in the matrix of mitochondria, it is likely that even a significant decrease in GSH levels may not have an impact on H_2O_2 detoxification by GSH-dependent enzymes.
- ***Glutathione-S-transferase.*** Mitochondria utilize GSH in two major ways: as a recyclable electron donor and as a consumable in conjugation reactions [84]. The latter are catalyzed by any of the several isoforms of glutathione-S-transferase (GST) present in mitochondria [85]. These enzymes protect mitochondria from various toxins including products of lipid peroxidation such as 4-hydroxynonenal by adding a GSH molecule to the toxin; GSH is consumed and has to be replenished by uptake from the cytosol. A sufficiently large intramitochondrial pool of GSH ensures an efficient operation of the GST-based detoxifying system.
- ***Glutathione reductase.*** Reduced GSH can either scavenge superoxide and hydroxyl radical nonenzymatically or serve as an electron-donating substrate to several enzymes involved in ROS detoxification. In either case, GSH is oxidized to GS-SG that cannot be exported to cytosol [86] and has to be reduced back to GSH in the mitochondrial matrix. The reduction is catalyzed by a specific enzyme, glutathione reductase (GSR), which is present in the matrix of mitochondria [87]. This enzyme utilizes intramitochondrial NADPH as a source of

reducing equivalents. In turn, mitochondrial NADPH can be regenerated by two major pathways: (1) the substrate-dependent reduction by dehydrogenases of mitochondrial matrix and (2) the hydride ion transfer reaction utilizing intramitochondrial NADH to reduce $NADP^+$. These NADPH regeneration pathways link mitochondrial ability to defend themselves against exogenously or endogenously generated ROS with their bioenergetic skill and oxidative capacity. As a result, ROS detoxification dissipates energy derived from oxidizable substrates.

- ***Glutathione peroxidase.*** Classical glutathione peroxidase (GPx1) is a widely investigated mitochondrial enzyme that utilizes GSH for the reduction of H_2O_2 to H_2O. This selenoenzyme can be detected in various cellular compartments including the mitochondrial matrix and intermembrane space. Apparently, GPx1 is involved in protecting tissues and mitochondria against acute oxidative stress rather than in providing a major defense against low-level endogenous mitochondrial ROS production.
- ***Peroxiredoxins and other oxins.*** Peroxiredoxins are peroxidases reducing H_2O_2 and lipid hydroperoxides [88]. Two isoforms of peroxiredoxins (Prx3 and Prx5) have been found in mammalian mitochondria. Both are regenerated in their active form by disulfide oxidoreductase thioredoxin (Trx2), a part of the mitochondrial thioredoxin system. Trx2 is reduced by thioredoxin reductase (TrxR2) that utilizes intramitochondrial NADPH as the source of reducing equivalents. Therefore, the efficient operation of Prx3 and Prx5 is dependent on efficient mitochondrial bioenergetics, similar to the GSH-linked systems (*vide supra*).

Consequences of ROS overproduction

Reactive oxygen and reactive nitrogen species can cause damage to all cellular macromolecules, including nucleic acids, proteins, carbohydrates, and lipids. Membrane lipids are major targets of ROS, and lipid peroxidation may lead to membrane dysfunction and alterations in cell permeability. Mitochondria are particularly vulnerable to oxidative damage because they are constantly exposed to high levels of ROS (Fig. 13) [36]. Mitochondrial DNA has been in fact shown to undergo oxidative damage.

In addition to lipid peroxidation, protein oxidation and nitration result into altered function of many metabolic enzymes in the mitochondrial matrix as well as of those comprising the electron transport chain. A particularly relevant protein that loses function upon oxidation is SOD, which would further compromise antioxidant capacity and lead to further oxidative stress. The following effects can all arise from ROS and RNS overproduction escaping detoxifying mechanisms of the cell.

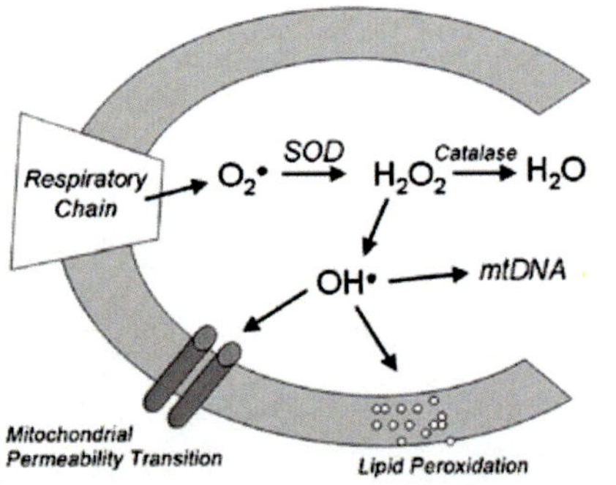

Fig. 13 Mitochondrial damage caused by reactive oxygen and nitrogen species. Free radicals generated by the electron transport chain can result in oxidative damage to mitochondrial DNA and proteins, lipid peroxidation, and opening of the mitochondrial permeability transition pore (Reprinted with permission from Ref. [36])

- Mitochondrial DNA is more vulnerable than nuclear DNA to oxidative damage because it is situated much closer to the site of ROS generation, and mitochondria have much more limited base excision repair mechanisms if compared with nuclei.
- Mitochondrial proteins can be modified by nitration or carbonylation. Protein oxidation and nitration result in altered function of many metabolic enzymes in the mitochondrial ETC.
- Oxidation of the adenine nucleotide translocator (ANT) impairs the influx of ADP into the matrix for ATP synthesis.
- MnSOD loses function upon oxidation, which further compromises antioxidant capacity and lead to further oxidative stress.
- ROS, especially $OH^{\cdot}$ can lead directly to peroxidation of lipids and the production of highly reactive aldehyde species such as 4-hydroxy-2,3,trans-nonenal (HNE), which can then result in secondary detrimental effects. Cardiolipin, the major phospholipid on the mitochondria inner membrane, is particularly susceptible to peroxidation because of its high degree of unsaturation.

Increasing evidence suggests that ROS play a key role in promoting cytochrome *c* release from mitochondria [89]. Cytochrome *c* is normally bound to the inner mitochondrial membrane by association with cardiolipin [90]. Peroxidation of cardiolipin leads to dissociation of cytochrome *c* and its release through the outer mitochondrial membrane into the cytosol [91]. The mechanism by which cytochrome *c* is released through the outer membrane is not clear. One mechanism may involve mitochondrial permeability transition (MPT), with swelling of the mitochondrial matrix and rupture of the outer membrane (14) [36]. ROS may promote MPT by causing oxidation of thiol groups on the adenine nucleotide translocator, which is believed to form part of the MPT pore [92]. Cytochrome *c* release may also occur via MPT-independent mechanisms and may involve an

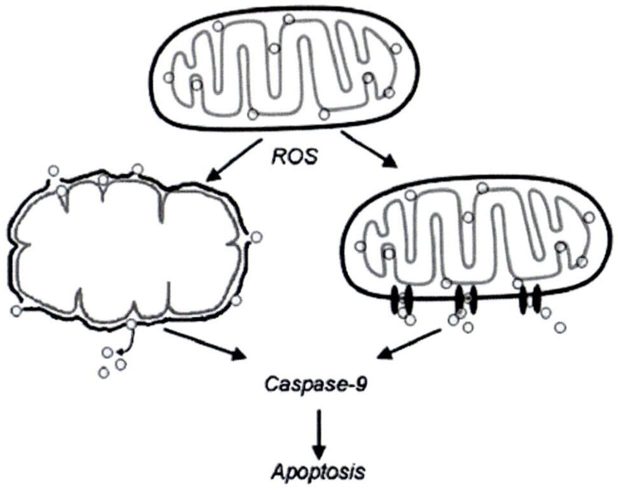

Fig. 14 Cytochrome *c* release from mitochondria. Cytochrome *c* (*gray sphere*) is normally associated with cardiolipin on the inner mitochondrial membrane. Cytochrome *c* is dissociated upon oxidation of cardiolipin and is believed to be released out of mitochondria either by mitochondrial permeability transition resulting in mitochondrial swelling and rupture of the outer membrane, or by channels formed by oligomerization of Bax. In the cytoplasm, cytochrome *c* activates caspase-9 and promotes apoptosis. Reprinted with permission from Ref. [36]

oligomeric form of Bax (Bcl-2-associated X protein, a protein of the Bcl-2 gene family, promoting apoptosis) [93] (Fig. 14). Cytochrome *c* in the cytoplasm triggers the activation of caspase-9, which triggers the caspase cascade and ultimately leads to apoptosis [94, 95].

Acute exposure to relatively high levels of oxidants, especially in the presence of calcium, can induce the mitochondrial permeability transition (MPT), uncouple oxidative phosphorylation with catastrophic effects on mitochondrial energetics, and contribute to cytotoxicity via necrosis and/or apoptosis (through release of cytochrome *c* or other factors). Induction of the MPT involves the calcium-mediated opening of a pore in the inner mitochondrial membrane that allows free diffusion of all solutes smaller than 1,500 Da in and out of the mitochondrial matrix. The MPT is attributed to a voltage-gated, cation-permeable channel, whose opening is favored by several factors including depolarization, intramatrix Ca^{2+}, and oxidizing agents and whose closing is favored by protons (low matrix pH) and adenine nucleotides.

Under proapoptotic conditions, two Bcl-2 family proteins, Bax and Bak, associate with the voltage-dependent anion channel component of the permeability transition (PT) pores on the outer membrane of the mitochondria. This calcium-dependent process allows the release of cytochrome *c* from the intermembrane space of the mitochondria into the cytoplasm (Fig. 15) [96].

The initial release of cytochrome *c* into the cytoplasm can result in its association with the IP3 receptor which acts as a calcium channel on the outer membrane of the endoplasmic

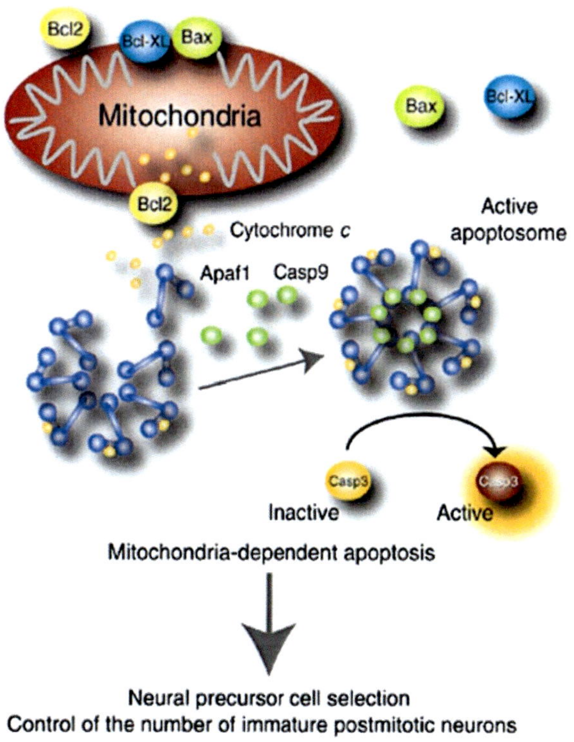

Fig. 15 Autophagosome formation and apoptosome-mediated cell death. Bcl2-like pro-apoptotic and antiapoptotic proteins regulate cytochrome *c* release from mitochondria. Cytosolic cytochrome *c* binds Apaf1 and induces the recruitment of the initiator caspase 9 (Casp9) on the active apoptosome. The active apoptosome, in turn, activates caspase 3 (Casp3), which mediates cell destruction. The relative roles of the two pathways in neurodevelopment are indicated. Notably, they are both required for neural tube closure (Adapted from Ref. [96])

reticulum. Subsequent release of calcium ions into the cytoplasm can induce apoptosis. Cytochrome *c* also participates in the cytosolic caspase proteolytic cascade of apoptosis as a component of the apoptotic protease activating factor (Apaf). The association with Apaf yields the formation of the apoptosome protein complex which can recruit and activate pro-caspase 9. Cytochrome *c* binds to Apoptosis Activation Factor 1 (Apaf-1) which is constitutively expressed in the cytoplasm. This complex cleaves off caspase-9 precursor's third domain. This allows the molecule to change conformation, and bind to another cleaved caspase-9 precursor, forming a homodimer. Caspase-9 is autocatalytic: it activates other caspase-9 s by cleaving off their third domain. This is known as the "Caspase Cascade." Caspase-9 also activates caspase-3. Caspase-3 and -9 activate other caspases, cleave enzymes, and activate endonucleases.

Neurodegeneration and oxidative stress

All aerobic cells suffer oxidative damage, yet the mammalian brain is often said to be especially sensitive. One reason is its high O_2 consumption; in adult humans, the brain accounts for only a few percent of body weight, but about 20 % of basal O_2 consumption. A major reason for the high O_2 uptake is the vast amounts of ATP needed to maintain neuronal intracellular ion homoeostasis in the face of all the openings and closings of ion channels associated with propagation of action potentials and neurosecretion. Thus interrupting mitochondrial function in neurons by toxins, or failing to supply O_2 or substrates for energy production, produces rapid damage. In particular, the high Ca^{2+} traffic across neuronal membranes means that interference with Ca^{2+} sequestration (e.g., by oxidative stress-dependent damage to plasma membrane Ca^{2+} exporters or to Ca^{2+} pumps in the endoplasmic reticulum) and/or disruption of the ATP supply produces especially rapid rises in intracellular free Ca^{2+}, that interfere with mitochondrial function and increase superoxide formation.

It has already been anticipated that cellular oxidative injury is implicated in aging and a wide array of clinical disorders including ischemia–reperfusion injury, neurodegenerative diseases, diabetes, and inflammatory diseases such as atherosclerosis, arthritis, and hepatitis. Generation of reactive oxidants, including ROS, appears to be increased in damaged mitochondria, and in cells with compromised mitochondrial function. Conversely, acute exposure to relatively high levels of oxidants, especially in the presence of calcium, can induce the mitochondrial permeability transition (MPT), uncouple oxidative phosphorylation with catastrophic effects on mitochondrial energetics, and contribute to cytotoxicity via necrosis and/or apoptosis (through release of cytochrome *c* and/or other factors).

The MPT plays a central role in both necrotic and apoptotic cell death. Induction of the MPT involves the calcium-mediated opening of a pore in the inner mitochondrial membrane that allows free diffusion of all solutes smaller than 1,500 Da in and out of the mitochondrial matrix. The MPT is attributed to a voltage-gated, cation-permeable channel, whose opening is favored by several factors including depolarization, intramatrix Ca^{2+}, and oxidizing agents and whose closing is favored by protons (low matrix pH) and adenine nucleotides. Oxidation of pyridine nucleotides favors pore opening whereas their reduction favors pore closing.

The various neurodegenerative diseases have different symptoms, affect different parts of the brain, and have different causes, but, as stated in the introductory chapter, they share many common features:

1. Impaired mitochondrial function [97],
2. Increased oxidative damage [98],
3. Defects in the ubiquitin–proteasome system [99],
4. The presence of abnormal, aggregated proteins [100],
5. Impaired metal ion homeostasis.

Oxidative damage is manifested as increases in lipid peroxidation end-products, DNA (and often RNA) base oxidation products, and oxidative protein damage. The protein aggregates frequently contain proteins that are nitrated, bear carbonyl residues, have attached aldehydes such as 4-hydroxy-2-*trans*-nonenal or acrolein and, sometimes, carry advanced glycation end-products. Most likely, any one of the above-mentioned events can initiate neuronal cell death and rapidly recruit the others [101].

Despite the heterogeneity of neurological disorders, a most relevant role played by mitochondrial involvement has progressively emerged from the huge body of investigations and studies [102]. As a matter of fact, mitochondria are key regulators of cell survival and death, have a central role in aging, and have recently been found to interact with many of the specific proteins implicated in genetic forms of neurodegenerative diseases.

Mitochondria are under the control of two genomes. Human mitochondria contain multiple copies of a 16.5 kb, double-stranded, circular DNA molecule (mtDNA) which is maternally inherited, although in rare circumstances there is paternal inheritance and recombination. mtDNA codes for 13 polypeptides, all of which are components of the respiratory chain. Although most mitochondrial proteins are encoded by nuclear DNA, mtDNA defects can cause numerous diseases, many of which are associated with neuronal degeneration. Human neurodegenerative diseases therefore can be caused by either mtDNA or nuclear DNA mutations which have effects on mitochondrial proteins.

The most important risk factor for neurodegenerative diseases is aging. A major theory of aging is that mutations in mtDNA and oxidative damage may contribute to the aging process. It has long been known that mitochondrial deletions and point mutations accumulate with normal aging. It has been speculated that mutations that accumulate with age might lead to impaired energy generation as well as increased amounts of ROS, resulting in cellular damage. Somatic mutations in the control region of mtDNA have been associated with human aging. There is strong evidence that mutations in mtDNA, which are acquired during normal aging, can contribute to the aging process. They, in turn, may lead to increased generation of ROS. mtDNA mutations and oxidative stress may contribute to aging. Somatic mutations accumulate in mtDNA with age. Inducing mtDNA mutations by disabling the proofreading activity of mtDNA POLG accelerates aging-related pathology in transgenic mice. Certain mtDNA polymorphisms are associated with increased longevity, possibly by reducing membrane potential and decreasing the generation of ROS. Overexpression of ROS-scavenging enzymes manganese superoxide dismutase (MnSOD), methionine sulphoxide reductase A (MSRA), or catalase within mitochondria prolongs lifespan. Knockout of p66SHC, a protein that promotes ROS generation and mitochondrial apoptosis, also prolongs lifespan. Complex IV and complex V activities decline with aging, and knockdown of complex V activity causes oxidative damage to nuclear

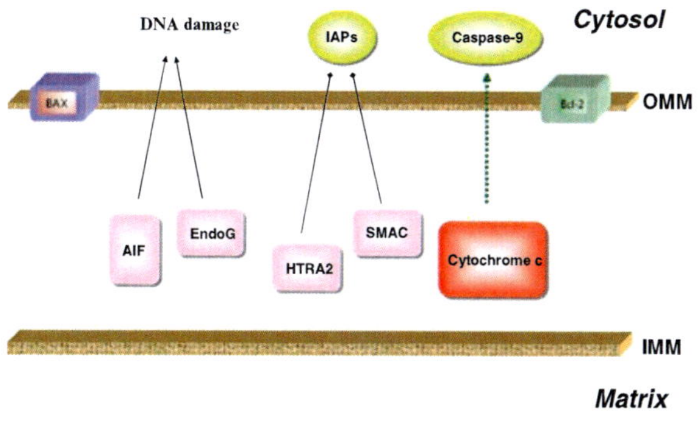

Fig. 16 Role of mitochondria in apoptosis. Several intermembrane space proteins are pro-apoptotic if released into the cytosol. Cytochrome *c* activates caspase-9. SMAC (second mitochondrial activator of caspases) and HTRA2 inhibit cytosolic inhibitor of apoptosis proteins (IAPs). HTRA2 is a serine protease that might function to remove denatured proteins within mitochondria, but degrades IAPs when released from mitochondria. Inhibiting HTRA2's normal quality control function or enhancing its IAP-degrading activity could both promote cell death. Apoptosis inhibitor factor (AIF) and endonuclease G (endo G) translocate to the nucleus and induce chromatin condensation and DNA fragmentation. Release of these proteins into the cell is modulated by recruitment of BAX (which is proapoptotic) or Bcl-2 (anti-apoptotic) to the outer mitochondrial membrane (OMM). Numerous extracellular and intracellular signals converge to regulate mitochondrial apoptosis

DNA, which probably results in decreased gene expression with aging. How mitochondria may exert their role in apoptosis and aging is shown in Fig. 16.

As already debated, net production of ROS and RNS is an additive and/or alternative important mechanism by which mitochondria are thought to contribute to aging. Mitochondria contain multiple electron carriers capable of producing ROS, as well as an extensive network of antioxidant defenses. Mitochondrial insults, including oxidative damage itself, can cause an imbalance between ROS production and removal, resulting in net ROS production. The importance to aging of net mitochondrial ROS production is supported by observations that enhancing mitochondrial antioxidant defenses can increase longevity. In *Drosophila*, overexpression of the mitochondrial antioxidant enzymes manganese superoxide dismutase (MnSOD) [103] and methionine sulphoxide reductase [104] prolong lifespan. Moreover, overexpression of catalase experimentally targeted to mitochondria increases lifespan in an already long-lived mouse strain [105]. The role of mitochondria in ROS metabolism is shown in Fig. 17 that also takes previous discussion into account [102]. Discussion of the main involvement of mitochondrial dysfunction in all forms of neurodegenerative disorders would be very long and probably tedious. We limit ourselves to assess the main features of mitochondria and oxidative stress in the two main pathologies, namely Alzheimer's and Parkinson's diseases.

Mitochondria and Alzheimer's disease

AD is characterized clinically by progressive cognitive decline, and pathologically by the presence of senile plaques composed primarily of Aβ and neurofibrillary tangles made up mainly of hyperphosphorylated tau. About 5–10 % of cases are familial, occurring in an early onset, autosomal-dominant manner. Three proteins are known to be associated with such familial cases:

- Amyloid precursor protein (APP), which is cleaved sequentially by β- and γ-secretases to produce Aβ;
- Presenilins 1 and 2 (PS1 and PS2), one or other of which is a component of each γ-secretase complex.

There is extensive literature supporting a role for mitochondrial dysfunction and oxidative damage in the pathogenesis of AD. Oxidative damage occurs early in the AD brain, before the onset of significant plaque pathology [106]. Oxidative damage also precedes Aβ deposition in transgenic APP mice [107], with upregulation of genes relating to mitochondrial metabolism and apoptosis occurring even earlier and co-localizing with the neurons undergoing oxidative damage [108]. Moreover, oxidative damage and mitochondrial dysfunction probably contribute causally to AD-related pathology.

Several pathways connecting oxidative stress and AD pathology have recently been uncovered. Oxidative stress may activate signaling pathways that alter APP or tau processing. For example, oxidative stress increases the expression of β-secretase [109] and increases aberrant tau phosphorylation [110].

Several recent reports suggest that many of the proteins implicated in AD pathogenesis have direct physical involvement with mitochondria or mitochondrial proteins (Fig. 18) [102]. Namely:

- APP has been found to have a dual endoplasmic reticulum/mitochondrial-targeting sequence, and in transfected cells and transgenic mice overexpressing APP it clogged the mitochondrial protein importation machinery, causing mitochondrial dysfunction and impaired energy metabolism [111];
- Aβ binds to a mitochondrial-matrix protein termed Aβ-binding alcohol dehydrogenase (ABAD) [112]. Blocking the interaction of Aβ and ABAD with a "decoy peptide" suppressed Aβ-induced apoptosis and free-radical generation in neurons. Conversely, overexpression of ABAD in transgenic APP-mutant mice exaggerated neuronal oxidative stress and impaired memory.
- Aβ also inhibits α-ketoglutarate dehydrogenase activity in isolated mitochondria [113], and deficiency of α-ketoglutarate dehydrogenase as well as of cytochrome oxidase activities has been observed in the brain and other tissues in AD.
- Aβ also interacts with the serine protease HTRA2 [114].

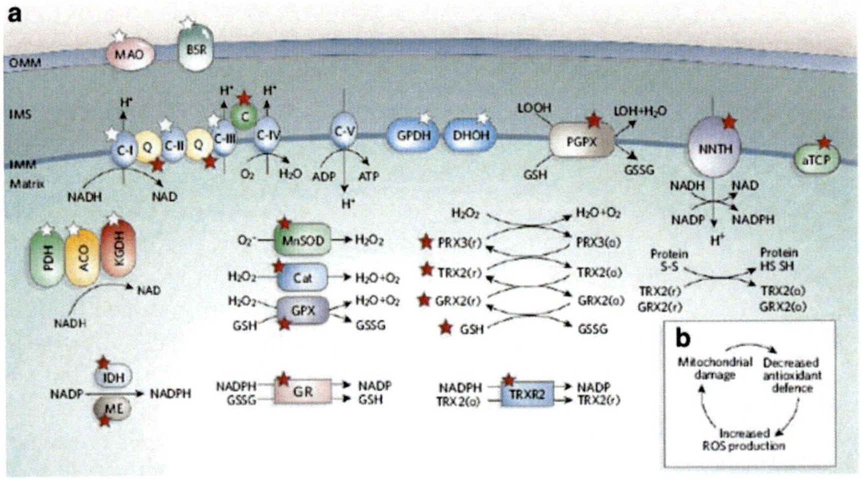

Fig. 17 Role of mitochondria in reactive oxygen species metabolism. The processes and components involved in ROS generation (*white stars*) and antioxidant defense (*red stars*). Mitochondria are the primary cellular consumers of oxygen and contain numerous redox enzymes capable of transferring single electrons to oxygen, generating the ROS superoxide. Mitochondrial enzymes so far known to generate ROS include the tricarboxylic acid (TCA) cycle enzymes aconitase (ACO) and α-ketoglutarate dehydrogenase (KGDH); the electron-transport chain (ETC) complexes I, II, and III; pyruvate dehydrogenase (PDH) and glycerol-3-phosphate dehydrogenase (GPDH); dihydroorotate dehydrogenase (DHOH); the monoamine oxidases (MAO) A and B; and cytochrome *b*5 reductase (B5R). The transfer of electrons to oxygen, generating superoxide, is more likely when these redox carriers are abundantly charged with electrons and the potential energy for transfer is high, as reflected by a high mitochondrial membrane potential. ROS generation is decreased when available electrons are few and potential energy for the transfer is low. Mitochondria also contain an extensive antioxidant defense system to detoxify the ROS generated by the reactions described earlier. Both the membrane-enclosed and soluble compartments are protected. Nonenzymatic components of the system include α-tocopherol (aTCP), coenzyme Q10 (Q), cytochrome *c* (C), and glutathione (GSH). Enzymatic components include manganese superoxide dismutase (MnSOD), catalase (Cat), glutathione peroxidase (GPX), phospholipid hydroperoxide glutathione peroxidase (PGPX), glutathione reductase (GR); peroxiredoxins (PRX3/5), glutaredoxin (GRX2), thioredoxin (TRX2), and thioredoxin reductase (TRXR2). The regeneration of GSH (through GR) and reduced TRX2 (through TRXR2) depends on NADPH, which is derived from substrates (through isocitrate dehydrogenase, IDH, or malic enzyme, ME) or the membrane potential (through nicotinamide nucleotide transhydrogenase, NNTH). So, like ROS generation, antioxidant defenses are also tied to the redox and energetic state of mitochondria. *GSSG* glutathione disulphide, *LOH* lipid hydroxide, *LOOH* lipid hydroperoxide, *o* oxidized state, *r* reduced state. In structurally and functionally intact mitochondria, a large antioxidant defense capacity balances ROS generation, and there is little net ROS production. Mitochondrial damage with decrease of antioxidant defense capacity is a prerequisite for net ROS production. Once this occurs, a vicious cycle (*inset* ***b***) can ensue whereby ROS can further damage mitochondria, causing more free-radical generation and loss or consumption of antioxidant capacity. For example, the Fe–S cluster in aconitase is easily inactivated by superoxide, the iron is released, and this induces hydroxyl radical production (Reprinted with permission from Ref. [102])

- All components of the γ-secretase complex have also been localized to mitochondria, where they form an active γ-secretase complex [115].

Mitochondria and Parkinson's disease

PD is characterized clinically by progressive rigidity, bradykinesia and tremor, and pathologically by loss of pigmented neurons in the substantia nigra and the presence of Lewy bodies, distinctive cytoplasmic inclusions that immunostain for α-synuclein and ubiquitin.

Mitochondria were first implicated in PD because MPTP (1-methyl 4-phenyl-1,2,3,6-tetrahydropyridine), whose metabolite MPP^+ inhibits complex I of the mitochondrial electron-transport chain, caused parkinsonism in designer drug abusers. The mechanism of toxicity in complex I inhibition models is most likely to involve oxidative stress [116]. Complex I inhibition and oxidative stress are relevant to naturally occurring PD when complex I deficiency and glutathione depletion are found in the *substantia nigra* [117]. Many of the genes associated with PD also implicate mitochondria in disease pathogenesis. So far, mutations or polymorphisms in mtDNA and nine nuclear genes have been identified as causing PD or affecting PD risk: α-synuclein, parkin, ubiquitin carboxy-terminal hydrolase L1, DJ-1, phosphatase and tensin homologue (PTEN)-induced kinase 1 (PINK1), leucine-rich-repeat kinase 2 (LRRK2), the nuclear receptor NURR1, HTRA2, and tau. Of the nuclear genes, α-synuclein, parkin, DJ-1, PINK1, LRRK2, and HTRA2 directly or indirectly involve mitochondria.

Mutations in α-synuclein are associated with autosomal dominant familial PD. α-Synuclein is a major component of Lewy bodies, and the primary effect of α-synuclein mutations is likely to be an increased formation of oligomeric or fibrillar aggregates. However, there seem to be close interrelationships between abnormal protein accumulation or degradation, oxidative stress, and mitochondrial dysfunction. In

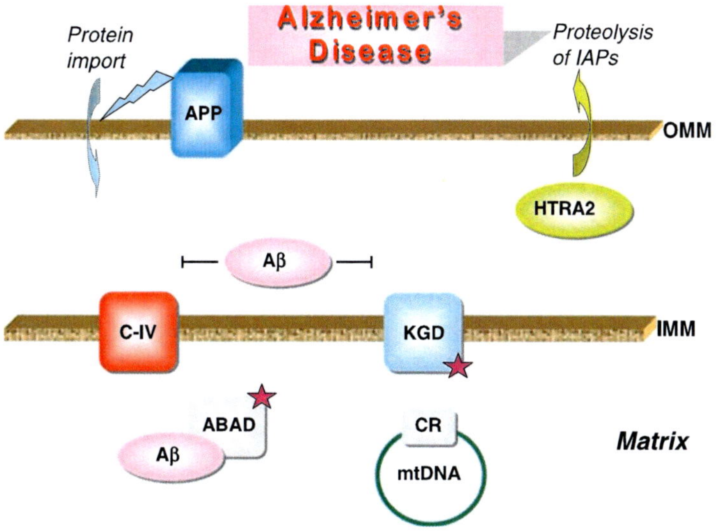

Fig. 18 In AD, mitochondrial ROS generation and inhibition of energy metabolism increase Aβ levels in cells and transgenic mice, and Aβ can interact with mitochondria and cause mitochondrial dysfunction. Aβ inhibits complex IV and α-ketoglutarate dehydrogenase (KGD), and binds Aβ-binding alcohol dehydrogenase (ABAD). Both KGD and ABAD produce ROS. Amyloid precursor protein (APP) may be targeted to the OMM and interfere with protein import. Mitochondria have also been reported to contain active γ-secretase complexes, which are involved in cleaving APP to form Aβ and contain presenilin 1, which increases the proteolytic activity of HTRA2 towards IAPs. AD patients have on average more somatic mutations in the mtDNA control region than control subjects (Adapted from Ref. [102])

transgenic mice, overexpression of α-synuclein impairs mitochondrial function, increases oxidative stress, and enhances nigral pathology induced by MPTP51. Moreover, in a recent study of mice overexpressing A53T mutant α-synuclein, degencrating mitochondria were immunostained for α-synuclein, raising the possibility that mutant α-synuclein might damage mitochondria directly [118].

Mutations in parkin are associated with autosomal recessive juvenile PD. Parkin encodes a ubiquitin E3 ligase, and the primary abnormality, therefore, is in the ubiquitin–proteasome system. However, as above, there seem to be close interrelationships between the ubiquitin–proteasome system, oxidative stress, and mitochondrial dysfunction. On one hand, parkin deficiency or mutations lead to oxidative stress and mitochondrial dysfunction; in fact, leukocytes from individuals with parkin mutations have a selective impairment in complex I activity [119]. Parkin can associate with the outer mitochondrial membrane and prevent mitochondrial swelling, cytochrome *c* release, and caspase activation, and this protective effect is abrogated by proteasome inhibitors and parkin mutations [120]. Parkin has also been localized to mitochondria in proliferating cells, where it has been shown to associate with mitochondrial transcription factor A and to enhance mitochondrial biogenesis [121]. On the other hand, mitochondrial dysfunction and oxidative stress can affect parkin function and exacerbate parkin mutations. S-nitrosylation of parkin, an oxidative modification, impairs its ubiquitin-ligase activity and compromises its protective function [122].

Mutations in *DJ-1* are also associated with autosomal recessive juvenile PD. DJ-1 interacts with α-synuclein [123], parkin [124], and PINK1 [125]. The overall function of DJ-1 seems to be to protect against cell death, especially that induced by oxidative stress. It can act as a redox sensor: oxidative stress causes a critical cysteine residue (C106) to be acidified, which leads to its relocalization to mitochondria. C106 mutations prevent this mitochondrial relocalization and impair the cell's response to oxidative stress and mitochondrial damage [126].

Mutations in *PINK1* represent a third form of autosomal recessive juvenile PD. PINK1 is a kinase localized to mitochondria [127], and, like DJ-1, seems to protect against cell death.

Mutations in *LRRK2* are the most common known cause of familial late-onset PD, and also account for 1–2 % of sporadic late-onset PD cases. On the basis of its sequence, LRRK2 is predicted to have a ROC–COR GTPase domain, a MAPK kinase domain and WD40 domains. It has been shown that LRRK2 is a kinase, that two disease-associated mutations, including the most common G2019S mutation, augment the kinase activity, and that ~10 % of LRRK2 is associated with mitochondria [128]. All these events can be summarized in Fig. 19 [102].

Targeting mitochondria

It has been previously discussed that, under normal conditions, mitochondria are protected from damage by ROS via several interacting antioxidant systems, but as soon as antioxidant protection is overpowered, oxidative

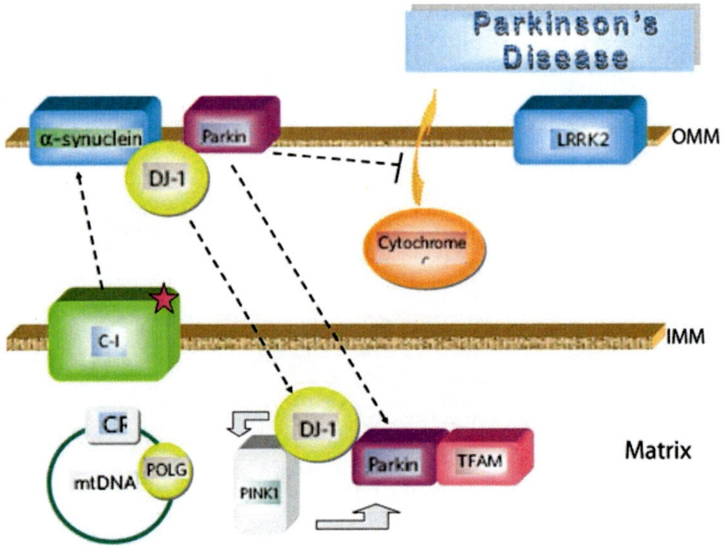

Fig. 19 Complex I activity is decreased in PD, and inhibition of complex I by MPTP or rotenone causes parkinsonism. Mutations in mtDNA-encoded complex I subunits, 12SrRNA, and POLG also cause parkinsonism. Many genes associated with PD also implicate mitochondria in disease pathogenesis. α-Synuclein immunostaining is seen in degenerating mitochondria from mice overexpressing A53T α-synuclein. α-Synuclein overexpression impairs mitochondrial function and enhances the toxicity of MPTP. Parkin associates with the OMM and protects against cytochrome *c* release. It may also associate with mitochondrial-transcription factor A (TFAM) and enhance mitochondrial biogenesis. When oxidized, DJ-1 translocates to mitochondria and protects the cell from oxidative-stress-induced cell death. The mitochondrial kinase PINK1 protects against apoptosis, an effect that is reduced by PD-related mutations or kinase inactivation. Physical associations have been reported between DJ-1 and α-synuclein, DJ-1 and parkin, and DJ-1 and PINK1, and there is genetic evidence that DJ-1, PINK1, and parkin function sequentially in the same pathway. About 10 % of the kinase LRRK2 is localized to mitochondria, and PD-related mutations augment its kinase activity. A mutation in HTRA2 was found in ~1 % of sporadic PD patients. Overexpression of the mutant impaired normal HTRA2 protease activity and HTRA2 knockout results in striatal degeneration and parkinsonism (Adapted from Ref. [102])

stress starts damaging nucleic acids, proteins, and lipids in mitochondria, resulting into loss of enzyme function in the ETC and eventually leading to mitochondrial dysfunction and impairment of ATP production [19, 129]. Endogenous antioxidant systems can also be damaged by protein oxidation, and peroxidation of cardiolipin leads to the dissociation of cytochrome *c* (compromising the function of cytochrome *c* oxidase), reduced ATP production, and further increased generation of ROS.

In these last years, the search for new protective remedies against damage caused by excessive free radical formation in mitochondria has accelerated. Similar to our body's own natural defenses against ROS, research has been primarily focused on molecules combining antioxidant utilities with recycling capacities. Large doses of antioxidants proved ineffective at preventing oxidative damage in animal disease models, presumably because the antioxidants and proteins such as manganese superoxide dismutase (MnSOD) cannot penetrate cell membranes effectively and therefore do not reach the relevant sites of ROS and RNS generation.

Since oxidative damage to mitochondria is central to the pathology of neurodegeneration, antioxidants specifically targeted at mitochondria have been considered as a helpful aid in treating the diseases. The desired effect of a drug or gene targeted at mitochondria in organs can be achieved only if the bioactive molecule is taken up by the required organ or cell type or both and accumulates in the desired subcellular location (in this case, mitochondria). The general strategy [130] consists in ameliorating the cell ROS scavenging potential by using antioxidants that:

(a) Are delivered specifically to mitochondria,
(b) Act where needed in the mitochondria,
(c) Pharmacologically or genetically increase endogenous expression of mitochondrial antioxidant systems.

One of the first ideas has been to target antioxidants selectively to mitochondria by conjugating an antioxidant to lipophilic cations that accumulate within mitochondria, driven by the mitochondrial membrane potential. For example, MitoQ consists of the lipophilic triphenylphosphonium (TPP) cation attached to the ubiquinone antioxidant moiety of the endogenous antioxidant co-enzyme Q10 [131]. The lipophilic TPP cation enables MitoQ to be taken up rapidly through the plasma and mitochondrial membranes without the requirement for a carrier, and the large membrane potential (negative inside) across the mitochondrial inner membrane causes MitoQ to accumulate several hundred-fold within mitochondria [131–133]. Within mitochondria, the MitoQ adsorbs to the matrix surface of

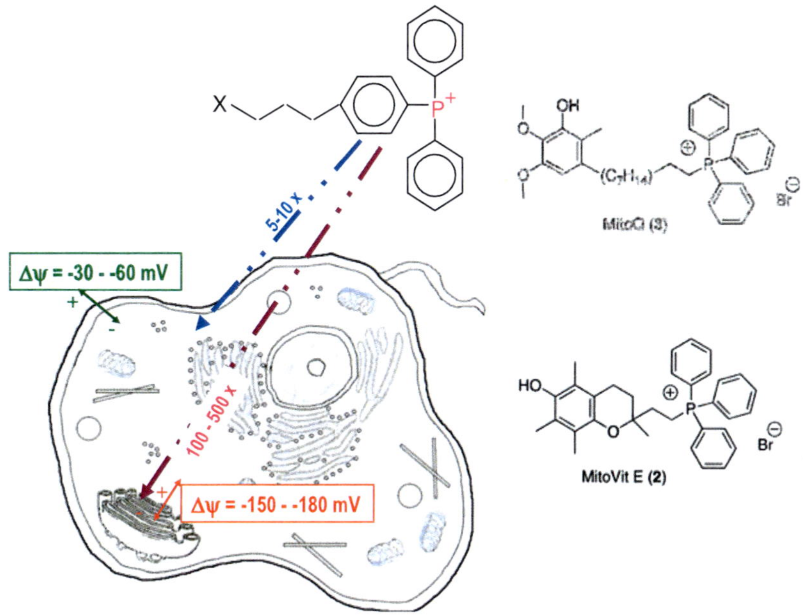

Fig. 20 If an antioxidant is attached to triphenylphosphonium, it accumulates several hundred-fold within mitochondria in cells and selectively blocks mitochondrial oxidative damage and mitochondrial redox signaling. Targeted antioxidants include derivatives of the endogenous antioxidants ubiquinol (MitoQ) and α-tocopherol (MitoVitE) (Redrawn from Ref. [130])

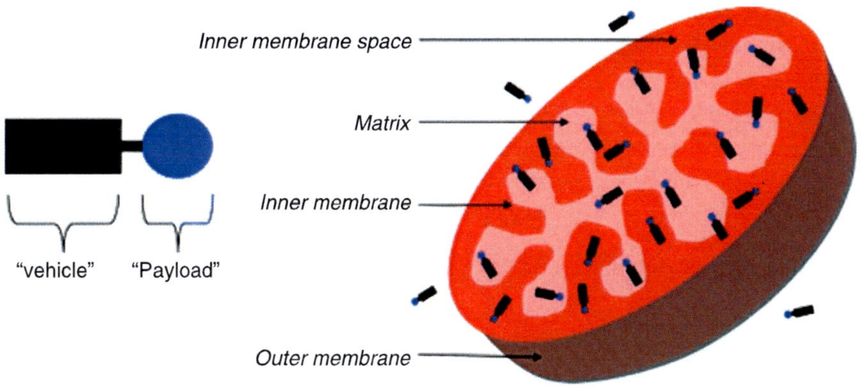

Fig. 21 Concept of dual function agents that use a vehicle to deliver an ROS scavenging payload into mitochondria (Reprinted with permission from Ref. [5])

the inner membrane and is recycled to the active ubiquinol antioxidant by the respiratory chain (Fig. 20). Antioxidants that accumulate within the matrix provide better protection from oxidative injury than untargeted antioxidants. MitoQ has been tested in a number of animal models of disease [134, 135] and has been also developed as a pharmaceutical for oral use in humans [136].

Other compounds have been conjugated to TPP. MitoVitE, a TPP-conjugated form of tocopherol, like MitoQ, protects mitochondria and whole cells from oxidative stress induced by several processes (Fig. 20). Other investigators have favored conjugating plastoquinone, a plant quinone needed for photosynthesis, to TPP to form a molecule named SkQ, that has been shown to protect cells against oxidative stress in vitro [137].

Another attractive solution has been suggested based on attaching a molecule with antioxidant properties onto a vehicle that can penetrate both the cellular and outer mitochondrial membranes and thereby deliver the medicine to a site where it can scavenge ROS and ameliorate oxidative damage (Fig. 21) [5]. Since the mitochondrial membrane spans across the mentioned negative potential, most agents have a positively charged moiety that takes advantage of electrostatic forces in locating its target.

Precursor proteins synthesized in the cytosol are usually required to encompass a specific amino acid sequence that is recognized by an import pathway [138] in order to be properly imported into mitochondria. While precursor proteins are prone to misfolding and aggregation, cytosolic chaperone proteins maintain them in an import-competent form [139]. The processed protein is then bound by translocases of the outer and inner membranes (TOM and TIM), which transport the target across the lipid bilayers [140].

Protein import and recognition is generally directed by an N-terminal or, less frequently, a C-terminal signal sequence consisting of about 20–30 amino acid residues, which are cleaved by mitochondrial processing peptidase (MPP) either during import or once inside the mitochondrial matrix.

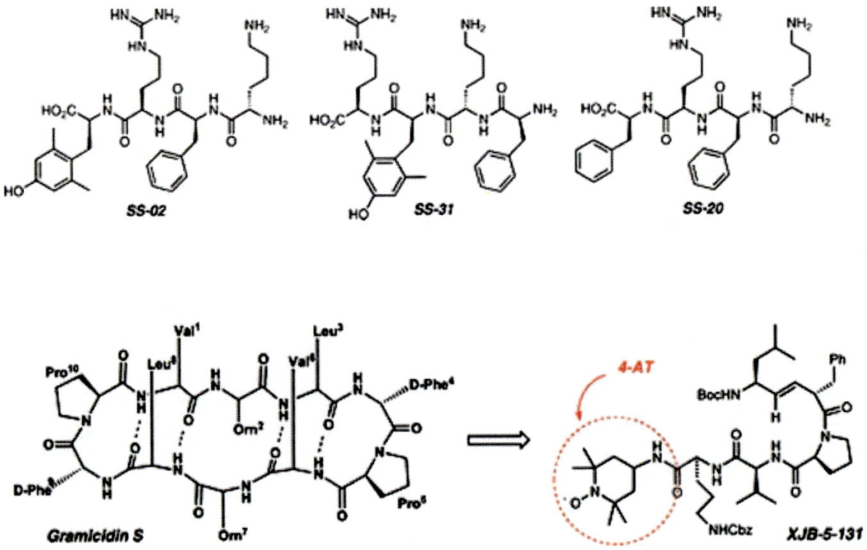

Fig. 22 Structures of active and control SS peptides and of the cyclodecapeptide antibiotic, microbial lipid targeting gramicidin S, and the designed **XJB-5-131**, which delivers the ROS scavenging unit 4-AT to mitochondrial membranes

A common primary structure has not been observed, but a common secondary structure as well as certain basic (arginine), hydrophobic (alanine, leucine), and polar residues (serine) might be present. Proteins such as cytochrome *c* and superoxide dismutases are imported with minimal processing since they contain the necessary recognition elements as part of their primary sequence [141, 142]. The N-terminal regions are postulated to fold into amphiphilic helices. It is proposed that this amphiphilicity in combination with localized positive charges emanating from basic residues are the two main features required for successful protein import [143]. Electrostatic interactions are thought to occur between the positive charges found on the helix and the negative charges of the TOM receptors [144].

Several cell-permeable mitochondrial targeting peptides with attached antioxidants have been conceived and tested, including SS peptides [145], which feature a 2′,6′-dimethyltyrosine (Dmt) payload, as well as the XJB peptide mimetics [146], which deliver 4-amino-TEMPO (4-AT), a stable nitroxide radical.

SS tetrapeptides represent a series of mitochondria-targeting antioxidant peptides that feature a common structural motif of alternating aromatic and basic residues (Fig. 22). The antioxidant properties of **SS-02** and **SS-31** are likely to originate from their dimethyltyrosine (Dmt) residues [147]. The basic residues provide for localization in the inner mitochondrial membrane, and the Dmt phenol moieties of **SS-02** and **SS-31** are likely responsible for chemically reducing reactive oxygen species and peroxide bonds. Tetrapeptide **SS-20**, where Dmt is substituted with a phenylalanine residue, was devised as a control and, in agreement with the hypothesis, demonstrated no ROS scavenging ability.

XJB peptides and peptide mimetics are based on the sequence of the membrane-active gramicidin S (GS) antibiotics (Fig. 22); their antioxidant properties stem from the attachment to the stable free radical, 4-amino-TEMPO (4-AT) [146], which allows to use ESR to measure distribution of the spin label and detect oxidative stress in the local cellular environment [148].

The final statement of Hoye et al. [5] can be adequately cited here as the most suitable conclusion to the issues we have tried to review here: "Mitochondrial targeting of ROS scavengers or compounds that interfere with the unique biochemistry in mitochondria is a relatively novel and highly relevant approach in drug discovery for the treatment of degenerative diseases and acute conditions derived from surging ROS and RNS. No pharmaceutical agents specifically designed to deliver a therapeutic compound to mitochondria have yet reached the market, but it is likely that a significant number of essential nutrients, including ascorbic acid, selenium, vitamins E and Q10, carotenoids, etc., fulfill at least part of their function by controlling the surge of reactive by-products of the oxidative phosphorylation process and ATP generation in mitochondria. Since age-related conditions are rapidly becoming a major source of a declining quality of life in a graying population, we can only hope that the near future will show the emergence of a new class of effective therapies that involve mitochondrial survival strategies. This is an exciting development for synthetic chemists who are being challenged with the discovery of innovative approaches to deliver functional small organic compounds as well as larger biomolecules across cell membranes to specific intracellular targets".

Acknowledgments We acknowledge Prof. Henryk Kozlowski for his continuous support and suggestions. We also want to apologize for all the reports we have not mentioned. The number of records found by any motor of research is incredibly huge and it is practically impossible to mention all the work done, especially in the last 5–10 years. Among the literature reports, many review articles can be found [2, 149–163].

References

1. Taylor JP, Hardy J, Fischbeck KH (2002) Toxic proteins in neurodegenerative disease. Science 296:1991–1995
2. Patel VP, Chu CT (2011) Nuclear transport, oxidative stress, and neurodegeneration. Int J Clin Exp Pathol 4:215–229
3. Harman D (1972) The biologic clock: the mitochondria? J Am Geriatr Soc 20:145–147
4. Fernández-Checa JC, Fernández A, Morales A, Marí M, García-Ruiz C, Colell A (2010) Oxidative stress and altered mitochondrial function in neurodegenerative diseases: lessons from mouse models. CNS Neurol Disord Drug Targets 9:435–454
5. Hoye AT, Davoren JE, Wipf P, Fink MP, Kagan VE (2008) Targeting mitochondria. Acc Chem Res 41:87–97
6. Baffy G (2005) Uncoupling protein-2 and non alcoholic fatty liver disease. Frontiers Biosci 10:2082–2096
7. Sheu S-S, Nauduri D, Anders MW (2006) Targeting antioxidants to mitochondria: a new therapeutic direction. Biochim Biophys Acta Mol Basis Dis 1762:256–265
8. Wood PM (1988) The potential diagram for oxygen at pH 7. Biochem J 253:287–289
9. Freinbichler W, Colivicchi MA, Stefanini C, Bianchi L, Ballini C, Misini B, Weinberger P, Linert W, Varešlija D, Tipton KF, Della Corte L (2011) Highly reactive oxygen species: detection, formation, and possible functions. Cell Mol Life Sci 68:2067–2079
10. D'Autreaux B, Toledano MB (2007) ROS as signalling molecules: mechanisms that generate specificity in ROS homeostasis. Nat Rev Mol Cell Biol 8:813–824
11. Shukla V, Mishra SK, Pant HC (2011) Oxidative stress in neurodegeneration. Adv Pharmacol Sci 2011:572634, published online 2011 September 21. doi: 10.1155/2011/572634
12. Walling C (1975) Fenton's reagent revisited. Acc Chem Res 8:125–131
13. Lloyd RV, Hanna PM, Mason RP (1997) The origin of the hydroxyl radical oxygen in the Fenton reaction. Free Radic Biol Med 22:885–888
14. Freinbichler W, Tipton KF, Della Corte L, Linert W (2009) Mechanistic aspects of the Fenton reaction under conditions approximated to the extracellular fluid. J Inorg Biochem 103:28–34
15. Merkofer M, Kissner R, Hider RC, Brunk UT, Koppenol WH (2006) Fenton chemistry and iron chelation under physiologically relevant conditions: electrochemistry and kinetics. Chem Res Toxicol 19:1263–1269
16. Kremer ML (2000) Is *OH the active Fenton intermediate in the oxidation of ethanol? J Inorg Biochem 78:255–257
17. Winterbourn CC (2008) Reconciling the chemistry and biology of reactive oxygen species. Nat Chem Biol 4:278–286
18. Andreyev AY, Kushnareva YE, Starkov AA (2005) Mitochondrial metabolism of reactive oxygen species. Biochem Mosc 70:200–214
19. Turrens JF (2003) Mitochondrial formation of reactive oxygen species. J Physiol 552:335–344
20. Balaban RS, Nemoto S, Finkel T (2005) Mitochondria, oxidants, and aging. Cell 120:483–495
21. Cadenas E, Davies KJ (2000) Mitochondrial free radical generation, oxidative stress, and aging. Free Radical Biol Med 29:222–230
22. Raha S, Robinson BH (2000) Mitochondria, oxygen free radicals, disease and ageing. Trends Biochem 25:502–508
23. Adam-Vizi V, Chinopoulos C (2006) Bioenergetics and the formation of mitochondrial reactive oxygen species. Trends Pharmacol Sci 27:639–645
24. Saugstad OD (1998) Chronic lung disease: the role of oxidative stress. Biol Neonate 74(suppl 1):21–28
25. Kennedy G, Spence VA, McLaren M, Hill A, Underwood C, Belch JJ (2005) Oxidative stress levels are raised in chronic fatigue syndrome and are associated with clinical symptoms. Free Radic Biol Med 39:584–589
26. Baynes JW (1991) Role of oxidative stress in development of complications in diabetes. Diabetes 40:405–412
27. Giugliano D, Ceriello A, Paolisso G (1996) Oxidative stress and diabetic vascular complications. Diabetes Care 19:257–267
28. Murphy MP (2009) How mitochondria produce reactive oxygen species. Biochem J 417:1–13
29. Jensen PK (1966) Antimycin-insensitive oxidation of succinate and reduced nicotinamide-adenine dinucleotide in electron-transport particles. I. pH dependency and hydrogen peroxide formation. Biochim Biophys Acta 122:157–166
30. Loschen G, Flohe L, Chance B (1971) Respiratory chain linked H_2O_2 production in pigeon heart mitochondria. FEBS Lett 18:261–264
31. Loschen G, Azzi A, Richter C, Flohe L (1974) Superoxide radicals as precursors of mitochondrial hydrogen peroxide. FEBS Lett 42:68–72
32. Forman HJ, Kennedy JA (1974) Role of superoxide radical in mitochondrial dehydrogenase reactions. Biochem Biophys Res Commun 60:1044–1050
33. Weisiger RA, Fridovich I (1973) Superoxide dismutase: organelle specificity. J Biol Chem 248:3582–3592
34. Szeto HH (2006) Mitochondria-targeted peptide antioxidants: novel neuroprotective agents. AAPS J 8:E521–E531
35. Muller FL, Liu Y, Van RH (2004) Complex III releases superoxide to both sides of the inner mitochondrial membrane. J Biol Chem 279:49064–49073
36. Szeto HH (2006) Cell-permeable, mitochondrial-targeted, peptide antioxidants. AAPS J 8:E277–E283
37. Alvarez B, Radi R (2003) Peroxynitrite reactivity with amino acids and proteins. Amino Acids 25:295–311
38. Greenacre SA, Ischiropoulos H (2001) Tyrosine nitration: localisation, quantification, consequences for protein function and signal transduction. Free Radic Res 34:541–581
39. Skulachev VP (1996) Role of uncoupled and non-coupled oxidations in maintenance of safely low levels of oxygen and its one-electron reductants. Q Rev Biophys 29:169–202
40. Sawyer DT, Valentine JS (1981) How super is superoxide? Acc Chem Res 14:393–400
41. Nishino H, Ito A (1986) Subcellular distribution of OM cytochrome b-mediated NADH-semidehydroascorbate reductase activity in rat liver. J Biochem (Tokyo) 100:1523–1531
42. Whatley SA, Curti D, Das Gupta F, Ferrier IN, Jones S, Taylor C, Marchbanks RM (1998) Superoxide, neuroleptics and the ubiquinone and cytochrome b5 reductases in brain and lymphocytes from normals and schizophrenic patients. Mol Psychiatry 3:227–237
43. Hauptmann N, Grimsby J, Shih JC, Cadenas E (1996) The metabolism of tyramine by monoamine oxidase A/B causes oxidative damage to mitochondrial DNA. Arch Biochem Biophys 335:295–304
44. Kunduzova OR, Bianchi P, Parini A, Cambon C (2002) Hydrogen peroxide production by monoamine oxidase during ischemia/reperfusion. Eur J Pharmacol 448:225–230
45. Maurel A, Hernandez C, Kunduzova O, Bompart G, Cambon C, Parini A, Frances B (2003) Age-dependent increase in hydrogen peroxide production by cardiac monoamine oxidase A in rats. Am J Physiol Heart Circ Physiol 284:H1460–H1467

46. Kumar MJ, Nicholls DG, Andersen JK (2003) Oxidative a-ketoglutarate dehydrogenase inhibition via subtle elevations in monoamine oxidase B levels results in loss of spare respiratory capacity: implications for Parkinson's disease. J Biol Chem 278:46432–46439
47. Loffler M, Becker C, Wegerle E, Schuster G (1996) Catalytic enzyme histochemistry and biochemical analysis of dihydroorotate dehydrogenase/oxidase and succinate dehydrogenase in mammalian tissues, cells and mitochondria. Histochem Cell Biol 105:119–128
48. Koza RA, Kozak UC, Brown LJ, Leiter EH, MacDonald MJ, Kozak LP (1996) Sequence and tissue-dependent RNA expression of mouse FAD-linked glycerol-3-phosphate dehydrogenase. Arch Biochem Biophys 336:97–104
49. Kwong LK, Sohal RS (1998) Substrate and site specificity of hydrogen peroxide generation in mouse mitochondria. Arch Biochem Biophys 350:118–126
50. Zhang L, Yu L, Yu CA (1998) Generation of superoxide anion by succinate-cytochrome c reductase from bovine heart mitochondria. J Biol Chem 273:33972–33976
51. Gardner PR (2002) Aconitase: sensitive target and measure of superoxide. Methods Enzymol 349:9–23
52. Vasquez-Vivar J, Kalyanaraman B, Kennedy MC (2000) Mitochondrial aconitase is a source of hydroxyl radical. An electron spin resonance investigation. J Biol Chem 275:14064–14069
53. Tretter L, Adam-Vizi V (2004) Generation of reactive oxygen species in the reaction catalyzed by α-ketoglutarate dehydrogenase. J Neurosci 24:7771–7778
54. Starkov AA, Fiskum G, Chinopoulos C, Lorenzo BJ, Browne SE, Patel MS, Beal MF (2004) Mitochondrial α-ketoglutarate dehydrogenase complex generates reactive oxygen species. J Neurosci 24:7779–7788
55. Skulachev VP (1988) Membrane bioenergetics. Springer, New York
56. Cadenas E, Boveris A, Ragan CI, Stoppani AO (1977) Production of superoxide radicals and hydrogen peroxide by NADH-ubiquinone reductase and ubiquinol-cytochrome c reductase from beef-heart mitochondria. Arch Biochem Biophys 180:248–257
57. Genova ML, Ventura B, Giuliano G, Bovina C, Formiggini G, Parenti Castelli G, Lenaz G (2001) The site of production of superoxide radical in mitochondrial Complex I is not a bound ubisemiquinone but presumably iron-sulfur cluster N2. FEBS Lett 505:364–368
58. Kang D, Narabayashi H, Sata T, Takeshige K (1983) Kinetics of superoxide formation by respiratory chain NADH-dehydrogenase of bovine heart mitochondria. J Biochem (Tokyo) 94:1301–1306
59. Kudin AP, Bimpong-Buta NY, Vielhaber S, Elger CE, Kunz WS (2004) Characterization of superoxide-producing sites in isolated brain mitochondria. J Biol Chem 279:4127–4135
60. Krishnamoorthy G, Hinkle PC (1988) Studies on the electron transfer pathway, topography of iron-sulfur centers, and site of coupling in NADH-Q oxidoreductase. J Biol Chem 263:17566–17575
61. Hinkle PC, Butow RA, Racker E, Chance B (1967) Partial resolution of the enzymes catalyzing oxidative phosphorylation. XV. Reverse electron transfer in the flavin-cytochrome beta region of the respiratory chain of beef heart submitochondrial particles. J Biol Chem 242:5169–5173
62. Starkov AA, Fiskum G (2003) Regulation of brain mitochondrial H_2O_2 production by membrane potential and NAD(P)H redox state. J Neurochem 86:1101–1107
63. Herrero A, Barja G (2000) Localization of the site of oxygen radical generation inside the complex I of heart and nonsynaptic brain mammalian mitochondria. J Bioenerg Biomembr 32:609–615
64. Sipos I, Tretter L, Adam-Vizi V (2003) Quantitative relationship between inhibition of respiratory complexes and formation of reactive oxygen species in isolated nerve terminals. J Neurochem 84:112–118
65. Trumpower BL (1990) The protonmotive Q cycle. Energy transduction by coupling of proton translocation to electron transfer by the cytochrome bc1 complex. J Biol Chem 265:11409–11412
66. Dionisi O, Galeotti T, Terranova T, Azzi A (1975) Superoxide radicals and hydrogen peroxide formation in mitochondria from normal and neoplastic tissues. Biochim Biophys Acta 403:292–300
67. Boveris A, Cadenas E, Stoppani AO (1976) Role of ubiquinone in the mitochondrial generation of hydrogen peroxide. Biochem J 156:435–444
68. Crofts AR, Barquera B, Gennis RB, Kuras R, Guergova-Kuras M, Berry EA (1999) Mechanism of ubiquinol oxidation by the bc_1 complex: different domains of the quinol binding pocket and their role in the mechanism and binding of inhibitors. Biochemistry 38:15807–15826
69. Zhang Z, Huang L, Shulmeister VM, Chi YI, Kim KK, Hung LW, Crofts AR, Berry EA, Kim SH (1998) Electron transfer by domain movement in cytochrome bc1. Nature 392:677–684
70. Turrens JF (1997) Superoxide production by the mitochondrial respiratory chain. Biosci Rep 17:3–8
71. Lass A, Forster MJ, Sohal RS (1999) Effects of coenzyme Q10 and a-tocopherol administration on their tissue levels in the mouse: elevation of mitochondrial a-tocopherol by coenzyme Q10. Free Radic Biol Med 26:1375–1382
72. Packer L, Weber SU, Rimbach G (2001) Molecular aspects of α-tocotrienol antioxidant action and cell signalling. J Nutr 131:369S–373S
73. Imai H, Nakagawa Y (2003) Biological significance of phospholipid hydroperoxide glutathione peroxidase (PHGPx, GPx4) in mammalian cells. Free Rad Biol Med 34:145–169
74. Gardner PR, Raineri I, Epstein LB, White CW (1995) Superoxide radical and iron modulate aconitase activity in mammalian cells. J Biol Chem 270:13399–13405
75. Hackenbrock CR, Chazotte B, Gupte SS (1986) The random collision model and a critical assessment of diffusion and collision in mitochondrial electron transport. J Bioenerg Biomembr 18:331–368
76. McCord JM, Fridovich I (1970) The utility of superoxide dismutase in studying free radical reactions. II. The mechanism of the mediation of cytochrome c reduction by a variety of electron carriers. J Biol Chem 245:1374–1377
77. Mailer K (1990) Superoxide radical as electron donor for oxidative phosphorylation of ADP. Biochem Biophys Res Commun 170:59–64
78. Pereverzev MO, Vygodina TV, Konstantinov AA, Skulachev VP (2003) Cytochrome c, an ideal antioxidant. Biochem Soc Trans 31:1312–1315
79. Chelikani P, Fita I, Loewen PC (2004) Diversity of structures and properties among catalases. Cell Mol Life Sci 61:192–208
80. Wahllander A, Soboll S, Sies H, Linke I, Muller M (1979) Hepatic mitochondrial and cytosolic glutathione content and the subcellular distribution of GSH-S-transferases. FEBS Lett 97:138–140
81. Mårtensson J, Lai JC, Meister A (1990) High-affinity transport of glutathione is part of a multicomponent system essential for mitochondrial function. Proc Natl Acad Sci USA 87:7185–7189
82. Griffith OW, Meister A (1985) Origin and turnover of mitochondrial glutathione. Proc Natl Acad Sci USA 82:4668–4672

83. Chen Z, Lash LH (1998) Evidence for mitochondrial uptake of glutathione by dicarboxylate and 2-oxoglutarate carriers. J Pharmacol Exp Ther 285:608–618
84. Dringen R (2000) Metabolism and functions of glutathione in brain. Prog Neurobiol 62:649–671
85. Raza H, Robin MA, Fang JK, Avadhani NG (2002) Multiple isoforms of mitochondrial glutathione S-transferases and their differential induction under oxidative stress. Biochem J 366:45–55
86. Olafsdottir K, Reed DJ (1988) Retention of oxidized glutathione by isolated rat liver mitochondria during hydroperoxide treatment. Biochim Biophys Acta 964:377–382
87. Zoccarato F, Cavallini L, Alexandre A (2004) Respiration-dependent removal of exogenous H_2O_2 in brain mitochondria: inhibition by Ca^{2+}. J Biol Chem 279:4166–4174
88. Wood ZA, Schroder E, Robin Harris J, Poole LB (2003) Structure, mechanism and regulation of peroxiredoxins. Trends Biochem Sci 28:32–40
89. Petrosillo G, Ruggiero FM, Pistolese M, Paradies G (2001) Reactive oxygen species generated from the mitochondrial electron transport chain induce cytochrome c dissociation from beef-heart submitochondrial particles via cardiolipin peroxidation: possible role in the apoptosis. FEBS Lett 509:435–438
90. Petrosillo G, Ruggiero FM, Paradies G (2003) Role of reactive oxygen species and cardiolipin in the release of cytochrome c from mitochondria. FASEB J 17:2202–2208
91. Shidoji Y, Hayashi K, Komura S, Ohishi N, Yagi K (1999) Loss of molecular interaction between cytochrome c and cardiolipin due to lipid peroxidation. Biochem Biophys Res Commun 264:343–347
92. Vieira HL, Belzacq AS, Haouzi D, Bernassola F, Cohen I, Jacotot E, Ferri KF, El Hamel C, Bartle LM, Melino G, Brenner C, Goldmacher V, Kroemer G (2001) The adenine nucleotide translocator: a target of nitric oxide, peroxynitrite, and 4-hydroxynonenal. Oncogene 20:4305–4316
93. Ott M, Robertson JD, Gogvadze V, Zhivotovsky B, Orrenius S (2002) Cytochrome c release from mitochondria proceeds by a two-step process. Proc Natl Acad Sci USA 99:1259–1263
94. Liu X, Kim CN, Yang J, Jemmerson R, Wang X (1996) Induction of apoptotic program in cell-free extracts: requirement for dATP and cytochrome c. Cell 86:147–157
95. Li P, Nijhawan D, Budihardjo I, Srinivasula SM, Ahmad M, Alnemri ES, Wang X (1997) Cytochrome c and dATP-dependent formation of Apaf-1/caspase-9 complex initiates an apoptotic protease cascade. Cell 91:479–489
96. Cecconi F, Piacentini M, Fimia GM (2008) The involvement of cell death and survival in neural tube defects: a distinct role for apoptosis and autophagy. Cell Death Differ 15:1170–1177
97. Reddy PH, Beal MF (2005) Are mitochondria critical in the pathogenesis of Alzheimer's disease? Brain Res Rev 49:618–632
98. Halliwell B (2001) Role of free radicals in the neurodegenerative diseases: therapeutic implications for antioxidant treatment. Drugs Aging 18:685–716
99. McNaught KS, Olanow CW, Halliwell B, Isacson O, Jenner P (2001) Failure of the ubiquitin–proteasome system in Parkinson's disease. Nat Rev Neurosci 2:589–594
100. Bence NF, Sampat RM, Kopito RR (2001) Impairment of the ubiquitin–proteasome system by protein aggregation. Science 292:1552–1555
101. Halliwell B (2006) Oxidative stress and neurodegeneration: where are we now? J Neurochem 97:1634–1658
102. Li MT, Beal MF (2006) Mitochondrial dysfunction and oxidative stress in neurodegenerative diseases. Nature 443:787–795
103. Sun J, Folk D, Bradley TJ, Tower J (2002) Induced overexpression of mitochondrial Mn-superoxide dismutase extends the life span of adult *Drosophila melanogaster*. Genetics 161:661–672
104. Ruan H, Thang HD, Chen ML, Joiner ML, Sun G, Brot N, Weissbach H, Heinemann SH, Iverson L, Wu CF, Hoshi T (2002) High-quality life extension by the enzyme peptide methionine sulfoxide reductase. Proc Natl Acad Sci USA 99:2748–2753
105. Schriner SE, Linford NJ, Martin GM, Treuting P, Ogburn CE, Emond M, Coskun PE, Ladiges W, Wolf N, Van Remmen H, Wallace DC, Rabinovitch PS (2005) Extension of murine life span by overexpression of catalase targeted to mitochondria. Science 308:1909–1911
106. Nunomura A, Perry G, Aliev G, Hirai K, Takeda A, Balraj EK, Jones PK, Ghanbari H, Wataya T, Shimohama S, Chiba S, Atwood CS, Petersen RB, Smith MA (2001) Oxidative damage is the earliest event in Alzheimer disease. J Neuropathol Exp Neurol 60:759–767
107. Pratico D, Uryu K, Leight S, Trojanoswki JQ, Lee VM (2001) Increased lipid peroxidation precedes amyloid plaque formation in an animal model of Alzheimer amyloidosis. J Neurosci 21:4183–4187
108. Reddy PH, McWeeney S, Park BS, Manczak M, Gutala RV, Partovi D, Jung Y, Yau V, Searles R, Mori M, Quinn J (2004) Gene expression profiles of transcripts in amyloid precursor protein transgenic mice: up-regulation of mitochondrial metabolism and apoptotic genes is an early cellular change in Alzheimer's disease. Hum Mol Genet 13:1225–1240
109. Tamagno E, Parola M, Bardini P, Piccini A, Borghi R, Guglielmotto M, Santoro G, Davit A, Danni O, Smith MA, Perry G, Tabaton M (2005) β-Site APP cleaving enzyme up-regulation induced by 4-hydroxynonenal is mediated by stress-activated protein kinases pathways. J Neurochem 92:628–636
110. Lovell MA, Xiong S, Xie C, Davies P, Markesbery WR (2004) Induction of hyperphosphorylated tau in primary rat cortical neuron cultures mediated by oxidative stress and glycogen synthase kinase-3. J Alzheimers Dis 6:659–671
111. Anandatheerthavarada HK, Biswas G, Robin MA, Avadhani NG (2003) Mitochondrial targeting and a novel transmembrane arrest of Alzheimer's amyloid precursor protein impairs mitochondrial function in neuronal cells. J Cell Biol 161:41–54
112. Lustbader JW, Cirilli M, Lin C, Xu HW, Takuma K, Wang N, Caspersen C, Chen X, Pollak S, Chaney M, Trinchese F, Liu S, Gunn-Moore F, Lue L-F, Walker DG, Kuppusamy P, Zewier ZL, Arancio O, Stern D, Yan SS, Wu H (2004) ABAD directly links Aβ to mitochondrial toxicity in Alzheimer's disease. Science 304:448–452
113. Casley CS, Canevari L, Land JM, Clark JB, Sharpe MA (2002) β-amyloid inhibits integrated mitochondrial respiration and key enzyme activities. J Neurochem 80:91–100
114. Park HJ, Seong YM, Choi JY, Kang S, Rhim H (2004) Alzheimer's disease-associated amyloid β interacts with the human serine protease HtrA2/Omi. Neurosci Lett 357:63–67
115. Hansson CA, Frykman S, Farmery MR, Tjernberg LO, Nilsberth C, Pursglove SE, Ito A, Winblad B, Cowburn RF, Thyberg J, Ankarcrona M (2004) Nicastrin, presenilin, APH-1, and PEN-2 form active γ-secretase complexes in mitochondria. J Biol Chem 279:51654–51660
116. Sherer TB, Betarbet R, Testa CM, Seo BB, Richardson JR, Kim JH, Miller GW, Yagi T, Matsuno-Yagi A, Greenamyre JT (2003) Mechanism of toxicity in rotenone models of Parkinson's disease. J Neurosci 23:10756–10764
117. Schapira AH, Cooper JM, Dexter D, Jenner P, Clark JB, Marsden CD (1989) Mitochondrial complex I deficiency in Parkinson's disease. Lancet 1:1269
118. Martin LJ, Pan Y, Price AC, Sterling W, Copeland NG, Jenkins NA, Price DL, Lee MK (2006) Parkinson's disease α-synuclein transgenic mice develop neuronal mitochondrial degeneration and cell death. J Neurosci 26:41–50

119. Muftuoglu M, Elibol B, Dalmızrak Ö, Ercan A, Kulaksız G, Ögüs H, Dalkara T, Özer N (2004) Mitochondrial complex I and IV activities in leukocytes from patients with parkin mutations. Mov Disord 19:544–548
120. Darios F, Corti O, Lücking CB, Hampe C, Muriel M-P, Abbas N, Gu W-J, Hirsch EC, Rooney T, Ruberg M, Brice A (2003) Parkin prevents mitochondrial swelling and cytochrome c release in mitochondria-dependent cell death. Hum Mol Genet 12:517–526
121. Kuroda Y, Mitsui T, Kunishige M, Shono M, Akaike M, Azuma H, Matsumoto T (2006) Parkin enhances mitochondrial biogenesis in proliferating cells. Hum Mol Genet 15:883–895
122. Chung KK, Thomas B, Li X, Pletnikova O, Troncoso JC, Marsh L, Dawson VL, Dawson TM (2004) S-nitrosylation of parkin regulates ubiquitination and compromises parkin's protective function. Science 304:1328–1331
123. Meulener MC, Graves CL, Sampathu DM, Armstrong-Gold CE, Bonini NM, Giasson BI (2005) DJ-1 is present in a large molecular complex in human brain tissue and interacts with α-synuclein. J Neurochem 93:1524–1532
124. Moore DJ, Zhang L, Troncoso J, Lee MK, Hattori N, Mizuno Y, Dawson TM, Dawson VL (2005) Association of DJ-1 and parkin mediated by pathogenic DJ-1 mutations and oxidative stress. Hum Mol Genet 14:71–84
125. Tang B, Xiong H, Sun P, Zhang Y, Wang D, Hu Z, Zhu Z, Ma H, Pan Q, Xia J-h, Xia K, ZhangPan Z (2006) Association of PINK1 and DJ-1 confers digenic inheritance of early-onset Parkinson's disease. Hum Mol Genet 15:1816–1825
126. Canet-Aviles RM, Wilson MA, Miller DW, Ahmad R, McLendon C, Bandyopadhyay S, Baptista MJ, Ringe D, Petsko GA, Cookson MR (2004) The Parkinson's disease protein DJ-1 is neuroprotective due to cysteine-sulfinic acid-driven mitochondrial localization. Proc Natl Acad Sci USA 101:9103–9108
127. Silvestri L, Caputo V, Bellacchio E, Atorino L, Dallapiccola B, Valente EM, Casari G (2005) Mitochondrial import and enzymatic activity of PINK1 mutants associated to recessive parkinsonism. Hum Mol Genet 14:3477–3492
128. West AB, Moore DJ, Biskup S, Bugayenko A, Smith WW, Ross CA, Dawson VL, Dawson TM (2005) Parkinson's disease-associated mutations in leucine-rich repeat kinase 2 augment kinase activity. Proc Natl Acad Sci USA 102:16842–16847
129. James AM, Murphy MP (2002) How mitochondrial damage affects cell function. J Biomed Sci 9:475–487
130. Galley HF (2010) Bench-to-bedside review: targeting antioxidants to mitochondria in sepsis. Crit Care 14:230
131. Smith RAJ, Porteous AM, Coulter CV, Murphy MP (1999) Selective targeting of an antioxidant to mitochondria. Eur J Biochem 263:709–716
132. Kelso GF, Porteous CM, Coulter CV, Hughes G, Porteous WK, Ledgerwood EC, Smith RA, Murphy MP (2001) Selective targeting of a redox-active ubiquinone to mitochondria within cells: antioxidant and antiapoptotic properties. J Biol Chem 276:4588–4596
133. Dhanasekaran A, Kotamraju S, Kalivendi SV, Matsunaga T, Shang T, Keszler A, Joseph J, Kalyanaraman B (2004) Supplementation of endothelial cells with mitochondria-targeted antioxidants inhibit peroxide-induced mitochondrial iron uptake, oxidative damage, and apoptosis. J Biol Chem 279:37575–37587
134. Adlam VJ, Harrison JC, Porteous CM, James AM, Smith RA, Murphy MP, Sammut IA (2005) Targeting an antioxidant to mitochondria decreases cardiac ischemia-reperfusion injury. FASEB J 19:1088–1095
135. Graham D, Huynh NN, Hamilton CA, Beattie E, Smith RA, Cochemé HM, Murphy MP, Dominiczak AF (2009) Mitochondria-targeted antioxidant MitoQ10 improves endothelial function and attenuates cardiac hypertrophy. Hypertension 54:322–328
136. Antipodean Pharmaceuticals, Inc. homepage. http://www.antipodeanpharma.com/
137. Skulachev VP, Anisimov VN, Antonenko YN, Bakeeva LE, Chernyak BV, Erichev VP, Filenko OF, Kalinina NI, Kapelko VI, Kolosova NG, Kopnin BP, Korshunova GA, Lichinitser MR, Obukhova LA, Pasyukova EG, Pisarenko OI, Roginsky VA, Ruuge EK, Senin II, Severina II, Skulachev MV, Spivak IM, Tashlitsky VN, Tkachuk VA, Vyssokikh MY, Yaguzhinsky LS, Zorov DB (2009) An attempt to prevent senescence: a mitochondrial approach. Biochim Biophys Acta 1787:437–461
138. Duby G, Boutry M (2002) Mitochondrial protein import machinery and targeting information. Plant Sci 162:477–490
139. Eilers M, Hwang S, Schatz G (1988) Unfolding and refolding of a purified precursor protein during import into isolated mitochondria. EMBO J 7:1139–1145
140. Neupert W (1997) Protein import into mitochondria. Annu Rev Biochem 66:863–917
141. Matsuura S, Arpin M, Hannum C, Margoliash E, Sabatini DD, Morimoto T (1981) In vitro synthesis and posttranslational uptake of cytochrome c into isolated mitochondria: Role of a specific addressing signal in the apocytochrome. Proc Natl Acad Sci USA 78:4368–4372
142. Rosenblum JS, Gilula NB, Lerner RA (1996) On signal sequence polymorphisms and diseases of distribution. Proc Natl Acad Sci USA 93:4471–4473
143. Von Heijne G (1986) Mitochondrial targeting sequences may form amphiphilic helices. EMBO J 5:1335–1342
144. Brix J, Dietmeier K, Pfanner N (1997) Differential recognition preproteins by the purified cytosolic domains of the mitochondrial import receptors Tom20, Tom22, and Tom70. J Biol Chem 272:20730–20735
145. Zhao K, Zhao G-M, Wu D, Soong Y, Birk AV, Schiller PW, Szeto HH (2004) Cell-permeable peptide antioxidants targeted to inner mitochondrial membrane inhibit mitochondrial swelling, oxidative cell death, and reperfusion injury. J Biol Chem 279:34682–34690
146. Wipf P, Xiao J, Jiang J, Belikova NA, Tyurin VA, Fink MP, Kagan VE (2005) Mitochondrial targeting of selective electron scavengers: synthesis and biological analysis of hemigramicidin-TEMPO conjugates. J Am Chem Soc 127:12460–12461
147. Winterbourn CC, Parsons-Mair HN, Gebicki S, Gebicki JM, Davies MJ (2004) Requirements for superoxide-dependent tyrosine hydroperoxide formation in peptides. Biochem J 381:241.248
148. Ban S, Nakagawa H, Suzuki T, Miyata N (2007) Novel membrane-localizing TEMPO derivatives for measurement of cellular oxidative stress at the cell membrane. Bioorg Med Chem Lett 17:1451–1454
149. Uttara B, Singh AV, Zamboni P, Mahajan RT (2009) Oxidative stress and neurodegenerative diseases: a review of upstream and downstream antioxidant therapeutic options. Curr Neuropharmacol 7:65–74
150. Malkus KA, Tsika E, Ischiropoulos H (2009) Oxidative modifications, mitochondrial dysfunction, and impaired protein degradation in Parkinson's disease: how neurons are lost in the Bermuda triangle. Mol Neurodegener 4:24. doi:10.1186/1750-1326-4-24
151. Migliore L, Coppedè F (2009) Environmental-induced oxidative stress in neurodegenerative disorders and aging. Mutat Res 674:73–84

152. Roberts RA, Laskin DL, Smith CV, Robertson FM, Allen EMG, Doorn JA, Slikkerk W (2009) Nitrative and oxidative stress in toxicology and disease. Toxicol Sci 112:4–16
153. Takashima A (2009) Amyloid-β, Tau, and dementia. J Alzheimer's Dis 17:729–736
154. Moreira PI (2010) Mitochondrial dysfunction and oxidative stress in Alzheimer's disease. Eur Neurol Rev 5:17–21
155. Douglas PM, Dillin A (2010) Protein homeostasis and aging in neurodegeneration. J Cell Biol 190:719–729
156. Squier TC. Oxidative stress and protein aggregation during biological aging
157. Gidalevitz T, Kikis EA, Morimoto RI (2010) A cellular perspective on conformational disease: the role of genetic background and proteostasis networks. Curr Opin Struct Biol 20:23–32
158. Voisine C, Pedersen JS, Morimoto RI (2010) Chaperone networks: tipping the balance in protein folding diseases. Neurobiol Dis 40:12–20
159. Kikis EA, Gidalevitz T, Morimoto RI (2010) Protein homeostasis in models of aging and age-related conformational disease. Adv Exp Med Biol 694:138–159
160. Khalil M, Teunissen C, Langkammer C (2011) Iron and neurodegeneration in multiple sclerosis. Multiple Sclerosis Int 2011. Article ID 606807, doi:10.1155/2011/606807
161. Farooqui T, Farooqui AA (2011) Lipid-mediated oxidative stress and inflammation in the pathogenesis of parkinson's disease. Parkinson's Dis 2011. Article ID 247467. doi:10.4061/2011/247467
162. Barsukova AG, Bourdette D, Forte M (2011) Mitochondrial calcium and its regulation in neurodegeneration induced by oxidative stress. Eur J Neurosci 34:437–447
163. Numakawa T, Matsumoto T, Numakawa Y, Richards M, Yamawaki S, Kunugi, H (2011) J Toxicol 2011. Article ID 405194, doi:10.1155/2011/405194

The chemical mechanism of oxidative stress by copper(II) and iron(III) ions in several neurodegenerative disorders

Yuzo Nishida

Abstract

The variety of factors and events involved in neurodegeneration renders the subject a major challenge. Neurodegenerative disorders include a number of different pathological conditions, which share similar critical metabolic processes, such as protein aggregation and oxidative stress, both of which are associated with the involvement of transition metal ions. In this artical, amyotrophic lateral sclerosis (ALS), Parkinson's disease (PD), and prion disease are discussed, with the aim of identifying the common trends underlying these devastating neurological conditions by elucidating the chemical mechanism of the oxidative stress due to copper(II) and iron(III) ions.

Keywords

ALS • SOD • Hydrogen peroxide • NTBI • Binuclear iron(III) species

Introduction

Neurodegeneration is a complex and multifaceted process leading to many chronic disease states. A conventional definition implies a progressive neuronal death, which usually affects a specific population of nerve cells, the vulnerability of which determines the clinical manifestations of a particular neurodegenerative disease.

A classification of neurodegeneration can be achieved on the basis of the principal neuropathological changes, characterized by the presence of abnormal protein components, which accumulate in the brain. There is extensive evidence for the association between protein aggregation and neurodegeneration in many neurodegenerative disorders [1–5]. Interestingly, metals such as iron and copper appear to play an important role in oxidative stress and are therefore likely to provide a link between the two pathological processes of protein aggregation and oxidative damage, as demonstrated below.

Y. Nishida
Medical Research Institute, Kanazawa Medical University,
1-1 Daigaku, Uchinada, Kahoku, Ishikawa 920-0263, Japan
e-mail: nsd-2210@kanazawa-med.ac.jp

Friedreich's ataxia, the most common hereditary ataxia accounting for approximately 50% of all cases of hereditary ataxia, is an autosomal recessive degenerative disease, characterized by the development of muscle weakness. It is caused by the expansion of a GAA triplet located within the first intron of the frataxin gene that encodes frataxin, a mitochondrial protein that plays a role in iron homeostasis. Deficiency of frataxin results in mitochondrial iron accumulation, defects in specific mitochondrial enzymes, enhanced sensitivity to oxidative stress, and eventually cell death [6, 7].

Post-mortem studies in Parkinson's disease (PD) brains indicate that a wide range of molecules undergo oxidative damage, including lipids, proteins, and DNA. In fact, significant neurochemical, physical, histochemical, and biochemical evidence confirm the hypothesis that oxidative stress generates the cascade of events, which is responsible of the preferential degeneration of melanized dopaminergic neurons in the substantia nigra pars compacta (SNc) in PD [8–10]. In Parkinsonian brains several phenomena have been observed, but we would like to emphasize the elevated level of iron in microglia, astrocytes, oligodendrocytes, and dopaminergic neurons of SNc, and also changes in the normal iron and antioxidant concentration in SNc of PD patients [11–13].

W. Linert and H. Kozlowski (eds.), *Metal Ions in Neurological Systems*,
DOI 10.1007/978-3-7091-1001-0_13, © Springer-Verlag Wien 2012

Transmissible spongiform encephalopathies (TSEs) or prion diseases form a group of fatal neurodegenerative disorders that have the unique property of being either infectious or sporadic or genetic in origin. They share many pathologic features with the other neurodegenerative disorders, including the importance of oxidative damage of the brain, accumulation of aggregated proteins, and neuronal cell loss. The exact pathogenic mechanism in TSEs remains uncertain, but it is believed that oxidative stress plays a central role, because in the last 10 years the role of metal ions, the copper(II) ion in particular, has been shown to have a critical function in the physiopathology of prion diseases and in the process leading to the abnormal form from the normal prion protein [4, 14].

The loss of neurons in amyotrophic lateral sclerosis (ALS), which is one of the most common neurodegenerative disorders after Alzheimer's disease (AD) and Parkinson's disease (PD), results from a complex interplay of oxidative injury, excitotoxic simulation, aggregation and/or dysfunction of critical proteins, and genetic factors [4, 15, 16]. Degeneration of cortical and spinal motor neurons is the typical feature of ALS occurring both sporadically and as a familial disorder, with inherited cases accounting for about 10 % of patients; in the latter cases participation of the mutant superoxide dismutase SOD enzyme, which contains the copper(II) ion, has been pointed out [15, 16].

Oxidative stress and transition metal ions

Oxidative stress has been defined as an imbalance between the production of oxidants and the ability of the cell to defend against them through a set of antioxidants and detoxifying enzymes. When this imbalance occurs, oxidatively modified molecules accumulate in the cellular compartment causing dysfunction. In the case of very sensitive cells such as neurons, the lack of control of defense systems may eventually lead to cell death.

As the plausible candidates for the oxidants described above, superoxide, hydroxyl radical, and nitric oxide have been postulated; all these molecules are referred to as reactive oxygen species (ROS). Among them, the hydroxyl radical ($OH^{\cdot}$), which may be generated by reaction between Fe (II) ion and hydrogen peroxide, was believed to be the most important; however, the formation of the hydroxyl radical seems to be completely negligible in the human body, because the concentration of free Fe(II) ion, i.e., $[Fe(H_2O)_6]^{2+}$, is very low.

It is clear that one of the main risk factors in most neurodegeneration is age and this should be directly linked to oxidative stress (lipid peroxidation, protein oxidation, DNA and RNA oxidation), which increases in the brain with age and plays a central role in the pathogenic mechanism of neurodegeneration. On the basis that abnormally high levels of iron in the brain have been demonstrated in a number of neurodegenerative disorders including PD and AD as described in the "Introduction," we concluded that the oxidative stress increasing with age should be closely related to the increased accumulation of iron levels in the brain with age. In order to elucidate the relationship between oxidative stress and accumulated iron ions, we investigated the chemical mechanism of H_2O_2 activation by several copper(II) and iron(III) complexes in vitro, and found that hydrogen peroxide can be activated to show reactivity similar to that of singlet oxygen ($^1\Delta_g$), and that this activated hydrogen peroxide should be a main risk factor for the oxidative stress to induce all the neurodegenerative disorders [17–20]. In addition, we also showed that the iron ions not associated with transferrin, generally termed as non-transferrin-bound iron (NTBI), facilely produce hydrogen peroxide in the presence of reducing agents [20].

On the basis of our results we have proposed that the intrinsic oxidant to induce the oxidative stress should be a copper(II) or an iron(III) complex containing hydrogen peroxide, and have given reasonable answers to the following key questions: why do iron levels increase abnormally in some regions of the brain, and what is the origin of toxicity due to iron ions accumulated in the brain [17, 18]?

Amyotrophic lateral sclerosis and mutant superoxide dismutase

ALS is a progressive paralytic disease characterized by selective degeneration of the upper and lower motor neurons [4, 21, 22]. Although ALS is predominantly a sporadic disease, ~10 % cases are inherited in an autosomal dominant manner [familial ALS (fALS)] and a subset of the fALS cases are caused by mutations in the SOD1 gene [15, 16, 23, 24]. The gene product of SOD1, cytoplasmic Cu,Zn-superoxide dismutase (SOD1), is a ubiquitously expressed enzyme that catalyzes the disproportionation reaction of superoxide radicals [Eqs. (1) and (2)] [25]. The crystal structure of the SOD(Cu/Zn) has already been determined, its dimeric structure being illustrated in Fig. 1 (PDB, 1spd_x). The copper and zinc ions are bridged by an anionic form of the imidazole ring of histidine.

$$O_2^- + Cu(II) \rightarrow O_2 + Cu(I) \quad (1)$$

$$Cu(I) + O_2^- + 2H^+ \rightarrow H_2O_2 + Cu(II) \quad (2)$$

There are several lines of evidence that SOD1 mutations result in a gain, rather than loss of function that causes ALS.

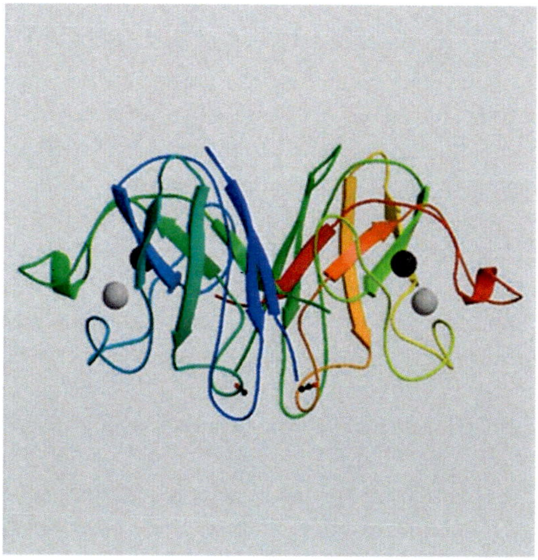

Fig. 1 Dimeric structure of the SOD molecule. Two copper and zinc ions are illustrated as *colored circles*

For instance, some fALS-associated mutant SOD1s retain full enzymatic activity [26]. In addition, SOD1 knockout mice lack ALS symptoms, whereas transgenic mice expressing the fALS-associated mutant G93A SOD1 develop ALS-like symptoms despite expression of endogenous mouse SOD1. Lastly, overexpression of human wild-type SOD1 fails to alleviate symptoms in this transgenic mouse model for ALS [27].

One hypothesis about the "gain of function" of SOD1 is that misfolding of the mutant alters the catalytic mechanism to allow production of oxidants such as peroxynitrite [28] and possibly hydrogen peroxide [29]. Another major hypothesis is that toxicity is caused by intracellular aggregation of SOD1. SOD1 inclusion bodies, which also react with anti-ubiquitous antibodies, are a common pathological finding in motor neurons and neighboring astrocytes of ALS patients [30]. Although SOD1 aggregates may be inherently toxic or cause motor neuron toxicity by sequestering chaperons and blocking proper functioning of the proteasome, origin of toxicity by SOD1 aggregates has not been elucidated.

Lindberg et al. [31] compared the folding behavior of 15 ALS-associated SOD mutants with survival times ranging from 1 to 17 years after the onset of the first symptom. They found that the folding pattern of the ALS-associated mutants shows different and characteristic features that can be linked to their respective disease progression rates. Common to all mutations is a shift of the folding equilibrium toward the denatured monomer; however, the magnitudes and the mechanistic origins of this shift are found to vary. In order to rationalize the above facts, we must consider the effects by oxidative stress due to the copper(II) ion; this point is omitted in the study by Lindberg et al. In 1997, Yim et al. [32] reported that an fALS mutant (Gly93Ala = G93A)

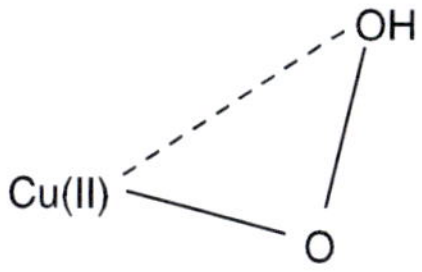

Scheme 1

exhibits an enhanced free radical-generating activity, while its dismutation activity is identical to that of the wild-type enzyme. They reported that the free radical-generating activity of the mutant, as measured by a spin-trapping method at low H_2O_2 concentration, is enhanced relative to that of the wild type and G93A, wild type < G93A < A4 V, but the reason for the above fact has not been clarified.

To understand ALS pathogenesis, we must clarify how altering the SOD molecule can induce cell injury. To carry out such an investigation, we have started to clarify the origin of the gain of function by the mutant SOD molecule. The reaction mechanism of SOD1 enzyme has been investigated by many authors. Very recently Nishida et al. pointed out the importance of formation of a copper(II)–OOH species (Scheme 1) as an intermediate in the second step [Eq. (2)] above, and this hydrogen peroxide produced is immediately removed from the wild-type enzyme because of the negligible interaction among hydrogen peroxide, the copper(II) ion, and the surrounding organic groups.

Origin of gain of function in mutant SOD enzyme and the unique reactivity of copper(II)–hydroperoxide adduct

In order to obtain a comprehensive solution for the correlation between the structural change in mutant SOD and pathogenesis of fALS, we studied the reactivity of copper(II)–OOH, proposed as an important intermediate in the SOD reaction. For this purpose, we synthesized many copper(II) compounds with ligands containing an *N,N*-bis(2-pyridinylmethyl)amine moiety as illustrated in Fig. 2 [33]. The structural features of all the copper(II) compounds are essentially similar to each other (as an example, the crystal structure of $[Cu(bdpg)Cl]^+$ is illustrated in Fig. 3). In the presence of hydrogen peroxide, formation of a peroxide adduct as shown at the right side of Fig. 3 was anticipated, and this was consistent with the results of the reaction with cyclohexane [34]; we also found that the reactivity of the peroxide adduct of the copper(II) compound is highly dependent on the R of the ligand system (Fig. 2), i.e., slight structural change around the copper(II) ion due to the different R moieties strongly controls the reactivity of the copper(II)–peroxide adduct.

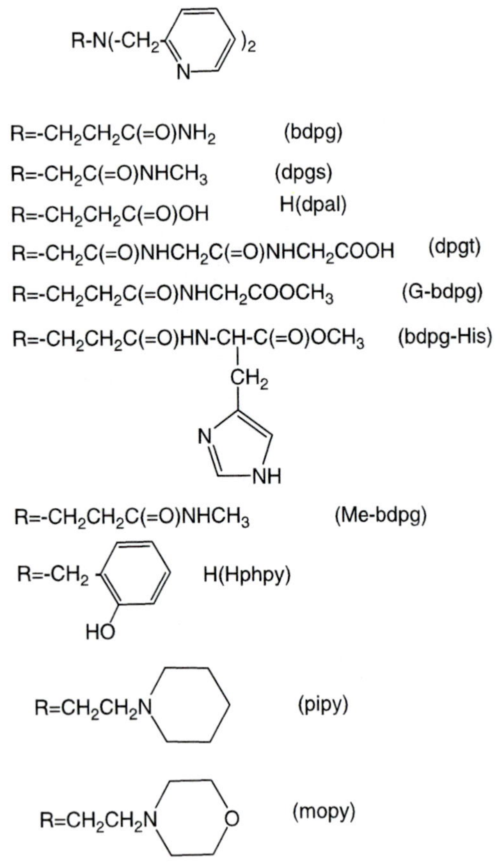

Fig. 2 Chemical structures of the ligands used in our study

We measured the ESR spectra of solutions containing a copper(II) complex and spin-trapping reagent, such as PBN (α-phenyl-*N*-*t*-butylnitrone) and TMPN (2,2,6,6-tetramethyl-4-piperidinol), specific reagents for OH· radical and singlet oxygen ($^1\Delta_g$) (Scheme 2), respectively [35].

No ESR signal due to the formation of a PBN radical was detected when the copper(II) complexes with tpa (tris(2-pyridinylmethyl)amine) or bdpg were mixed with H_2O_2 and PBN. However, strong peaks due to nitrone radical formation of the corresponding TMPN (Scheme 2) were detected in the solution with the Cu(tpa) complex, but not with the Cu(bdpg) complex.

The comparison between the Cu(pipy)Cl^+ and the Cu(mopy)Cl^+ is very interesting [35]. Structural features of the two compounds are essentially the same, the only difference being that the oxygen atom of the morpholine ring of the Cu(mopy)Cl^+ complex is replaced by a $-CH_2-$ in the Cu(pipy)Cl^+ complex. In the case of Cu(pipy)Cl^+, no formation of the nitrone radical was observed in the presence of hydrogen peroxide; in contrast, high activity for the radical formation by the Cu(mopy)Cl^+ complex was detected as illustrated in Fig. 4. Similar high activity for TMPN radical formation was also observed for the copper(II) complex [Cu(Hphpy)Cl]$^+$. In this case, similar to the Cu(mopy)Cl^+ complex, the addition of the H_2O_2 to the copper(II) solution does not induce the change in ESR spectrum due to the copper(II) ion; but the addition of TMPN leads to the dramatic change in the ESR signals attributed to the copper(II) species (i.e., the change of hyperfine structure values due to the copper atom). These are all comprehensively elucidated on the assumption that the complex formation of copper(II), hydrogen peroxide, and TMPN occurs only when three reagents are present in the solution (Fig. 5), and the unique reactivity of hydrogen peroxide observed is detected only when the intermediate is formed in the solution.

The above facts indicate that the reactivity of Cu(II)–OOH is controlled by the structural properties of the intermediate (Fig. 5), i.e., by the chemical interactions among copper(II)–OOH species, peripheral groups, and substrate [17, 18]. It should be noted here that although hydrogen peroxide was believed to be relatively inert and not toxic to cells, our present results clearly show that some copper(II) chelates can activate the hydrogen peroxide to exhibit high reactivity similar to that of singlet oxygen ($^1\Delta_g$).

In order to get further information about the reactivity of a copper(II)–OOH species, we measured ESI mass spectra of the solutions of copper(II) compounds and hydrogen peroxide. When hydrogen peroxide was added to Cu(Me-bdpg)Cl solution (for Me-bdpg, see Fig. 2), the formation of [Cu(bdpg)Cl], not [Cu(dpal)], was detected by ESI–MS [36]. This clearly indicates that Cu(II)–OOH species can cleave the peptide at the C– N bond oxidatively, not hydrolytically, because the hydrolytic cleavage may give Cu(dpal) species from the Cu(Me-bdpg) compound (Scheme 3).

We also found that some copper(II) complexes exhibit high activity to oxygenate the methionine residue of amyloid beta-peptide(1–40) at the sulfur atom [37, 38], and decompose several proteins in the presence of hydrogen peroxide [39]. All these facts may suggest that the gain of function of the mutant SOD is due to formation of a long-lived highly reactive copper(II)–OOH as an intermediate in the process of SOD reaction. The chemical structures around the copper(II) in the mutant SOD are slightly changed, and this gives an unexpected effect on the reactivity of copper(II)–OOH as observed in our papers. In the mutant SOD the C–N bond cleavage by Cu(II)–OOH may give great changes in the surface of SOD, leading to destabilization of the dimer contact of the SOD enzyme [40]. Thus, it is quite likely that formation and existence of a highly reactive Cu(II)–OOH species is an intrinsic origin of oxidative stress in the pathogenesis of fALS, which may be consistent with the recent studies on the destabilization of the dimer contact of the SOD enzyme [41, 42].

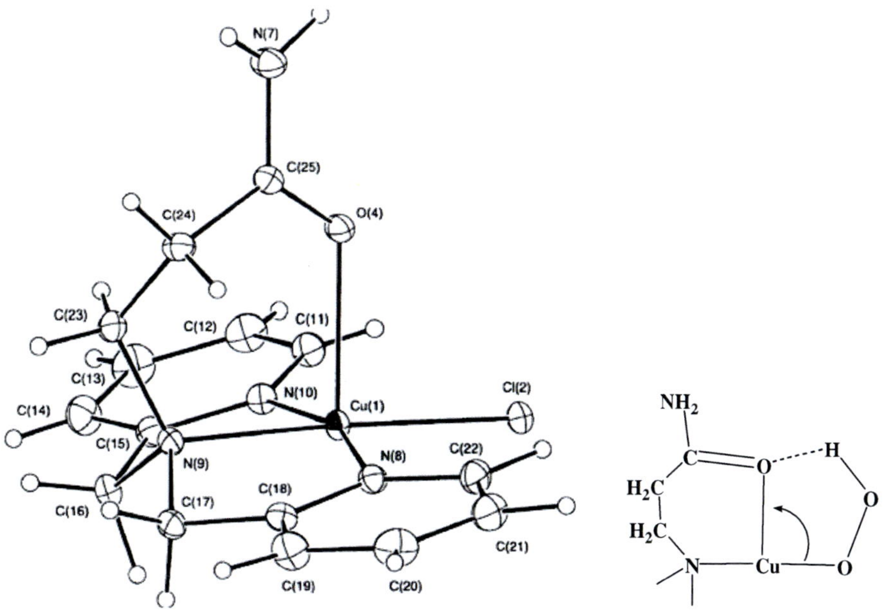

Fig. 3 *Left* crystal structure of $[Cu(bdpg)Cl]^+$ [34]. *Right* assumed structure of the peroxide adduct $[Cu(bdpg)(OOH)]^+$

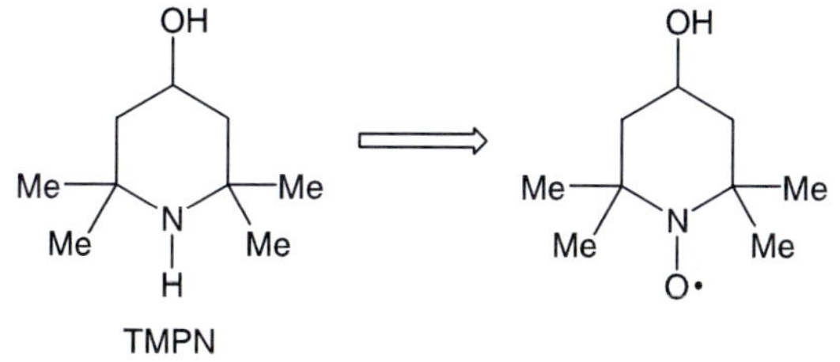

Scheme 2

Dissociation of dimeric SOD molecule into monomers

As stated before, it is widely recognized that protein aggregations are common pathological features of many neurological disorders, including Huntington's, Alzheimer's, and Parkinson's diseases, and that SOD1 aggregates may be inherently toxic or cause motor neuron toxicity by sequestering chaperons and blocking proper functioning of the proteasome.

In 2004, Rakhit et al. [43] reported that SOD1, normally a dimeric enzyme, dissociates to monomers prior to aggregation for both wild-type and mutant proteins. They used the dynamic light scattering (DLS) method to detect the dissociation of dimeric SOD to monomers. Very recently we reported that the capillary electrophoresis method (CE) is very suitable to investigate the conformational change of the proteins and aggregation states of the proteins in solution [19].

We observed that a drastic decrease of the peak strength due to the dimeric SOD molecule occurs when copper(II)/ascorbic acid solution was added to the SOD molecule [44]; our experimental system was the same as that reported by Rakhit et al. This clearly shows that the dissociation of the dimeric SOD molecule can be readily detected by the CE method. We also found that the presence of excess hydrogen peroxide induces the dissociation of the dimeric structure of the wild-type SOD molecule, because a drastic decrease of the peak height due to the dimeric structure was observed (Fig. 6) [45]. Thus, it seems quite likely that the oxidant in the system, the copper(II)/ascorbic acid solution, used by Rakhit et al. should be hydrogen peroxide, and that sporadic ALS should be closely related to the presence of excess hydrogen peroxide [19]; the same discussion may be applied to the elucidation of sporadic prion diseases (see later).

By using antibody methods to rapidly purify SOD1 and coupling this with mass spectrometry, Sato et al. [16] measured the relative accumulated levels of wild-type and mutant SOD1 in erythrocytes of 29 SOD1-mutated fALS patients. They observed that the patients with undetectable SOD1 mutant had the shortest disease durations. Although age at disease onset was found to be uncorrelated with the amount of mutant SOD1, the evidence convincingly shows a strong inverse correlation between disease duration and mutant accumulation. Said another way, an accelerated disease course is found for mutants that are less stable. This surprising discovery implies that it is the misfolded unstable forms of SOD1 mutants that contribute to toxicity underlying disease progression, and that despite its apparent importance for progression, SOD1 mutant stability is not correlated with disease onset. Thus, dissociation of the dimeric SOD1 molecule to misfolded monomers should be an essential important process for APS pathogenesis. As it has become apparent that hydrogen peroxide plays an important role in the formation of SOD1 monomers [45], we

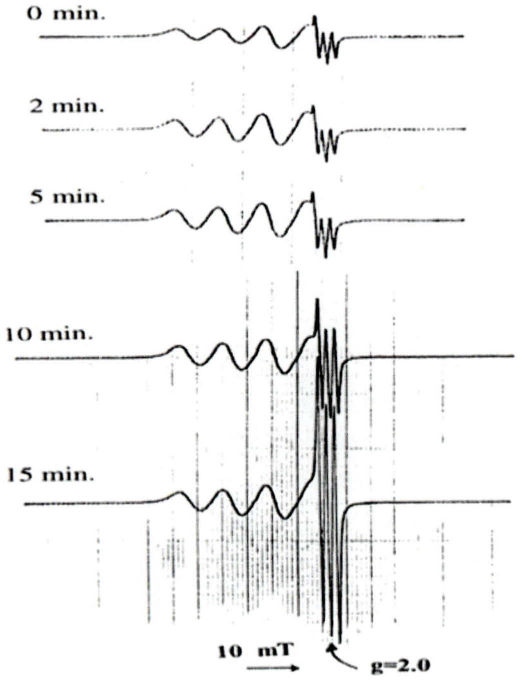

Fig. 4 Time course of ESR spectra of the solutions containing [Cu(mopy)Cl]$^+$, H_2O_2, and TMPN

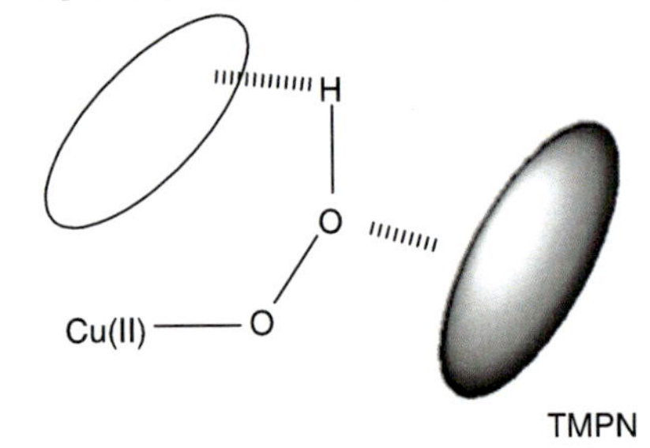

Fig. 5 Assumed intermediate among copper(II) chelate, H_2O_2, and TMPN

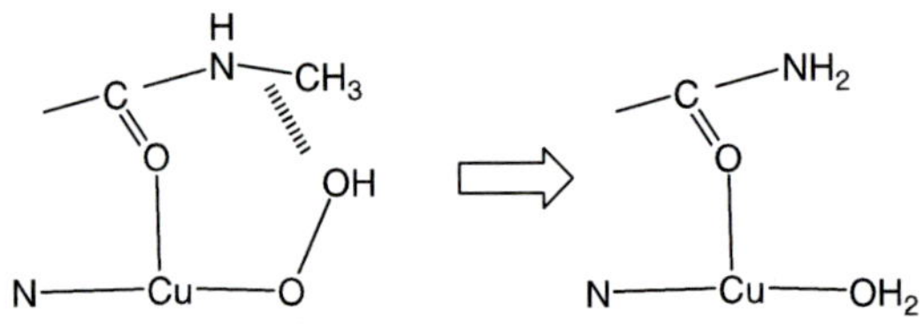

Scheme 3

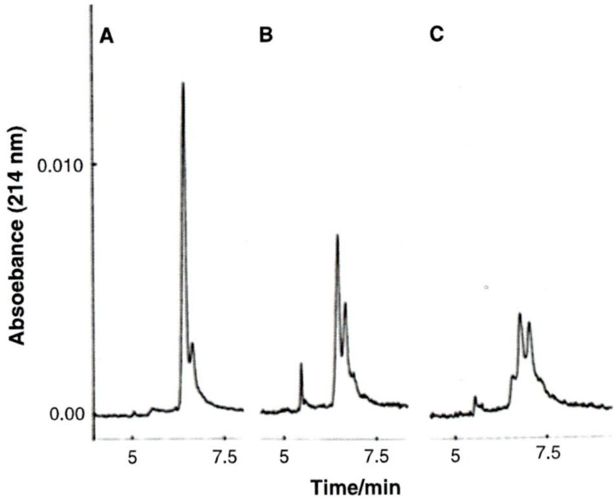

Fig. 6 CE profiles of the solutions. *A* Wild-type SOD (1 mg/cm^3), *B* measured immediately after H_2O_2 was added, *C* measured 60 min after addition of H_2O_2

should pay attention to the formation of excess hydrogen peroxide in the human body, especially due to the reaction between a dimeric iron(III) species and the glutathione cycle and other related systems [18, 20].

Oxidative stress by copper(II)–OOH in sporadic prion diseases

Between 1980 and roughly 1996, about 750,000 cattle infected with BSE (bovine spongiform encephalopathy, one of the TSEs) were slaughtered for human consumption in Great Britain, and at present it is accepted that the central event in TSEs is the post-translational conversion of the normal cellular prion protein (PrP^C) into an abnormal isoform of called scrapie PrP (PrP^{Sc}) that has a high β-sheet content [46]. It is generally recognized that PrP^C is a copper-containing protein (at most four copper ions are present within the octarepeat region located in the unstructured N-terminus). Analysis of recombinant mouse and chicken PrP^C led to the discovery of an important gain of function following the formation of the PrP^C copper complex; PrP^C has been shown to contribute directly to cellular SOD activity.

The misfolded prions (PrP^{Sc}) ultimately kill neurons and leave the brain riddled with holes, like a sponge. In addition to PrP^{Sc}, another protease-resistant PrP of 27–30 kDa, which is called PrP27-30, was extracted from affected brains. It should be noted here that PrP27-30 is derived from only PrP^{Sc} (not from PrP^C), and no difference in amino acid sequence between PrP^C and PrP^{Sc} has been identified. On the basis of these facts we may assume that the chemical environment around the copper ion in the PrP^{Sc} should be different from those in the PrP^C; this situation is similar to the difference observed between environments around copper(II) ions in the wild-type and mutant SOD enzyme. Thus, it is most likely that the gain of function in the PrP^{Sc} due to a "highly reactive" Cu(II)–OOH formation may occur as described for the mutant SOD molecule, which leads to the

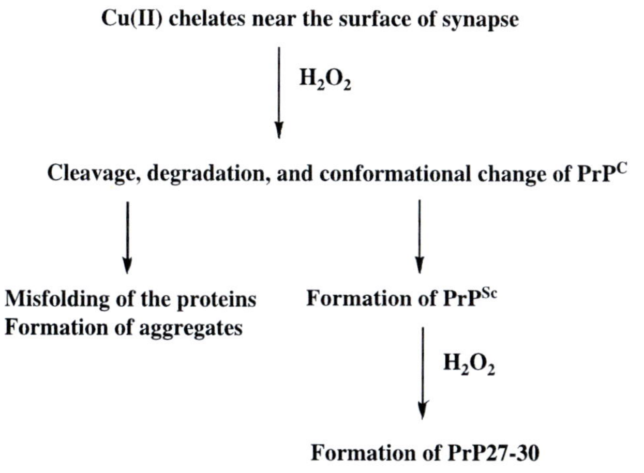

Scheme 4

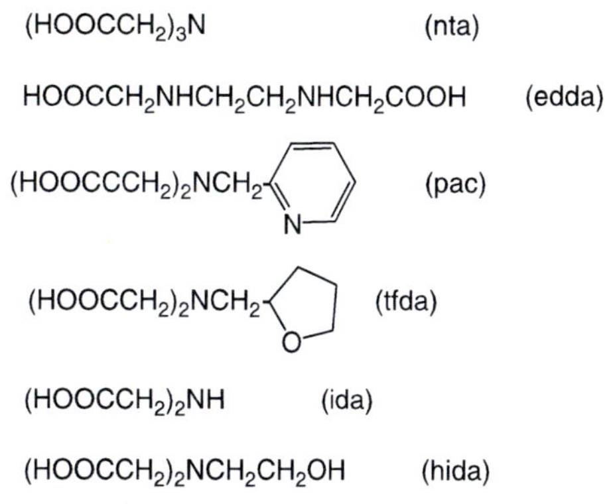

Fig. 7 Several chelates used in our study

cleavage of the peptide bonds around the copper ion (at about site 90), giving dangerous PrP27-30; the latter protein may behave like the misfolded SOD monomer. In addition to this, it seems quite likely that the copper(II) ions in PrP^{c} and also PrP^{Sc} may react with hydrogen peroxide to yield a Cu(II)–OOH species, which may exert serious effects on the PrP^{C} such as oxygenation at methionine residues, conformational change (i.e., formation of PrP^{Sc}), and degradation of protein in the presence of hydrogen peroxide (Scheme 4). Several experimental facts observed for the native prion proteins [47–54] seem to be consistent with our discussions. All these findings support our proposal that hydrogen peroxide, which may derive from the SOD function of PrP^{Sc} and/or from NTBI [20], should be one of the serious origins of the oxidative stress in sporadic prion diseases.

Structural characteristics of iron(III) chelates that induce tissue damage and renal carcinoma: the chemical origin of iron toxicity

It is generally recognized that iron, the most abundant transition metal ion in mammalian systems, is a necessary trace element and is required for normal metabolic processes spanning molecular oxygen transport, respiratory electron transfer, DNA synthesis, and drug metabolism. In fact, iron deficiency leads to the deficiency of neurotransmitters such as dopamine and serotonin in the brain, inducing several mental diseases such as Parkinson's disease, depression, and schizophrenia [18]. Thus, the ancient Greeks concocted potions of iron filings dissolved in vinegar, hoping that drinking this liquor would empower them with the properties of the element.

In the human body, we have no real mechanism for the elimination of excess iron and, as a result, cells continuously store excess absorbed iron in a complex with the protein ferritin. Under conditions of iron excess some of it is shunted into another storage form known as hemosiderin [55, 56] in which the excess iron is deposited as ferrihydrate structures. Hemosiderin is typically insoluble, but when amino acid derivatives or small peptides are present in the plasma, the precipitated ferrihydrate in hemosiderin may dissolve forming an iron(III) chelate. These iron ions, or NTBI, are detected in the plasma of patients with thalassemia, hemochromatosis, and several neurodegenerative disorders, and are thought to play an important role in iron-induced cell damage as pointed out in the "Introduction." Despite numerous studies over the last 30 years since plasma NTBI was first postulated to exist, little is understood about the chemical composition of NTBI and the origin of toxicity due to NTBI at present. In this section, we discuss the chemical origin of iron toxicity in the human body due to NTBI, which is closely related with the accumulation of manganese ions (see Chapter 6).

Iron(III)-nta chelate as a renal carcinogen

Ferric nitrolotriacetate (Fe(III)-nta) is a well-known renal carcinogen (for nta, see Fig. 7), and Fe(III)-nta-injected animals have been used as a model of carcinogenesis [57–60]. When Fe(III)-nta is intraperitoneally injected into animals, lipid peroxidation and oxidative modification of proteins and DNA occur in renal proximal tubules, and tubular epithelial cells are damaged. Thiobarbituric acid reactive substance (TBARS) [61] has also been shown to increase in kidney, and increases in 4-hydroxy-2-nonenal (4-HNE)-modified proteins and 8-hydroxy-deoxyguanosine (8-OH-dG) were also demonstrated.

We have determined the crystal structures of several iron (III) compounds including nta, pac, and ida [62–64]. As shown in Fig. 8, the Fe(III)-(nta) complex is of a dimeric structure with oxo and carbonato bridges. It should be noted here that although the crystal structure of the Fe(III)-(pac)

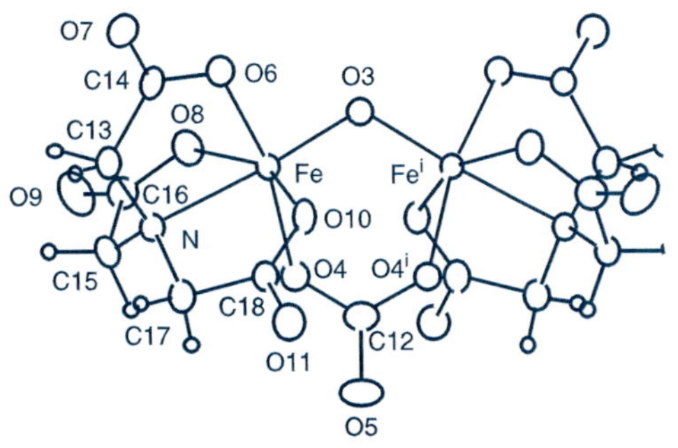

Fig. 8 Crystal structure of the $[Fe_2O(nta)_2(CO_3)]^{4-}$ ion

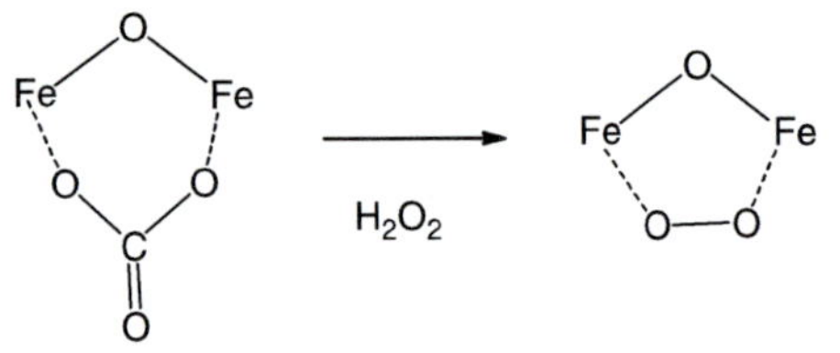

Scheme 5

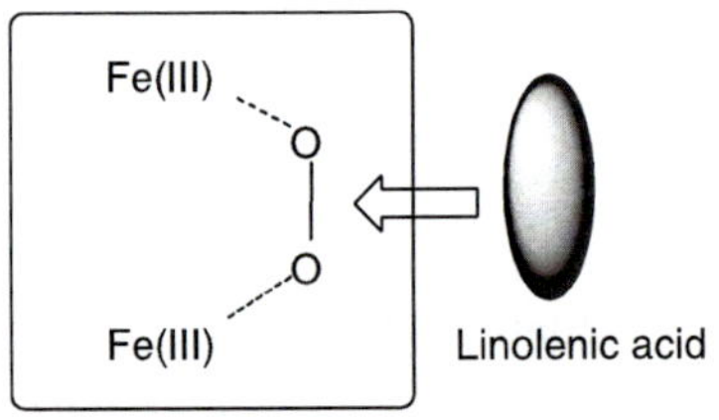

Scheme 6

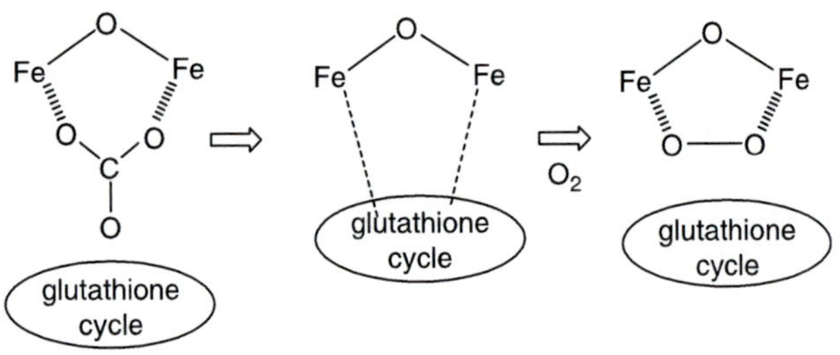

Scheme 7

complex is of a dimeric structure similar to that of the nta compound, tubular injuries by the Fe(III)-(pac) compound are negligible and the Fe(III)-(pac) compound does not induce renal carcinoma [20, 62].

On the basis of the spectral studies on the solution containing an iron(III) compound and hydrogen peroxide, we found that in the presence of hydrogen peroxide the Fe-(nta) complex facilely gives a (μ–η^1:η^1)-peroxodiiron (III)-(nta) species shown in Scheme 5 [20, 62, 65], which exhibits high reactivity similar to that of singlet oxygen ($^1\Delta_g$) [66–68]. However, no formation of (μ–η^1:η^1)-peroxodiiron(III) species occurs in the solution of Fe-(pac) complex in the presence of hydrogen peroxide. In the case of the edda compound, the carbonato ion in the dimeric unit is readily replaced by the hydrogen peroxide, leading to the formation of a dimeric species with a linear Fe–O–Fe core which does not give a (μ–η^1:η^1)-peroxodiiron(III) species.

Why does the tissue damage occur only in the vicinity of renal proximal tubules?

It is clear that the proximal tubules necrosis induced by artificial iron(III) chelates in rat kidneys is highly dependent on the chelate structure (Fig. 7), and injuries such as lipid peroxidation and protein oxidation are observed mainly in the renal proximal tubules [63, 64], but no injury was observed in the distal position. It should be noted here that the glutathione cycle is highly active in the renal proximal position [69], and this may suggest that the glutathione cycle promotes the iron(III)-induced injuries.

It has become apparent that some binuclear iron(III) compounds exhibit very unique reactivity towards reducing agents in the presence of oxygen. For example, the binuclear iron(III) complex with H(HPTB), $Fe_2(HPTB)(OH)(NO_3)_2$, exhibits high activity for oxygenation of linolenic acid in the presence of oxygen, and the two-electron transfer reaction to oxygen (formation of hydrogen peroxide) from TMPD (*N,N,N',N'*-tetramethyl-*p*-phenylenediamine) [70, 71]. This was elucidated on the assumption that a binuclear iron (III)–oxygen intermediate formation is promoted through interaction with the reducing agent, such as linolenic acid or TMPD (Scheme 6) [20].

On the basis of the above discussion, the tissue damage and renal carcinoma induced by the Fe(III)-(nta) chelate may be explained as follows (Scheme 7) [20, 63, 64]: when the binuclear Fe(III)-(nta) compound reacts with the glutathione cycle, the carbonato ions dissociate from the compound, leading to the interaction between two iron(III) atoms and the protein. At this stage, when oxygen is present, formation of peroxide ion is accelerated through the interaction with the glutathione cycle, and the peroxide adduct of the binuclear Fe(III)-(nta) complex formed shows high oxidative reactivity towards the proteins, leading to the tissue injuries and renal carcinoma. Thus, the tissue damage and renal carcinoma occur only in the renal proximal tubules where glutathione cycles are highly active. In the case of the Fe-(pac) complex, no formation of a (μ–η^1:η^1)-peroxodiiron (III) species occurs even in the presence of the glutathione cycle, which should be due to the stronger bonds in the Fe–O–Fe–CO_3 moiety of this complex [20].

Our conclusion on the active species inducing tissue injuries shown in Scheme 7 can explain comprehensively all the results concerning cell damage by the iron(III) chelates [20, 60, 63, 64]. In previous papers, the role of the hydroxyl radical in inducing tissue damage and renal carcinoma has been frequently pointed out, but this cannot explain the lower tissue-damaging activity exerted by the Fe(III)-(edda) chelate, nor the difference in the tissue damage exerted by the two alkoxo-bridged binuclear iron(III) complexes $Fe_2(hida)_2(H_2O)_2$ (non-toxic) and $Fe_2(HPTP)Cl_4$ (highly toxic), or by Fe(III)-(nta) and Fe(III)-(pac) chelates [20, 63, 64].

The examples of NTBI models

We have found that deposition of iron(III) hydroxide occurs readily on the aggregates of amyloid beta-peptide (1–40) [abbreviated as Aβ(1–40)] by the addition of zinc(II) chloride to the solution containing iron(III) compounds with nta, edda, and other amino acid derivatives [72]. This implies that iron(III) compounds with amino acids or peptides in plasma may be an intrinsic iron(III)-ion carrier to induce the high level accumulation of iron(III) ions in the amyloid deposits.

Recently it was suggested that the toxicity of Aβ and other amyloidogenic proteins lies not in the insoluble fibrils that aggregate but rather in the soluble oligomeric intermediates, indicating that the soluble oligomers may be more important pathologically than are the fibrillar deposits [73, 74]. The origin of the high toxicity of the soluble oligomeric intermediates should be due to the iron(III) species bound in the oligomers, which may contain a dimeric iron(III) species with oxo bridges; the soluble iron(III) species in the oligomers may readily oxidize the proteins in the presence of hydrogen peroxide, leading to the aggregation of Aβ proteins and/or degradation of surrounding proteins. Thus, it seems quite reasonable to assume that several iron (III) compounds with amino acids or peptides including amyloid β-peptide in plasma are possible candidates for NTBI models. The participation of the copper(II) ion in the toxicity of Aβ and other amyloidogenic proteins has been discussed by many authors [4, 75].

The above discussion also supports the estimation that the formation of insoluble amyloid plaques including iron(III) ions promoted by zinc(II) ions may be one of the important methods to protect against oxidative stress by soluble oligomeric iron(III) compounds with amino acids or peptides in vivo [76]. Recently, the contribution of D-serine in the pathogenesis of ALS and other neurodegenerative disorders including schizophrenia and Alzheimer's disease was been pointed out [77]. The studies on the formation of D-serine have led to the assumption that NTBI and hydrogen peroxide should play an important role in the formation and accumulation of D-serine through its high oxidative ability to racemize the L-amino acids.

References

1. Butterfield DA, Kanski J (2001) Mech Ageing Dev 122:945
2. Shastry BS (2003) Neurochem Int 43:1
3. Dobson MC (2003) Nat Drug Discov 2:154
4. Gesta A, Hider RC (2005) Br J Pharm 145:1041
5. Hider RC, Ma Y, Holgado FM, Gesta A, Roy S (2008) Biochem Soc Trans 36:1304
6. Babcock M, De Silva D, Oaks R, Davis-Kalpan S, Jiralerspong S, Montermini L, Pandolfo M, Kalpan J (1997) Science 276:1709
7. Bradley JL, Blake JC, Chamberlain S, Thomas PK, Cooper JM, Schapira AHV (2000) Hum Mol Genet 9:275
8. Dexter DT, Wells R, Lees AJ (1989) J Neurochem 52:381
9. Sanchez-Ramos J, Overvick E, Ames BN (1994) Neurodegeneration 3:197
10. Alam Z, Daniel SE, Lees AJ, Marsden DC, Jenner P, Halliwell B (1997) J Neurochem 69:1326
11. Mizuno Y, Ohta S, Tanaka M, Takayama S, Suzuki K, Sato T, Oya H, Ozawa T, Kagawa Y (1989) Biochem Biophys Res Commun 163:1450
12. Lan J, Jiang DH (1997) J Neural Transm 104:469
13. Jelliger KA (1999) Drugs Aging 14:115
14. Wong B, Chen SG, Colucci M, Xie Z, Pan T, Liu T, Li R, Gambetti P, Sy M, Brown DR (2001) J Neurochem 78:1400
15. Rosen DR, Siddique T, Figlewicz DA, Sapp P, Hentati A, Donaldson D, Goto J, O'Regan JP, Deng HX, Rahmani Z, Krizus A, Yasek DM, Cayabyab A, Gaston SM, Bergere R, Tanzi RE, Halperin JJ, van den Bergh R, Hung WY, Bird T, Deng G, Mulder DW, Smyth C, Laing NG, Soriano E, Pericak-Vance MA, Haines J, Rouleau GA, Gusella JS, Horvitz HR, Brown RH (1993) Nature 362:59
16. Sato T, Nakanishi T, Yamamoto Y, Anderson PM, Ogawa Y, Fukuda K, Zhou Z, Aoike F, Sugai F, Nagano S, Hirata S, Ogawa M, Nakano R, Ohi T, Kato T, Nakagawa M, Hamasaki T, Shimizu A, Sakoda S (2005) Neurology 65:1954
17. Nishida Y (2003) Z Naturforsch 58C:752
18. Nishida Y (2004) Med Hypothesis Res 1:227
19. Nishida Y (2007) TCIMAIL 135:2. http://www.tciamerica.com/tcimail/backnumber/article/135drE.PDF
20. Nishida Y (2009) TCIMAIL 141:2. http://www.tciamerica.com/tcimail/backnumber/article/141drE.pdf
21. Smith RG, Appel SH (1995) Annu Rev Med 46:133
22. Brown RH II (1995) Cell 80:687
23. Deng HX, Hhentani A, Tainer JA, Iqbal Z, Cayabyab A, Hung WY, Getzoff ED, Hu P, Herzfeldt B, Roos RP, Warner C, Deng G, Soriano E, Smyth C, Parge HE, Ahmed A, Roses AD, Hallewell RA, Pericak-Vance MA, Siddique T (1993) Science 261:1047
24. Pazos MW, Goto JJ, Rabizadeh S, Gralla EB, Loe JA, Lee MK, Valenstein JS, Bredesen DE (1996) Science 271:15
25. Fridovich I (1995) Annu Rev Biochem 64:97
26. Borchelt DR, Lee MK, Slunt HS, Guarnieri M, Xu ZS, Wong PC, Brown RH Jr, Price DL, Sisodia SS, Cleaveland DW (1994) Proc Natl Acad Sci USA 91:8292
27. Brujin L, Houseweart MK, Kato S, Anderson SD, Ohama E, Reaume AG, Scott RW, Cleaveland DW (1998) Science 281:1851
28. Estevez AG, Crow JP, Sampson JB, Reiter C, Zhuang Y, Richardson GJ, Tarpey MM, Barbeito L (1999) Science 286:2498
29. Yim MB, Kang JH, Yim HS, Kwak HS, Chock PB, Stadman ER (1996) Proc Natl Acad Sci USA 93:5709
30. Watanabe M, Hobera MD, Culotta VC, Price DL, Wong PC, Rothstein JD (2001) Neurobiol Dis 8:933
31. Lindberg MJ, Bystrom R, Bokna N, Andersen PM, Oliveber M (2005) Proc Nat Acad Sci USA 102:9754

32. Yim HS, Kang JH, Chock PB, Stadman ER, Yim MB (1997) J Biol Chem 272:8861
33. Kobayashi T, Okuno T, Suzuki T, Kunita M, Ohba S, Nishida Y (1998) Polyhedron 17:1553
34. Okuno T, Ohba S, Nishida Y (1998) Polyhedron 16:3765
35. Nishino S, Kobayashi T, Kunita M, Ito S, Nishida Y (1999) Z Naturforsch 54C:94
36. Nishino S, Kunita M, Kani Y, Ohba S, Matsuhsima H, Tokii T, Nishida Y (2000) Inorg Chem Commun 3:145
37. Nishino S, Nishida Y (2001) Inorg Chem Commun 4:86
38. Nishino S, Nishida Y (2005) Synth Reac Inorg Metal-org Nano-Metal Chem 35:677
39. Nishino S, Kishita A, Nishida Y (2001) Z Naturforsch 56C:1144
40. Cioni P, Pesce A, Morozzo B, Castelli S, Falconi M, Parrilli L, Bolognesi M, Strambini G, Desideri A (2003) J Mol Biol 326:1351
41. Matsumoto O, Fridovich I (2002) Proc Natl Acad Sci USA 99:9010
42. Yamanaka K, Cleveland DW (2005) Neurology 65:1859
43. Rakhit R, Crow JP, Lepock JR, Kondejewski LH, Cashman NR, Chakrabartty A (2004) J Biol Chem 279:15499
44. Abe K, Nishida Y (2008) Z Naturforsch 63C:151
45. Chiba Y, Sutoh Y, Nishida Y (2006) Z Naturforsch 61C:273
46. Caughe B (2001) Trends Biochem Sci 25:235
47. Requena JR, Dimitrova ND, Legname G, Teijira S, Prusiner SB, Levine RL (2004) Arch Biochem Biophys 432:188
48. MaMahon HEM, Mange NishidaN, Creminon C, Casanova D, Lehmann S (2001) J Biol Chem 276:2286
49. Requena R, Groth RD, Legname G, Sradtman ER, Prusiner SB, Revine RL (2001) Proc Natl Acad Sci USA 98:7170
50. Watt NT, Taylor DR, Gillott A, Thomas DA, Perera WS, Hooper NM (2005) J Biol Chem 280:35914
51. Tabler BJ, Turnbull S, Fullwood NJ, German M, Allsop D (2005) Biochem Soc Trans 33:548–590
52. Tabler BJ, Agnaf OMAE, Turnbull S, German MJ, Paleologou KE, Hayashi Y, Kooper LJ, Fullwood NJ, Allsop D (2005) J Biol Chem 280:35789
53. Watt NT, Hopper NM (2005) Biochem Soc Trans 33:1123
54. Fernaeus S, Reis K, Bedecs K, Land T (2005) Neurosci Lett 389:133
55. Harrison PM, Arosio A (1996) Biochim Biophys Acta 1275:161
56. Chasteen D, Harrison PM (1999) J Struct Biol 126:182
57. Okada S, Midorikawa O (1982) Jpn Arch Intern Med 29:485
58. Li JL, Okada S, Hamazaki S, Ebina Y, Midorikawa O (1987) Cancer Res 47:1867
59. Kawabata T (1997) Carcinogenesis 18:1389
60. Toyokuni S, Sagripanti JL (1993) Carcinogenesis 14:223
61. Halliwell B, Gutteridge JMC (1985) Free radicals in biology and medicine. Oxford University Press, London
62. Nishida Y, Goto A, Akamatsu T, Ohba S, Fujita T, Tokii T, Okada S (1994) Chem Lett 1994:641
63. Mizuno S, Kawabata T, Sutoh Y, Nishida Y, Okada S (2006) Biometals 19:675
64. Nishida Y, Ito Y, Satoh T (2007) Z Naturforsch 62C:608
65. Nishida Y, Ito S (1995) Polyhedron 14:2301
66. Nishida Y, Takeuchi M (1987) Z Naturforsch 42B:52
67. Chen MS, White MC (2007) Science 318:783
68. Okuno T, Ito S, Ohba S, Nishida Y (1997) J Chem Soc Dalton Trans 1997:3547
69. Okada S, Minamiyama Y, Hamazaki S, Toyokuni S, Sotomatsu A (1993) Arch Biochem Biophys 301:138
70. Nishida Y, Takeuchi M, Oishi N, Kida S (1985) Inorg Chim Acta 96:115
71. Nishida Y, Nasu M, Akamatsu T (1992) J Chem Soc Chem Commun 1992:93
72. Okawamukai Y, Sutoh Y, Nishida Y (2006) Synth React Inorg Metal-org Nano-Metal Chem 36:373
73. Kayed R, Head E, Thompson JL, Mcintire TM, Milton SC, Cotman CW, Glabe CG (2003) Science 300:486
74. Klein WL, Stine WB Jr, Teplow DB (2004) Neurobiol Aging 25:569
75. Streltsov VA, Titmaus SJ, Epa VC, Barnham KJ, Mastrers CL, Varghese JN (2008) Biophys J 95:3447
76. Gauthier MA, Eibl JK, Crispo JA, Ross GM (2008) Neurotox Res 14:317
77. Sasabe J, Chiba T, Yamada M, Okamoto K, Nishimoto I, Matsuoka M, Aiso S (2007) EMBO J 19:4149

Metal ion roles and responses in the CNS under toxic organophosphonate exposure: traces of understanding and various open questions

Olga G. Tsay, Kibong Kim, and David G. Churchill

Abstract

In this short review article, we list fresh questions regarding how metal ions are operating (or differently operating) in CNS compartments in the presence of concentrations of toxic organophosphonates (Nerve agents) [$O = PR^1R^2R^3$]. Clearly, AChE active site serine residue phosphonylation is of primary and acute concern, but there may be a trigger for later onset (long ranging effects) events that may be acute and symptomatic but noncritical. Based on a biological inorganic chemistry approach, we have tried to position important questions herein starting from what is known in the primary literature, namely about zinc, iron, calcium, and magnesium ions.

Keywords

Nerve agent • Brain • CNS • Metals • Zinc • Manganese • Inorganic neurochemistry • Enzymes • Organophosphonate • Toxicology

Introduction

> what the critical variables are that determine that some OP compounds have these profound central actions and some do not would appear to be a fruitful area for further basic research. [1]

Organophosphonate (OP) nerve agents and pesticides continue to be molecules of significance in environmental contamination and remediation research fields. Of course, environmental aspects impact human health. Such chemical agents weigh heavily in neurotoxicity. Since the brain and CNS are central in neurology, we are underscoring here that there is still a vast amount of important scientific information to learn about with respect to brain chemistry, and more specifically, OP activity and metal species in the brain [1]. The quote above refers to the fact that not all nerve agents have the same activity. Nerve agent AChE acute toxicity/antagonism has been the dominant theme for decades, and for good reason: the buildup of acetylcholinesterase is acutely toxic to nerve functioning. In recent years, this avenue has been considered for its therapeutic potential as well, for instance with Alzheimer's disease [2–4]. Whereas agent toxicity is well documented, there is a paucity of reports describing how metalions in various CNS compartments might act on this toxicity. This lack is clear from Boolean searches involving relevant keywords (Fig. 1). There are many hits for "metals" in the "brain" but then including a third term "posphon*," or "nerve agent," yields no hits [5, 6]. This gives us clear reasoning to extend current research; at the same time, a variety of related research questions arise which extend to the biochemistry and pharmacology of agents, and to the chemistry of "antidotes" as well. Two main themes can be created: OP exposure may invoke or depress (normal) expression of metals; secondly, its related compounds may interact directly with metals. As a corollary to the points, oxime treatments may also have an impact; these are discussed as well. Clearly, there are various roles of a metal ion: it can bind or react with the agent, the enzyme, or a metal and agent can bind or react with each other.

D.G. Churchill (✉)
Department of Chemistry, Korea Advanced Institute of Science and Technology (KAIST), 373-1 Guseong-dong, Yuseong-gu, Daejeon 305-701, Republic of Korea
e-mail: dchurchill@kaist.ac.kr

W. Linert and H. Kozlowski (eds.), *Metal Ions in Neurological Systems*,
DOI 10.1007/978-3-7091-1001-0_14,

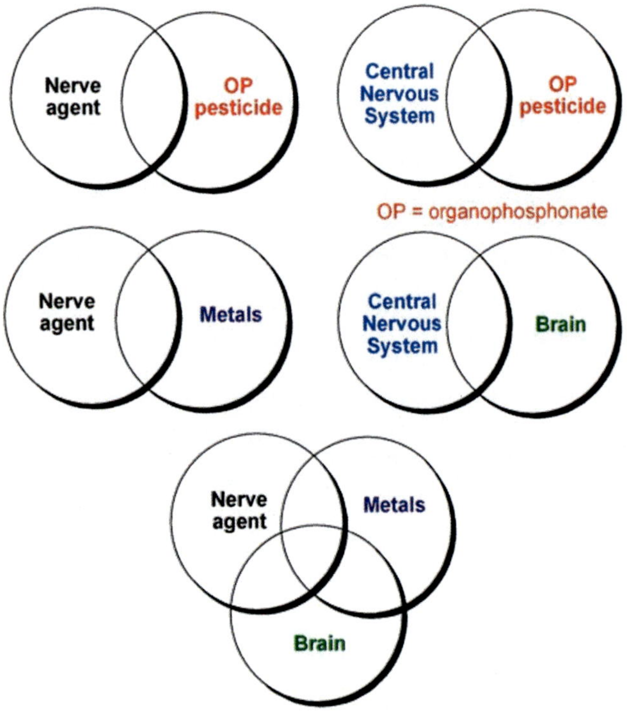

Fig. 1 Venn diagrams of some search terms of concepts explored herein

A variety of metals are found in the brain at relatively high concentrations. Various transition metals, while being in enzymes, are also labile species: Cu^{2+}, Fe^{2+}/Fe^{3+}. Not much is known about brain metals and NAs/organophosphonates. First, we will try to find out what is already known about brain metals and NAs/organophosphonates and what animal models tell us. In 1955, Augustinsson reported the effects of metal ions on enzymes extracted from kidney (pig) and blood serum (human) [7]. Phosphorylate-phosphatases were charged with Sarin. While this is a relatively old report, continued work regarding metal ions seems to be absent perhaps due to the general difficulty in working with nerve agents despite the increased understanding and importance that metals now have in neurology. There is also the general question about the limitations of animal models. Importantly, in this article, enzymes are inhibited by some metal ions and less by others. Reports by Matsuda involved brain tissues [8]; however, it is obviously difficult to clearly extrapolate to humans from animal models [9]. Also since metals are used in muscle contraction and OP exposure may lead to convulsions, it is a particularly interesting conceptual tug-of-war to sometimes determine cause and effect.

Calcium (II)

We will first discuss calcium (II) reports. In vivo fluctuations in intracellular [Ca^{2+}] are known and can be monitored. Activation of muscarinic receptors by nerve agents leads to the hydrolysis of phosphatidylinositol 4,5-bisphosphate and results in the formation of a second messenger inositol 1,4,5-triphosphate (IP3), which, in turn, transmits the signals from the receptor into the cell by releasing Ca^{2+} from non-mitochondrial intracellular stores [10]. Elevation of free intracellular Ca^{2+} leads to necrosis or apoptosis. Calcium levels were studied in rat brain (cortex) after exposure with malaoxon. Increase of Ca^{2+} concentration was observed for both convulsing and nonconvulsing rats by 20 % in 1.0 h after malaoxon treatment, and by 25 % and 15 %, respectively, at 4.0 h. In convulsing rats, within 72 h, increased levels of Ca^{2+} remained constant while in nonconvulsing rats Ca^{2+} concentration declined to normal levels, although the redistribution pathway(s) of free extra Ca^{2+} in tissue is not known. However,based on the timeline, changes in morphology appeared to be a result of this increased [Ca^{2+}] [11].

It is also known that Soman-induced seizure-related brain damage results from glutamate excitotoxicity. Excessive stimulation of *N*-methyl-D-aspartate (NMDA) ionotropic glutamate receptor causes elevation in intracellular Na^+ and especially calcium ion concentration which leads to depolarization of postsynaptic membranes. Calcium ion overload in neurons initiates activation of several potentially damaging enzymes, such as oxygenases, phospholipases, and nitric oxide synthase (NOS), which produce reactive oxygen species (ROS) [12]. Therefore, the role of calcium imbalance in neuronal injury is important as schematically proposed by Ballough et al. [13].

DeLorenzo and his group studied calcium ion concentration dynamics in hippocampal neurons upon diisopropyl fluorophosphate (DFP) exposure. It was shown that treatment by DFP produces significant increases in calcium ion levels that lasted for almost 336 h after the brain was constantly in seizure (*status epilepticus*). Approximately 50 % of the neurons in DFP-treated animals had calcium levels greater than 500 nM, whereas control values showed lower levels (time of experiment: 1 h after *status epilepticus*). These results confirmed that calcium concentrations in neurons (hippocampus) were raised in *status epilepticus* by DFP induction. Time-dependent experiments showed that the concentration of Ca^{2+} remained still higher (378.06 ± 46.21 nM) than those values found in control trials (248.12 ± 23.12 nM) after 1 month; a Ca^{2+} concentration plateau was observed for over 168 h. In this study, the elevation in calcium concentration was mediated and maintained by calcium entry from the NMDA receptor system; experimental findings indicate that the activation of the NMDA receptor during *status epilepticus* plays an important role in the development of the calcium plateau [14]. A disruption of calcium homeostasis was also reported by Hamilton and Posavad [15]. An increase in calcium ion content was observed after during 24 h and remained high

24 h at the LD_{50} level (lethal dose 50) of Soman and at least 168 h at concentration of 4 times the LD_{50}.

The effect of Soman administration on intracellular calcium uptake by microsomes or mitochondria from rabbit heart muscle was also investigated [16]. It was shown that acute exposure by Soman at concentrations of 1.0–100 μM did not give any change in Ca^{2+} uptake. But after ~168 h, Soman administration is correlated with a marked reduction of $^{45}Ca^{2+}$ uptake by both microsomes and mitochondria. This suggests that calcium uptake is dependent on repeated Soman administration and relevant applications to brain chemistry can be made. Here, other *S*-block metal ions can also be considered. While many such species are spectator ions, they are often at high concentration; the general inertness of Mg^{2+}, Na^{+}, and K^{+} towards complexation or reactivity with OPs gives rise to the issue of whether these could participate in any important way to nerve agent interactions or aqueous chemistry [17]. The Mg^{2+} ion has been mentioned in the context of nerve agents and CNS previously [1].

Iron (II,III)

Pazdernik et al. showed that Soman-induced seizures increase the concentration of "catalytic" iron ions in tissue. Also, Metallothioneins and heme oxygenase-1 levels rise and cause a depletion in the concentrations of tissue glutathione. Free radical concentrations might increase; cellular damage may occur from free iron by way of the well-known Fenton-type chemistry. The highest values of increasing "catalytic" iron ions were observed 72 h after soman-induced seizure. In the *piriform cortex*, a 202 % increase in iron ions as observed, compared with the control; in the thalamus, there was a 130 % increase and in the hippocampus a 125 % increase. No changes in iron ion concentrations were observed in brain regions, wherein minimal damage occurred (*pariental cortex*, 96 % and *caudate* 96 %). In this study, the authors detected loosely bound iron and heme iron in tissue, but not ferritin-bound iron. From experimental data, it was proposed that the increase of "catalytic" iron ion concentration is coming from heme iron associated with cell lysis and microhemorrhages stemming from capillary rupture in the vasculature. Changes in redox chemistry in brain disrupt oxidative defense mechanisms and lead to oxidative damage of brain (cell lysis and tissue destruction). Also increase of concentrations of heme oxygenase-1 (OH-1), crucial for iron homeostasis, by 1.5- to 2.0-fold above control levels may be an indicator of an increase in delocalized iron ions and subsequent oxidative stress [18]. Changes in dopamine levels might impact iron levels as well [1].

Zinc (II)

Labile zinc pools are ubiquitous and essential in neurology; Zn^{2+} is a very mobile ion; it is considered a neurotransmitter and is under investigation as a species involved in the etiology of certain neurodegenerative diseases. It is known that zinc ion is co-localized and co-released with glutamate at the NMDA receptor. The protein metallothionein-3 (MT-3) plays a functional role in neuronal zinc deposition and release. So, the action of increasing concentrations of MT-1 mRNA in non-neuronal cells observed in brain with Soman-induced seizure may indicate a presence of excessive co-release of glutamate and zinc ion [19].

Some questions for further research and the related field of study

- Pharmacology and toxicology:
  - What are all the possible mechanisms of in vivo transport of *unreacted* nerve agent or pesticide into the brain from a remote (e.g., epidermal) tissue exposure?
- Bioinorganic chemistry:
  - What metalloenzymes are affected (inhibited or poisoned)?
  - Which metals are not related or involved?
  - Which chelation environments are most amenable to disruption/poisoning?
  - Which nerve agent or pesticide downstream byproducts are of most concern?
  - How do these downstream products (above) interact with labile metals?
- Pharmacology:
  - Are there new metal ion-related biomarkers that can be discovered for low-dose chronic exposure?
- Bioinorganic chemistry/pharmacology:
  - How will oximes administered as an antidote engage endogenous metals?
  - What more can we learn about this aqueous chemistry in vitro?
  - What concentrations of oximes build up in the cerebral spinal fluid (CSF)? [19]
- Pharmacology and environmental biochemistry:
  - Could there be some moderating or ameliorating or healthful effects to low-dose chronic OP exposure?
- Bioinorganic chemistry and toxicology:
  - Could phosphonates enter Ferritin or disrupt Ferritin stores?

Conclusion

In this brief review article, we tried to treat what little is known about how metal ions are pesticide operating (or differently operating) in the CNS in the presence of toxic organophosphonate (e.g., Nerve Agent) exposure. The metallobiochemistry of zinc, iron, calcium and magnesium ions is described. Clearly, the AChE active site serine residue phosphonylation is an effect of acute and primary concern; but there may be a host of other chemistry occurring that may be acute but noncritical/symptomatic. Some effects might be long term. We have tried to develop and position important study questions as a guide for important future studies. Based on the quote that prefaced this article, this is meant to be a good, but albeit challenging area for further research.

Acknowledgment The Molecular Logic Gate Laboratory headed by D.G.C. was supported financially for this work by the National Research Foundation of Korea (NRF) grant funded by the Korea government (MEST) (No. 2009–0070330 and 2010–0013660). We thank Officer Yoon Jeong Jang of the Chemical Defense Research Institute (Naegok-dong San 12-555, Seocho-gu, Seoul, Republic of Korea) for helpful discussions related to this research.

References

1. McDonough JH, Shih TM (1997) Neuropharmacological mechanisms of nerve agent-induced seizure and neuropathology. Neurosci Biobehav Rev 21:559–579, and references therein
2. Giacobini E (2004) Cholinesterase inhibitors: new roles and therapeutic alternatives. Pharmacol Res 50:433–440
3. Kuhl DE, Koeppe RA, Minoshima S, Snyder SE, Ficaro EP, Foster NL, Frey KA, Kilbourn MR (1999) In vivo mapping of cerebral acetylcholinesterase activity in aging and Alzheimer's disease. Neurology 52:691–699
4. Weinstock M (1999) Selectivity of cholinesterase inhibition. CNS Drugs 12:307–323
5. "Organophosphate" and "brain" and "metals" (25 hits); "organophosphate" and "neuron" (169); "organophosphonate" and "neuron" (0). 1 reference was found containing all of the concepts "metal", "brain" and "organophosphonate". SciFinder Scholar. Accessed Dec 2011
6. Kim K, Tsay OG, Atwood DA, Churchill DG (2011) Destruction and detection of chemical warfare agents. Chem Rev 111:5345–5403
7. Augustinsson KB, Heimburger G (1955) Enzymatic hydrolysis of organophosphorus compounds. VI. Effect of metallic ions on the phosphorylphosphatases of human serum and swine kidney. Acta Chem Scand 9:383
8. Matsuda Y, Nagao M, Takatori T, Niijima H, Nakajima M, Iwase H, Kobayashi M, Iwadate K (1998) Detection of the sarin hydrolysis product in formalin-fixed brain tissues of victims of the Tokyo subway terrorist attack. Toxicol Appl Pharmacol 150:310–20
9. Bajgar J (2004) A organophosphates/nerve agent poisoning: mechanism of action, diagnosis, prophylaxis, and treatment. Adv Clin Chem 38:151–216
10. Berridge MJ, Irvine RF (1984) Inositol trisphosphate, a novel second messenger in cellular signal transduction. Nature 312:315–321
11. Hirvonen MR, Paljärvi L, Naukkarinen A, Komulainen H, Savolainen KM (1990) Potentiation of malaoxon-induced convulsions by lithium: early neuronal injury, phosphoinositide signaling, and calcium. Toxicol Appl Pharmacol 104:276–289
12. Filbert M, Levine E, Ballough G (2005) Neuroprotection for nerve agent-induced brain damage by blocking delayed calcium overload: a review. J Med Chem Biol Radiol Def 3:1–21
13. Ballough GPH, Newmark J, Levine ES, Filbert MG (2008) Neuroprotection as a treatment for nerve agent survivors. In: Tuorinsky SD (ed) Medical aspects of chemical warfare, 1st edn. TMM publications, Washington, DC
14. Deshpande LS, Carter DS, Blair RE, DeLorenzo RJ (2010) Development of a prolonged calcium plateau in hippocampal neurons in rats surviving status epilepticus induced by the organophosphate diisopropylfluorophosphate. Toxicol Sci 116:623–631
15. Hamilton MG, Posavad C (1991) Alteration of calcium influx in rat cortical synaptosomes by soman. Neuroreport 2:273–276
16. Hu CY, Hsu CH, Robinson CP (1991) Effects of soman on calcium uptake in microsomes and mitochondria from rabbit aorta. J Appl Toxicol 11:293–296
17. Bandyopadhyay I, Kim MJ, Lee YS, Churchill DG (2006) Favorable pendant-amino metal chelation in VX nerve agent model systems. J Phys Chem A 110:3655–3661
18. Pazdernik TL, Emerson MR, Cross R, Nelson SR, Samson FE (2001) Soman-induced seizures: limbic activity, oxidative stress and neuroprotective proteins. J Appl Toxicol 21:s87–s94
19. Frederickson CJ (1989) Neurobiology of zinc and zinc-containing neurons. Int Rev Neurobiol 31:145–238

Heavy metals, behavior, and neurodegeneration: using *Caenorhabditis elegans* to untangle a can of worms

Samuel Caito and Michael Aschner

Abstract

The complexity of the vertebrate brain has made the study of neurodegenerative disease processes slow, difficult, and expensive. *Caenorhabditis elegans* offers viable in vivo model system for addressing numerous issues pertinent to neurodegenerative diseases. Differentiation and migration patterns in the nematode are well characterized, thus allowing for analysis of changes in nervous system in response to mutations and toxic insults. The full sequencing of the nematode genome and a high-density map of polymorphisms for the wild type nematode allows for mapping of gene mutations and linking of mechanisms of neurodegeneration to genetic susceptibility. In addition to the high level of gene conservation, the processes of synaptic release, trafficking and formation are also conserved between this invertebrate and mammalians. Given these advantages, *C. elegans* has been employed in numerous studies to address neurodegeneration and mechanisms of toxicity of a wide range of toxicants. In this chapter, we provide an overview on the system model and discuss contemporary insights derived from *C. elegans* on the involvement of metals in behavior and neurodegenerative diseases.

Keywords

Caenorhabditis elegans • Metals • Neurodegeneration • Behavior • Parkinson's disease • Alzheimer's disease • Methylmercury

Abbreviations

AD	Alzheimer's disease
APP	Amyloid precursor protein
DA	Dopamine
DAergic	Dopaminergic
DMT1	Divalent metal transporter 1
GABA	γ-aminobutyric acid
GST	Glutathione-S-transferase
LRRK2	Leucine-rich repeat kinase 2
MeHg	Methylmercury
MPT	Mitochondrial permeability transition
NGM	Nematode growth medium
Nrf2	Nuclear factor-2 erythroid 2-related factor-2
PD	Parkinson's disease
PHP	Psuedohyperphosphorylated
PINK1	PTEN-induced novel kinase 1
PTEN	Phosphatase and tensin homolog
ROS	Reactive oxygen species
SNpc	Substantia nigra pars compacta

M. Aschner (✉)
Division of Clinical Pharmacology and Pediatric Toxicology, Vanderbilt University Medical Center, 11425 MRB IV 2215-B Garland Avenue, Nashville, TN 37232-0414, USA

Center in Molecular Toxicology, Vanderbilt University Medical Center, Nashville, TN, USA

Center for Molecular Neuroscience, Vanderbilt University Medical Center, Nashville, TN, USA

The Kennedy Center for Research on Human Development, Vanderbilt University Medical Center, Nashville, TN, USA
e-mail: michael.aschner@vanderbilt.edu

W. Linert and H. Kozlowski (eds.), *Metal Ions in Neurological Systems*,
DOI 10.1007/978-3-7091-1001-0_15,

Introduction

As the population ages, there is an increased prevalence of neurodegenerative diseases, the two most common being Alzheimer's disease (AD) and Parkinson's disease (PD). While common mechanisms involved in neurodegenerative diseases have been identified, such as altered protein processing, oxidative stress, death and degeneration of specific neurons and brain regions, initiating events and underlying causes remain elusive. Genetics have identified genes that contribute to the development of familial AD and PD; however, the majority of cases are sporadic and idiopathic in origin. Epidemiological studies have found environmental toxicants, such as herbicides, pesticides, and heavy metals, are associated with an increased risk of AD or PD [1, 2]. Due to the large numbers of people who are exposed to these contaminants, the influence of environmental contributors to the development of neurodegenerative diseases is gaining wider attention.

Metals are persistent environmental contaminants that have been associated with neurodegeneration, as well as developmental and behavioral deficits [3–5]. Certain metals, such as Zn, Mn, and Fe are essential for cell viability and function; however, excessive levels of metals can be deleterious. Metal dysregulation in neurodegenerative diseases may occur as one ages. Neurodegenerative diseases are primarily found in older individuals, and it is known that there are morphological alterations in brain structure in the elderly [6]. Concurrent with altered structure, a number of studies in rodents have demonstrated reduced transport of Mn, Fe, Cu, and Zn in aged animals, allowing for accumulation of metals in the brain [7]. Exposure to low levels of metals over the course a lifetime may also contribute to brain metal dysregulation, especially if these metals are not removed from the brain as efficiently in aged individuals. Metals can generate free radicals and reactive oxygen species (ROS) through Fenton chemistry, leading to protein and DNA oxidation and lipid peroxidation. Oxidation of macromolecules can damage cellular organelles and generate more ROS to create a vicious cycle. Brain metal accumulation has been associated with severe neuronal damage and AD [8]. Postmortem tissue from patients with confirmed neurodegenerative diseases has revealed elevated levels of metals in affected brain regions [9]. Although it is unclear whether elevated metals in the brain are a contributing mechanism or a consequence of the disease, altered metal homeostasis may be an accelerant to the neurodegenerative process through interactions with accumulated proteins [10, 11].

The nematode *Caenorhabditis elegans* has emerged as a powerful model system for neuroscience research due to its well characterized nervous system and genetic manipulability. *C. elegans* are small (~1.5 mm) round soil-dwelling worms with a short lifespan of 3 weeks and a rapid life cycle of 3 days, that can be maintained cheaply in a humid environment containing atmospheric oxygen and a food source of bacteria grown on agar plates. Because of the short lifetime and low cost of upkeep, *C. elegans* are a perfect model for high-throughput screens of chemicals for a variety of parameters, such as toxicity, lifespan, and reproduction/brood size. A single adult hermaphrodite is capable of producing ~300 progeny. Upon hatching (L1 phase), worms proceed through three subsequent larval stages (L2 after 12 h, L3 after 8 h, and L4 after 10 h). The simplicity of the worm's body comprising less than 1,000 cells has allowed for the mapping of every cell's development and lineage [12]. As an adult, an hermaphrodite is able to self-fertilize, or may be used for mating with an adult male, a rare sex with a frequency of 0.5 %, allowing for crossing worms of different genotypes. The genome of *C. elegans* has been fully sequenced and has 60–80 % homology with mammals, including ~40 % of human disease-related genes [13, 14]. The genetic architecture of *C. elegans* genes is known and there are standardized protocols for knocking down genes through RNAi and introducing DNA through injections. Due to the ease of genetic manipulation, scientists have created numerous strains of knockout alleles, transgenic overexpression alleles, and genes fused with reporters such as mCherry and GFP to study a wide range of cellular processes. These advantages have revealed insights into signaling pathways and gene–gene and gene–environment interactions that have not been observed in more complex systems. In this review, we highlight recent research using *C. elegans* that provide insight on the involvement of metals in behavior and neurodegenerative diseases.

C. elegans nervous system

Adult hermaphrodite *C. elegans* contain 302 neurons which are grouped into ganglia with their processes organized into nerve cord bundles and commissures [15]. There are 56 support cells which function similarly to glial cells. Males have an additional 79 neurons and 36 support cells, which are involved in male mating behavior [16]. Neurons undergo processes of outgrowth, cell migration, apoptosis, and developmental plasticity using mechanisms conserved in mammals. All synapses have been characterized by serial electron microscopy [17], and the neurons may be characterized into motor neurons, sensory neurons, and interneurons. *C. elegans* have conserved neurochemistry with mammals, utilizing glutamate, acetylcholine, γ-aminobutyric acid (GABA), dopamine (DA), serotonin, and an estimated 150 neuropeptides. Nematodes and mammals also share conserved mechanisms to produce, package into vesicles, traffic, and release neurotransmitters.

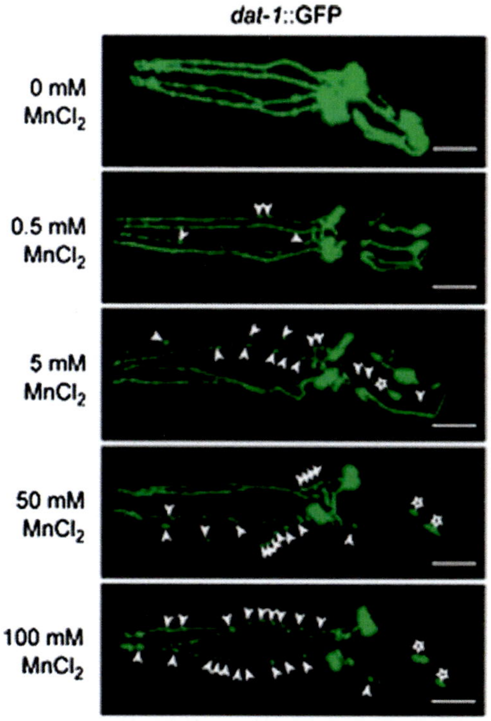

Fig. 1 Degeneration of dopaminergic neurons after Mn exposure. Mn dose-dependently causes CEP and ADE degeneration in *C. elegans*. Dopaminergic neurons express GFP under the dat-1 promoter. *Arrowheads* indicate neuronal processes exhibiting abnormal discontinuous GFP signal, while *stars* indicate the position of degenerated neuronal cell bodies (from Ref. [21])

C. elegans displays complex behaviors including locomotion (sinusoidal crawling and C-shaped swimming/thrashing), foraging, feeding, defecation, and avoidance or movement toward chemicals, pheromones, odorants, temperatures, and food sources (reviewed in Ref. [18–20]. Due to its transparency, GFP fusion strains have been utilized to visualize specific neurons and synapses in living worms. Cell death may be observed by counting total number of cells expressing GFP. Changes in morphology, such as vacuolated neurons, nerve cord breaks or puncta can be observed in dying or degenerating cells (Fig. 1), making *C. elegans* a powerful system for studying neurodegenerative mechanisms and genetics.

However, there are important considerations that must be taken into account in using C. *elegans* as a model system for neuroscience research. Neurons in *C. elegans* are not myelinated, which is not amiable for studying diseases characterized by demyelination, such as multiple sclerosis [22]. Most neurodegenerative diseases have an inflammation or immune system component, which cannot be studied in the worm as *C. elegans* has a rudimentary immune system that does not contain an adaptive immune response or major inflammatory pathways, such as NF-κB. Additionally, vascular contribution to disease cannot be examined in nematodes, as there is no circulatory system or blood–brain barrier.

Worm models in metal neurotoxicity research

C. elegans is an attractive model for studying effects of metals on neural functioning. Outside of the lab, *C. elegans* live in soil, and are exposed to a variety of compounds and contaminants not essential for life. Worms contain many of the transporters and stress response genes critical for xenobiotic and metal detoxification, including metallothionines [23, 24], transporters involved in metal homeostasis [25–27], heat shock proteins [28], and genes involved in glutathione homeostasis [29]. Additionally, *C. elegans* possess phytochelatins, small heavy metal binding peptides synthesized from glutathione which are also present in plants and fungi [30]. Nematodes are the only animals identified to use these peptides. *C. elegans* has served as a tool for toxicogenomic analyses for genes and pathways involved in resistance to Cd toxicity. Cui et al. have identified 290 genes differentially expressed after 4 or 24 h exposure to 100 μM Cd. Metallothionines and other genes known to be involved in metal detoxification were identified, but the majority of genes were novel, including transporters, cytochrome P450s, and MAPK pathways [26]. It is unknown how many of these genes are similarly regulated after exposure to other metals. Due to the conserved nature of *C. elegans*' genes, many of the proteins identified may be similarly regulated in higher organisms following Cd exposure.

Essential for their survival, *C. elegans* are able to sense noxious stimuli, such as metals, alkaloids, and acidic pH, and rapidly reverses direction. This response requires ASH nocioceptive neurons, which also respond to mechanical stimuli, such as light touch to the nose [31, 32]. The withdrawal response to metals also uses information from the ASK and ASE neurons of the head and the phasmid neurons in the tail to decide which direction to move [33]. Interestingly, when worms are repeatedly exposed to a noxious repellant, such as 10 mM Cu, they will adapt to the stimuli and will not be repelled as a worm sensing the repellant for the first time [34]. Repulsion behavior has been shown to occur through the activation of *gpc-1*, a G-protein-coupled receptor, as deletion of *gpc-1* resulted in loss in sensory adaptation [34]. *C. elegans* will also reduce food intake when exposed to sublethal concentrations of metals, such as Cu, Pb and Cd [35]; however, it is not known which neurons are involved in this behavior. ASE neurons are involved in chemotaxis to water-soluble attractants. *C. elegans* exposed to Pb, Cu, Ag, or Cr results in reduced chemotaxis to NaCl, cAMP, and biotin [36]. Pb, Cu, Ag, and Cr exposure altered morphology of the ASE neurons, observing puncta of cell bodies decreased lengths of sensory endings and altered dendrite morphology, which may explain the reduced chemotaxis [36].

Metals alter nematode neurobehavior

Exposure to Pb and methylmercury lead to altered behavior in mammals and developmental changes in the brain, such as learning deficiencies, attention deficits, and aggression (reviewed in Ref. [3, 37]). Several worm behaviors are decreased in response to metals. Similar to mammals, developmentally, younger larvae were more sensitive to Pb, Hg, Cd, and Cr, showing decreased thrashing and body bends in L2–L3 stages than in adults [38]. As pathways for learning are complex in vertebrate models, *C. elegans* provides a simpler platform for investigating neurobehavioral pathways affected by metal exposures. To evaluate memory and learning plasticity, a thermotaxis tracking assay has been developed where young adult worms are incubated overnight at 20 °C on nematode growth medium plates (NGM) with bacteria washed off the plate and transferred to an unseeded plate at 20 °C for 2 min. Individual animals are then placed on a new unseeded NGM plate for an incubation period of 0–18 h. A frozen vial containing acetic acid is then placed in the center of the plate at 26 °C for 90 min to create a temperature gradient ranging from 17 to 25 °C. In this assay, worms learn to associate food with a given growth temperature during the conditioning period and will isothermal track on the unseeded plates to the growth temperature to find food. The percentage of worms that track to the growth temperature reflects the ability of the worms to learn. By varying the incubation time, one can evaluate memory. Reduced learning and memory has been observed in worms exposed to Al, Pb, Hg, Cu, Zn, or Ag [39, 40]. The thermosensation assay is a valuable tool in *C. elegans* research which has the potential to reveal insights on memory and learning and modifiers of memory loss. *C. elegans* either pretreated or posttreated with 200 μg/ml vitamin E had less Al- or Pb-induced reduction in memory [39], suggesting that the metal-induced oxidative stress contribution to learning and memory impairments may be reversible with antioxidants. Exposure to Ni reduces the lifespan of *C. elegans* and slows the development and number of progeny [41]. Examining associative learning using the thermosensation assay revealed cross-generational effects of Ni; Ni exposure reduced learning not only in the exposed worms but also in their progeny [41]. As the progeny were not exposed to Ni, these data suggest heritable changes in associative learning pathways caused by metal exposure. Ni has been found to affect heterochromatin by inhibiting histone acetylation [42], perhaps leading to heritable deficits. Further work is needed to tease out mechanisms and pathways of metal-induced alterations in behavior.

Oxidative stress, neurodegeneration, and cell death

Neurons are metabolically active cells, which due to high oxygen utilization and cellular respiration levels are vulnerable to oxidative stress. ROS can be produced by the mitochondria during respiration; however, cells contain antioxidant defense systems for protection. When there is an increase in the cellular ROS burden, such as after metal exposure, cells will accumulate oxidized DNA, protein, and lipids, and, if the damage is too extensive, die. In neurodegenerative diseases there are not only increased levels of metals present, but increased markers of oxidative stress, such as oxidatively damaged DNA [43, 44]. Both necrotic and apoptotic cell death pathways have been observed from metal-induced oxidative stress. Necrosis is the process of cell death characterized by plasma membrane rupture and dilation of cellular organelles [45]; that in neurons may be due to ROS formation and increased lipid peroxidation after metal exposure [46]. ROS contributes to mitochondrial permeability transition (MPT) induction, which can trigger apoptosis. Induction of the MPT is caused by opening of a pore, called the permeability transition pore (PTP) in the mitochondrial inner membrane. Apoptosis, or programmed cell death, is characterized by chromatin condensation and fragmentation, cell shrinkage, blebbing of plasma membrane, and formation of apoptotic bodies containing nuclear or cytoplasmic material. There are several pathways identified to activate apoptosis, form extracellular signals through death receptors, to intracellular pathways activated by cytochrome c release from the mitochondria, via the MPT. *C. elegans* has played a crucial role in the initial characterization of programmed cell death in development and in neurons, through the discovery of *ced-3*, ortholog to caspases in vertebrates [47, 48]. Therefore it is not surprising that worms are being used to investigate cell death pathways involved in metal exposure.

Although metals have been shown to induce apoptosis in cell lines, mechanisms for metal-induced apoptosis *in vivo* are not well characterized. To elucidate mechanisms of Cd-induced apoptosis, Wang et al. exposed mutant lines to Cd (1, 10, 50, and 100 μM) for the DNA damage response checkpoint genes HUS1 (*hus-1*) and p53 (*cep-1*), and genes involved in the JNK or p38 signaling cascades, and apoptotic germ cells were stained by AO vital staining [49]. Mutants for hus-1 and cep-1 did not attenuate the Cd-induced apoptosis, however mutants for the JNK (*nsy-1*, *mek-1*, and *jnk-1*) and p38 (*sek-1*, *pmk-1*, and *pmk-3*) MAPK signaling pathways greatly reduced apoptotic cells in *C. elegans* [49].

Studies have shown that p38 MAPK can activate caspase-8 independently of activation of Fas/FADD, leading to loss of membrane potential and caspase-9 and caspase-3 activation [50]. JNK has been shown to phosphorylate 14-3-3 protein, which leads to dissociation of proapoptotic Bcl-2-family member Bax, allowing for translocation of Bax to mitochondria, cytochrome c release and apoptosis [51]. Apoptosis characterized by activation of JNK and p38 MAPK and independent of *hus-1* and *cep-1* is not unique to Cd exposure, as exposure to Cu, Co, or Ni have been characterized similarly [52–54], suggesting that metals activate apoptotic pathways through similar mechanisms.

Surprisingly, cell death has been observed in few *C. elegans* neurodegeneration models as a result of metal exposure. Loss of neurons has been observed following Pb and Hg exposure in worms expressing GFP-labeled GABAergic neurons (oxIs12 line) [38]. The GABAergic neurodegeneration observed also included dorsal and ventral cord gaps. Cell loss and neurodegeneration was greater in L1-3 stage worms after metal exposure than in L4 stage or young adult worms, suggesting a developmental susceptibility of worms to these metals. Deficits in synaptic function after exposure was also observed in worms exposed to Hg or Pb and Aldicarb or Levamisole, cholinesterase inhibitor and cholinergic receptor antagonist respectively [38]. Whether the loss of neurons was due to apoptosis or necrosis was not investigated.

Alzheimer's disease

Alzheimer's disease is the most common neurodegenerative disease and form of dementia worldwide, and the number of cases diagnosed is expected to quadruple in the next 40 years [55]. The disease is characterized by memory loss, confusion, irritability, language breakdown, and social withdrawal. AD is a progressive disease diagnosed in people over 65, suggesting a role for aging in its development. AD patients gradually lose their independence, dependent on care givers, often becoming bedridden and losing the ability to feed themselves. While AD is a terminal disease, death results from an external factor, such as pneumonia. The underlying pathophysiology of AD is complex and includes neurofibrillary tangles of β-amyloid and hyperphosphorylated, aggregated tau protein. Amyloid precursor protein (APP) is cleaved by β- and γ-secretases, resulting in deposition of dense, extracellular plaques of Aβ around neurons [56]. Tau is a microtubule-associated protein which stabilizes microtubules, providing support and acting as a transport system necessary for neuronal survival. It is normally phosphorylated, but in AD hyperphosphorylation of tau leads to disassembly of the support and transport system, forming tangles with the microtubules [57]. Additionally, metals are highly enriched in senile plaques, including Cu, Zn, Fe, and Al [58]. How the pathophysiology contributes to the disease phenotype is not entirely understood.

C. elegans models have been developed to investigate genetic influences on AD development. The majority of AD cases are sporadic, however early onset AD shows autosomal dominant familial inheritance. Familial AD comprises around 0.1 % of AD cases and is due to mutations in either APP, amyloid precursor-like proteins (APLP) 1 and 2, or presenilins 1 and 2, components of the γ-secretase protease complex [59]. *C. elegans* contains one ortholog to APP, *apl-1*, which encodes a single-pass transmembrane domain protein. APL-1 is also similar to APLP1 and APLP2 since neither proteins contain the Aβ peptide [60]. Deletion of *apl-1* is lethal, while knockdown of *apl-1* with RNAi reduces the worm's body size. Interestingly, overexpression of *apl-1* results in interference of motor neuron functions controlling swimming and crawling rates as compared to wild type worms [60]. Since the Aβ peptide is not present in worms, strains have been developed to express human Aβ under muscle-specific promoters, which results in aggregation of Aβ and paralysis [61, 62]. In worms overexpressing Aβ in muscle cells, synaptic transmission has been found to be defective due to mislocalization of nicotine-sensitive acetylcholine receptor ACR-16 [63]. Cholinergic neurons are thought to be most vulnerable in AD [64], suggesting that Aβ aggregation may result in cholinergic neuron dysfunction. Defects in cholinergic signaling have also been observed in animals with mutated *sel-12*, an ortholog to presenilins 1 and 2. When *sel-12* is mutated, there are deficits in thermotaxis dependent on the AFD sensory neurons and the AIZ and AIY interneurons [64]. In this model, AIY interneurons show axonal abnormalities, which would affect its cholinergic signaling. The *C. elegans* ortholog to MAPT, the gene that encodes the tau protein, is *ptl-1*. Mutations and deletions of *ptl-1* however does not model tau-related pathology. Transgenic worm lines have been created that express human wild type, mutant, and pseudohyperphosphorylated (PHP) tau. The PHP tau line has impaired motor neuron development associated with age-dependent uncoordinated movement phenotype [65]. Other tau lines have been made that model diseases other than AD, such as frontotemporal dementia with parkinsonism-17 (FTDP-17) [66].

Metals have been implicated in AD pathology. Senile plaques have been found to contain metals, with Cu levels elevated in rims of the plaques and Zn levels elevated in AD neuropil as compared to control [67]. Additionally, divalent metal transporter DMT1 co-localizes to Aβ plaques in post-mortem brains of AD patients [68]. DMT1 is an important transporter for various metals, including metals implicated in AD (Zn, Fe, and Cu), implicating the transporter in the development of the disease. It is unknown whether the elevated metals are a cause of AD or a consequence of the disease, but it has been found that Fe, Cu, and Zn can cause conformational changes in Aβ, stabilizing it in its toxic oligomeric form, as well as increase Aβ-induced oxidative stress [69].

AD *C. elegans* models have been important in investigating the relationship between metal exposure and AD-related neurotoxicity. Wan et al. have demonstrated that in worms that express human Aβ, there is increased Fe accumulation and oxidative stress [104], suggesting that Aβ plays a similar role in nematode metal homeostasis as in mammals. Whether Aβ and metal accumulation are beneficial or deleterious in AD is under debate [58]. In *C. elegans*, much work has been done on investigating the interplay between AD and Cu. APP and APP-like proteins have been shown to contain specific metal binding sites [70, 71], with histidines playing crucial roles in Cu toxicity [70]. This N-terminal binding domain not only binds Cu, but also strongly reduces it from Cu^{2+} to Cu^{+}, potentially allowing for the formation of damaging ROS [72]. There are Cu and Zn binding sites also found in APP's C-terminus that contains the Aβ peptide. Interestingly, overexpression of the Aβ C-terminal fragment that contains the metal binding domain results in significantly decreased Cu and Fe levels [71], suggesting Aβ and APP act as regulators of metal homeostasis. Aggregation of Aβ can potentially alter proper metal levels in AD. Surprisingly, White et al. have shown that worms are protected from Cu-mediated lipid peroxidation and neurotoxicity due to having tyrosine and lysine instead of histidines in the APL-1 Cu binding domain [73]. When the corresponding residues in human APP were mutated to tyrosine and lysine, Cu-mediated lipid peroxidation and toxicity were decreased in vitro. Furthermore, when the Cu binding domain of APL-1 was injected into rat dorsal hippocampus, it protected against Cu^{2+}-mediated toxicity [74]. In *C. elegans* strains overexpressing human Aβ in muscle, Cu^{2+} treatment results in enhanced amyloid aggregation [75]. Worms expressing Aβ have aggregates containing truncated $A\beta_{3-42}$, which has been shown to aggregate faster in vitro than full length $A\beta_{1-42}$ [76]. These transgenic worms expressing Aβ display a decreased sensitivity to $CuCl_2$ exposure compared to control worms [76], while Cu chelators histidine and clioquinol have been shown to diminish the formation of aggregates in worms expressing human Aβ [75]. Data also suggest that Cu accumulation may have beneficial effects in *C. elegans*. Cu^{2+} treatment has been shown to dampen Aβ-mediated synaptic dysfunction in *C. elegans* [63], and Luo et al. have found that cytoprotective genes, superoxide dismutase *sod-1*, peroxiredoxin prdx-2, *skn-1* (the worm ortholog to nuclear factor-2 erythroid 2-related factor-2, Nrf2), and heat shock proteins, *hsp-60* and *hsp-16*, are involved in the oxidative stress response to Cu in Aβ expressing worms [77]. These findings highlight the complex relationship between metal homeostasis and Aβ aggregation that require further attention.

Parkinson's disease

Parkinson's disease is the second most common neurodegenerative disease, and is characterized by the selective loss of dopaminergic (DAergic) cells in the substantia nigra pars compacta (SNpc) and locus ceruleus regions of the brain. Patients with PD show emotional and cognitive decline, bradykinesia, rigidity, tremors, and postural instability [78]. Symptoms start to show when 70–80 % of DAergic neurons are lost. In addition to cell loss, proteinaceous intracellular inclusions called Lewy bodies accumulate. There are also symptoms of PD not involving the SNpc, such as loss of sense of smell or gastrointestinal problems, however it is unknown how these features develop. PD affects 1–2 % of people over the age of 60, suggesting a role for aging in the etiology of the disease. The majority of PD cases are sporadic, while 10–20 % of cases have a genetic component. Familial PD may be due to autosomal dominant genes for α-synuclein and leucine-rich repeat kinase 2 (LRRK2) or autosomal recessive genes for parkin, DJ-1, and phosphatase and tensin homolog (PTEN)-induced novel kinase 1 (PINK1).

C. elegans are an attractive model organism for PD research. Rodent models of PD, either chemical-induced or genetic, are complex; however, there are only eight DAergic neurons in *C. elegans*, six anterior DAergic neurons (four cephalic—CEP and two anterior deirid—ADE neurons) and two posterior neurons (PDE). Dopamine controls specific, measurable behaviors in worms, including food searching behavior, defecation, and egg laying [79, 80]. All the necessary machinery for dopamine synthesis, storage, transport, and signaling are present in *C. elegans*. Many of the genes associated with familial PD have orthologs in worms, such as Parkin (*pdr-1*), DJ-1 (*djr-1*), LRRK2 (*lrk-1*), PINK1 (*pink-1*), but not α-synuclein. Proteinaceous inclusions that contain α-synuclein and ubiquitin are a feature of PD that has been observed in *C. elegans*. Knockdown of Parkin, an E3 ubiquitin ligase involved in proteasomal degradation of damaged proteins, ortholog pdr-1 in worms leads to reduced levels of high molecular weight ubiquitin-conjugated proteins [81, 82]. Overexpression of either wild type of mutant α-synuclein under the dopamine transporter (*dat-1*) promoter results in accumulation of α-synuclein in both cell bodies and neurites, and degeneration of DAergic dendrites [83]. Interestingly, if α-synuclein is expressed under a pan-neuronal promoter there is still selective DAergic neuronal and dendrite degeneration, but not if it is expressed under a motor neuron promoter, suggesting a specific sensitivity to DAergic neurons to α-synuclein toxicity [84]. Mammalian models of α-synuclein overexpression have not been successful in selectively degenerating DAergic neurons, which makes *C. elegans* a valuable tool. A number of mitochondrial inhibitors have been shown to cause PD-like symptoms in humans and rodents, suggesting a role for mitochondria in PD. Expression of wild type human α-synuclein in *C. elegans* increases vulnerability to mitochondrial complex-I inhibitors, such as rotenone, fenperoximate, pyridaben, and stigmatellin, showing an increase in mitochondrial fragmentation [82]. DAergic cell death has been reported in worms expressing α-synuclein and treated with MPP+ (1-methyl-4-phenyl-1,2,3,6-tetrahydropyridine metabolite) [85]. Knockdown of *pink-1* causes reduced mitochondrial cristae length in neuronal cells and affects axonal outgrowth of a pair of canal-associated neurons [86]. Mutants in *djr-1* have decreased oxygen consumption and survival after exposure to rotenone [82]. Genetic modulation of *pdr-1* and *lrk-1* also disrupts mitochondrial function in *C. elegans* [82, 86].

In addition to genetic components to PD, environmental exposures are thought to contribute to the disease development. Exposure to Mn, either from the environment or in occupational settings, can produce Parkinsonian-like symptoms. These symptoms include rigidity, tremor, gait disturbances and hypokinesia, and are attributed to selective interaction of Mn with the basal ganglia downstream of the nigrostriatal DAergic projection, which are areas that readily accumulate Mn [87]. Mn has been shown oxidize DA to produce highly toxic reactive metabolite leukoaminochrome o-semiquinone [88]. Benedetto et al. found that Mn toxicity in *C. elegans* relies on extracellular oxidation of DA. Mn causes a dose-dependent degeneration in DAergic neurons in *C. elegans* and that the neurodegeneration required the presence of the reuptake transporter, DAT-1, as neurodegeneration was not observed in dat-1 knockdown worms [21]. Worms lacking DAT-1 displayed higher susceptibility to Mn toxicity, whereas toxicity was prevented by the loss of tyrosine hydroxylase (TH)/CAT-2 function in the double knockout strain, *cat-2(e1112);dat-1(ok157)*. Knockdown of vesicular monoamine transporter (VMAT2)/CAT-1 in *cat-1(e1111)* mutants, in which DAergic neurons were unable to release DA at the synaptic cleft, resulted in increased tolerance to Mn [21]. The neurotoxic effects were specific to DAergic neurons, and absent from GABAergic, serotononergic or glutamatergic neurons [21]. As in mammals, Mn enters DAergic neurons in *C. elegans* through NRAMP/divalent metal transporters. In worms the NRAMP orthologs, SMF-1, SMF-2 and SMF-3 differ in location and functional roles [89]. Deletion of the *smf-1* gene has been shown to attenuate Mn-induced DAergic neurodegeneration [90]. *Smf-1* and *smf-3* loss-of-function mutations cause a significant decrease in Mn accumulation and increased survival [89].

The use of *C. elegans* has facilitated identification of genes that are involved in cellular response to Mn exposure. As metals induce oxidative stress, Benedetto et al. investigated whether SKN-1, a transcription factor that responds to oxidative stress, is involved in Mn exposure. Loss-of-function mutants for *skn-1* had increased sensitivity to Mn toxicity, whereas wild type worms expressing SKN-1::GFP showed activation of the transcription factor in the ASI neurons [21]. Generation of ROS by Mn occurs through Fenton chemistry, however since cells naturally produce ROS as defense mechanisms by specific enzymatic pathways, it has been investigated whether these enzymes are affected by Mn exposure. Loss of function of *bli-3* gene, a dual oxidase involved in di-tyrosine bond formation in the worm cuticle and pathogen-induced ROS production, caused increased resistance to Mn toxicity [21]. Furthermore, these mutants had no increase in ROS production from Mn exposures, implying that BLI-3 is required for ROS-mediated effects in Mn exposure [21]. It remains to be determined if dual oxidases are involved in mammalian responses to Mn and whether they would be a potential therapeutic target for Mn toxicity or PD.

Methylmercury (MeHg) exposure has also been implicated in PD. MeHg poisoning is characterized by severe neurological deficits due to brain lesions and disruptions of neurotransmitter systems [91], and both PD and MeHg poisoning present resting tremors and alterations in motor functioning [92, 93]. MeHg exposure occurs through seafood consumption, due to the global cycling and accumulation of Hg from industries that reaches the aquatic environment [94]. Several studies have shown an association between an increased prevalence to PD and fish consumption or occupational exposure to Hg [92, 95–98]. Furthermore, low serum Hg levels are associated with decreased risk for development of PD, and serum levels of Hg have been shown to increase post-PD diagnosis [99]. In *C. elegans*, MeHg causes several toxic effects, including decreased survival, developmental delay, and decreased pharyngeal pumping [100, 101]. MeHg exposure did not lead to neurodegeneration in *C. elegans*, suggesting the presence of protective mechanisms in neurons. Exposure to MeHg causes oxidative stress in *C. elegans*, as measured by the alterations in GSH levels and increased expression of heat shock proteins and glutathione-S-transferase (GST) [100]. This is not surprising as GST levels are controlled by Nrf2, which has been shown to be activated by MeHg in mammalian glial cells [102]. Knockout of Nrf2 in glial cells significantly reduces the upregulation of antioxidant genes, increasing MeHg toxicity [102]. In *C. elegans*, the reduction of *skn-1* gene expression led to DAergic neuron degeneration in 30 % of the worms exposed to 1 mM of MeHg, whereas no degeneration was observed in wild type animals [103]. Further studies are necessary to understand important metal-induced pathways that may be involved in the neurodegenerative process.

Conclusions

The ease of genetic manipulability of *C. elegans* and homology with mammalian systems offers a unique platform for studying neurodegenerative disease etiology and mechanisms. The *C. elegans* model has proved to be an invaluable tool in studying metal-induced neurodegeneration, apoptosis, and alterations in behavior. The short lifetime has proved useful in performing genetic screens for xenobiotics inducing neurodegeneration and the large RNAi libraries available have been utilized to find novel proteins and protein interactions involved in disease processes, providing a unique platform to explore gene–environment interactions. Although *C. elegans* cannot model all aspects of disease, such as inflammatory and vascular contributions, further research utilizing *C. elegans* will prove vital in defining roles for genetic and environmental contribution to neurodegenerative diseases.

Acknowledgments We are grateful for support by NIEHS R01ES07331, R01ES10563, the Center in Molecular Toxicology NIH grant P30ES00267, and the training program in Environmental Toxicology grant T32ES007028.

References

1. Langston JW (1998) Epidemiology versus genetics in Parkinson's disease: progress in resolving an age-old debate. Ann Neurol 44: S45–S52
2. Tyas SL, Manfreda J, Strain LA, Montgomery PR (2001) Risk factors for Alzheimer's disease: a population-based, longitudinal study in Manitoba, Canada. Int J Epidemiol 30:590–597
3. Kordas K (2010) Iron, lead, and children's behavior and cognition. Annu Rev Nutr 30:123–148
4. Koyashiki GA, Paoliello MM, Tchounwou PB (2010) Lead levels in human milk and children's health risk: a systematic review. Rev Environ Health 25:243–253
5. Neal AP, Guilarte TR (2010) Molecular neurobiology of lead (Pb (2+)): effects on synaptic function. Mol Neurobiol 42:151–160
6. Bertoni-Freddari C, Fattoretti P, Casoli T, Di Stefano G, Giorgetti B, Balietti M (2008) Brain aging: the zinc connection. Exp Gerontol 43:389–393
7. Tarohda T, Yamamoto M, Amamo R (2004) Regional distribution of manganese, iron, copper, and zinc in the rat brain during development. Anal Bioanal Chem 380:240–246
8. Butterfield DA, Reed T, Newman SF, Sultana R (2007) Roles of amyloid beta-peptide-associated oxidative stress and brain protein modifications in the pathogenesis of Alzheimer's disease and mild cognitive impairment. Free Radic Biol Med 43:658–677
9. Dexter DT, Jenner P, Schapira AH, Marsden CD (1992) Alterations in levels of iron, ferritin, and other trace metals in neurodegenerative diseases affecting the basal ganglia. The Royal Kings and Queens Parkinson's Disease Research Group. Ann Neurol 32(Suppl):S94–S100
10. Allsop D, Mayes J, Moore S, Masad A, Tabner BJ (2008) Metal-dependent generation of reactive oxygen species from amyloid proteins implicated in neurodegenerative disease. Biochem Soc Trans 36:1293–1298
11. Ricchelli F, Drago D, Filippi B, Tognon G, Zatta P (2005) Aluminum-triggered structural modifications and aggregation of beta-amyloids. Cell Mol Life Sci 62:1724–1733
12. Sulston JE, Horvitz HR (1977) Post-embryonic cell lineages of the nematode, *Caenorhabditis elegans*. Dev Biol 56:110–156
13. Kaletta T, Hengartner MO (2006) Finding function in novel targets: *C. elegans* as a model organism. Nat Rev Drug Discov 5:387–398
14. McDonald PW, Jessen T, Field JR, Blakely RD (2006) Dopamine signaling architecture in *Caenorhabditis elegans*. Cell Mol Neurobiol 26:593–618
15. Hall DH, Lints R, Altun Z (2006) Nematode neurons: anatomy and anatomical methods in *Caenorhabditis elegans*. Int Rev Neurobiol 69:1–35
16. Whittaker AJ, Sternberg PW (2009) Coordination of opposing sex-specific and core muscle groups regulates male tail posture during *Caenorhabditis elegans* male mating behavior. BMC Biol 7:33
17. Ward S, Thomson N, White JG, Brenner S (1975) Electron microscopical reconstruction of the anterior sensory anatomy of the nematode *Caenorhabditis elegans*.?2UU. J Comp Neurol 160:313–337
18. Ardiel EL, Rankin CH (2010) An elegant mind: learning and memory in *Caenorhabditis elegans*. Learn Mem 17:191–201
19. de Bono M, Maricq AV (2005) Neuronal substrates of complex behaviors in *C. elegans*. Annu Rev Neurosci 28:451–501
20. Sengupta P, Samuel AD (2009) *Caenorhabditis elegans*: a model system for systems neuroscience. Curr Opin Neurobiol 19:637–643
21. Benedetto A, Au C, Avila DS, Milatovic D, Aschner M (2010) Extracellular dopamine potentiates mn-induced oxidative stress, lifespan reduction, and dopaminergic neurodegeneration in a BLI-3-dependent manner in *Caenorhabditis elegans*. PLoS Genet 6
22. Teschendorf D, Link CD (2009) What have worm models told us about the mechanisms of neuronal dysfunction in human neurodegenerative diseases? Mol Neurodegener 4:38
23. Bofill R, Orihuela R, Romagosa M, Domenech J, Atrian S, Capdevila M (2009) *Caenorhabditis elegans* metallothionein isoform specificity–metal binding abilities and the role of histidine in CeMT1 and CeMT2. FEBS J 276:7040–7056
24. Swain SC, Keusekotten K, Baumeister R, Sturzenbaum SR (2004) *C. elegans* metallothioneins: new insights into the phenotypic effects of cadmium toxicosis. J Mol Biol 341:951–959
25. Bandyopadhyay J, Song HO, Park BJ, Singaravelu G, Sun JL, Ahnn J, Cho JH (2009) Functional assessment of Nramp-like metal transporters and manganese in *Caenorhabditis elegans*. Biochem Biophys Res Commun 390:136–141
26. Cui Y, McBride SJ, Boyd WA, Alper S, Freedman JH (2007) Toxicogenomic analysis of *Caenorhabditis elegans* reveals novel genes and pathways involved in the resistance to cadmium toxicity. Genome Biol 8:R122
27. Kim S, Selote DS, Vatamaniuk OK (2010) The N-terminal extension domain of the *C. elegans* half-molecule ABC transporter, HMT-1, is required for protein-protein interactions and function. PLoS One 5:e12938

28. Candido EP (2002) The small heat shock proteins of the nematode *Caenorhabditis elegans*: structure, regulation and biology. Prog Mol Subcell Biol 28:61–78
29. Lindblom TH, Dodd AK (2006) Xenobiotic detoxification in the nematode *Caenorhabditis elegans*. J Exp Zool A Comp Exp Biol 305:720–730
30. Vatamaniuk OK, Bucher EA, Ward JT, Rea PA (2001) A new pathway for heavy metal detoxification in animals. Phytochelatin synthase is required for cadmium tolerance in *Caenorhabditis elegans*. J Biol Chem 276:20817–20820
31. Kaplan JM, Horvitz HR (1993) A dual mechanosensory and chemosensory neuron in *Caenorhabditis elegans*. Proc Natl Acad Sci USA 90:2227–2231
32. Sambongi Y, Nagae T, Liu Y, Yoshimizu T, Takeda K, Wada Y, Futai M (1999) Sensing of cadmium and copper ions by externally exposed ADL, ASE, and ASH neurons elicits avoidance response in *Caenorhabditis elegans*. Neuroreport 10:753–757
33. Hilliard MA, Bargmann CI, Bazzicalupo P (2002) *C. elegans* responds to chemical repellents by integrating sensory inputs from the head and the tail. Curr Biol 12:730–734
34. Hilliard MA, Apicella AJ, Kerr R, Suzuki H, Bazzicalupo P, Schafer WR (2005) In vivo imaging of *C. elegans* ASH neurons: cellular response and adaptation to chemical repellents. EMBO J 24:63–72
35. Boyd WA, Cole RD, Anderson GL, Williams PL (2003) The effects of metals and food availability on the behavior of *Caenorhabditis elegans*. Environ Toxicol Chem 22:3049–3055
36. Xing X, Du M, Zhang Y, Wang D (2009) Adverse effects of metal exposure on chemotaxis towards water-soluble attractants regulated mainly by ASE sensory neuron in nematode *Caenorhabditis elegans*. J Environ Sci (China) 21:1684–1694
37. Johansson C, Castoldi AF, Onishchenko N, Manzo L, Vahter M, Ceccatelli S (2007) Neurobehavioural and molecular changes induced by methylmercury exposure during development. Neurotox Res 11:241–260
38. Xing X, Guo Y, Wang D (2009) Using the larvae nematode *Caenorhabditis elegans* to evaluate neurobehavioral toxicity to metallic salts. Ecotoxicol Environ Saf 72:1819–1823
39. Ye H, Ye B, Wang D (2008) Trace administration of vitamin E can retrieve and prevent UV-irradiation- and metal exposure-induced memory deficits in nematode *Caenorhabditis elegans*. Neurobiol Learn Mem 90:10–18
40. Zhang Y, Ye B, Wang D (2010) Effects of metal exposure on associative learning behavior in nematode *Caenorhabditis elegans*. Arch Environ Contam Toxicol 59:129–136
41. Wang D, Wang Y (2008) Nickel sulfate induces numerous defects in *Caenorhabditis elegans* that can also be transferred to progeny. Environ Pollut 151:585–592
42. Costa M, Sutherland JE, Peng W, Salnikow K, Broday L, Kluz T (2001) Molecular biology of nickel carcinogenesis. Mol Cell Biochem 222:205–211
43. Lyras L, Cairns NJ, Jenner A, Jenner P, Halliwell B (1997) An assessment of oxidative damage to proteins, lipids, and DNA in brain from patients with Alzheimer's disease. J Neurochem 68:2061–2069
44. Mecocci P, MacGarvey U, Kaufman AE, Koontz D, Shoffner JM, Wallace DC, Beal MF (1993) Oxidative damage to mitochondrial DNA shows marked age-dependent increases in human brain. Ann Neurol 34:609–616
45. Edinger AL, Thompson CB (2004) Death by design: apoptosis, necrosis and autophagy. Curr Opin Cell Biol 16:663–669
46. Lopez E, Arce C, Oset-Gasque MJ, Canadas S, Gonzalez MP (2006) Cadmium induces reactive oxygen species generation and lipid peroxidation in cortical neurons in culture. Free Radic Biol Med 40:940–951
47. Horvitz HR, Sternberg PW, Greenwald IS, Fixsen W, Ellis HM (1983) Mutations that affect neural cell lineages and cell fates during the development of the nematode *Caenorhabditis elegans*. Cold Spring Harb Symp Quant Biol 48(Pt 2):453–463
48. Yuan J, Shaham S, Ledoux S, Ellis HM, Horvitz HR (1993) The *C. elegans* cell death gene ced-3 encodes a protein similar to mammalian interleukin-1 beta-converting enzyme. Cell 75:641–652
49. Wang S, Tang M, Pei B, Xiao X, Wang J, Hang H, Wu L (2008) Cadmium-induced germline apoptosis in *Caenorhabditis elegans*: the roles of HUS1, p53, and MAPK signaling pathways. Toxicol Sci 102:345–351
50. Schrantz N, Bourgeade MF, Mouhamad S, Leca G, Sharma S, Vazquez A (2001) p38-mediated regulation of an Fas-associated death domain protein-independent pathway leading to caspase-8 activation during TGFbeta-induced apoptosis in human Burkitt lymphoma B cells BL41. Mol Biol Cell 12:3139–3151
51. Tsuruta F, Sunayama J, Mori Y, Hattori S, Shimizu S, Tsujimoto Y, Yoshioka K, Masuyama N, Gotoh Y (2004) JNK promotes Bax translocation to mitochondria through phosphorylation of 14-3-3 proteins. EMBO J 23:1889–1899
52. Cai K, Ren C, Yu Z (2010) Nickel-induced apoptosis and relevant signal transduction pathways in *Caenorhabditis elegans*. Toxicol Ind Health 26:249–256
53. Chong R, Ke-zhou C, Zeng-liang Y (2009) Induction of germline apoptosis by cobalt and relevant signal transduction pathways in *Caenorhabditis elegans*. Toxicol Mech Methods 19:541–546
54. Wang S, Wu L, Wang Y, Luo X, Lu Y (2009) Copper-induced germline apoptosis in *Caenorhabditis elegans*: the independent roles of DNA damage response signaling and the dependent roles of MAPK cascades. Chem Biol Interact 180:151–157
55. Holtzman DM, Morris JC, Goate AM (2011) Alzheimer's disease: the challenge of the second century. Sci Transl Med 3:77sr71
56. Crews L, Masliah E (2010) Molecular mechanisms of neurodegeneration in Alzheimer's disease. Hum Mol Genet 19: R12–R20
57. Li B, Chohan MO, Grundke-Iqbal I, Iqbal K (2007) Disruption of microtubule network by Alzheimer abnormally hyperphosphorylated tau. Acta Neuropathol 113:501–511
58. Bolognin S, Messori L, Zatta P (2009) Metal ion physiopathology in neurodegenerative disorders. Neuromolecular Med 11:223–238
59. Campion D, Dumanchin C, Hannequin D, Dubois B, Belliard S, Puel M, Thomas-Anterion C, Michon A, Martin C, Charbonnier F et al (1999) Early-onset autosomal dominant Alzheimer disease: prevalence, genetic heterogeneity, and mutation spectrum. Am J Hum Genet 65:664–670
60. Hornsten A, Lieberthal J, Fadia S, Malins R, Ha L, Xu X, Daigle I, Markowitz M, O'Connor G, Plasterk R et al (2007) APL-1, a Caenorhabditis elegans protein related to the human beta-amyloid precursor protein, is essential for viability. Proc Natl Acad Sci USA 104:1971–1976
61. Link CD (1995) Expression of human beta-amyloid peptide in transgenic *Caenorhabditis elegans*. Proc Natl Acad Sci USA 92:9368–9372
62. Wu Y, Wu Z, Butko P, Christen Y, Lambert MP, Klein WL, Link CD, Luo Y (2006) Amyloid-beta-induced pathological behaviors are suppressed by *Ginkgo biloba* extract EGb 761 and ginkgolides in transgenic *Caenorhabditis elegans*. J Neurosci 26: 13102–13113
63. Rebolledo DL, Aldunate R, Kohn R, Neira I, Minniti AN, Inestrosa NC (2011) Copper reduces A{beta} oligomeric species and ameliorates neuromuscular synaptic defects in a *C. elegans* model of inclusion body myositis. J Neurosci 31:10149–10158
64. Wittenburg N, Eimer S, Lakowski B, Rohrig S, Rudolph C, Baumeister R (2000) Presenilin is required for proper morphology and function of neurons in *C. elegans*. Nature 406:306–309

65. Brandt R, Gergou A, Wacker I, Fath T, Hutter H (2009) A *Caenorhabditis elegans* model of tau hyperphosphorylation: induction of developmental defects by transgenic overexpression of Alzheimer's disease-like modified tau. Neurobiol Aging 30:22–33
66. Kraemer BC, Zhang B, Leverenz JB, Thomas JH, Trojanowski JQ, Schellenberg GD (2003) Neurodegeneration and defective neurotransmission in a *Caenorhabditis elegans* model of tauopathy. Proc Natl Acad Sci USA 100:9980–9985
67. Lovell MA, Robertson JD, Teesdale WJ, Campbell JL, Markesbery WR (1998) Copper, iron and zinc in Alzheimer's disease senile plaques. J Neurol Sci 158:47–52
68. Zheng W, Xin N, Chi ZH, Zhao BL, Zhang J, Li JY, Wang ZY (2009) Divalent metal transporter 1 is involved in amyloid precursor protein processing and Abeta generation. FASEB J 23:4207–4217
69. Faller P (2009) Copper and zinc binding to amyloid-beta: coordination, dynamics, aggregation, reactivity and metal-ion transfer. ChemBioChem 10:2837–2845
70. Hesse L, Beher D, Masters CL, Multhaup G (1994) The beta A4 amyloid precursor protein binding to copper. FEBS Lett 349:109–116
71. Maynard CJ, Cappai R, Volitakis I, Cherny RA, White AR, Beyreuther K, Masters CL, Bush AI, Li QX (2002) Overexpression of Alzheimer's disease amyloid-beta opposes the age-dependent elevations of brain copper and iron. J Biol Chem 277:44670–44676
72. Multhaup G, Schlicksupp A, Hesse L, Beher D, Ruppert T, Masters CL, Beyreuther K (1996) The amyloid precursor protein of Alzheimer's disease in the reduction of copper(II) to copper(I). Science 271:1406–1409
73. White AR, Multhaup G, Galatis D, McKinstry WJ, Parker MW, Pipkorn R, Beyreuther K, Masters CL, Cappai R (2002) Contrasting, species-dependent modulation of copper-mediated neurotoxicity by the Alzheimer's disease amyloid precursor protein. J Neurosci 22:365–376
74. Cerpa WF, Barria MI, Chacon MA, Suazo M, Gonzalez M, Opazo C, Bush AI, Inestrosa NC (2004) The N-terminal copper-binding domain of the amyloid precursor protein protects against Cu^{2+} neurotoxicity in vivo. FASEB J 18:1701–1703
75. Minniti AN, Rebolledo DL, Grez PM, Fadic R, Aldunate R, Volitakis I, Cherny RA, Opazo C, Masters C, Bush AI et al (2009) Intracellular amyloid formation in muscle cells of Abeta-transgenic *Caenorhabditis elegans*: determinants and physiological role in copper detoxification. Mol Neurodegener 4:2
76. McColl G, Roberts BR, Gunn AP, Perez KA, Tew DJ, Masters CL, Barnham KJ, Cherny RA, Bush AI (2009) The *Caenorhabditis elegans* A beta 1-42 model of Alzheimer disease predominantly expresses A beta 3-42. J Biol Chem 284:22697–22702
77. Luo Y, Zhang J, Liu N, Zhao B (2011) Copper ions influence the toxicity of beta-amyloid(1-42) in a concentration-dependent manner in a *Caenorhabditis elegans* model of Alzheimer's disease. Sci China Life Sci 54:527–534
78. Lees AJ, Hardy J, Revesz T (2009) Parkinson's disease. Lancet 373:2055–2066
79. Sulston J, Dew M, Brenner S (1975) Dopaminergic neurons in the nematode *Caenorhabditis elegans*. J Comp Neurol 163:215–226
80. Weinshenker D, Garriga G, Thomas JH (1995) Genetic and pharmacological analysis of neurotransmitters controlling egg laying in *C. elegans*. J Neurosci 15:6975–6985
81. Mata IF, Lockhart PJ, Farrer MJ (2004) Parkin genetics: one model for Parkinson's disease. Hum Mol Genet 13 Spec No 1: R127–R133
82. Ved R, Saha S, Westlund B, Perier C, Burnam L, Sluder A, Hoener M, Rodrigues CM, Alfonso A, Steer C et al (2005) Similar patterns of mitochondrial vulnerability and rescue induced by genetic modification of alpha-synuclein, parkin, and DJ-1 in *Caenorhabditis elegans*. J Biol Chem 280:42655–42668
83. Kuwahara T, Koyama A, Gengyo-Ando K, Masuda M, Kowa H, Tsunoda M, Mitani S, Iwatsubo T (2006) Familial Parkinson mutant alpha-synuclein causes dopamine neuron dysfunction in transgenic *Caenorhabditis elegans*. J Biol Chem 281:334–340
84. Lakso M, Vartiainen S, Moilanen AM, Sirvio J, Thomas JH, Nass R, Blakely RD, Wong G (2003) Dopaminergic neuronal loss and motor deficits in *Caenorhabditis elegans* overexpressing human alpha-synuclein. J Neurochem 86:165–172
85. Wang YM, Pu P, Le WD (2007) ATP depletion is the major cause of MPP+ induced dopamine neuronal death and worm lethality in alpha-synuclein transgenic *C. elegans*. Neurosci Bull 23:329–335
86. Samann J, Hegermann J, von Gromoff E, Eimer S, Baumeister R, Schmidt E (2009) *Caenorhabditits elegans* LRK-1 and PINK-1 act antagonistically in stress response and neurite outgrowth. J Biol Chem 284:16482–16491
87. Aschner M, Guilarte TR, Schneider JS, Zheng W (2007) Manganese: recent advances in understanding its transport and neurotoxicity. Toxicol Appl Pharmacol 221:131–147
88. Graumann R, Paris I, Martinez-Alvarado P, Rumanque P, Perez-Pastene C, Cardenas SP, Marin P, Diaz-Grez F, Caviedes R, Caviedes P et al (2002) Oxidation of dopamine to aminochrome as a mechanism for neurodegeneration of dopaminergic systems in Parkinson's disease. Possible neuroprotective role of DT-diaphorase. Pol J Pharmacol 54:573–579
89. Au C, Benedetto A, Anderson J, Labrousse A, Erikson K, Ewbank JJ, Aschner M (2009) SMF-1, SMF-2 and SMF-3 DMT1 orthologues regulate and are regulated differentially by manganese levels in *C. elegans*. PLoS One 4:e7792
90. Settivari R, Levora J, Nass R (2009) The divalent metal transporter homologues SMF-1/2 mediate dopamine neuron sensitivity in *Caenorhabditis elegans* models of manganism and parkinson disease. J Biol Chem 284:35758–35768
91. Aschner M, Syversen T (2005) Methylmercury: recent advances in the understanding of its neurotoxicity. Ther Drug Monit 27:278–283
92. Biernat H, Ellias SA, Wermuth L, Cleary D, de Oliveira Santos EC, Jorgensen PJ, Feldman RG, Grandjean P (1999) Tremor frequency patterns in mercury vapor exposure, compared with early Parkinson's disease and essential tremor. Neurotoxicology 20:945–952
93. Kaur P, Aschner M, Syversen T (2007) Role of glutathione in determining the differential sensitivity between the cortical and cerebellar regions towards mercury-induced oxidative stress. Toxicology 230:164–177
94. Fitzgerald WF, Clarkson TW (1991) Mercury and monomethylmercury: present and future concerns. Environ Health Perspect 96:159–166
95. Fabrizio E, Vanacore N, Valente M, Rubino A, Meco G (2007) High prevalence of extrapyramidal signs and symptoms in a group of Italian dental technicians. BMC Neurol 7:24
96. Kirkey KL, Johnson CC, Rybicki BA, Peterson EL, Kortsha GX, Gorell JM (2001) Occupational categories at risk for Parkinson's disease. Am J Ind Med 39:564–571
97. Petersen MS, Halling J, Bech S, Wermuth L, Weihe P, Nielsen F, Jorgensen PJ, Budtz-Jorgensen E, Grandjean P (2008) Impact of dietary exposure to food contaminants on the risk of Parkinson's disease. Neurotoxicology 29:584–590
98. Petersen MS, Weihe P, Choi A, Grandjean P (2008) Increased prenatal exposure to methylmercury does not affect the risk of Parkinson's disease. Neurotoxicology 29:591–595
99. Gellein K, Syversen T, Steinnes E, Nilsen TI, Dahl OP, Mitrovic S, Duraj D, Flaten TP (2008) Trace elements in serum from patients with Parkinson's disease–a prospective case-control study: the Nord-Trondelag Health Study (HUNT). Brain Res 1219:111–115
100. Helmcke KJ, Aschner M (2010) Hormetic effect of methylmercury on *Caenorhabditis elegans*. Toxicol Appl Pharmacol 248:156–164

101. Helmcke KJ, Syversen T, Miller DM 3rd, Aschner M (2009) Characterization of the effects of methylmercury on *Caenorhabditis elegans*. Toxicol Appl Pharmacol 240:265–272
102. Ni M, Li X, Yin Z, Jiang H, Sidoryk-Wegrzynowicz M, Milatovic D, Cai J, Aschner M (2010) Methylmercury induces acute oxidative stress, altering Nrf2 protein level in primary microglial cells. Toxicol Sci 116:590–603
103. Vanduyn N, Settivari R, Wong G, Nass R (2010) SKN-1/Nrf2 inhibits dopamine neuron degeneration in a *Caenorhabditis elegans* model of methylmercury toxicity. Toxicol Sci 118:613–624
104. Wan L, Nie G, Zhang J, Luo Y, Zhang P, Zhang Z, Zhao B (2011) beta-Amyloid peptide increases iron content and oxidative stress in human all and Caenorhabditis elegans models of Alzheimer's disease. Free Radic. Biol. Med. 50, 122–129

The effect of point mutations on copper(II) complexes with peptide fragments encompassing the 106–114 region of human prion protein

Csilla Kállay, Ildikó Turi, Sarolta Timári, Zoltán Nagy, Daniele Sanna, Giuseppe Pappalardo, Paolo de Bona, Enrico Rizzarelli, and Imre Sóvágó

Abstract

The tetrapeptides Ac-SKHM-NH_2, Ac-TKHM-NH_2, Ac-MKHS-NH_2, Ac-S(OMe)KHM-NH_2, and Ac-MKHS(OMe)-NH_2 and the nonapeptides Ac-KTNSKHMAG-NH_2 and Ac-KTNMKHSAG-NH_2 were synthesized and their copper(II) complexes were studied by potentiometric, UV–Vis, circular dichroism (CD), and electron paramagnetic resonance (EPR) spectroscopic methods. These peptides mimic the 109–112 and 106–114 residues of the sequence of human prion protein. The imidazole-N donor atoms of histidyl residues were found to be the primary metal binding sites of all peptide fragments. This binding mode provides a good possibility for the cooperative deprotonation and metal ion coordination of two amide functions preceding histidine. The (N_{im},N^-,N^-)-bonded species predominate in the pH range 5.5–7.0 and the free coordination sites of these species make possible the metal binding of weakly coordinating side chains. The comparison of the potentiometric and spectroscopic results revealed the stabilizing role of the oxygen donors of seryl, threonyl, or methoxyseryl residues of Ac-SKHM-NH_2, Ac-TKHM-NH_2, Ac-S(OMe)KHM-NH_2, and Ac-KTNSKHMAG-NH_2 containing the mutations in position 109. These interactions were, however, not observed in the peptides containing the specific amino acids in other locations of the peptide sequence.

Keywords

Bioinorganic chemistry • Metal complexes • Peptides • Prion proteins

Introduction

Peptides are versatile and effective ligands and their coordination chemistry has been thoroughly reviewed [1–4]. It is clear from these compilations that the anchoring role of the terminal amino and/or histidyl residues is the most common characteristic of the complex formation processes of small peptides. The proteins responsible for the development of various forms of neurodegenerative disorders are generally rich in histidyl residues and this fact gave a big impetus to studies on the complex-forming ability of the peptide fragments of these proteins [5, 6]. The side chain imidazole nitrogen atoms proved to be the primary ligating sites of the terminally protected peptides containing internal histidyl residues. We reported the copper(II) binding affinity of the peptide fragments of human prion protein (HuPrP) containing histidyl residues outside the octarepeat domain (His96, His111) [7–11]. The predominance of 3N complexes with (N_{im},N^-,N^-) coordination mode was observed in the pH range 6–8 with all peptides studied. In the case of the peptide fragments containing His111, the -MKHM- sequence of the ligands made possible a weak interaction of the thioether function of Met109 residues in a narrow pH range. Slight increase of pH, however, resulted in the deprotonation and

I. Sóvágó (✉)
Department of Inorganic and Analytical Chemistry, University of Debrecen, Debrecen 4010, Hungary
e-mail: sovago@delfin.unideb.hu

W. Linert and H. Kozlowski (eds.), *Metal Ions in Neurological Systems*,
DOI 10.1007/978-3-7091-1001-0_16,

metal ion coordination of the third amide nitrogen and 4N-coordinated complexes were formed with all peptides.

Thioether sulfur atoms of methionine are generally the primary metal binding sites for soft metal ions, e.g., platinum(II) [12] or palladium(II) [13], but their interaction is very weak or negligible with the 3d transition elements [14]. However, both potentiometric and spectroscopic data reported for the tetrapeptide fragments of human and chicken prion protein, HuPrP109–112 (Ac-MKHM-NH_2) and ChPrP122–125 (Ac-FKHV-NH_2), respectively, provided an indication for the existence of a weak Cu–S (thioether) bond in the 3N complexes of Ac-MKHM-NH_2. Similar data were obtained for the corresponding nonapeptides, HuPrP106–114 (Ac-KTNMKHMAG-NH_2) and ChPrP119–127 (Ac-KTNFKHVAG-NH_2), supporting the assumption that the simultaneous metal binding of Met109 and His111 sites (or the XaaMetXaaHisXaa sequences) provide a specific arrangement for metal ion coordination (where Xaa is any amino acid without coordinating side chain).

These results suggest that the presence of other amino acids with weakly coordinating side chains in the same position as Met109 may also enhance the stability or alter the structure of peptide complexes. Alcoholic –OH groups of seryl and threonyl residues are also considered as low affinity metal binding sites [15]. In the case of dipeptides with C-terminal seryl residues, a slight stability enhancement of complexes was only observed and explained by the indirect effect of the –OH group via a bridging water molecule [16, 17]. The stabilizing role of N-terminal seryl residues was reported to be more effective [18] than those of the C-terminal ones and on the basis of electron paramagnetic resonance (EPR) measurements even the metal ion promoted deprotonation and coordination of alcoholic –OH groups were suggested to occur in bis(ligand) and dinuclear complexes above pH 8–9 [19]. The highest degree of stability enhancement from –OH groups was obtained for peptide derivatives containing the α-hydroxymethyl-L-seryl (HmS) residues. In this case, the extra stabilization was observed in the copper(II), nickel(II), and zinc(II) complexes too, and especially effective ligands were obtained if both HmS and His residues were present in the peptide sequence [20–23].

In this paper we report the synthesis and studies of the copper complexes of the serine mutants of human prion peptide fragments, including the nonapeptides Ac-KTNSKHMAG-NH_2 and KTNMKHSAG-NH_2 and the corresponding tetrapeptides Ac-SKHM-NH_2 and Ac-MKHS-NH_2. For the unambiguous clarification of the role of alcoholic –OH groups the methoxyserine (S-OMe) and threonine derivatives of the tetrapeptides, Ac-S(OMe)KHM-NH_2, Ac-MKHS(OMe)-NH_2, and Ac-TKHM-NH_2, were also synthesized and studied.

Results and discussion

All tetrapeptides studied in this work have two protonation sites and the corresponding p*K* values are included in Table 1. The protonation reactions of the imidazole-N atoms of histidyl and the ε-amino groups of lysyl residues are well separated and take place in the pH ranges 5.5–7.0 and 9.5–11.0, respectively. It is also clear from Table 1 that the differences in the p*K* values of the various tetrapeptides are very small, suggesting that the replacement of one amino acid with another without coordinating side chain does not significantly affect the acid–base properties of the peptides. In the case of nonapeptides two lysyl residues are present in the sequence, but their protonations occur under similar conditions, as is reported for the tetrapeptides. Both lysyl residues are, however, present in their protonated forms in most of the copper(II) complexes, resulting in different stoichiometries and charges of the same coordination modes formed with tetra- and nonapeptides. As a consequence, the equilibrium data obtained for the nonapeptides and their complexes are included in a separate table (Table 2).

The equilibrium and structural characterization of the copper(II) complexes of the tetra- and nonapeptide fragments of human and chicken prion proteins have been published by us elsewhere [7, 8]. It is clear from Tables 1 and 2 that the same species were formed with the mutants containing seryl or methoxyseryl residues instead of Met109 or Met112. The same effect was obtained for the peptide Ac-TKHM-NH_2 containing a threonyl residue instead of Met109. Even the pH-dependent metal ion speciation of the systems is very similar to those reported for the native peptide fragments [7, 8]. This is further supported by Fig. 1, where the metal ion speciation of the copper(II) complexes of Ac-SKHM-NH_2 and Ac-MKHS-NH_2 is compared.

Three major binding modes of the ligands can be identified in all systems and defined in terms of the number of coordinated nitrogen donor atoms. The $[CuHL]^{3+}$ complexes of tetrapeptides and $[CuH_2L]^{4+}$ complexes of nonapeptides are formed in rather low concentrations in all cases and their stoichiometries can be best explained by the monodentate binding of side chain imidazole-N donor atoms. The EPR spectroscopic parameters (see Table 3) and the lack of circular dichroism (CD) activity of these complexes are in good agreement with the Cu–N_{im} coordination mode. The log *K* values in the last rows of Tables 1 and 2 provide further support for this assumption.

In the case of tetrapeptides, the species $[CuH_{-1}L]^+$ predominates in the pH range 5.0–7.0 and it is formed in a cooperative process when the loss of two amide protons results in the formation of a 3N (N_{im},N^-,N^-)-coordinated complex. The species $[CuL]^{2+}$ with (N_{im},N^-) binding sites

Table 1 p*K* values and stability constants of the copper(II) complexes of the terminally protected tetrapeptides (Ac-$[X]_4$-NH_2)

Ligand $[X]_4$	S(OMe)KHM	SKHM	TKHM	MKHS(OMe)	MKHS	MKHM (HuPrP109–112) [7]	FKHV (ChPrP122–125) [7]
p*K*(Im)	6.27(1)	6.23(1)	6.18(2)	6.30(1)	6.29(1)	6.22	6.29
p*K*(Lys)	10.31(1)	10.33(2)	10.25(2)	10.33(1)	10.29(2)	10.28	10.28
$[CuHL]^{3+}$	13.70(2)	13.52(2)	14.29(4)	13.96(7)	13.84(3)	13.98	13.88
$[CuH_{-1}L]^{+}$	2.29(1)	3.34(2)	3.65(2)	2.53(4)	2.34(1)	2.70	2.02
$[CuH_{-2}L]$	–6.37(1)	–4.41(3)	–4.65(2)	–6.58(6)	–6.73(2)	–6.26	–6.50
$[CuH_{-3}L]^{-}$	–16.87(1)	–15.19(4)	–15.32(3)	–16.73(7)	–16.89(2)	–16.32	–16.69
p*K*(12)	5.71	5.09	5.32	5.72	5.75	5.64	5.93
p*K*(3)	8.66	7.75	8.30	9.11	9.07	8.96	8.52
p*K*(4)	10.50	10.78	10.67	10.15	10.16	10.06	10.19
log *K*(Cu + HL)	3.39	3.19	4.04	3.63	3.55	3.70	3.60

$I = 0.2$ mol dm^{-3} KCl, $T = 298$ K, standard deviations are in parentheses

Table 2 p*K* values and stability constants of the copper(II) complexes of the terminally protected nonapeptides (Ac-$[X]_9$-NH_2)

Ligand $[X]_9$	KTNSKHMAG (M109S)	KTNMKHSAG (M112S)	KTNMKHMAG HuPrP(106–114) [8]	KTNFKHVAG ChPrP(119–127) [8]
p*K*(Im)	6.07(1)	6.18(1)	6.24	6.22
p*K*(Lys1)	9.83(2)	9.88(2)	9.91	9.89
p*K*(Lys2)	10.69(2)	10.62(2)	10.54	10.62
$[CuH_2L]^{4+}$	23.75(4)	23.57(3)	23.54	24.19
$[CuL]^{2+}$	13.58(2)	12.27(1)	12.39	12.39
$[CuH_{-1}L]^{+}$	6.10(2)	4.73(2)	4.56	5.34
$[CuH_{-2}L]$	–3.91(2)	–5.38(3)	–5.45	–4.65
$[CuH_{-3}L]^{-}$	–14.60(3)	–15.88(4)	–16.11	–15.25
p*K*(12)	5.09	5.65	5.58	5.90
p*K*(3)	7.48	7.54	7.83	7.05
p*K*(4)	10.01	10.11	10.01	9.99
p*K*(5)	10.69	10.50	10.66	10.60
log *K* (Cu + H_2L)	3.23	3.07	3.09	3.68

$I = 0.2$ mol dm^{-3} KCl, $T = 298$ K, standard deviations are in parentheses

can also be present in very low concentrations, but its formation overlaps very much with those of $[CuHL]^{3+}$ and $[CuH_{-1}L]^{+}$ and neither the equilibrium nor the spectroscopic parameters can be unambiguously determined for $[CuL]^{2+}$ complexes of tetrapeptides. According to a literature survey the cooperative deprotonation of the first two amide nitrogens seems to be a common feature of peptides containing histidyl residues in internal positions [4].

UV–Vis and CD spectra of copper(II) complexes have been recorded at many different pH values and the use of the PSEQUAD program made it possible to calculate the spectra of all individual species. These data are collected in Tables 4 and 5.

Figure 2 is used to compare the CD spectra of the species $[CuH_{-1}L]^{+}$, $[CuH_{-2}L]$, and $[CuH_{-3}L]^{-}$ of the Ser-mutated peptides Ac-SKHM-NH_2 (a) and Ac-MKHS-NH_2 (b), whereas Fig. 3 shows the same spectra of their methoxyserine counterparts Ac-S(OMe)KHM-NH_2 (a) and Ac-MKHS(OMe)-NH_2 (b).

Several important conclusions can be drawn from the comparison of the four sets of CD spectra shown in Figs. 2 and 3. First, it is evident that the CD spectra of Ac-SKHM-NH_2 are completely different from those of the other three ligands. On the other hand, the parameters of the absorption and CD spectra of the copper(II)–Ac-SKHM-NH_2 and copper(II)–Ac-TKHM-NH_2 systems are almost the same (Table 4). Moreover, the CD spectra of the 4N complexes (species $[CuH_{-3}L]^{-}$) of Ac-MKHS-NH_2 and Ac-MKHS(OMe)-NH_2 are also very similar to each other. A comparison with the literature data on similar peptides reveals that the last two spectra correspond well to those of any other histidine peptide with the same (N_{im},N^-,N^-,N^-) coordination mode in the form of (6,5,5)-membered chelate rings [7–11, 24]. The unusual CD spectra of the copper(II)–Ac-SKHM-NH_2 and copper(II)–Ac-TKHM-NH_2 systems suggest the existence of a different binding mode in their copper(II) complexes. The residue Met109 is replaced by serine or threonine in these peptides and this can provide a good chance for a weak

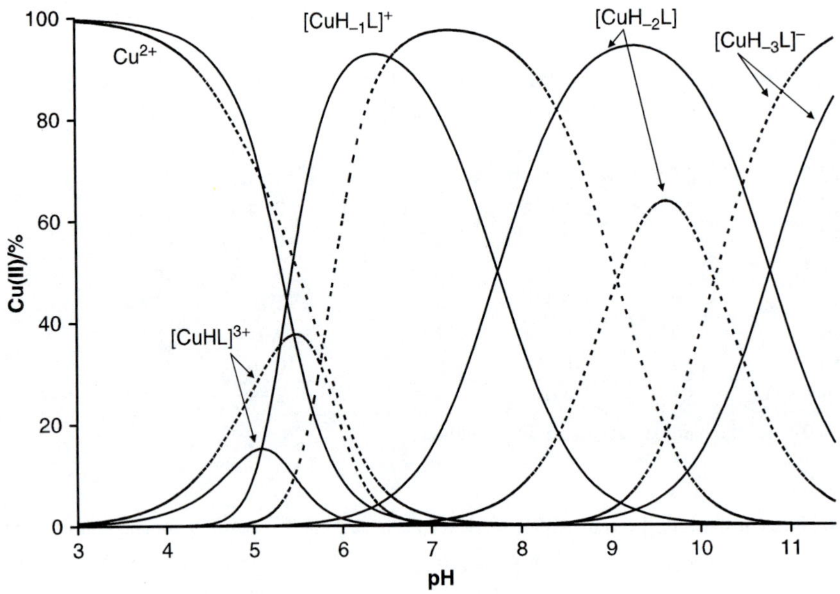

Fig. 1 Species distribution of the complexes formed in the copper(II)–Ac-SKHM-NH_2 (*solid line*) and copper(II)–Ac-MKHS-NH_2 system (*dotted line*) ($c_{Cu(II)} = c_L = 3 \times 10^{-3}$ mol dm^{-3}) as a function of pH

Table 3 EPR spectroscopic parameters of the tetra- and nonapeptide fragments of prion protein

Ligand	Species	$|A_{\parallel}|$ (10^{-4} cm^{-1})	$g_{\parallel}$
KTNFKHVAG [8]	1N	135	2.366
	3N	169	2.231
	4N	194	2.201
KTNMKHMAG [8]	1N	133	2.366
	3N	174	2.220
	4N	198	2.198
KTNSKHMAG (M109S)	1N	141	2.364
	3N	193	2.220
	4N	204	2.204
KTNMKHSAG (M112S)	1N	141	2.363
	3N	167	2.228
	4N	199	2.196
S(OMe)KHM	1N	–	–
	3N	175	2.229
	4N	197	2.198
MKHS(OMe)	1N	–	–
	3N	169	2.223
	4N	195	2.190
TKHM	1N	141	2.368
	3N	188	2.227
	4N	199	2.218

binding of the protonated alcoholic –OH group. In accordance with this expectation, slightly enhanced stability constants can be calculated for the $[CuH_{-1}L]^+$ species of Ac-SKHM-NH_2 and Ac-TKHM-NH_2. The increase of stability is also reflected in the low values of the deprotonation of the first two amide functions (see pK(12) values in Table 1) and in the slightly reduced pH range of complexation as shown by Fig. 1.

Both potentiometric and spectroscopic data support the notion that the deprotonation reactions of the Met109Ser and Met109Thr mutated peptides are different from all the others. In the case of Ac-SKHM-NH_2 and Ac-TKHM-NH_2 the species $[CuH_{-2}L]$ and $[CuH_{-3}L]^-$ have the same CD and absorption spectra, whereas that of $[CuH_{-1}L]^+$ is different from any of the other tetrapeptides (see Table 4). The similarities in the spectra of $[CuH_{-2}L]$ and $[CuH_{-3}L]^-$ suggest that the deprotonations of the non-coordinated lysyl side chains are the major processes during the transformation of $[CuH_{-2}L]$ to $[CuH_{-3}L]^-$. As a consequence, the low pK(3) value and the different CD spectra of $[CuH_{-1}L]^+$ can be best explained by the deprotonation and metal ion coordination of the third amide functions of the peptides in the pH range of 7.0–8.0. On the contrary, both Fig. 1 and spectroscopic data suggest the overlap of amide and lysyl deprotonation reactions in the copper(II) complexes of the other tetrapeptides. The low pK values of amide deprotonation reactions of Ac-SKHM-NH_2 and Ac-TKHM-NH_2 strongly suggest that the weak axial interaction of the alcoholic –OH groups of serine and threonine facilitates the metal binding of the third amide nitrogen. Alternately, the conformational change caused by a hydrogen bond between the Ser/Thr–OH and Lys–NH_2 groups may also be responsible for the promotion of amide binding.

Table 4 UV–Vis and circular dichroism spectral parameters of the 3N- and 4N-coordinated copper(II) complexes of tetrapeptides

Ligand $[X]_4$	Binding sites	S(OMe)KHM	SKHM	TKHM	MKHS(OMe)	MKHS	MKHM HuPrP(109–112) [7]	FKHV ChPrP(122–125) [7]
UV–Vis, λ_{max} (nm)/ε (dm^3 mol^{-1} cm^{-1})	N_{im},N^-,N^-	610/89	592/68	591/57	630/117	628/94	629/123	606/98
	N_{im},N^-,N^-,N^-	520/104	554/80	551/83	525/112	522/92	521/127	525/142
CD, λ_{max} (nm)/$\Delta\varepsilon$ (dm^3 mol^{-1} cm^{-1})	N_{im},N^-,N^-	635/–0.28 525/+0.29	605/–0.70	592/–0.66	665/+0.13 535/+0.36	655/+0.12 537/+0.40	660/+0.15 530/+0.33	605/–0.35 510/+0.40
	N_{im},N^-,N^-,N^-	665/+0.32 570/–0.50sh 495/–0.66	550/–0.54 480/+0.19	558/–0.57 479/+0.24	640/+0.92 495/–1.42	638/+1.08 495/–1.56	652/+0.92 494/–1.56	650/+1.00 500/–1.33

Table 5 UV–Vis and circular dichroism spectral parameters of the 3N- and 4N-coordinated copper(II) complexes of nonapeptides

	Ligand $[X]_9$				
	Binding sites	KTNSKHMAG (M109S)	KTNMKHSAG (M112S)	KTNMKHMAG HuPrP(106–114) [8]	KTNFKHVAG ChPrP(119–127) [8]
UV–Vis, λ_{max} (nm)/ε (dm^3 mol^{-1} cm^{-1})	N_{im},N^-,N^-	592/64	616/107	616/77	610/72
	N_{im},N^-,N^-,N^-	562/95	532/120	532/108	538/117
CD, λ_{max} (nm)/$\Delta\varepsilon$ (dm^3 mol^{-1} cm^{-1})	N_{im},N^-,N^-	602/–0.91 332/+0.69 254/+3.20	552/+0.30 475/–0.20 371/+0.32 321/–0.58 251/+5.47	760/–0.22 535/+0.39 385/+0.11 330/–0.61 249/+8.29	615/–0.19 525/+0.42 350/–0.77 248/+7.21 224/–7.69
	N_{im},N^-,N^-,N^-	558/–0.51 486/+0.08 266/+2.32 223/–15.21	624/+0.82 498/–1.14 355/–0.14 316/+0.94 257/+6.03	631/+1.05 495/–1.22 317/+1.27 292/+0.16 256/+7.82	645/+1.27 500/–1.35 325/+0.97 293/–0.75 260/+7.14 224/+25.4

The results obtained for the nonapeptides provide further support for these assumptions. It was demonstrated in our previous publication [8] that the complex formation processes of the nonapeptide fragments of prion protein are very similar to those of the short tetrapeptide fragments. The presence of two uncoordinated lysyl side chains, however, results in the different stoichiometries of the same coordination modes; e.g., the (N_{im},N^-,N^-)-bonded species corresponds to the stoichiometries $[CuH_{-1}L]^+$ and $[CuL]^{2+}$ for Ac-FKHV-NH_2 and Ac-KTNFKHVAG-NH_2, respectively. Another difference between the copper(II) complexes of tetra- and nonapeptides is reflected in the pK(3) values. All of these data in Table 2 are in the range of 7–8, supporting the complete separation of amide and lysyl ammonium deprotonation reactions. In agreement with this expectation the last two deprotonation reactions are not accompanied by any spectral changes and the spectroscopic parameters are the same for the species $[CuH_{-1}L]^+$, $[CuH_{-2}L]$, and $[CuH_{-3}L]^-$ having the same 4N coordination modes. This is best illustrated by CD and EPR spectra of the nonapeptides depicted in Figs. 4 and 5. The comparison of Figs. 2b and 4b reveals the high similarity in the coordination modes of Ac-MKHS-NH_2 and Ac-KTNMKHSAG-NH_2. In the case of the nonapeptide, the (N_{im},N^-,N^-,N^-) coordination mode exists in the species $[CuH_{-1}L]^+$ and the further two deprotonation reactions are connected to the lysyl side chains without any change in the coordination sphere of the metal ion. This is also supported by Fig. 5 in which the EPR spectra of the copper(II)–Ac-KTNMKHSAG-NH_2 system are plotted at four different pH values. The species $[CuL]^{2+}$ predominates at pH 6.55 and its spectral parameters correspond well to those of the common (N_{im},N^-,N^-)-coordinated species. This spectrum is completely changed by pH 9.05, supporting the involvement of the third amide group in metal binding. Further increase of pH, however, does not affect the EPR and CD parameters. On the other hand, the comparison of Fig. 4a, b reveals a big difference in the conformation of the two nonapeptides containing the Ser residues at positions 109 and 112, respectively. There is, however, an almost complete agreement in the spectroscopic data of the corresponding tetra- and nonapeptides Ac-SKHM-NH_2 and Ac-KTNSKHMAG-NH_2, supporting the involvement of the seryl–OH group in metal binding for both Met109Ser mutated peptides. The metal ion coordination of the hydroxyl groups will not change the number of coordinated nitrogen donor atoms around the metal ions; thus, the EPR spectroscopic parameters for the $[CuH_{-1}L]^+$ to $[CuH_{-3}L]^-$ species of the two nonapeptides are not much different. The major difference is reflected in the parameters of $[CuL]^{2+}$ where the

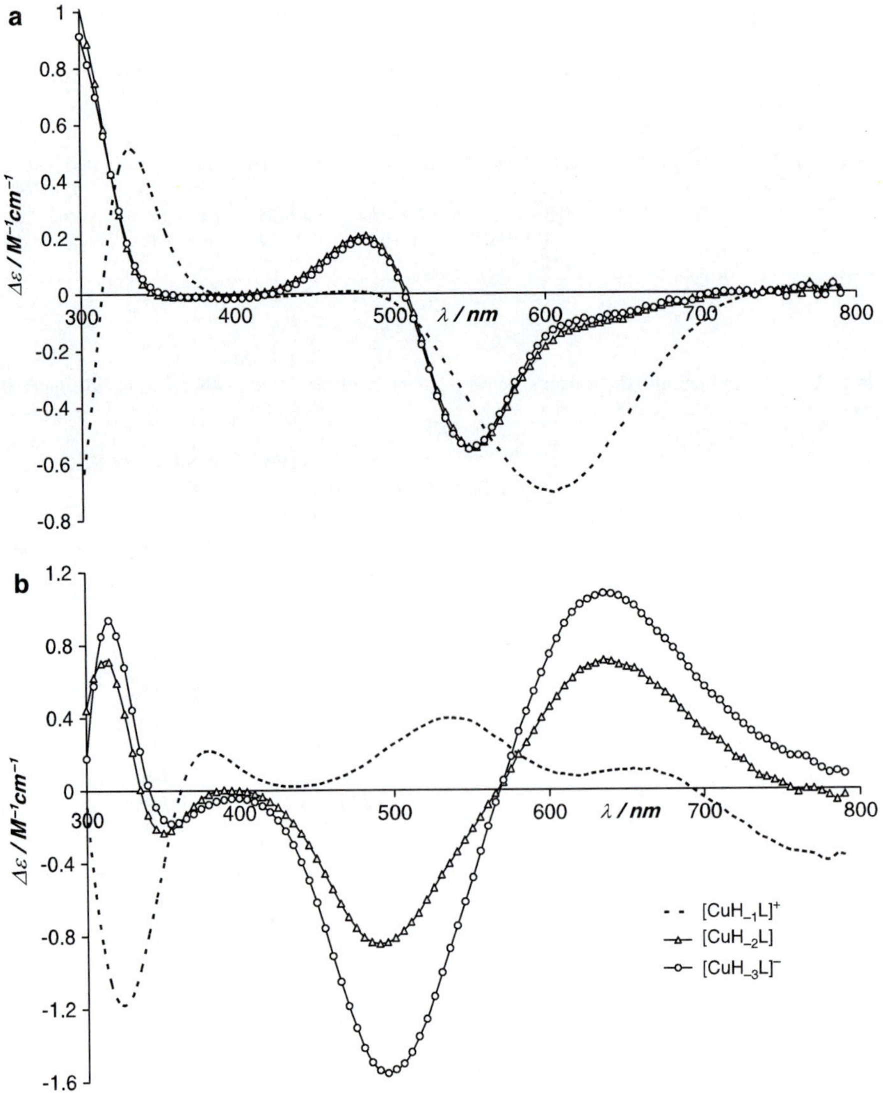

Fig. 2 Circular dichroism spectra of the complexes formed in the copper(II)–Ac-SKHM-NH_2 (**a**) and copper(II)–Ac-MKHS-NH_2 (**b**) systems

common (N_{im},N^-,N^-) coordination mode is changed to (N_{im}, N^-,N^-,OH). The $g_{\|}$ values of the unsaturated (N_{im},N^-,N^-)-bonded species in Table 3 correspond well to the metal ion coordination of three nitrogen donors in all peptides, but the $|A_{\|}|$ values are generally in the range 160–170 × 10^{-4} cm^{-1}. In the case of Ac-KTNSKHMAG-NH_2 this value is, however, much higher, $|A_{\|}| = 193 \times 10^{-4}$ cm^{-1}. It is a common feature of EPR spectra that the increase in the number of coordinated donor atoms is accompanied by an increase of the hyperfine splitting constant [25] and, as a consequence, these data support the presence of a seryl–OH donor function in the equatorial plane.

Experimental

Materials

Stock solutions of copper(II) ions were prepared from analytical grade reagents ($CuCl_2 \cdot 2H_2O$) and their concentrations were checked gravimetrically via the precipitation of oxinate. The other solutions (KOH, HCl, KCl, potassium hydrogen phthalate) were also prepared from analytical grade reagents.

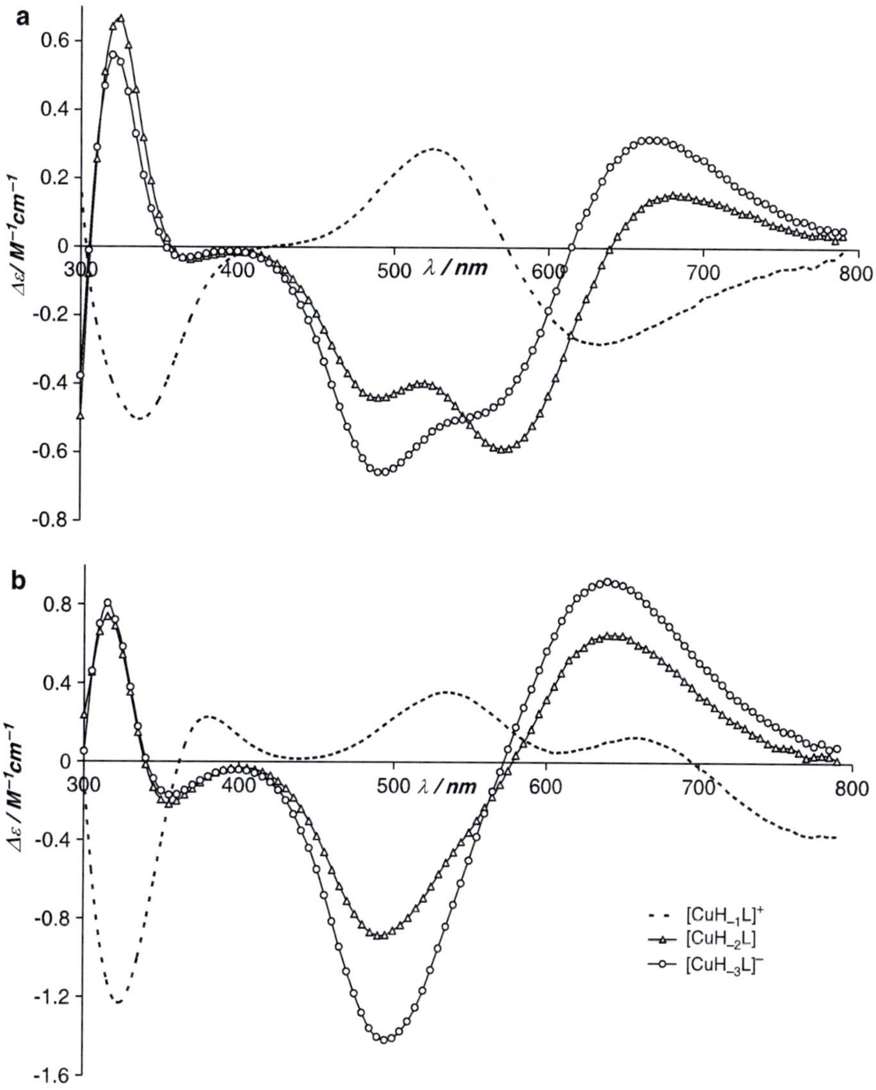

Fig. 3 Circular dichroism spectra of the complexes formed in the copper(II)–Ac-S(OMe)KHM-NH_2 (**a**) and copper(II)–Ac-MKHS(OMe)-NH_2 (**b**) systems

Synthesis of prion fragments and mutants

The tetra- and nonapeptide fragments of human and chicken prion proteins were prepared by solid-phase peptide synthesis and details of the procedure and purification of the peptides are described in our previous publications [7–10]. A similar procedure was used for the synthesis of the peptides containing the Ser or Ser(OMe) and Thr residues.

Potentiometric measurements

The pH-potentiometric titrations in the pH range 2.5–11.0 were performed on 3-cm^3 samples in the concentration range 1×10^{-3}–4×10^{-3} mol dm^{-3} at metal ion to ligand ratios between 1:1 and 1:2. The measurements were made with a MOLSPIN pH meter equipped with a 6.0234.100 combined electrode (Metrohm) and a MOL-ACS microburette controlled by computer.

The titrations were performed with carbonate-free stock solution of potassium hydroxide of known concentration. During the titration argon was bubbled through the samples to ensure the absence of oxygen and carbon dioxide and for stirring of the solutions. All pH-potentiometric measurements were carried out at a constant ionic strength of 0.2 M KCl and at constant temperature (298 K). The number of experimental points was around 50–70 data (cm^3, pH) for each titration curve. The pH readings were converted into hydrogen ion concentration as described earlier [26]. Protonation constants of the ligands and the overall stability constants (log β_{pqr}) of the complexes were calculated by means of general computational programs, PSEQUAD [27] and SUPERQUAD [28] using Eqs. (1) and (2).

$$pM + qH + rL \rightleftharpoons M_pH_qL_r \qquad (1)$$

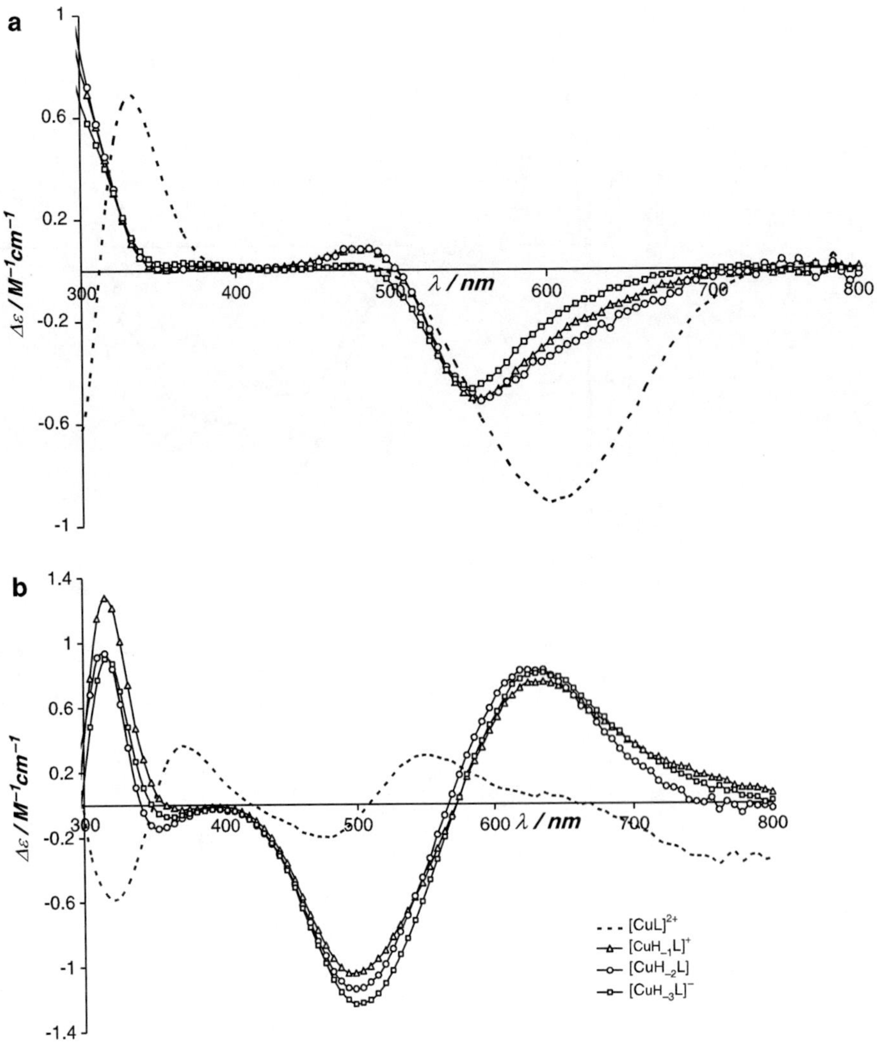

Fig. 4 Circular dichroism spectra of the complexes formed in the copper(II)–Ac-KTNSKHMAG-NH_2 (**a**) and copper(II)–Ac-KTNMKHSAG-NH_2 (**b**) systems

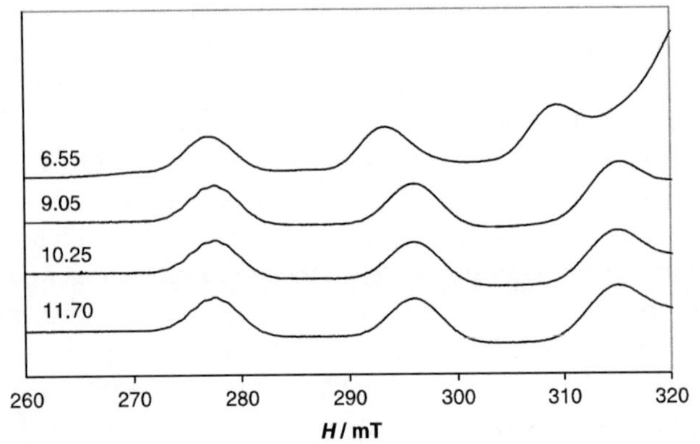

Fig. 5 EPR spectra of the copper(II)–Ac-KTNMKHSAG-NH_2 system at 1:1 metal ion to ligand ratio ($[Cu^{2+}] = 2 \times 10^{-3}$ mol dm^{-3}) at four different pH values

$$\beta_{pqr} = \frac{[M_pH_qL_r]}{[M]^p[H]^q[L]^r} \qquad (2)$$

Spectroscopic studies

UV–Vis spectra of the copper(II) complexes were recorded on a Perkin-Elmer Lambda 25 double beam spectrophotometer in the same concentration range as used for pH-potentiometry.

The EPR continuous wave spectra were recorded at the X-band at 120 K, using a Bruker EMX spectrometer. Copper

(II) stock solution was prepared from $CuSO_4 \cdot 5H_2O$ enriched with ^{63}Cu to get better resolution of EPR spectra. Metallic copper (99.3 % ^{63}Cu and 0.7 % ^{65}Cu) was purchased from JV Isoflex (Moscow, Russia) for this purpose and converted into the sulfate.

CD spectra of copper(II) complexes were recorded on a JASCO J-810 spectropolarimeter using 1- or 10-mm cells in the 200- to 800-nm range in the same concentration range as used for potentiometry. CD spectra of the individual species were calculated by the same general program (PSEQUAD) as used for the evaluation of potentiometric measurements.

Acknowledgments This work was supported by the MTA(Hungary)-CNR(Italy) bilateral program and the research grant OTKA-NKTH 77586, OTKA 72956 and PD 72321 and TAMOP 4.2.1/B-09/1/KONV-2010-0007 (Hungary).

References

1. Sigel H, Martin RB (1982) Chem Rev 82:385
2. Sóvágó I (1990) In: Burger K (ed) Biocoordination chemistry, metal complexes of peptides and derivatives. Ellis Horwood, Chichester, p 135
3. Kozlowski H, Bal W, Dyba M, Kowalik-Jankowska T (1999) Coord Chem Rev 184:319
4. Sóvágó I, Ősz K (2006) Dalton Trans 3841
5. Brown DR, Kozlowski H (2004) Dalton Trans 1907
6. Millhauser GL (2004) Acc Chem Res 37:79
7. Jószai V, Nagy Z, Ősz K, Sanna D, Di Natale G, La Mendola D, Pappalardo G, Rizzarelli E, Sóvágó I (2006) J Inorg Biochem 100:1399
8. Di Natale G, Grasso G, Impellizzeri G, La Mendola D, Micera G, Mihala N, Nagy Z, Ősz K, Pappalardo G, Rigó V, Rizzarelli E, Sanna D, Sóvágó I (2005) Inorg Chem 44:7214
9. Grasso D, Grasso G, Guantieri V, Impellizzeri G, La Rosa C, Milardi D, Micera G, Ősz K, Pappalardo G, Rizzarelli E, Sanna D, Sóvágó I (2006) Chem Eur J 12:537
10. Ősz K, Nagy Z, Pappalardo G, Di Natale G, Sanna D, Micera G, Rizzarelli E, Sóvágó I (2007) Chem Eur J 13:7129
11. Di Natale G, Ősz K, Nagy Z, Sanna D, Micera G, Pappalardo G, Sóvágó I, Rizzerelli E (2009) Inorg Chem 48:4239
12. Hahn M, Wolters D, Sheldrick WS, Hulsbergen FB, Reedijk J (1999) J Biol Inorg Chem 4:412
13. Bóka B, Nagy Z, Várnagy K, Sóvágó I (2001) J Inorg Biochem 83:77
14. Várnagy K, Bóka B, Sóvágó I, Sanna D, Marras P, Micera G (1998) Inorg Chim Acta 275–276:440
15. Pettit LD, Robbins RA (1995) In: Berthon G (ed) Metal-peptide complex formation, in handbook of metal–ligand interactions in biological fluids, vol 1. Dekker, New York, p 636
16. Gergely A, Farkas E (1982) J Chem Soc Dalton Trans 381
17. Bühl M (2000) J Inorg Biochem 80:137
18. Farkas E, Kiss T (1989) Polyhedron 8:2463
19. Szabó-Plánka T, Árkosi Z, Rockenbauer A, Korecz L (2001) Polyhedron 20:995
20. Kowalik-Jankowska T, Stasiak M, Leplawy MT, Kozlowski H (1997) J Inorg Biochem 66:193
21. Mlynarz P, Gaggelli N, Panek J, Stasiak M, Valensin G, Kowalik-Jankowska T, Leplawy ML, Latajka Z, Kozlowski H (2000) J Chem Soc Dalton Trans 1033
22. Mlynarz P, Kowalik-Jankowska T, Stasiak M, Leplawy ML, Kozlowski H (1999) J Chem Soc Dalton Trans 3673
23. Mlynarz P, Bal W, Kowalik-Jankowska T, Stasiak M, Leplawy ML, Kozlowski H (1999) J Chem Soc Dalton Trans 109
24. Kállay C, Várnagy K, Malandrinos G, Hadjiliadis N, Sanna D, Sóvágó I (2006) Dalton Trans 4545
25. Nishida Y, Hayashida K, Kida S (1980) J Coord Chem 10:101
26. Irving H, Miles G, Pettit LD (1967) Anal Chim Acta 38:475
27. Zékány L, Nagypál I (1985) In: Leggett D (ed) Computational methods for the determination of formation constants. Plenum, New York, p 291
28. Gans P, Sabatini A, Vacca A (1985) J Chem Soc Dalton Trans 1195

Interactions of pyridinecarboxylic acid chelators with brain metal ions: Cu(II), Zn(II), and Al(III)

Éva Sija, Annalisa Dean, Tamás Jakusch, Valerio B. Di Marco, Alfonso Venzo, and Tamás Kiss

Abstract

The interactions of Cu(II), Zn(II), and Al(III) with 1,6-dimethyl-4-hydroxy-3-pyridinecarboxylic acid (DQ716) and 2,6-dimethyl-3-hydroxy-4-pyridinecarboxylic acid (DT726), possible chelating agents in Alzheimer's disease, were investigated in aqueous solution. The proton dissociation constants of the ligands, the stability constants, and the coordination modes of the metal complexes formed were determined by pH-potentiometric, UV–vis spectrophotometric, and ^{1}H NMR methods. The nitrogen of the pyridine ring changes the proton affinity of the carboxylate and phenolate moieties and these pyridine derivatives form stronger complexes with Cu(II), Zn(II), and Al(III) than salicylic acid. Interactions of the ligands with human serum albumin as their potential transporter in blood were investigated at physiological pH through ultrafiltration by UV-Vis and fluorescence spectroscopy.

Keywords

Potentiometry • Carboxylate ligands • Ultrafiltration • Fluorescence spectroscopy • Chelation therapy • Human serum albumin

Introduction

A number of age-related degenerative diseases, such as Alzheimer's disease (AD), are becoming progressively more important, especially in Europe and the USA, as a result of the significant increase of life expectancy observed nowadays. Enough experimental evidence has been gathered so far which closely correlates AD with abnormal protein folding (misfolding), driving peptides towards their β-sheet conformation. In turn, in this kind of conformation the peptides have the ability to aggregate giving rise to many different types of supramolecular structures, such as amyloid deposits. The main protein component found in amyloid deposits is a 4-kDa amyloid-β-protein (Aβ) [1, 2] that is generated from a much larger ~100- to 130-kDa amyloid precursor protein (APP) by the proteolytic activity of β- and γ-secretases [3].

Aβ displays high affinity towards Cu(II) and Zn(II) at pH 7.4 [4]. These metal ions can promote the formation of Aβ deposits and oligomers in the brain. Elevated levels of Cu(II), Zn(II), and Al(III) have been already proven in deposits and oligomers [5, 6].

Two main forms of Aβ can be distinguished by electrophoresis: a shorter form, composed of 40 amino acids (Aβ_{40}), and a longer one, composed of 42 amino acids (Aβ_{42}) [7]. Aβ_{42} is more prone to aggregation and more toxic to neurons than Aβ_{40} [8]. In the presence of Cu(II) and H_2O_2, the in vitro Aβ_{42} monomers form cross-linked oligomers that are resistant to proteolysis [9]. The oligomeric forms of Aβ_{42} are proven to be potentially very toxic [10], as they cause further generation of reactive oxygen species and favour membrane depolarization, protein, DNA, and RNA oxidations, and lipid peroxidation.

V.B. Di Marco (✉)
Department of Chemical Sciences, University of Padova, via Marzolo 1, 35131 Padua, Italy
e-mail: valerio.dimarco@unipd.it

T. Kiss (✉)
Department of Inorganic and Analytical Chemistry, University of Szeged, 440Szeged 6701, Hungary
e-mail: tkiss@chem.u-szeged.hu

W. Linert and H. Kozlowski (eds.), *Metal Ions in Neurological Systems*,
DOI 10.1007/978-3-7091-1001-0_17, © Springer-Verlag Wien 2012

Metal-induced precipitation of Aβ can be reversed by chelation [11, 12]. However, chelating agents can disrupt the oxidative activity and break the interaction of these metals with Aβ [13]. These and other observations render chelation therapy a very challenging task for future pharmacological treatments of AD [14, 15]. An ideal chelating agent for AD should be sufficiently small to cross the blood–brain barrier, and it should be non-toxic. Moreover, it should exert specific and moderate, rather than indiscriminate and massive, chelation of excess metals [15]. Ligands with intermediate affinity and appreciable metal selectivity are capable of disrupting a few relevant metal–protein interactions rather than inducing generalized, and thus highly toxic, metal depletion like in the classic chelation therapy. For this reason, such ligands can be referred to as metal–protein attenuating compounds (MPACs) or metal-targeting compounds. One of the targets of an MPAC should be the inhibition of Aβ oligomerization and Aβ-related generation of free radicals that can prevent reactive metals from participating in potentially harmful redox reactions [16].

Several attempts were made to obtain efficient MPACs. The antibiotic clioquinol (5-chloro-7-iodo-8-hydroxyquinoline, CQ) has moderate affinity towards Cu(II) and Zn(II), and it was proven to inhibit metal-induced Aβ aggregation and generation of reactive oxygen species in vitro [17]. Clinical studies were stopped because of the difficulties in preventing diiodo-8-hydroxyquinoline contamination in large-scale chemical synthesis of CQ, but the positive effects of CQ encouraged scientists to design new possible MPACs for therapeutic purposes in AD (e.g. see Ref. [18]). Various other compounds have been prepared and tested in vitro and partly in vivo in different laboratories around the world to remove metal ions from the amyloid oligomers and thus to prevent formation of plaques or to achieve at least a partial resolubilisation of the already formed amyloids [19–21].

Besides the toxicity and metal binding strength, other properties should also be investigated for the ligands to be used as an MPAC or as classic chelating agent. Information about the fate of these new potential drugs in the biological systems is needed. As they are expected to be transported in the blood stream, their interactions with plasma proteins might be important. The most probable transporter molecule in serum is albumin, therefore the interaction with this protein should be investigated. The binding strengths and the sites can be determined by studying competition with different site markers such as warfarin or dansyl-glycine [22, 23].

In this work we focus on a previously poorly investigated class of ligands, hydroxypyridinecarboxylic acids (HPs), which have been proposed by us for the "classic" chelation therapy of iron and aluminium [24]. HPs have several positive properties: a low molecular weight, a negligible or even no toxicity, high affinity towards Fe(III) and Al(III), and absence of redox-cycling under in vivo conditions. Here

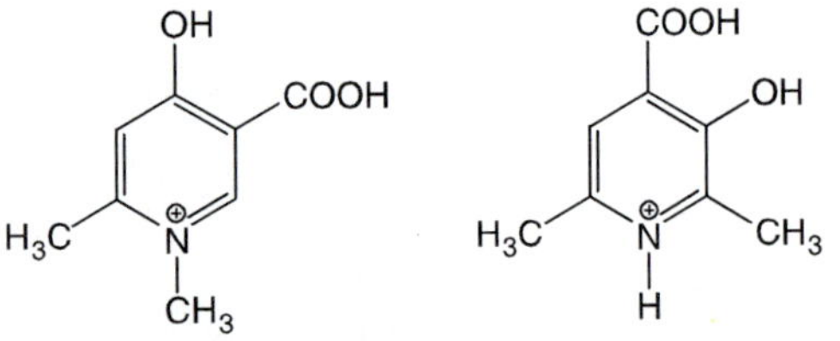

DQ716 DT726

Fig. 1 1,6-Dimethyl-4-hydroxy-3-pyridinecarboxylic acid (DQ716) and 2,6-dimethyl-3-hydroxy-4-pyridinecarboxylic acid (DT726)

we investigate if HPs can also be proposed as new potential MPACs that could regulate the metal ion distribution in the neurodegenerative disordered brain. In the present study, we consider 1,6-dimethyl-4-hydroxy-3-pyridinecarboxylic acid (DQ716) and 2,6-dimethyl-3-hydroxy-4-pyridinecarboxylic acid (DT726) (Fig. 1). DQ716 is the most promising HP synthesized so far for the classic chelation therapy, and it was proposed for a pharmacologic experimentation [24]. DT726 is under investigation for this aim, and preliminary data indicate very similar positive properties as for DQ716 [25]. The complexation properties of DQ716 and DT726 with Cu(II) and Zn(II) in aqueous solution were determined by means of pH-potentiometric titrations. Also the Al(III)-DT726 complexes were studied, employing potentiometry, UV–vis, and NMR, with the aim to evaluate DT726 as a classic chelating agent. The interactions of DQ716 and DT726 with human serum albumin (HSA) were studied by ultrafiltration and by UV–vis and fluorescence spectroscopy.

Results and discussion

Potentiometric results

Potentiometric titration of each ligand allowed the determination of some pK_A values, which are reported in the upper part of Tables 1 and 2. The pK_{A1} of DQ716 was too low for a pH-potentiometric determination, and it was therefore obtained by UV–vis. The high acidity for the most protonated forms (H_2L for DQ716, H_3L for DT726) is in agreement with values previously observed for other HPs examined so far [24]. As was justified in previous work [24, 26], the first and the last pK_A of both ligands can be assigned to the carboxylic COOH and to the phenolic OH, respectively, and the intermediate pK_A of DT726 (pK_{A2}) to the pyridinic NH. It is worth noting that there is an approximately five orders of magnitude difference in the acidity of the phenolic OH of pyridine (DT726) and of *N*-methylpyridine (DQ716 derivatives), as a result of the positive charge of the methylated nitrogen. The medium has a small but not negligible effect on the acidity of both ligands: pK_A values decrease by ca. 0.2 log units upon

Table 1 pK_A values of DQ716 and DT726, and stability constants of Cu(II) and Zn(II) complexes at 25 °C in aqueous KCl 0.2 mol/dm^3

Species	pK_A or log β (uncertainty[a])		
	Salicylic acid[b]	DQ716[c]	DT726[d]
H_3L	–	–	1.21 (5)
H_2L	2.78	<1	6.23 (3)
HL	13.15	6.38 (2)	11.58 (6)
CuLH	–	–	16.08 (2)
CuL	10.62	6.63 (2)	7.58 (13)
$CuLH_{-1}$	–	0.84 (5)	–
CuL_2H_2	–	–	29.89 (24)
CuL_2H	–	–	22.18 (15)
CuL_2	18.45	12.03 (2)	13.26 (12)
$Cu^{2+} + H_3L = Cu(HL) + 2H^+$	–	–	–2.94
$Cu^{2+} + H_2L = CuL + 2H^+$	–5.31	–0.04	–
$Cu(HL) + H_3L = Cu(HL)_2 + 2H^+$	–	–	–5.20
$CuL + H_2L = CuL_2 + 2H^+$	–8.1	–1.27	–7.11
K_d[e] (mol/dm^3)	1.34×10^{-5}	5.48×10^{-9}	3.42×10^{-5}
ZnL	6.85	3.91 (5)	6.56 (23)
ZnL_2	–	6.92 (2)	12.12 (28)
ZnL_3	–	8.84 (15)	16.37 (29)
$Zn^{2+} + H_2L = ZnL + 2H^+$	–9.08	–2.76	–6.23
$ZnL + H_2L = ZnL_2 + 2H^+$	–	–3.65	–7.23
$ZnL_2 + H_2L = ZnL_3 + 2H^+$	–	–4.75	–8.54
K_d[e] (mol/dm^3)	7.94×10^{-2}	1.32×10^{-4}	4.45×10^{-3}

Values for salicylic acid are reported for comparison. Data are taken from Refs. [42, 43].
[a]Three times the standard deviation.
[b]L is the negatively double charged species (L^{2-}).
[c]L is the negatively singly charged species (L^-).
[d]L is the negatively doubly charged species (L^{2-}).
[e]$K_d = [M]_{free} \sum[H_x L]/\sum[M_wH_yL_z]$ computed at pH 7.4 for $C_M^0 = 25$ μmol/dm^3, $C_L^0 = 50$ μmol/dm^3.

Table 2 pK_A values of DQ716 and DT726, and stability constants of Al(III) complexes at 25 °C in aqueous NaCl 0.6 mol/kg

Species	pK_A or log β (uncertainty)		
	Salicylic acid[a]	DQ716[b]	DT726
H_3L	–	–	0.66 (18)[c]
H_2L	2.79	0.37 (15)	6.25 (5)
HL	13.4	6.30 (5)	11.44 (3)
AlLH	–	–	18.64 (3)
AlL	13.22	7.66 (3)	–
AlL_2H_2	–	–	35.82 (6)
AlL_2H	–	–	29.44 (18)
AlL_2	23.73	14.38 (1)	–
AlL_2H_{-1}	16.60	–	–
AlL_3H_3	–	–	51.10 (6)
AlL_3H_2	–	–	44.85 (7)
AlL_3H	–	–	37.42 (9)
AlL_3	–	19.37 (1)	29.46 (16)
$AL_2L_2H_{-2}$	17.9	–	–
$Al^{3+} + H_3L = Al(HL)^+ + 2H^+$	–	–	0.68
$Al^{3+} + H_2L = AlL^+ + 2H^+$	–2.71	1.00	–
$Al(HL)^+ + H_3L = Al(HL)_2 + 2H^+$	–	–	–2.14
$AlL^+ + HL = AlL_2 + 2H^+$	–5.42	0.06	–

Values for salicylic acid are reported for comparison.
[a]pK_A of free salicylic acid and log β values of Al(III)-salicylic acid [44] refer to a KCl 0.2 mol/dm^3 medium.
[b]pK_A values of DQ716 in NaCl 0.6 mol/kg, and stability constants for Al(III)/DQ716, were obtained in a previous work [24].
[c]Value obtained by UV–vis.

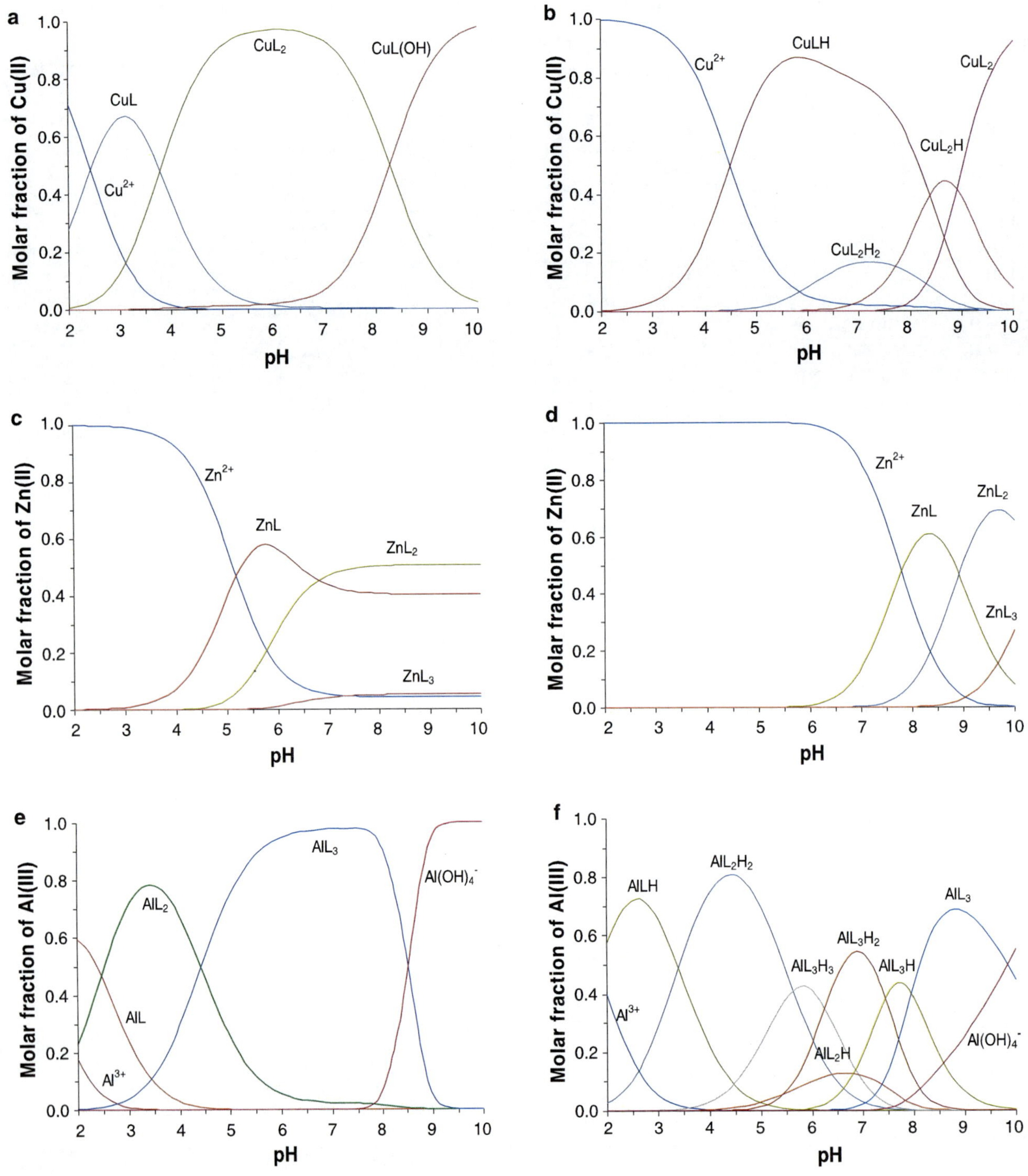

Fig. 2 Distribution diagrams of the most important Cu(II) (**a**, **b**), Zn(II) (**c**, **d**), and Al(III) (**e**, **f**) species in the presence of DQ716 (**a**, **c**, **e**) and DT726 (**b**, **d**, **f**); $C_M{}^0 = 5 \times 10^{-4}$, $C_L{}^0 = 2 \times 10^{-3}$ mol/dm^3 or mol/kg

increasing the ionic strength from 0.2 mol/dm^3 KCl to 0.6 mol/kg NaCl.

The stoichiometry and the values of the stability constants of the metal–ligand complexes identified in solution are reported in Tables 1 and 2. The representative distribution diagrams are shown in Fig. 2.

As seen from the structural formulae of DQ716 and DT726, both ligands contain a carboxylic and a phenolic function in a reciprocal *ortho* position, i.e. they can be considered as salicylic acid derivatives. Accordingly, their metal ion binding properties were compared to those of salicylate. Differences in the acidity of the functional groups and the resultant proton displacement constants are compared in rows 10, 11, and 16–18 in Table 1 and 17–20 in Table 2. It is seen from the data that both ligands form salicylate-type (COO^-, O^-) mono, bis, and tris complexes depending on the maximum coordination number of the metal ions, and the pyridine-N may be protonated or deprotonated depending on the pH. If the pH ranges of metal complex formation and deprotonation of the pyridine-NH^+ overlap, then protonated complexes are formed, which deprotonate with increasing pH (see Al(III)-ligand systems and Cu(II)-DT726). A comparison of the proton displacement constants reveals that the presence of the pyridinic N in the benzene ring increases the metal binding ability of the salicylate chelating function. This effect appears to be significantly more pronounced when the pyridinic N is methylated.

Although the basic binding mode of the ligands is the same, the speciation showed by the three metal ions and by the two ligands is rather different. As regards Cu(II) + DQ716, only three complexes were observed. CuL_2 is the main species at slightly acidic and neutral pH values, whereas $CuLH_{-1}$ predominates at higher pH. As the bound ligand has no free protons, $CuLH_{-1}$ can be only the CuL(OH) mixed hydroxo species, in which one of the coordinated water molecules liberates a proton. The pK_A of CuL (5.79) is larger than the pK_A of free Cu(II) in water (ca. 7.2) probably because of an electronic effect exerted by the ligand on the metal centre. A similar speciation is observed for Cu(II) + DT726, but here the complexed ligand is able to deprotonate as it binds the metal ion in its monoprotonated (HL) form resulting in Cu(LH) and $Cu(LH)_2$ species. It is most likely that the pyridinic NH^+ of the ligand remains free. Then $Cu(LH)_2$ deprotonates in a stepwise way with pK values 7.71 and 8.92, significantly higher than that of the non-coordinated pyridinic NH^+ ($pK = 6.23$), indicating that assignment of the deprotonation constant of the ligand can not be made unambiguously.

Zn(II) forms weak complexes, as they start to have a significant concentration only at pH larger than 3.5–4 (for DQ716) or even at pH larger than 6 (for DT726). Only the simple ZnL, ZnL_2, and ZnL_3 complexes were detected in solution, ZnL_3 being a minor species as it forms at the start of hydroxide precipitation.

Al(III) forms strong complexes with both ligands. For Al (III) + DQ716, a complete discussion can be found in Ref. [24]. DT726 forms a large number of Al(III) complexes, especially in the neutral pH range. The fully coordinated complex AlL_3H_3 is able to deprotonate stepwise at the pyridine rings, producing AlL_3H_2, AlL_3H, and AlL_3. pK_A values are 6.25, 7.43, and 7.96, respectively. The first pK_A is similar to that of free DT726, and the other two increase following the negative charge produced by deprotonation, and in agreement with statistical considerations. The deprotonation of AlL_2H_2 to give AlL_2H ($pK_A = 6.38$) might occur at the pyridinic nitrogen and/or at a coordinated water molecule (see above).

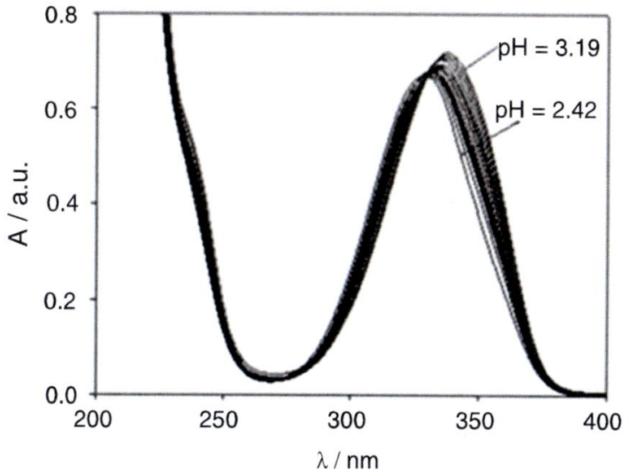

Fig. 3 UV–vis spectra for Al(III)-DT726 solutions at pH = 2.42, 2.48, 2.55, 2.58, 2.62, 2.65, 2.70, 2.74, 2.79, 2.82, 2.85, 2.88, 2.91, 2.95, 2.98, 3.03, 3.07, 3.12, and 3.19. Path cell = 1 cm. Other details are reported in the "Experimental" section

UV–vis of Al(III)-DT726

UV spectra for Al(III)-DT726 solutions at several pH values are depicted in Fig. 3. Metal complexation affects the typical pyridine ring absorption at around 340 nm, producing a red shift. The following values were obtained for log β (AlLH) and log β (AlL_2H_2): 18.39 (0.03) and 35.67 (0.03). These data are close (0.1–0.3 log units) to the corresponding values obtained by potentiometric titrations (see Table 2). The differences can be attributed to the intrinsic uncertainty of these measurements, as the spectral shifts induced by Al(III) complexation are small.

NMR of Al(III)-DT726

The ^1H NMR spectra of the free ligand at pD 2.0 show the presence of two singlets, one in the aliphatic zone (6H, $CH_3(2)$ + $CH_3(6)$, δ 2.60) and one in the aromatic zone (1H, H(3), δ 7.81). For pyridine ring numbering see the top of Fig. 4. In principle the two methyl groups are not chemically equivalent because DT726 is not symmetric, but they appear at the same chemical shift. Only at very basic pD values can the two signals be distinguished (spectra not shown), likely because of the deprotonation of the phenolic OH which inductively affects $CH_3(2)$ more than $CH_3(6)$. Upon increasing the pD from 2.0 to 9.4, all signals shift upfield. In agreement with the pK_A values of DT726, the maximum shift is observed at neutral pD: for example, H(5) proton resonates at δ 7.81, 7.80, 7.70, 7.47, and 7.37 ppm at pD 2.0, 3.8, 5.6, 6.8, and 9.4, respectively.

In the presence of Al(III), the resonances of the free ligand are still observed, together with new peaks due to the complexes at all investigated pD values. Figure 4 displays spectra of Al(III)-DT726 solutions.

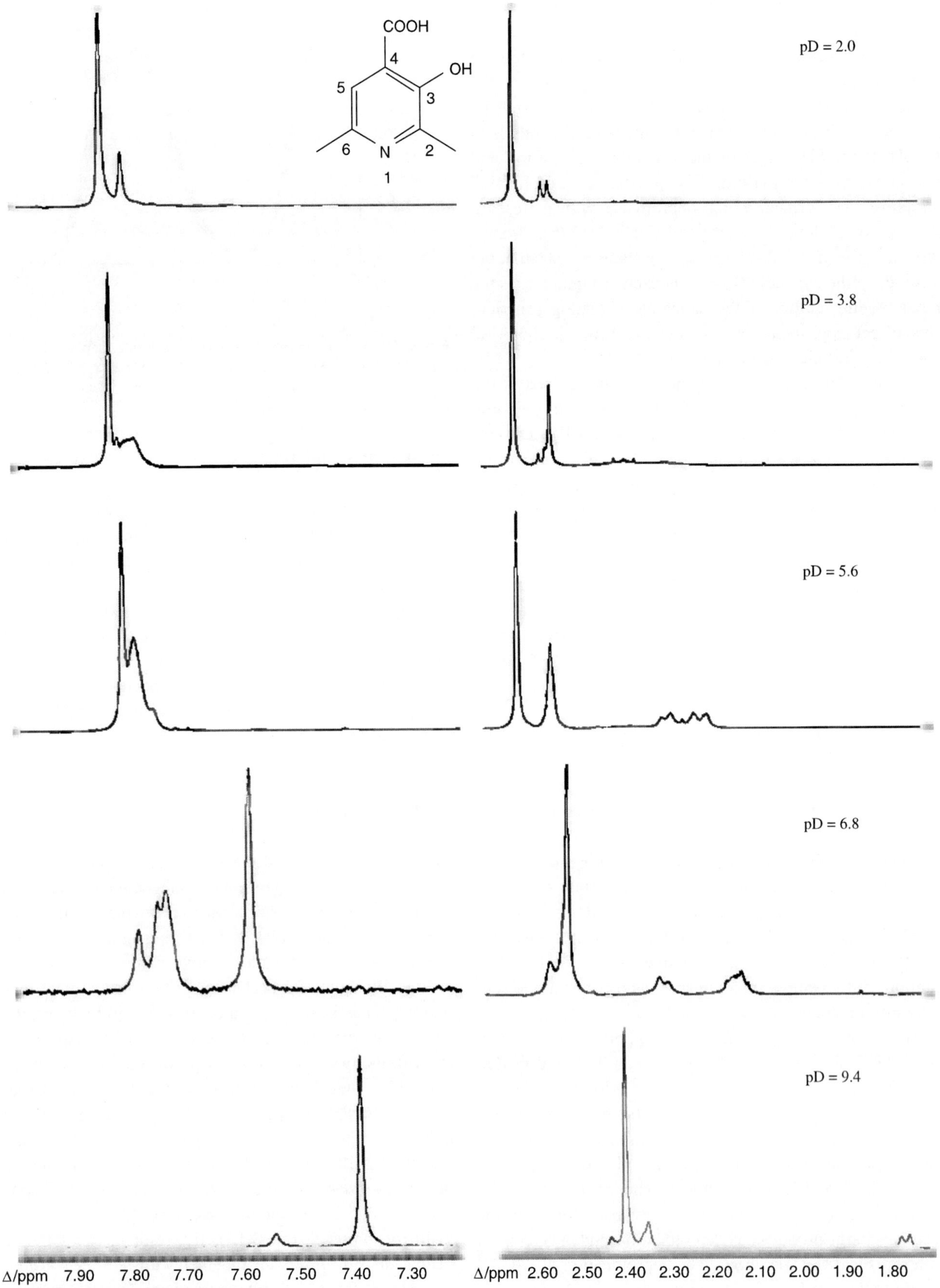

Fig. 4 ^{1}H NMR spectra of D_2O solutions containing Al(III) ($C_{Al}{}^0 = 2.0 \times 10^{-3}$ mol/kg) and DT726 ($C_L{}^0 = 8.0 \times 10^{-3}$ mol/kg) at pD = 2.0, 3.8, 5.6, 6.8, and 9.4. Only the aromatic and the aliphatic zone of the spectra are shown

At pD = 2.0 three new narrow signals appear, one in the aromatic region and two in the aliphatic zone. These new peaks can be assigned to AlLH, which according to potentiometric data predominates at this pD. The $CH_3(2)$ and $CH_3(6)$ peaks of the complexed ligand are more resolved than the corresponding signals of the free ligand, probably because of an inductive effect exerted by the metal ion, which is closer to $CH_3(2)$.

At pD values larger than 2.0, further new signals can be observed, most of which are broad. At pD 3.8 the copresence of AlLH, AlL_2H_2, and AlL_3H_3 is expected on the basis of the potentiometric data. Therefore, the signals should be assigned to these species. A very interesting feature can be observed in the spectrum at pD 5.6, where the complex AlL_3H_3 should predominate. Here, a band composed of at least six to seven singlets appear centred at δ 2.20. Our hypothesis is that this band is composed of eight signals, which have the same intensity, and which can altogether be assigned to the $CH_3(2)$ and the $CH_3(6)$ of AlL_3H_3. DT726 is asymmetric, thus it may chelate the metal, forming AlL_3H_3 in two different spatial configurations. If the three ligands chelate the metal ion in the same configurations, a symmetrical diastereoisomer (AlL_3H_3-sym) forms, and the protons are chemically equivalent. If one ligand chelates differently from the other two, an unsymmetrical diastereoisomer (AlL_3H_3-unsym) forms, and the protons should give rise to different signals in the NMR spectrum. As two different protons are there ($CH_3(2)$ and $CH_3(6)$), eight signals are expected. The signals have about the same intensity: this can be explained by supposing that the formation microconstants of AlL_3H_3-sym and AlL_3H_3-unsym are very similar, so that their ratio in solution is the statistical 1:3 one. Similar ^{1}H NMR patterns have been observed previously for the AlL_3-type complexes of other HPs [24 and references therein]. These signals separate into two groups (having the same intensity) when the pH is increased to pD 6.8 and (even more) to pD 9.4, probably because the deprotonation of the pyridine ring (leading to AlL_3H_2, AlL_3H, and AlL_3) differentiates $CH_3(2)$ from $CH_3(6)$. In the aromatic zone, a multiplet composed of at least three peaks forms, this shape being due to four H(3) signals of AlL_3H_3 according to the previous hypothesis.

NMR data were treated also in a semiquantitative way, by integrating the signals of the free ligand and those of the complexed ligand. Because of partial overlap among these signals, the calculation was performed only at pD 2.0 and 3.8. At pD 2.0, the signals of AlLH account altogether for 20 % of the total NMR signal: this value agrees well with the corresponding number, 24.7 %, which can be computed from potentiometric data, and which represents the ratio between complexed and total ligand. At pD 3.8, the two numbers (NMR integration ratio and potentiometric ratio) are 43 and 49.4 %, respectively. The semiquantitative agreement among two different techniques suggests the absence of systematic errors in our results.

Table 3 Conditional overall and stepwise stability constants (log β', log K') of HSA–ligand (L = DT726) complexes calculated from ultrafiltration-UV measurements

	Log β'_n	Log K'
(HSA)-L	4.83 (9)	4.83
(HSA)-L_2	9.1 (3)	4.27
(HSA)-L_3	13.1 (1)	4.0
(HSA)-L_4	16.1 (6)	3

Measurements were performed at pH 7.40 (HEPES 0.10 M) and $T = 25$ °C. Uncertainties in the stability constants (as ±3 SD values) are given in parentheses.

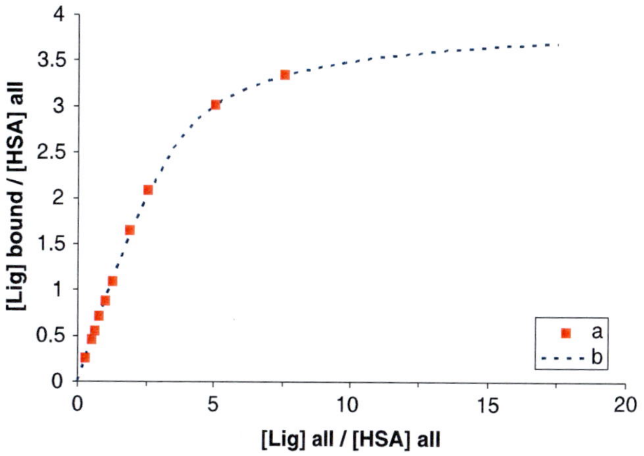

Fig. 5 Equilibrium concentration of the bound ligand/total HSA versus free ligand/total HSA (ligand = DT726): **a** (*squares*) measured by UV spectrophotometry; **b** (*dashed line*) calculated with the conditional binding constants given in Table 3

HSA–ligand interaction through ultrafiltration

In order to determine the maximum number of DT726 molecules binding to HSA and the strength of the interaction, evaluation using centrifugal devices with ultrafiltration coupled with UV spectrophotometric was performed. The stoichiometry and the conditional stability constants calculated from the distribution data are listed in Table 3.

Albumin is able to bind up to four DT726 molecules with high affinity at pH 7.40. In order to check the correctness of the calculation, the "formation function" obtained by using the measured equilibrium concentrations of the free and bound ligands together with those calculated from the conditional binding constants determined is presented in Fig. 5.

On the other hand, DQ716 does not exhibit a measurable tendency to bind to HSA, and for this reason constants for HAS-DQ716 species could not be determined. One possible explanation of the different behaviour of the two ligands might lie in the different charges carried by the pyridine

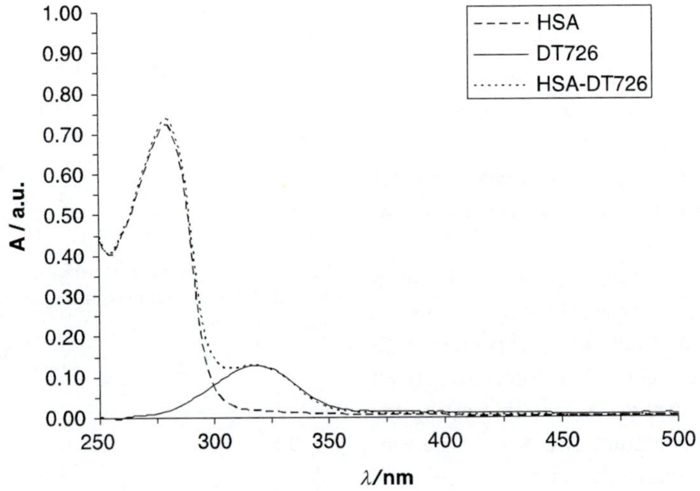

Fig. 6 Absorption spectra of DT726, HSA, and DT726 in the presence of HSA. $C_{HSA}{}^{0} = 2 \times 10^{-5}$ mol/dm^{3}, $C_{DT726}{}^{0} = 2 \times 10^{-5}$ mol/dm^{3} in 0.1 mM HEPES buffer 7.4 and 25 °C

ring of the two ligands at pH 7.40: for DT726 the pyridine ring is neutral, for DQ716 it is positively charged. Thus, its affinity to the hydrophobic binding site of HSA decreases.

There are two major drug binding regions on the albumin molecule. One is the warfarin site (warfarin and azopropazone binding area), also called site I. The other is the diazepam site (indole and benzodiazepine binding area), also called site II [27, 28]. Site I was shown to prefer large heterocyclic and negatively charged compounds, whereas site II is the preferred site for small aromatic carboxylic acids. Hydrophobic interactions seem to be important at both sites, and thus further details about the actual binding sites of our molecules can be obtained only by competition measurements of the different site markers. Such studies are in progress in our laboratories.

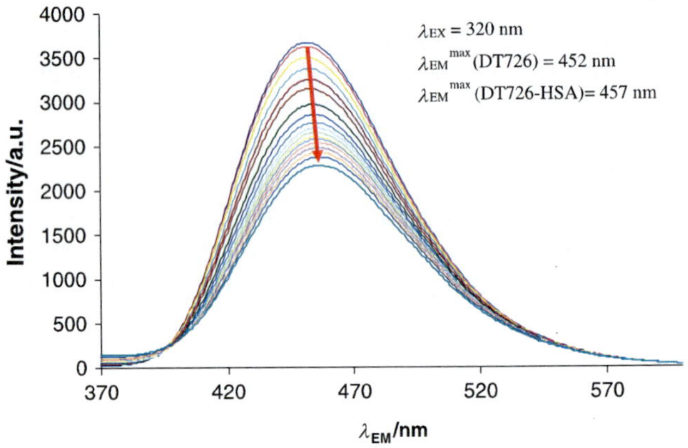

Fig. 7 Fluorescence emission spectra of DT726 as a function of HSA concentration in 20 mM HEPES buffer pH 7.4 and 25 °C. The HSA concentration was increased from 0 to 16 μM and at the same time the concentration of the ligand decreased from 1.5 to 1.2 μM in the direction of the *arrow*

HSA–ligand interaction through fluorescence

Fluorescence spectroscopy has been the most widely used spectroscopic technique for monitoring drug binding to plasma albumin because of its sensitivity, accuracy, rapidity, and ease of use [29].

The absorption of DT726 is characterized by a strong band at 320 nm but it has no absorption in the range of 390–600 nm. The HSA and HSA-DT726 adduct have no absorption at 320 nm and in the examined emission range (see Fig. 6).

A set of emission spectra of DT726 at various concentrations of HSA are shown in Fig. 7. The maximum of emission wavelength shifts from 452 to 457 nm in the presence of HSA. This shift indicates a small structural change (micro-environment) of HSA upon interaction with DT726 that can be rationalized in terms of interactions between DT726 and HSA in the ground state and formation of a complex.

Other information that is shown by the spectra, when HSA is added to the solution containing DT726, is that the fluorescence intensity of DT726 decreased continuously with increasing concentration of HSA. The fluorescence quenching spectra of DT726 with varying concentrations of HSA are shown in Fig. 7. The observed behaviour suggests that complexes are formed between DT726 and HSA, which are responsible for the fluorescence quenching of DT726.

The calculated molar emission spectrum of the ligand alone agrees very well with the experimental one. At the same time the emission intensity maximum of DT726 in the presence of HSA is shifted to 457 nm indicating the interaction between the two components (Fig. 8).

The calculated log β is 5.80 $\pm$ 0.01. The values reported in the literature [30–32] for the interaction of HSA with various antibiotic drugs are of the order of 10^{3}–10^{5}. It has to be noted that this value is one order of magnitude higher than that obtained by ultrafiltration (see above), and that the interaction

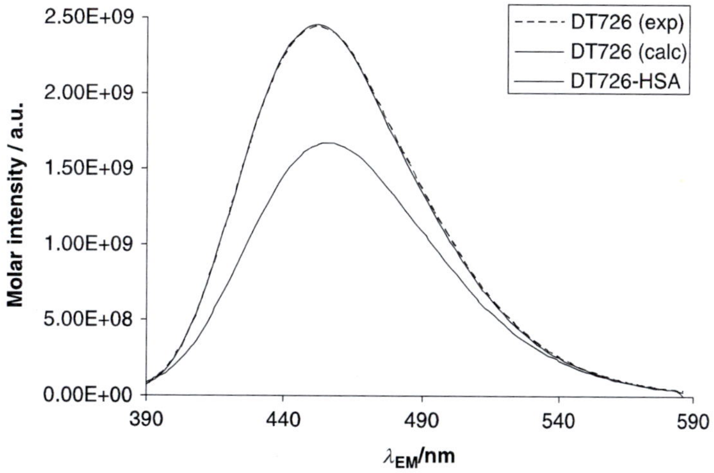

Fig. 8 Calculated and experimental molar emission spectra of DT726 (which coincide) and DT726-HSA adduct

of only one DT726 could be detected by fluorescence, instead of four ligands as found by ultrafiltration. This difference can be explained by the three orders of magnitude difference in the concentration range of the measurement: ultrafiltration characterises the system in the millimolar range, whereas fluorescence works at the micromolar range and was not able to detect the formation of polynuclear species.

Discussion of the chelation strength towards metal ions

The metal binding strength of the various chelator molecules towards the metal ions can be compared in different ways. Biologists prefer to use K_D, the dissociation constant, which gives the extent of dissociation at given experimental conditions ($C_M{}^0$, $C_L{}^0$, pH) and can be calculated from the thermodynamic stability constants (see footnote of Table 1). The K_D values have been calculated for the Cu(II) and Zn(II) complexes, as we have reasonably good indicator values for their potential applicability as metal ion modulators in AD [11].

Another parameter used to characterised the complexation strength of different ligands towards a given metal ion (preferably used by chemists) is the pM graph (pM = –log [M_{free}]) as a function of pH. The larger the pM, the stronger is the binding strength of the ligand. pM graphs of our ligands for the metal ions studied are displayed in Fig. 9.

The pM graphs demonstrate that both HPs form stronger complexes than salicylic acid does with all considered metal ions and over all the pH range. The stability increase with respect to salicylic acid can be attributed to the nitrogen of the pyridine ring: its positive charge reduces the proton affinity of the carboxylate and phenolate moieties, without affecting (or affecting to a minor degree) their metal ion affinity, in the same way as happens when passing from catechol to hydroxypyridinones [33]. In previous complexation work regarding Fe(III) and Al(III) [24, 34] we observed that the methylation of the pyridine ring of HPs increases the metal ion affinity even more, because of the electron donor properties of the methyl group(s), and we found that the stability increase is maximum if the methyl group is at position 1 (i.e. bound to nitrogen). The pM graphs in Fig. 9 show that DQ716 forms stronger complexes with Cu(II), Zn (II), and Al(III) than DT726 does: this result further supports our previous finding.

Application as metal–protein attenuating compounds

The dissociation constants for 1:1 Cu(II)-Aβ and Zn(II)-Aβ complexes reported in the literature [11], computed at pH 7.4 as K_D values, vary with the type of buffer used in the measurements and the methods employed for their determination. Faller and Hureau [11] suggest that a good MPAC should have K_D values of around 1–10 pmol/dm^3 and 0.1–10 μmol/dm^3 for Cu(II) and Zn(II), respectively. Table 1 shows that both DQ716 and DT726 have larger values than these, i.e. these ligands chelate Cu(II) and Zn(II) too weakly and cannot efficiently remove the metal ions from Aβ. For this reason, it is likely that DQ716 and DT726 cannot be proposed as MPACs for the therapy of AD.

Nevertheless, the weakness of the Cu(II) and Zn(II) complexes formed by DQ716 and DT726, together with the strength of their Al(III) (and Fe(III)) complexes), confirms that they can be selective drugs for the classic chelation therapy of iron and aluminium. In principle they are able to

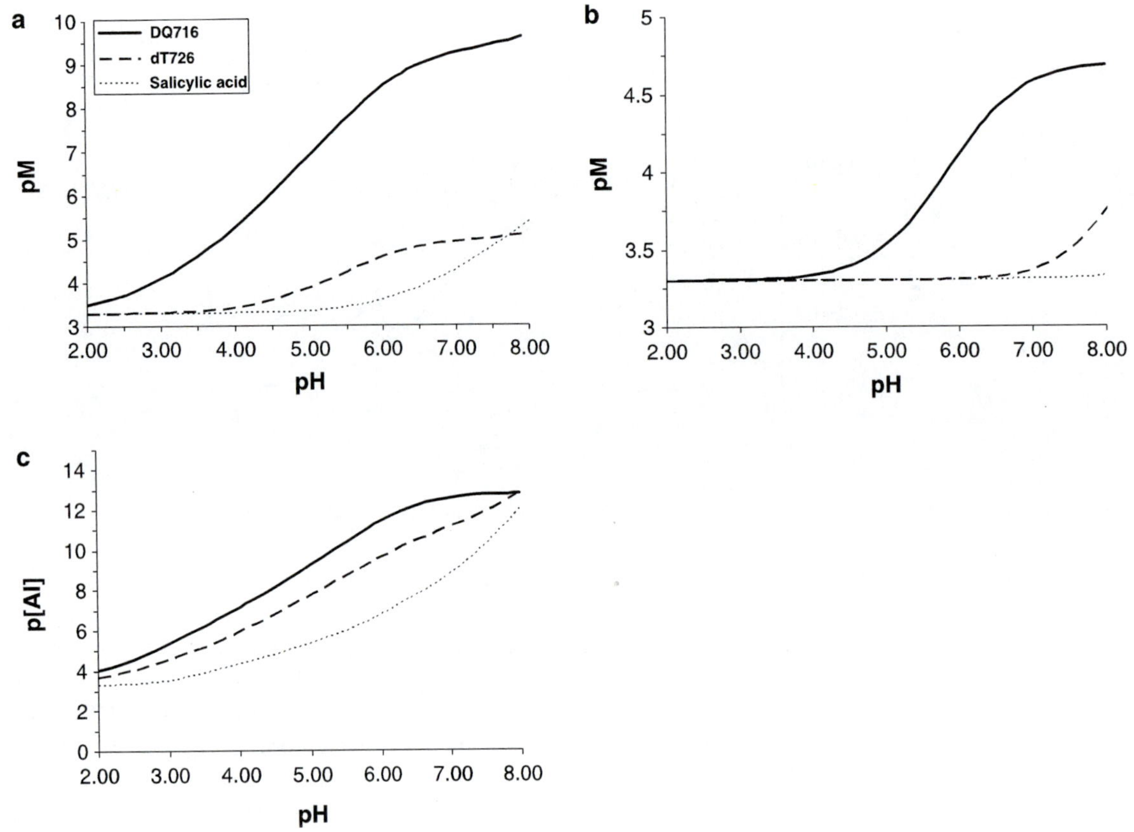

Fig. 9 pM versus pH at $C_M{}^0 = 2 \times 10^{-3}$, $C_L{}^0 = 5 \times 10^{-4}$ mol/dm^3 or mol/kg, M = Cu(II) (**a**), Zn(II) (**b**), Al(III) (**c**)

remove iron and aluminium from the body without significantly affecting copper and zinc metabolism.

Experimental

Compounds

The pH-metric titrations were performed with 0.2 mol/dm^3 KOH solution or 0.1 mol/kg NaOH prepared from KOH (Merck) and NaOH pellets (Fluka). The base was standardised against HCl solutions prepared from 36% HCl (Merck). The 0.1 mol/dm^3 metal ion solutions were prepared from $AlCl_3·6H_2O$ (Fluka), $CuCl_2$, and $ZnCl_2$ (Aldrich), and their exact concentrations were determined gravimetrically via the oxinates. DQ716 [24] and DT726 [25] were obtained by synthesis, and their solutions were standardised against base.

Solutions for UV–vis were the same as for potentiometric titrations. Solutions for ^{1}H NMR measurements were prepared by dissolving weighed amounts of DT726 and $AlCl_3$ (Carlo Erba) in D_2O (Aldrich, 99.9% atom D). The internal reference was $Me_3SiCH_2CH_2COOH$ (TSP, Aldrich).

Solutions for fluorescence measurements were prepared using the ligands, HEPES (Sigma), and HSA (Sigma). HSA solutions were freshly prepared just before the measurements.

Equilibrium measurements

The stability constants of the proton and metal complexes of the ligands were determined by pH-potentiometric titrations of 20-cm^3 samples. A Radiometer pHM 84 instrument equipped with a Metrohm combined electrode (type 6.0234.110) and Metrohm 715 Dosimat burette was used for the pH-metric measurements. The samples were in all cases thermostatted at 25.0 ± 0.1 °C, and completely deoxygenated by bubbling purified argon for ca. 15 min before the measurements. Argon was also passed over the solutions during the titrations. The electrode system was calibrated according to Irving et al. [35] and the pH-metric readings could therefore be converted into hydrogen ion concentration. About 200 titration points were used for each system. The accepted fitting of the titration curves was always less than 0.01 cm^3. Duplicate titrations were performed. The reproducibility of the titrations was within 0.005 pH units. The pH-metric titrations were performed in the pH range 2.0–11.0 or until precipitation occurred in the samples. The ligand concentrations varied in the range

1.00×10^{-3}–4.00×10^{-3} mol/dm^3. The metal ion to ligand ratios was in the range 1:1–1:4. After the mixing of Al(III) and DT726 in an acidic solution (pH ca. 3), the measured pH drifted and reached a constant value only after ca. 1 h, suggesting a low complex formation rate. The kinetics of the Al(III) + DT726 reaction became "normal" (equilibration after ca. 1 min) at pH values above 3. These findings are in agreement with previous results with other HPs [24 and references therein]. Experimental details regarding the handling of the slow kinetics during the titrations are reported elsewhere [36].

For Cu(II)- and Zn(II)-related titrations, the ionic strength of the solutions was adjusted to 0.2 mol/dm^3 KCl, and a 0.2 mol/dm^3 carbonate-free KOH solution was the titrant. For Al(III)-related titrations, medium was a 0.6 mol/kg NaCl aqueous solution, and the titrant was a 0.1 mol/kg carbonate-free NaOH solution. A 0.6 mol/kg NaCl ionic strength (=0.594 mol/dm^3) was used to allow a rigorous comparison among Al(III) data which have always been obtained in this medium ([24] and references therein). The water ionization constant, pK_w, is 13.76 ± 0.01 in KCl 0.2 mol/dm^3, and 13.69 ± 0.01 in NaCl 0.6 mol/kg.

The general equilibrium reaction is: $p\mathrm{M} + q\mathrm{L} + r\mathrm{H} = \mathrm{M}_p\mathrm{L}_q\mathrm{H}_r$, where M denotes the metal ion and L the non-protonated ligand molecule (see footnote of Table 1). Charges are omitted for simplicity. The corresponding concentration stability constants $\beta_{pqr} = [\mathrm{M}_p\mathrm{L}_q\mathrm{H}_r]/[\mathrm{M}]^p[\mathrm{L}]^q[\mathrm{H}]^r$ were calculated using the PSEQUAD computer program [40]. Calculations were always made on the experimental results obtained before precipitation. pK_A values were calculated from the protonation constants obtained by PSEQUAD and refer to the general reaction: $\mathrm{H}_h\mathrm{L} = \mathrm{H}_{h-1}\mathrm{L} + \mathrm{H}$.

UV–vis spectra were recorded at 25 °C using a Perkin-Elmer Lambda 20 spectrophotometer operating in a NaCl 0.6 mol/kg medium. For free DT726, solutions were prepared dissolving 1.39×10^{-4} mol/kg of DT726 and varying the pH between 0.3 and 1.1. Here the pH was computed from the stoichiometric concentration of HCl, because the [H^+] modifications produced by the other species were negligible under these conditions. For Al(III) + DT726, two solutions were prepared. In the first, 1.08×10^{-3} mol/kg of Al(III) and 1.06×10^{-3} mol/kg of DT726 were dissolved, the pH was between 2.4 and 3.2. In the second, 5.53×10^{-4} mol/kg of Al (III) and 1.08×10^{-3} mol/kg of DT726 were dissolved, the pH was between 2.9 and 4.5. Calculations of the stability constants were performed at the wavelength displaying the maximum absorbance variation with pH (338 and 350 nm for free DT726; 310, 328, 341, and 350 nm for Al(III) + DT726).

^{1}H NMR spectra were obtained at 25 °C using a Bruker DRX-400 spectrometer operating at 400.13 MHz. Chemical shift values are given in δ units with reference to internal TSP. Suitable integral values for the proton signals were obtained by a prescan delay of 10 s. The assignment of the proton resonances was performed by standard chemical shift correlations and NOESY measurements when necessary. Spectra were collected in D_2O solutions containing free DT726, or Al(III) + DT726. The pH was measured with a Crison 5014 combined glass electrode previously calibrated in buffered aqueous solutions at pH = 4 and pH = 7. The values of pD were computed by adding 0.41 pH units to the pH meter readings, to correct for isotopic and solvent effects due to the use of D_2O instead of H_2O [37].

Fluorescence measurements

Fluorescence spectra were measured with an Hitachi F4500 fluorescence spectrophotometer using quartz cuvettes of 1-cm optical path and operating at 25 °C. For the measurements of the free ligand (DT726), 2.0 cm^3 of a 1.5 or a 3 μmol/dm^3 ligand solution were placed into the cell. Titrations of DT726 solutions in 20 mM HEPES buffer (pH 7.4) were performed adding 80 μM HSA stock solution directly to the cuvette. The concentration of DT726 changed between 1.5–1.2 and 3.0–2.4 μmol/dm^3. The fluorescence emission spectra were registered after 5 min. An excitation wavelength of 320 nm was selected and the fluorescence spectra were recorded in the range of 390–600 nm. The fluorescence intensities were corrected for DT726 absorption of exciting light [38]:

$$F_{\mathrm{corr}} = F_{\mathrm{obs}} \times 10^{-\varepsilon c l} \qquad (1)$$

where F_{cor} and F_{obs} are the fluorescence intensities corrected and observed, respectively; ε is the molar absorption coefficient at 320 nm; l = 1.0 cm (the optical path); and c the DT726 concentration. The molar absorption coefficient was determined by UV–vis spectrometry, that is calculated as 4,792 M^{-1} cm^{-1}. The intensity of fluorescence used in this paper is the corrected fluorescence intensity [39].

The calculations were performed with the aid of the PSEQUAD computer program [40, 41]. We supposed that the two emitting forms are the DT726 and DT726-HSA adduct.

Ultrafiltration

Samples were separated by ultrafiltration through 10-kDa membrane filters (Microcon YM-10 centrifugal filter unit, Amicon, Millipore) according to a standard procedure [41]. In the protein–ligand interaction studies, all 0.50-cm^3 samples contained 200 μM HSA and the carrier ligand in 0.10 M HEPES buffer at pH 7.4. The ligand concentrations were varied from 50 μM to 1 mM. Time-dependent measurements were carried out to determine the time needed to reach equilibrium between the reacting partners. An incubation time of 20 min was found to be enough for attaining true equilibrium. The samples were then transferred to filter tubes and centrifuged for 10 min at 10,000 rpm (25 °C). After the separation, the low molecular mass (LMM) fractions were obtained. The non-bound ligand was determined by UV spectrophotometry. Control samples containing the carrier ligand and HEPES at the same concentrations but without the protein were used to determine the accurate analytical concentration of the ligand. No adsorption of the drug on the filter was observed. The binding constants were calculated with the computer program PSEQUAD [40] from the total and free concentration data pairs obtained at different protein–ligand ratios.

Acknowledgments The work was made in the frame of the CNR-HAS Bilateral Research Program, and was also supported by the Hungarian Research Funds (TÁMOP 4.2.1/B. and OTKA K77833). This paper was also supported by the János Bolyai Research Scholarship of the Hungarian Academy of Sciences (TJ). The project named "TÁMOP-4.2.1/B-09/1/KONV-2010-0005 – Creating the Center of Excellence at the University of Szeged" is supported by the European Union and co-financed by the European Regional Fund.

References

1. Glenner GG, Wong CW (1984) Biochem Biophys Res Commun 120:885
2. Masters CL, Simms G, Weinman NA, Multhaup G, McDonald BL, Beyreuther K (1985) Proc Natl Acad Sci U S A 82:4245
3. Checler F (1995) J Neurochem 65:1431
4. Atwood CS, Scarpa RC, Huang X, Moir RD, Jones WD, Fairlie DP, Tanzi RE, Bush AI (2000) J Neurochem 75:1219
5. Lovell MA, Robertson JD, Teesdale WJ, Campbell JL, Markesbery WR (1998) J Neurol Sci 158:47
6. Bush IA (2003) Trends Neurosci 26:207
7. Klafki HW, Wiltfang J, Staufenbiel M (1996) Anal Biochem 237:24
8. Hureau C, Faller P (2009) Biochimie 91:1212
9. Atwood CS, Perry G, Zeng H, Kato Y, Jones WD, Ling KQ, Huang X, Moir RD, Wang D, Sayre LM (2004) Biochemistry 43:560
10. Fang CL, Wu WH, Liu Q, Sun X, Ma Y, Zhao YF, Li YM (2010) Regul Pept 163:1
11. Faller P, Hureau C (2009) Dalton Trans 2009:1080–1094
12. Atwood CS, Moir RD, Huang X, Bacarra NME, Scarpa RC, Romano DM (1998) J Biol Chem 273:12817
13. Cuajungco MP, Faget KY, Huang X, Tanzi RE, Bush AI (2000) Ann N Y Acad Sci 920:292
14. Crichton RR, Ward RJ (2006) Metal-based neurodegeneration: from molecular mechanisms to therapeutic strategies. Wiley, Chichester
15. Bolognin S, Drago D, Messori L, Zatta P (2009) Med Res Rev 29:547
16. Bush IA, Tanzi RE (2008) Neurotherapeutics 5:421
17. Cherny RA, Atwood CS, Xilinas ME, Gray DN, Jones WD, McLean CA, Barnham KJ, Volitakis I, Fraser FW, Kim YS (2001) Neuron 30:665
18. Adlard PA, Cherny RA, Finkelstein DI, Gautier E, Robb E, Cortes M, Volitakis I, Liu X, Smith JP, Perez K, Laughton K, Li QX, Charman SA, Nicolazzo JA, Wilkins S, Deleva K, Lynch T, Kok G, Ritchie CW, Tanzi RE, Cappai R, Masters CL, Barnham KJ, Bush AI (2008) Neuron 59:43
19. Hindo SS, Mancino AM, Braymer JJ, Liu Y, Vivekanandan S, Ramamoorthy A, Lim MH (2009) J Am Chem Soc 131:16663
20. Lakatos A, Zsigó É, Hollender D, Nagy NV, Fülöp L, Simon D, Bozsó Zs, Kiss T (2010) Dalton Trans 39:1302
21. Green DE, Bowen ML, Scott LE, Storr T, Merkel M, Bohmerle K, Thompson KH, Patrick BO, Schugar HJ, Orvig C (2010) Dalton Trans 39:1604
22. Sudlow G, Birkett DJ, Wade DN (1975) Clin Exper Pharmacol Physiol 2:129
23. Muller N, Lapicque F, Drelon E, Netter P (1994) J Pharm Pharmacol 46:300
24. Dean A, Ferlin MG, Brun P, Castagliuolo I, Yokel RA, Badocco D, Pastore P, Venzo A, Bombi GG, Di Marco VB (2009) Dalton Trans 2009:1815–1824
25. Dean A, Di Marco VB (in preparation)
26. Di Marco VB, Yokel RA, Ferlin MG, Tapparo A, Bombi GG (2002) Eur J Inorg Chem 2002:2648–2655
27. Sudlow G, Birkett DJ, Wade DN (1975) Mol Pharmacol 11:824
28. Sudlow G, Birkett DJ, Wade DN (1976) Mol Pharmacol 12:1052
29. Seedher N, Bhatia S (2006) Pharmacol Res 54:77
30. Bi S, Song D, Tian Y, Zhou X, Liu Z, Zhang H (2005) Spectrochim Acta A 61:629
31. Chengnong Y, Jinqiang T, Dan X, Yi L, Zuting P (2006) Chin J Anal Chem 34:796
32. Ahmad B, Parveen S, Khan RH (2006) Biomacromolecules 7:1350
33. Scarrow RC, Riley PE, Abu-Dari K, White DL, Raymond KN (1985) Inorg Chem 24:954
34. Di Marco VB, Dean A, Ferlin MG, Yokel RA, Li H, Venzo A, Bombi GG (2006) Eur J Inorg Chem 2006:1284–1293
35. Irving HM, Miles MG, Pettit LD (1967) Anal Chim Acta 38:475
36. Di Marco VB, Tapparo A, Bertani R, Bombi GG (1999) Ann Chim (Rome) 89:535
37. Covington AK, Paabo M, Robinson RA, Bates RG (1968) Anal Chem 40:700
38. Whalley CV, Rankin SM, Hoult JRS, Jessup W, Leake DS (1990) Biochem Pharmacol 39:1743
39. Epps DE, Raub TJ, Caiolfa V, Chiari A, Zamai M (1999) J Pharm Pharmacol 51:41
40. Zékány L, Nagypál I, Peintler G (1991) Technical software distribution, Baltimore
41. Enyedi ÉA, Farkas E, Dömötör O, Santos MA (2011) J Inorg Biochem 105:326
42 Kurz H (1986) In: Reidenberg MM, Erill S (eds) Drug protein binding. Praeger, New York, p 70
43 Martell AE, Smith RM (1982) Critical stability constants, vol 5. Plenum, New York
44 Kiss T, Atkári K, Jezowska-Bojczuk M, Decock P (1993) J Coord Chem 29:81

Spatially resolved imaging methods to probe metals in the brain: from subcellular to organ level

Sylvain Bohic, Dominic Hare, Alexia Daoust, Peter Cloetens, and Emmanuel L. Barbier

Abstract

Very little is known about the subcellular distribution of metal ions in cells. Some metals such as zinc, copper, and iron are essential and play an important role in the cell metabolism. Dysfunctions in this delicate housekeeping may be at the origin of major diseases. There is also a prevalent use of metals in a wide range of diagnostic agents and drugs for the diagnosis or treatment of a variety of disorders. This is becoming more and more of a concern in the field of nanomedicine with the increasing development and use of nanoparticles, which are suspected of causing adverse effects on cells and organ tissues and particularly the brain. Various analytical methods are developing into well-suited submicrometer analytical tools for addressing new problems when studying the role of metals in the brain. In the present review, we describe the possibilities and some current applications offered by three major techniques that cover imaging of metals from subcellular to organ level, namely synchrotron-based X-ray microspectroscopy, laser ablation inductively coupled plasma-mass spectrometry (LA-ICP-MS) and, less thoroughly addressed, the capabilities of magnetic resonance imaging (MRI) for *in vivo* imaging of metals in the brain.

Keywords

Metal ions • Brain • Synchrotron • Mass spectrometry • Magnetic resonance imaging

Introduction

It is known that living systems depend for survival on their ability to accumulate and use some elements, particularly metal ions, to define a certain composition that is held constant within a given homeostatic state. Several essential metal ions participate in the control of numerous metabolic and signaling pathways. Their coordination chemistry and redox properties confer them a propensity to randomly coordinate and catalytically react inside the cell with protein sites other than those tailored for that purpose. Indeed, about one-third of all structurally characterized proteins are metalloproteins and bound metal ions or co-factors play a pivotal role in the structure–function relationship of proteins and other bio-molecules [1]. In addition, these cellular essential metals are also potentially toxic. Their uptake is thus tightly regulated by intracellular transport and compartmentalization. To avoid toxic reactions involving metals, a number of sophisticated networks of trafficking pathways are available.

The brain is an exceptional tissue in that it constitutively concentrates biometals for its functional metabolism but is

S. Bohic (✉)
Inserm U-836, Grenoble Institut des Neurosciences, Team 6: Rayonnement Synchrotron et Recherches Médicales, ESRF, 6 rue Jules Horowitz, BP220, 38043 Grenoble, France

European Synchrotron Radiation Facility, ESRF, BP220, Grenoble, France

Université Joseph Fourier 1, Grenoble Institut des Neurosciences, Grenoble 38042, France

Elemental Bio-imaging Facility, University of Technology, Sydney, Broadway, New South Wales, Australia
e-mail: bohic@esrf.fr

W. Linert and H. Kozlowski (eds.), *Metal Ions in Neurological Systems*,
DOI 10.1007/978-3-7091-1001-0_18, © Springer-Verlag Wien 2012

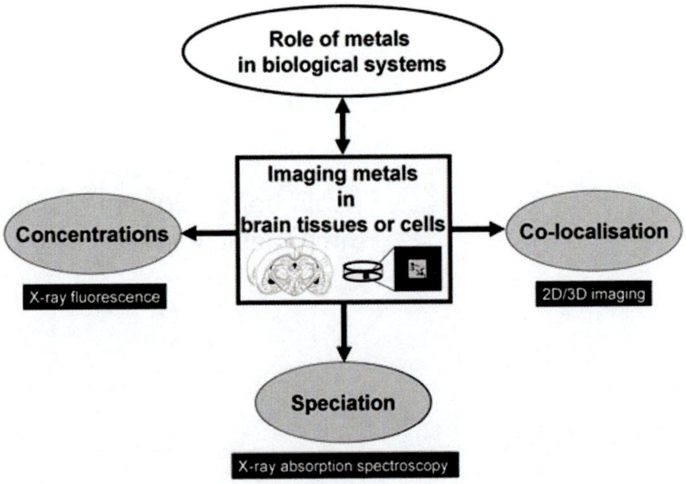

Fig. 1 Scheme that depict the synchrotron microspectroscopy approaches in the study of metals in the brain, namely metal ions concentrations (dose), elemental distribution (co-localization), and chemical state (speciation)

particularly prone to oxidative stress. Indeed, abnormalities of metal ion biochemistry in neural tissue arise mainly by protein aggregation mediated by metal ions and oxidative reaction catalysed by redox-active metals. More generally, toxicity of essential or non-essential metals does not depend only on their total local concentration but also depends on their chemical form. Therefore, the brain chemical microenvironment has to be strictly controlled, the blood–brain barrier being a key element of this regulation [2]. Besides the well-known and abundant alkali metals Na, K, and Ca, which control cell excitability, signal transduction, synaptic transmission and plasticity, or gene expression, transition metals are essential for normal brain development, myelination, or are involved in neurotransmitter synthesis and also regulation of synaptic transmission. One abundant example is zinc (more specifically Zn^{2+}), which plays a dynamic role in both the physiology and pathophysiology of brain function. Cortical neurons release ionically exchangeable copper (Cu) and Zn during depolarization and neurotransmission [3, 4]. The other main transition metals like iron (Fe) [5], manganese (Mn) [6], Cu [7], or selenium (Se) [8] are also of importance in the brain functioning. Next, certain metals (aluminum, arsenic, cadmium, mercury, lead, uranium, etc.) with no known benefit for human physiology are of health concern and may trigger neurotoxic responses.

Metals have a heterogeneous distribution throughout the brain and in spite of all the progress made, still, our ignorance on metals' trafficking in neurons' cell bodies, axons or synapses, their distribution, concentration, i.e., their homeostasis in the different regions of the brain and associated nerve cells, remains vast. Indeed, biological metal-relevant information mainly relies on their concentration measured in brain homogenates [9–11]. Various types of neuronal cells are then found mixed and contamination from adjacent tissues cannot be avoided when dissecting small brain nuclei. Quantitative elemental imaging and speciation of metal ions, in cells and their subcellular compartments are challenging but important objectives of modern analytical chemistry. Recent advances in spatially resolved analytical probe can overcome the aforementioned drawbacks (Fig. 1).

Different methods are used to look at the distribution of metals within tissues and cells. Histochemical methods are semi-quantitative but have proved to be useful to localize various metals like iron or zinc in tissues and applied to central nervous system (CNS) [12–14]. Optical fluorescence imaging with metal-responsive small molecule offer the potential to probe pathophysiological consequences of the cell biology of free metal ions in a spatially and time-resolved manner. This imaging method will not be discussed here but a comprehensive literature review of applications in the field of metals in neurobiology can be found elsewhere [15, 16]. These methods do not allow quantitative multi-elemental analysis and prevent the possibility to study element inter-relationships. Modern techniques in electron microscopy like scanning transmission electron microscopy with energy-dispersive X-ray analysis (STEM-EDX) or energy-filtered transmission electron microscopy (EFTEM) provide also some information on the distribution of metals at the ultrastructural level [17]. Their sensitivities are, however, limited to detect trace element, a fortiori when ultrathin sample sections are required (about 30 nm) and restricted size of the area probed. Similar sample preparation is needed to perform nano-secondary ion mass spectrometry (nano-SIMS), which is fully complementary to electron microscopy and provides high sensitivity isotopic analysis in addition to the atomic composition. However, quantification is difficult due to matrix effects that influence the secondary ion yield and depends on the chemical environment of the surface being sampled. Particle-induced X-ray emission (PIXE) technique is a fully quantitative method and has a good sensitivity at the micron scale and even at sub-micron resolution, as recently demonstrated [18]. All these techniques are complementary and should be selected

Table 1 Major spatially resolved methods for imaging metals in biology

	STEM-EDX	STEM-EELS/ EFTEM	Dynamic SIMS	X-PEEM	SXRF [micro-XANES]	PIXE	LA-ICP-MS	MRI
Selectivity	Atomic composition	Atomic composition, Chemical	Atomic composition, Isotopic	Atomic composition, Chemical	Atomic composition [Chemical]	Atomic composition	Atomic composition, Isotopic	Indirect, Paramagnetic metals
Spatial resolution (nm)	1	1	50–100	20	50–1,000 [150–1,000]	500–1,000	5,000–20,000	>20,000
Analytical depth (μm)	0.1–1	<0.5 @ 200 keV/<0.1 @ 100 keV	<0.1	<0.05	10–1,000	10–100	200	>10,000
Field of view	<10 μm^2	<10 μm^2	<0.1 cm^2	10–100 μm^2	Sub mm^2 to cm^2	Sub mm^2	Sub cm^2 to 200 cm^2	>cm^2
Detection limit (mg/kg)	>1,000	>1,000	0.1	100–1,000	<0.1 [10–100]	1–10	<0.1	Not known
Quantification	Semi-quantitative	Semi-quantitative	Semi-quantitative	Semi-quantitative	Quantitative [poor]	Quantitative	Quantitative	Semi-quantitative
Samples	Thin section (resin-embedded material, frozen-hydrated)	Thin section (resin-embedded material, frozen-hydrated)	Semi-thin or thin section of resin-embedded material	Thin section of resin-embedded material; cryofixed and freeze-dried	Frozen hydrated; dried or embedded material	Freeze-dried samples/ embedded material	10–40 μm Native cryosections	*In vivo, ex vivo*
Environment	High-vacuum	High-vacuum	Ultra high-vacuum	Ultra high-vacuum	In air or high-vacuum	High-vacuum	In air	In air, in liquid

EDX X-ray energy-dispersive spectrometry, *EELS* electron energy loss spectrometry, *EFTEM* energy filtered transmission electron microscopy, *LA-ICP-MS* laser ablation inductively coupled plasma mass spectrometry, *PIXE* particle-induced X-ray emission, *SIMS* secondary ion mass spectrometry, *SR-XRF* synchrotron-radiation X-ray fluorescence, *STEM* scanning transmission electron microscopy, *XANES* X-ray Absorption Near-Edge Spectroscopy, *X-PEEM* X-ray photoelectron emission microscopy, *MRI* magnetic resonance imaging

depending upon the research objectives, type of sample, spatial resolution and sensitivity, required throughput and even the availability of the technique. As a guide for the reader, key performance indicators of the most commonly used metal imaging techniques are compiled in Table 1.

Here, we provide a brief review of the possibilities and some current applications in neurosciences offered by three major techniques that cover imaging of metals from subcellular to organ level, namely synchrotron-based X-ray microspectroscopy, laser ablation inductively coupled plasma-mass spectrometry (LA-ICP-MS), and, less thoroughly addressed, the capabilities of magnetic resonance imaging (MRI) for *in vivo* imaging of metals in the brain.

Synchrotron-based X-ray fluorescence for chemical imaging of the brain at mesoscale

The X-ray fluorescence (XRF) method depends on fundamental principles that are common to several other instrumental methods involving interactions between electron beams, accelerated protons, or X-rays. These allow to analyze the elemental composition of the sample. Besides the determination of the local composition, beams can be focused and sample raster-scanned to map elemental distribution. The advantage of the use of high brilliance, high-energy synchrotron as an excitation source for XRF analysis, coupled with recent advances in manufacturing technologies of X-ray-focusing optics, is that high resolution and high-sensitivity XRF probe can be achieved. The synchrotron radiation is naturally collimated and has low divergence associated to high-incident flux capabilities thus contributing to lower radiation damages when compared to electron or proton probes. The physics of X-ray interaction with material is well described and simpler than for the other methods, resulting in more straightforward quantification procedures. This is particularly true for the analysis of high atomic number elements diluted in a matrix of low atomic number elements (thin biological sample like a cell or tissue section containing mostly carbon and other light elements). In this case, absorption and secondary enhancement effects can be neglected and the intensity of XRF is directly proportional to the concentration of elements within the sample [19]. Typical areal detection limits and absolute detection limits are in the order of ng/cm^2. The attogram range detection level may now be achieved at sub-100 nm level on synchrotron nanoprobe [20–22]. Although the high flux delivered by synchrotron sources greatly contributed to this high sensitivity of XRF probe, the possibility to use

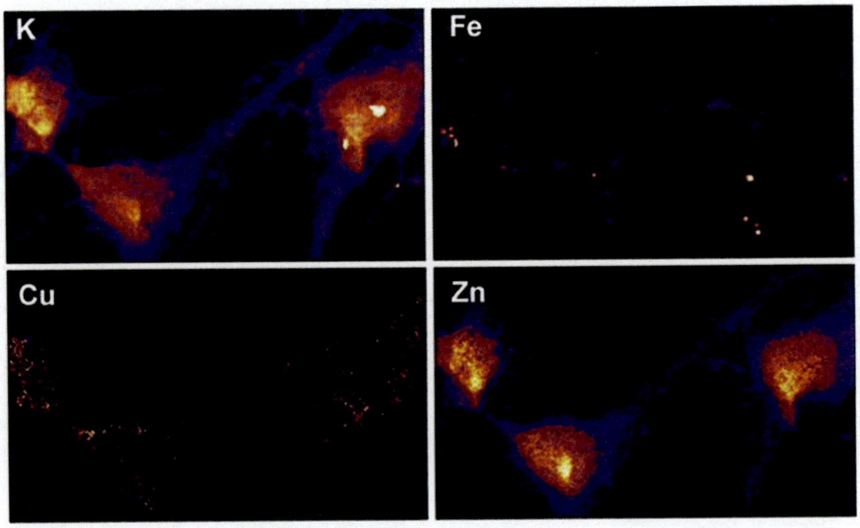

Fig. 2 Two-dimensional (2D) elemental mappings of cortical neurons. These representative 2D mappings of potassium (K), iron (Fe), copper (Cu), and zinc (Zn) in isolated mouse cortical neurons were obtained at the European Synchrotron Radiation Facility (ESRF) ID22NI nanoimaging synchrotron beamline using incident photon energy of 17.5 keV, a stepsize of 150 nm and a dwell time of 300 ms

monochromatic radiation and the intrinsic high degree of polarization yields an important reduction in the spectral background from scattered radiation, resulting in improved limits of detection. The other asset of the synchrotron source is the energy tunability that enables X-ray absorption spectroscopy to probe the chemical local environment of an element within subcellular compartment [22–24]. These improvements contributed to the increase use of synchrotron micro/nano-analysis for a more comprehensive view of the role of metals in the cellular physiopathology by studying from the tissular to subcellular scale ("mesoscale") the heterogeneous distribution of metals.

The large penetration depth of hard X-rays (ranging from a few microns to a few mm for a biological matrix depending on the excitation energy) offers possibilities for simple sample preparation procedures and more versatile in-situ observations in controlled environments. Whole cells can be studied avoiding cell fractionation and other preparation steps that might modify the chemical species. Cells are generally grown onto appropriate XRF substrates (thin inert polymer or silicon nitride membranes free of contaminants) while thin brain tissues cryosections can be deposited onto micrometer-thick polycarbonate film stretch over a plastic frame holder [22, 25]. Laser microdissection slides can also be used [26]. Thorough washing of the cell is required to remove extracellular medium that would interfere with the analysis. Cells are first washed through phosphate buffer but a few second rinsing with an ice-cold solution such as ammonium acetate or pure water is often necessary. Further, cells are cryo-immobilized through fast freezing methods to preserve the elemental information of diffusible or displaceable ions but also of cell morphology and intracellular structures [27]. Some analytical setup allow vitrified sample to be analyzed at cryogenic temperature with cells or tissues in their near-native hydrated state [24, 28, 29]. Synchrotron cryo-scanning facilities are now in development worldwide but, up to now, freeze-drying at low temperature under vacuum was the method of choice to best preserve cells for synchrotron micro/nano-analysis [22].

On another hand, human brain tissue collections rely on archived tissue samples that are most of the time chemically fixed. Although the effects of chemical fixation and long-term fixed tissue storage are still under debate [30–33], there are evidences that chemical fixation, dehydration using organic solvents and drying, often causes redistribution and the washing-out of ions in cellular structures. Even if all procedures are carried out carefully, this still raises the question about the redistribution and loss of elements caused by chemical fixation methods. Thus, fresh frozen tissues should be considered as far as possible for imaging of metals and probe their chemical state.

An increasing number of applications demonstrated the wide potential of synchrotron-based X-ray fluorescence (SRXRF) to explore the distribution of metals in various physiopathological conditions on *in vitro* models of neuronal cells (Fig. 2), in brain of small animals and in human post-mortem brain tissues (Fig. 3). Cellular models are adequate to investigate many aspects of neurodegenerative disorders. Ortega et al. provide the first evidence of iron-dopamine co-localization in neuro-vesicles and demonstrated that when dopamine production is obstructed, the iron in the vesicles drastically decreases [20]. The subcellular chemical imaging capabilities of the synchrotron nanoprobe greatly contributed to reveal this new function of dopamine vesicles in iron storage that is of critical importance to understand the molecular mechanisms involved in Parkinson's disease. Recently, in support to electrophysiological recording and imaging experiments with fluorescent Zn^{2+} probe, SRXRF could show a higher Zn content in the nuclear region indicating that the intracellular distribution of this metal was influenced by the over-expression of TRPC6 channels, a plasma membrane cationic channels widely expressed in the brain [34]. As chemical images provided are largely static, these examples highlighted the importance of

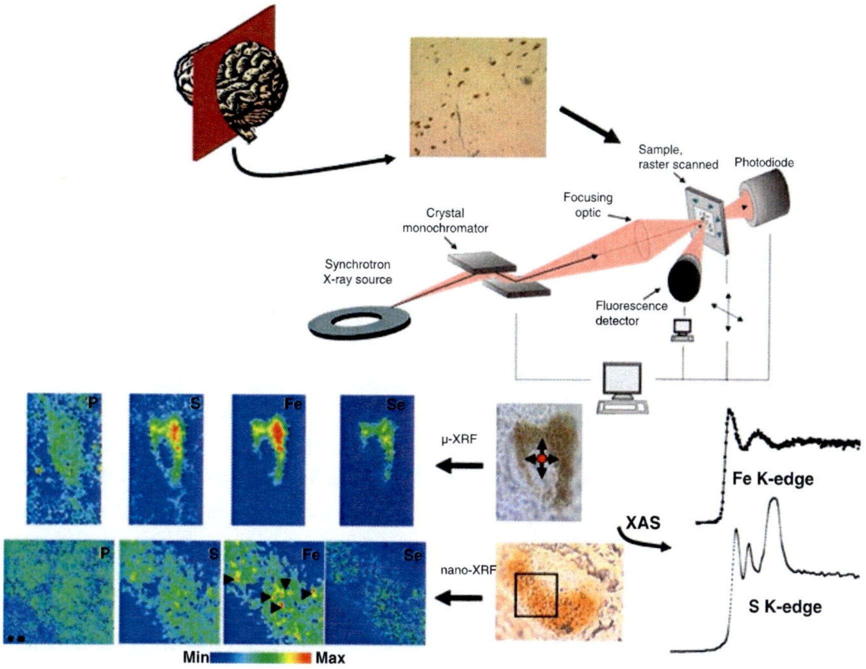

Fig. 3 Scheme showing the different steps involved in the synchrotron X-ray microspectroscopic analysis of pigmented neurons of the human substantia nigra. The 8 μm thick brain tissue sections are mounted on a thin polymer film. Iron (Fe) and sulfur (S) intracellular speciation (XAS) are performed using micro X-ray absorption spectroscopy at the sulfur and iron K-edge at the European Synchrotron Radiation Facility (ESRF). Using X-ray fluorescence (XRF), ESRF microprobe, and nanoprobe, neuromelanin-containing neurons were raster scanned to generate a pixel-by-pixel spectral image providing topographical elemental maps at micrometer and sub-100 nm level, respectively. *Arrow heads* show identified iron-rich microdomains within neuromelanin aggregates, scale bar 1 μm

combining SRXRF approach to conventional laboratory molecular, biological, and biochemical techniques to have a better understanding of cellular metal homeostasis that accompany processes of development, differentiation, senescence, and stress responses, etc. This is also well exemplified by the work of Duong et al. who were able to observe that neuroglobin ameliorates the neuronal response to experimental hypoxia-reoxygenation injury through maintenance of cellular bioenergetics, inhibition of calcium (Ca^{2+}) influx, a reduction in cellular uptake of Fe, Cu, and Zn at the expense of sulfur, chlorine, and potassium and an enhancement of cell viability through inhibiting necrosis and apoptosis [35]. Another important result was obtained through correlative imaging using Cu fluorescent sensor and SRXRF [36]. In this work, the authors discovered that neuronal cells move significant pools of Cu from their cell bodies to peripheral processes upon their activation and are also dependent on calcium release. Although SRXRF cannot identify which cuproprotein is implicated in these processes, it is suggested that ATP7A is involved in neuronal process guidance [36]. The importance of metals was also reported to study the chemical composition and distribution of metabolites during biogenesis in the process of myelination [37]. Specific element accumulation sites in myelin at the subcellular level were observed [37]. Subcellular chemical imaging is also well suited to explore the toxicity mechanisms of some metals as demonstrated for Mn within dopaminergic neuron-like cells [38].

SRXRF can cover large sample area scanning and it is thus found ideally suited to image trace elements within animal brain tissues cryosections. The development of models of neurodegenerative diseases based on transgenic animals opens the possibility to study the link between metal dyshomeostasis and neurodegenerative processes. This has been applied in animal models of Alzheimer's disease [39, 40] or in the case of Syrian hamsters intracerebrally infected with the scrapie strain ME7-H [41]. In the latter, the authors found that the copper-binding prion protein (PrP) in its pathological form PrP^{Sc} was diffusely dispersed in the scrapie-infected brains, and that the copper distribution differed from that in the control animals. Also of interest, the potent relationship between copper hot spot and pathological deposit of PrP^{Sc} needs to be investigated [41]. Specific pathways for metal regulation can also be investigated, such as in the work of Linkous et al. who use ZNT3 knockout mice and show that ZNT3 protein is essential to control the actual Zn concentration in mossy fiber synaptic vesicles. This confirms the metal transporting role of the ZNT3 protein in the brain [42].

Increasing number of studies explore the role of metals in the physiopathological mechanisms of neurological diseases through the analysis of post-mortem human brain tissues. Rapid scanning XRF has now been used to get quick anatomic views of global metal distribution in slices of whole brain [43]. This has been applied to neurodegenerative disorders such as Parkinson's disease [44] or spinocerebellar ataxia [45]. The role of metals has also been investigated in other disorders like Alzheimer [40], the amyotrophic lateral sclerosis [46, 47], or brain tumors [48]. Beyond a better understanding of the contributions of metals to these pathologies, these examples suggest that one could get the fingerprint of a particular disorder using a quantitative evaluation of trace elements.

After elemental mapping within cells or tissues from the CNS are generated and thus informed on the quantitative intracellular distribution of metals, points of interest can be chemically analyzed using X-ray near-edge spectroscopy (XANES). Indeed, the improvements in synchrotron microprobe as mentioned above opened the field of speciation at the cell level that result in the characterization of the local structure of the absorber elements (oxidation state, coordination, ligand) within subcellular compartments [22, 49]. Very little is known about the chemical form of some metals that are found in neurons. At first, XANES was used on fresh frozen pellet of cells with unfocused X-ray beams. For example, Gunter et al. studied Mn toxicity in PC12 cells differentiated to a phenotype similar to neurons through the action of nerve growth factor (NGF), but also in primary rat astrocytes and neuroteratocarcinoma cells (NT2) [50, 51]. These XANES data collected failed to show the formation of an active Mn^{3+} complex by oxidation of Mn^{2+} that could have been suspected to play a role in Mn neurotoxicity [52]. Recently, Mn was found to accumulate within the Golgi apparatus of PC12 dopaminergic cells, but until now direct speciation of Mn in the different subcellular compartment has not yet been reported [38]. The same group reported speciation of Fe in NGF-differentiated PC12 cells cultured in normal conditions and found Fe to be present as Fe^{3+} in their nucleus and cytoplasm [24]. Micro-XANES is still a challenging technique and has only been scarcely applied to subcellular analysis [22, 24, 53]. The recent advances in synchrotron instrumentation make it one of the very rare techniques to allow intracellular speciation. However, micro-XANES has been more applied to animal and post-mortem human brain tissues with a particular focus on Fe. Free iron has long been suspected to be implicated in several neurodegenerative diseases where abnormal Fe handling and redox transitions that may occur *in vivo* exacerbates the production of free radicals and further induce oxidative stress to neurons. Micro-XANES has been used to study chemical forms of mitochondrial iron in Friedreich's Ataxia [54]. The analysis of chemical states of Fe in the neuromelanin and in and around the substantia nigra pars compacta neurons show that Fe shifted toward Fe^{3+} in parkisonian cases, but showed mixed states of Fe^{2+} and Fe^{3+} in the parkinsonism–dementia complex cases and control cases [55]. This was also observed in brain tissues of monkeys injected with 1-methyl-4-phenyl-1,2,3,6-tetrahydropyridin (MPTP) [56]. Other studies confirmed that Fe in human parkisonian brain tissues was present only as Fe^{3+} in nigral neurons, with a spectral signature that closely match the one of ferritin [26, 57]. Copper oxidation state has also been explored in Parkisnon's disease, and shows that most of the Cu inside nigral neurons occurs in tetrahedral environment and probably as Cu^{2+} [58].

One of the limitations of these results is the use of human post-mortem tissues that only allows an evaluation of the end point of the disease and not necessarily the process by which neuronal cell death has occurred. Analysis at different point in the time course of the disease reproduced to some extent in animal models could be of interest to better understand which chemical forms of metals and to what extent they are involved in neurodegenerative processes.

Such researches also apply to metals without known physiological relevance such as lead (Pb), mercury (Hg), or cadmium. These heavy metals can enter our bodies by way of food, drinking water, and air, and they can produce toxicity by forming complexes with cellular compounds containing sulfur, oxygen, or nitrogen. The complexes inactivate enzyme systems or modify critical protein structures leading to cellular dysfunction and death [59]. The central neurotoxicity caused by Pb is caused by disruption of the intercellular junction that seals the capillary endothelium and chronic exposure to Pb affects numerous neurotransmitter systems [60]. SRXRF mapping of Pb distribution in brain tissues have been reported [61], but, on the overall, the ways heavy metals distributed into the CNS and under which chemical forms are poorly understood. Recently, an interesting study investigated the chemical forms of Hg in the brain. Korbas et al. shows that after high exposures to methylmercury, the selenium levels in the human cortical brain were significantly elevated and the mercury was present in at least two chemically distinct species, inert nano-particulate mercuric selenide, and mobile mercuric bis-thiolate [62]. Their studies suggested that selenium plays an essential role in the detoxification process of inorganic and organic mercury in the brain [23, 62]. The possibilities offered by synchrotron microspectroscopies open new avenues for directly investigating the molecular toxicology of heavy metals, including testing of new chelation therapy drugs.

Imaging of metals in brain regions using laser ablation-inductively coupled plasma mass spectrometry

Inductively coupled plasma-mass spectrometry (ICP-MS) is a highly sensitive technique for trace element analysis capable of achieving detection limits in the parts-per-trillion ($ng\ kg^{-1}$) while maintaining a linear response over 9 orders of magnitude. The major limitation of ICP-MS is the need for aqueous samples, requiring complex bulk digestion procedures that are prone to contamination and render spatial information invalid. Laser ablation (LA) was first coupled to ICP-MS in 1985 as an alternative sampling technique that allows for the direct analysis of solids [63], negating the need for dissolution of the sample in strong mineral acids. Initial uptake of LA-ICP-MS was mostly confined to geological and environmental sciences [64]. Increased interest in applying LA-ICP-MS techniques to biological samples, including the brain, has been responsible for an increased number of applications appear in the last 10 years, particularly those involving imaging [65–67].

In a typical imaging LA-ICP-MS experiment, a cut section mounted on a microscope slide is placed in an ablation chamber, which is in turn fixed on a movable *x–y–z* stage. Focused laser energy beams ranging from 4 to 250 μm in diameter are directed onto the sample surface, where material is ablated. These particles are swept to the ICP-MS by a carrier gas (usually argon, though helium has been suggested to provide superior performance for some applications [68]), where they are atomized and ionized by a high-temperature argon plasma. Ions pass through an interface region and enter a mass spectrometer (typically quadrupole-based) [69], where they are separated on the basis of their mass-to-charge (m/z) ratio and detected [70, 71]. Sensitivity is partially controlled by ablation conditions—larger beam diameters ablate more material at the cost of spatial information. Achieving robust quantitative data is a challenge in LA-ICP-MS, as a lack of suitable matrix-matched reference materials often results in a reliance on single-point calibrations.

Imaging of trace elements in biological samples is a process not dissimilar to that of proteins and peptides by organic mass spectrometry [72]. The motorized sample stage allows the tissue section to traverse the laser beam, with the resultant particles detected by the ICP-MS in a time-resolved manner. Lines of ablation are drawn across the section, with each line producing a single data file that can be combined into a hyperspectral image for all measured m/z values [73]. Quantitative information can be provided through representative ablation of a standard material [74].

Samples are best analyzed as a flat cut surface, such as a microtomed section. Optimal sample thickness is usually dependent on laser performance (high-energy beams can cause sample fracturing, particularly in very thin sections), though as a guide approximately 30 μm is an appropriate thickness. Contamination and metal redistribution during sampling should always be considered. For example, transfer from stainless steel blades may preclude analysis of iron and chromium. Samples should be air-dried in clean conditions prior to analysis. Feldmann et al. suggested using a cryogenically cooled ablation chamber might improve signal stability when ablating soft tissue [75]. Use of an internal standard element, typically ^{13}C, can compensate for signal drift and variation in laser power output [76, 77].

An increasing number of applications describing the use of LA-ICP-MS in imaging brain tissue have appeared in recent years. Early examples using high-resolution magnetic-sector mass analyzers were described by Becker et al., who conducted proof-of-principle imaging experiments on both mouse [78, 79] and human [80, 81] tissues. LA-ICP-MS imaging was used in conjunction with receptor autoradiography to profile Cu, Fe, Pb, and uranium distributions in human glioblastoma multiforme [82]. A similar method was used to observe Cu, Zn, phosphorus, and sulfur distribution in implanted F98 cell tumors in rat brains [83]. All these described examples determined metal concentrations through ablation of spiked tissue homogenates prior to imaging.

Jackson et al. used a series of marine-certified reference materials to quantify copper, iron, and zinc in rat brain sections [84]. The development of quantitative methods has allowed LA-ICP-MS imaging to examine more closely the role of redox-active metals in neurodegeneration pathology, particularly through the study of metal dyshomeostasis following neurotoxin administration. Increased nigral iron levels were confirmed by LA-ICP-MS analysis of brain sections taken from unilateral 6-hydroxydopamine (6-OHDA)-lesioned mice, with no obvious alterations in equivalent Zn, Cu, and Mn concentrations [85]. In parkinsonism stimulated by MPTP administration in mice, no corresponding nigral Fe increase was observed through imaging of brain sections, though significant decreases in copper levels were identified in the periventricular zone and fascia dentata. The potential for LA-ICP-MS imaging for targeted study of pathological features observed in neurodegeneration was demonstrated by Hutchinson et al. [86], who used europium-labeled antibodies to mark and image β-amyloid plaques in transgenic TASTPM mice, a commonly used model of Alzheimer's disease.

Contemporary LA-ICP-MS imaging development is progressing in two directions: larger scale and three-dimensional imaging, and imaging approaching single cell resolution. Imaging of larger samples has required improvements in LA-ICP-MS acquisition time, which is limited by both laser parameters and ICP-MS acquisition speeds [87]. Increased capacity for LA-ICP-MS imaging of larger sample areas has resulted in preliminary studies on the feasibility of producing three-dimensional volume reconstructions. Twelve serial sections taken from a 6-OHDA-lesioned mouse brain were used to produce three-dimensional models of Fe, Cu, Zn, and Mn as shown in Fig. 4 [87]. Progression of large-scale imaging is moving toward visualization of metal distribution throughout the whole mouse brain. On a smaller scale, Giesen et al. recently described an innovative method applying iodine-labeled antibodies as a means to discriminate between immunostained fibroblasts and surrounding tissue [89]. With a maximal practical resolution of 4 μm, LA-ICP-MS is capable of producing high-resolution images at the neuron level using similar methods. Application of the near-field effect [88] to

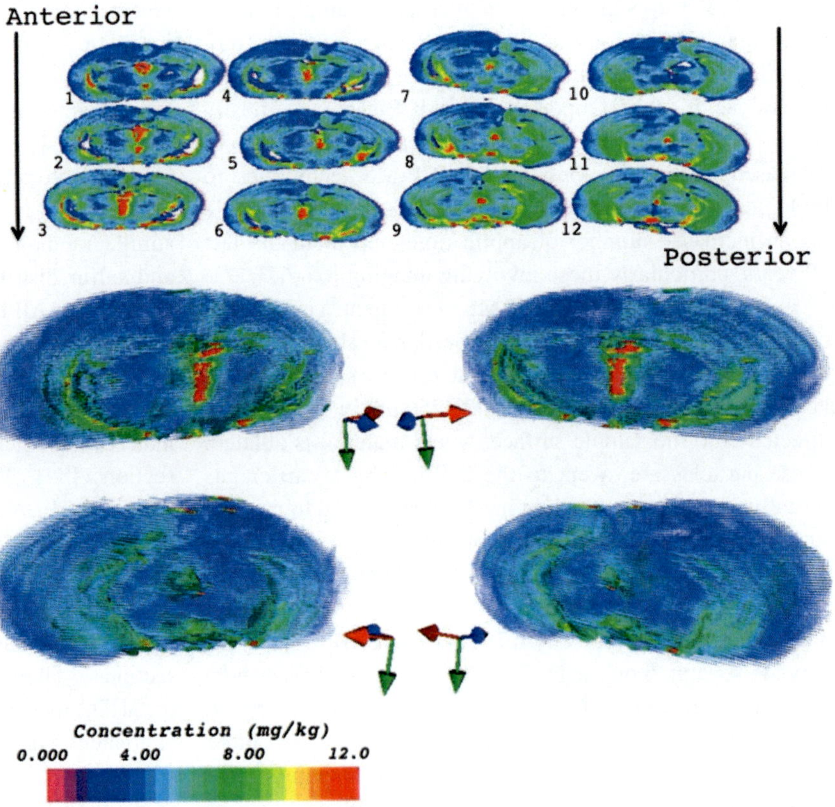

Fig. 4 Two dimensional images (*top*) and three-dimensional models of copper in a 6-OHDA-lesioned mouse brain at the level of the substantia nigra. Reproduced by permission of The Royal Society of Chemistry

LA-ICP-MS imaging, which decreases the incident laser beam diameter to as low as 200 nm, will further improve LA-ICP-MS imaging as a complementary technique to higher resolution synchrotron-based techniques [89].

Innovative quantitative methods have been proposed to improve accuracy and precision while reducing total analytical time, all of which are viable in the analysis of brain tissue. Modified ink jet printers have been used to deposit precise concentrations of standard solutions onto matrix-matched supports, producing quantitative data in good agreement with alternative calibration methods [92]. Homogenized brain tissue samples spiked with metal standards have been used to produce reasonable multi-point calibrations [90], as have whole blood and serum samples [91]. Austin et al. proposed a novel method for quantification where doped polymethylmethacrylate films were used in place of matrix-matched tissue standards (Fig. 5) [77].

Magnetic resonance imaging for *in vivo* imaging of paramagnetic metals

Even if it has not the specificity of the previously mentioned techniques, MRI may be used to indirectly detect some metals with a good sensitivity *in vivo* contrary to the aforementioned methods that required fixed samples. Biological applications of MRI usually rely on the signal from water protons. Other nuclei will not be mentioned in the following but in principle, they would behave like protons in the presence of metal. However, the presence of a paramagnetic or a diamagnetic metal in the vicinity of water protons affects the collected MR signal, thus enabling the detection of the metal's presence. A superparamagnetic (e.g., iron oxide particles) or a paramagnetic (e.g., Fe, Mn, or gadolinium) may interact with the MR signal from protons in two ways: a relaxivity effect and a susceptibility effects. A diamagnetic metal (e.g., Zn, Cu, or bismuth) causes only weak susceptibility effects.

Superparamagnetic and paramagnetic metals catalyze the spin relaxation (spin–spin and spin–network mechanisms) due to their unpaired electrons and thereby reduce the T_1 and T_2 relaxation time constants of tissue magnetization. A metal may be characterized by its relaxivities, i.e., its ability to decrease T_1 and T_2, denoted r_1 and r_2 (in mmol^{-1} s^{-1}). For a given concentration of metal [M] in a given solution, the apparent relaxation time may be obtained from:

$$1/T_i = (1/T_{i0}) + r_i\,[\mathrm{M}]$$

where $i = 1$ or 2, T_{i0} is the relaxation time of the solution in absence of metal, and r_i is the relaxivity of the metal M for the studied solution, for a given magnetic field and a given temperature.

The susceptibility effect arises from the modification of the magnetic field in the vicinity of the metal. If the net magnetization in the metal differs from that of its

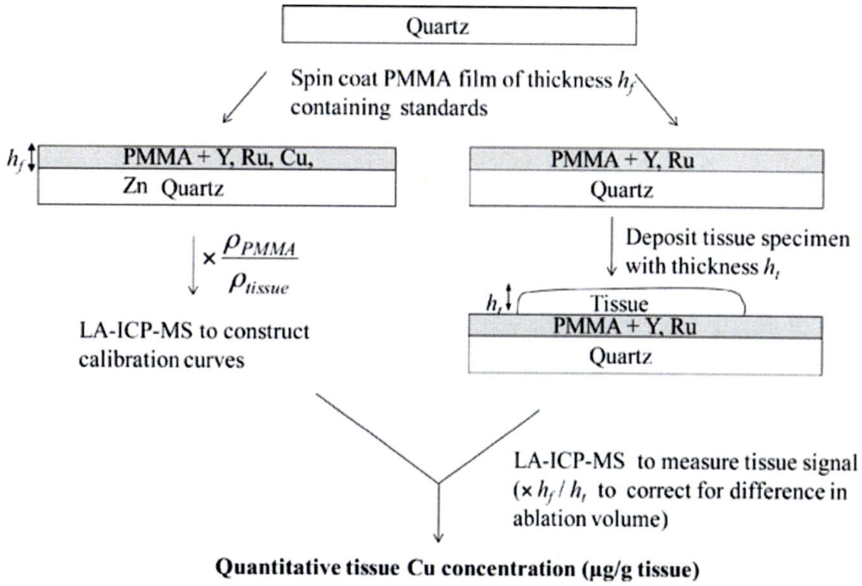

Fig. 5 Calibration workflow for quantifying metals in tissue using doped PMMA films [77]. Reproduced by permission of The Royal Society of Chemistry

environment (e.g., a paramagnetic metal in water, which is diamagnetic), a magnetic field gradient appears around the metal. This heterogeneity in the magnetic field induces a widening of the distribution of resonance frequencies around the metal (as a reminder, the resonance frequency is proportional to the intensity of the magnetic field). This widening of the resonance frequencies around the metal yields a faster dephasing of the transverse magnetization and thus a faster decay of the MR signal. The extent of this effect depends on the size of metal piece. For example, the susceptibility effects of some aesthetic orthodontic brackets may be seen up to the frontal lobe of the brain. For diamagnetic and paramagnetic materials, the net magnetization varies linearly with the applied magnetic field; for superparamegnetic materials, this relation is a S-shaped function (e.g., the net magnetization of iron-oxide particles saturates above about 0.5 T).

As reported by Walter, transcranial sonography can detect trace metal accumulation in deep brain structures with higher sensitivity than conventional MRI [92]. However, MRI has now covered a wide range of applications that benefit from sensitivity of MRI to various metals for their detection *in vivo* with much-improved spatial resolution and potential for temporal resolution. Many metals are commonly used in MRI, mostly as exogenous contrast agents. Within these intravascular contrast agents, the most common is gadolinium (Gd), used in a chelated form and injected into the blood stream. Both relaxivity and susceptibility effects are used to monitor various physiological properties such as vascular wall permeability or tissue perfusion. The use of exogenous Fe is rapidly growing, in the form of superparamagnetic contrast agents. These contrast agent are based on iron oxide crystals (mostly maghemite, γ-Fe_2O_3, and magnetite, Fe_3O_4), coated with a biocompatible material such as inorganic materials, monomeric stabilizer, or polymer stabilizers (e.g., dextran) [93]. The metabolisms of these particles depend on their size and on their coating. They may be used to assess perfusion at a higher spatial resolution than with Gd-Chelates [94] or to monitor brain inflammation: 24–48 h after injection, these particles are taken-up (phagocytosis) by macrophages and/or microglia in the tissues [95, 96]. An accumulation of iron oxide particles in the brain may be interpreted as a sign of inflammation [97].

Much progress has also been realized to bring MRI in the domain of cellular imaging. Iron oxide nanoparticules have been used to label cells, *ex vivo* or *in vivo*. *In vivo* labeling corresponds to the phagocytosis described above. As it is exemplified through Fig. 6, *ex vivo* labeling has been used for various cell types such as immune cells, tumor cells, stem cells, etc. [98–101]. *Ex vivo* labeling yields higher intracellular concentrations (up to 50 pg). It can be obtained via simple incubation with cells or using more refined techniques such as electroporation [102]. Single cell detections have been reported at magnetic loads as low as 0.2 pg of iron per cell [103]. Most of the cellular studies are based on superparamagnetic contrast agents and uses susceptibility effects to detect the contrast agent [104]. However, susceptibility effects yield a reduced signal, which is not ideal for a positive detection. For this reason, a few teams evaluate the use of new Gd-based contrast agents to obtain a signal increase based on the relaxivity effects [105, 106]. A very interesting emerging modality of MRI is the use of Mn mainly to better understand neuronal transports *in vivo*. Indeed, after its uptake by neurons, Mn is transported and crosses synapses and nicely enhanced the cytoarchitecture of the brain as depicted in Fig. 7. Manganese, a paramagnetic metal, enters cells as Mn^{2+} via calcium channels [107–109]. It has been proposed to use this

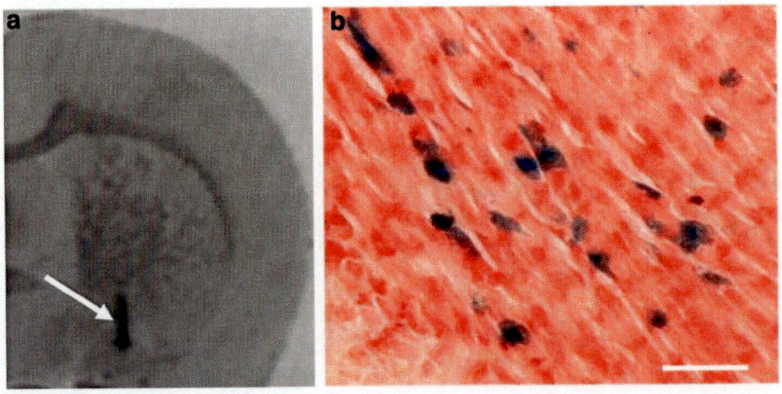

Fig. 6 *Ex vivo* characterization of the colonization of a tumor following intravenous injection of macrophages. (**a**) *Ex vivo* T2*-weighted MR image (7 T) of a rat brain bearing a C6 tumor, 7 days after implantation of the tumor and 3 days after administration of macrophages labeled with iron oxide particles. (**b**) Prussian blue detection of iron on adjacent coronal brain sections; Scale bar: 50 μm [119]

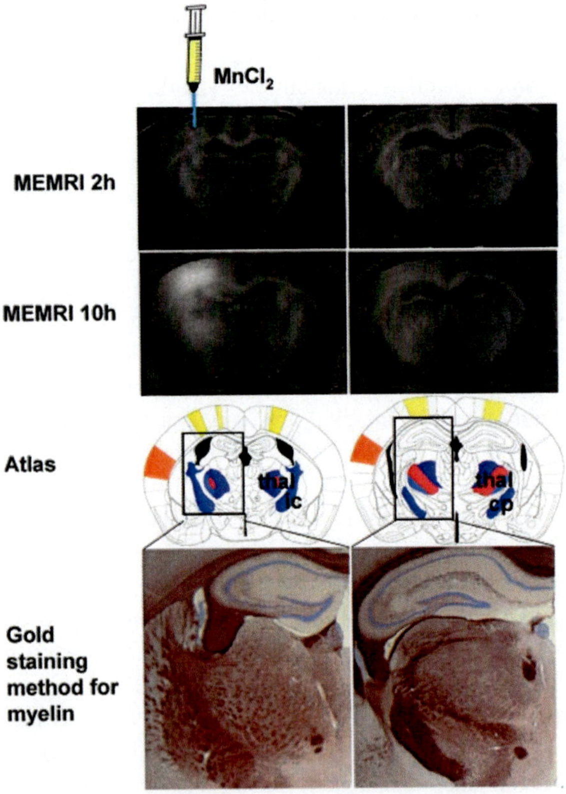

Fig. 7 T1-weighted MEMRI in mice, on a 7T MRI system. A single MnCl2 injection on the S1 cortex can improved the thalamo-cortical tract. This tract is visible by Gold staining. *Cp* cerebellum pedoncle, *ic* internal capsule, *thal* thalamus

property to study brain activity and neuronal transports. For example, after injection of Mn in the eye, the signal from the optic nerve is enhanced up to area V2 in the visual cortex [109, 110]. The speed of Mn transport seems to correspond to that of intracellular vesicles, i.e., in the range of 100 nm/s. A detection threshold of 5 μM of Mn has been proposed [110, 111] but is not clearly defined yet.

Finally, there is a clear trend toward the use of very high magnetic field with a still open question of the origin of contrasts in the brain and detection of endogenous metals such as iron. Indeed, as the strength of available magnetic fields for humans is increasing—7 T magnet are becoming available to a larger community of researchers—there is a growing interest to reach higher spatial resolutions (for rodent studies, field above 17 T are now available). Quantitative susceptibility mapping, an MR approach to estimate the tissue magnetic susceptibility based on MRI signal phase, or field-dependent relaxation rate increase, a technique based on the estimate of relaxation rates at two different magnetic fields, have recently emerged [112]. A few promising results have already been obtained. For example, Fukunaga et al. suggest that MRI would be sensitive to the variation of iron concentrations between the line of Gennari and the superficial layers in the visual cortex (a difference of 34 μg/g of iron) [113].

Conclusion

New developments in synchrotron microspectroscopy, LA-ICP-MS, and high-field MRI allow to get a better picture of the chemistry of metals in the brain from subcellular scale to the organ level, *in vitro*, *ex vivo*, and *in vivo*. As demonstrated in this review, the combined determination of the distribution/concentration/chemical state of metals inside the cell represents an essential step toward the precise understanding of the altered processes occurring at various specific cellular regions (organelles like mitochondria or lysosomes, the rough endoplasmic reticulum, golgi apparatus, vacuoles, etc.) and at different stages of the diseases. It is clear that, yet, no unique method can allow microscopic and spectroscopic analysis, at ultra-high spatial resolution, molecular resolution, and ultra-high sensitivity in 2D/3D in a time-resolved manner. Although this would be the ultimate analytical instrument, correlative imaging, which relies on best use of complementary metals imaging techniques, could partially reach this goal. The use of both light and electron microscopies already combines invaluable molecular and ultrastructural information [114]. NanoSIMS and electron microscopy both combine ultrastructural and chemical information and were found useful to better depict the subcellular iron distribution within brain tissues of patients suffering from Alzheimer's disease [115, 116]. Recent combined use of SRXRF and MRI to image iron in brains of patients suffering Parkinson's and Alzheimer's disease also appeared [117, 118]. The fast growing use of correlative imaging will without doubt provide new insight in the complexity of the cellular machinery, the role of metals therein and the implication of metal ions in the development of some brain disorders.

Acknowledgements The European Synchrotron Radiation Facility is acknowledged for provision of synchrotron radiation beam time. This work was supported by the Partenariats Hubert Curien (PHC)-FAST and the Australian Department of Innovation, Industry, Science and Research International Science Linkage FAST programs.

References

1. Zastrow ML, Peacock AFA, Stuckey JA, Pecoraro VL (2011) Nat Chem
2. Schmitt C, Strazielle N, Richaud P, Bouron A, Ghersi-Egea JF (2011) J Neurochem 117:747
3. Sensi SL, Paoletti P, Koh JY, Aizenman E, Bush AI, Hershfinkel M (2011) J Neurosci 31:16076
4. Sensi SL, Paoletti P, Bush AI, Sekler I (2009) Nat Rev Neurosci 10:780
5. Zecca L, Youdim MB, Riederer P, Connor JR, Crichton RR (2004) Nat Rev Neurosci 5:863
6. Sloot WN, Gramsbergen JB (1994) Brain Res 657:124
7. Zheng W, Monnot AD (2011) Pharmacol Ther
8. Zhang S, Rocourt C, Cheng WH (2010) Mech Ageing Dev 131:253
9. Tarohda T, Yamamoto M, Amamo R (2004) Anal Bioanal Chem 380:240
10. Tarohda T, Ishida Y, Kawai K, Yamamoto M, Amano R (2005) Anal Bioanal Chem 383:224
11. Pushie MJ, Pickering IJ, Martin GR, Tsutsui S, Jirik FR, George GN (2011) Metallomics 3:206
12. Pearse AGE (1985) Histochemistry, theoretical and applied. Churchill livingston, Edinburgh
13. Danscher G (1991) Prog Histochem Cytochem 23:273
14. Lopez-Garcia C, Varea E, Palop JJ, Nacher J, Ramirez C, Ponsoda X, Molowny A (2002) Microsc Res Tech 56:318
15. Que EL, Domaille DW, Chang CJ (2008) Chem Rev 108:1517
16. Domaille DW, Que EL, Chang CJ (2008) Nat Chem Biol 168
17. Leapman RD (2004) Curr Opin Neurobiol 14:591
18. Barberet P, Daudin L, Gordillo N, Sorieul S, Simon M, Seznec H, Idarraga I, Incerti S, Balana A, Moretto Ph (2011) Nucl Instrum Methods Phys Res B 269:2163
19. Bohic S, Simionovici A, Ortega R, Heymann D, Schroer C, Snigirev A (2001) Nucl Instrum Methods Phys Res B 181:728
20. Ortega R, Cloetens P, Deves G, Carmona A, Bohic S (2007) PLoS One 2:e925
21. Bohic S, Cotte M, Salome M, Fayard B, Kuehbacher M, Cloetens P, Martinez-Criado G, Tucoulou R, Susini J (2011) J Struct Biol
22. Ortega R, Deves G, Carmona A (2009) J R Soc Interface 6(Suppl 5):S649
23. Korbas M, Macdonald TC, Pickering IJ, George GN, Krone PH (2011) ACS Chem Biol
24. Bacquart T, Deves G, Carmona A, Tucoulou R, Bohic S, Ortega R (2007) Anal Chem 79:7353
25. Carter EA et al (2010) Mol Biosyst 6:1316
26. Bohic S, Murphy K, Paulus W, Cloetens P, Salome M, Susini J, Double K (2008) Anal Chem 80:9557
27. Al-Amoudi A, Norlen LP, Dubochet J (2004) J Struct Biol 148:131
28. Matsuyama S (2010) X-ray Spectrom 39:260
29. Ortega R et al (2009) Toxicol Lett 188:26
30. James SA et al (2011) Anal Bioanal Chem 401:853
31. Hackett MJ et al (2011) Analyst 136:2941
32. Chwiej J, Szczerbowska-Boruchowska M, Lankosz M, Wojcik S, Falkenberg G, Stegowski Z, Setkowicz Z (2005) Spectrochim Acta Part B At Spectrosc 60:1531
33. Schrag M, Dickson A, Jiffry A, Kirsch D, Vinters HV, Kirsch WM (2010) BioMetals 23:1123
34. Gibon J, Tu P, Bohic S, Richaud P, Arnaud J, Zhu M, Boulay G, Bouron A (2011) Biochim Biophys Acta 1808:2807
35. Duong TT, Witting PK, Antao ST, Parry SN, Kennerson M, Lai B, Vogt S, Lay PA, Harris HH (2009) J Neurochem 108:1143
36. Dodani SC, Domaille DW, Nam CI, Miller EW, Finney LA, Vogt S, Chang CJ (2011) Proc Natl Acad Sci USA 108:5980
37. Ducic T, Quintes S, Nave KA, Susini J, Rak M, Tucoulou R, Alevra M, Guttmann P, Salditt T (2011) J Struct Biol 173:202
38. Carmona A, Deves G, Roudeau S, Cloetens P, Bohic S, Ortega R (2010) ACS Chem Neurosci 1:194
39. Miller LM, Wang Q, Telivala TP, Smith RJ, Lanzirotti A, Miklossy J (2006) J Struct Biol 155:30
40. Leskovjan AC, Lanzirotti A, Miller LM (2009) NeuroImage 47:1215
41. Kühbacher M, Weseloh G, Thomzig A, Bertelsmann H, Falkenberg G, Radtke M, Riesemeier H, Kyriakopoulos A, Beekes M, Behne D (2005) X-ray Spectrom 34:112
42. Linkous DH, Flinn JM, Koh JY, Lanzirotti A, Bertsch PM, Jones BF, Giblin LJ, Frederickson CJ (2008) J Histochem Cytochem 56:3
43. Gh Popescu BF, Nichol H (2011) CNS Neurosci Ther 17:256
44. Popescu BF et al (2009) Phys Med Biol 54:651
45. Popescu BF, Robinson CA, Chapman LD, Nichol H (2009) Cerebellum 8:340
46. Kastyak MZ, Szczerbowska-Boruchowska M, Adamek D, Tomik B, Lankosz M, Gough KM (2010) Neuroscience 166:1119
47. Tomik B et al (2006) Neurochem Res 31:321
48. Szczerbowska-Boruchowska M, Lankosz M, Adamek D (2011) J Biol Inorg Chem 16:1217
49. Penner-Hahn JE (2005) Coord Chem Rev 249:161
50. Gunter KK, Aschner M, Miller LM, Eliseev R, Salter J, Anderson K, Gunter TE (2006) Neurobiol Aging 27:1816
51. Gunter KK, Aschner M, Miller LM, Eliseev R, Salter J, Anderson K, Hammond S, Gunter TE (2005) Free Radic Biol Med 39:164
52. Gunter TE, Gavin CE, Aschner M, Gunter KK (2006) Neurotoxicology 27:765
53. Lobinski R, Moulin C, Ortega R (2006) Biochimie 88:1591
54. Popescu BF, Pickering IJ, George GN, Nichol H (2007) J Inorg Biochem 101:957
55. Yoshida S, Ide-Ektessabi A, Fujisawa S (2003) Struct Chem 14:85
56. Ide-Ektessabi A, Rabionet M (2005) Anal Sci 21:885

57. Chwiej J, Adamek D, Szczerbowska-Boruchowska M, Krygowska-Wajs A, Wojcik S, Falkenberg G, Manka A, Lankosz M (2007) J Biol Inorg Chem 12:204
58. Chwiej J, Adamek D, Szczerbowska-Boruchowska M, Krygowska-Wajs A, Bohic S, Lankosz M (2008) J Trace Elem Med Biol 22:183
59. Chang LW, Verity MA (1995) Mercury neurotoxicity: effects and mechanisms. In: Chang LW, Dyers RS (eds) Handbook of neurotoxicology. Dekker, New York, p 31
60. Ibrahim D, Froberg B, Wolf A, Rusyniak DE (2006) Clin Lab Med 26:67
61. Zoeger N, Streli C, Wobrauschek P, Jokubonis C, Pepponi G, Roschger P, Bohic S, Osterode W (2005) Powder Diffraction 20:158
62. Korbas M, O'Donoghue JL, Watson GE, Pickering IJ, Singh SP, Myers GJ, Clarkson TW, George GN (2010) ACS Chem Neurosci 1:810
63. Gray AL (1985) Analyst 110:551
64. Mokgalaka N, Gardea-Torresdey J (2006) Appl Spectrosc Rev 41:131
65. Becker JS, Zoriy M, Matusch A, Wu B, Salber D, Palm C (2010) Mass Spectrom Rev 29:156
66. Becker JS, Matusch A, Palm C, Salber D, Morton KA (2010) Metallomics 2:104
67. Smeal J, Wilkins CL (2011) Appl Spectrosc Rev 46:425
68. Gunther D, Heinrich CA (1999) J Anal At Spectrom 14:1363
69. Potter D (2008) J Anal At Spectrom 23:690
70. Durrant SF (1999) J Anal At Spectrom 14:1385
71. Durrant SF, Ward NI (2005) J Anal At Spectrom 20:821
72. Caprioli RM, Farmer TB, Gile J (1997) Anal Chem 69:4751
73. Hare D, Burger F, Austin C, Fryer F, Grimm R, Reedy B, Scolyer RA, Thompson JF, Doble P (2009) Analyst 134:450
74. Hare D, Tolmachev S, James A, Bishop D, Austin C, Fryer F, Doble P (2010) Anal Chem 82:3176
75. Feldmann J, Kindness A, Ek P (2002) J Anal At Spectrom 17:813
76. Kindness A, Sekaran CN, Feldmann J (2003) Clin Chem 49:1916
77. Austin C, Fryer F, Lear J, Bishop D, Hare D, Rawling T, Kirkup L, McDonagh A, Doble P (2011) J Anal At Spectrom 26:1494
78. Zoriy MV, Dehnhardt M, Matusch A, Becker J (2008) Spectrochim Acta Part B At Spectrosc 63:375
79. Becker JS, Dobrowolska J, Zoriy M, Matusch A (2008) Rapid Commun Mass Spectrom 22:2768
80. Becker JS, Zoriy MV, Pickhardt C, Palomero-Gallagher N, Zilles K (2005) Anal Chem 77:3208
81. Zoriy MV, Becker JS (2007) Int J Mass Spectrom 264:175
82. Dehnhardt M, Zoriy MV, Khan Z, Reifenberger G, Ekstrom TJ, Sabine Becker J, Zilles K, Bauer A (2008) J Trace Elem Med Biol 22:17
83. Becker JS, Zoriy MV, Pickhardta C, Zilles K (2005) J Anal At Spectrom 20:912
84. Jackson B, Harper S, Smith L, Flinn J (2006) Anal Bioanal Chem 384:951
85. Hare D, Reedy B, Grimm R, Wilkins S, Woltakis I, George JL, Cherny RA, Bush AI, Finkelstein DI, Doble P (2009) Metallomics 1:53
86. Hutchinson RW, Cox AG, McLeod CW, Marshall PS, Harper A, Dawson EL, Howlett DR (2005) Anal Biochem 346:225
87. Lear J, Hare D, Adlard P, Finkelstein D, Doble P (2012) J Anal At Spectrom 27:159
88. Stockle R, Setz P, Deckert V, Lippert T, Wokaun A, Zenobi R (2001) Anal Chem 73:1399
89. Giesen C, Waentig L, Mairinger T, Drescher D, Kneipp J, Roos PH, Panneab U, Jakubowski N (2011) J Anal At Spectrom 26:2160
90. Sela H, Karpas Z, Zoriy M, Pickhardt C, Becker JS (2007) Int J Mass Spectrom 261:199
91. Pugh JAT, Cox AG, McLeod CW, Bunch J, Whitby B, Gordon B, Kalber T, White E (2011) J Anal At Spectrom 26:1667
92. Bellis DJ, Santamaria-Fernandez R (2010) J Anal At Spectrom 25:957
93. Corot C, Robert P, Idee JM, Port M (2006) Adv Drug Deliv Rev 58:1471
94. Barbier EL, Lamalle L, Decorps M (2001) J Magn Reson Imaging 13:496
95. Saleh A, Schroeter M, Jonkmanns C, Hartung HP, Modder U, Jander S (2004) Brain 127:1670
96. Desestret V, Brisset JC, Moucharrafie S, Devillard E, Nataf S, Honnorat J, Nighoghossian N, Berthezene Y, Wiart M (2009) Stroke 40:1834
97. Weinstein JS, Varallyay CG, Dosa E, Gahramanov S, Hamilton B, Rooney WD, Muldoon LL, Neuwelt EA (2010) J Cereb Blood Flow Metab 30:15
98. Valable S, Barbier EL, Bernaudin M, Roussel S, Segebarth C, Petit E, Remy C (2007) NeuroImage 37(Suppl 1):S47
99. Wilhelm C, Gazeau F (2008) Biomaterials 29:3161
100. Frank JA, Miller BR, Arbab AS, Zywicke HA, Jordan EK, Lewis BK, Bryant LH Jr, Bulte JW (2003) Radiology 228:480
101. Arbab AS, Bashaw LA, Miller BR, Jordan EK, Bulte JW, Frank JA (2003) Transplantation 76:1123
102. Walczak P, Kedziorek DA, Gilad AA, Lin S, Bulte JW (2005) Magn Reson Med 54:769
103. Smirnov P, Poirier-Quinot M, Wilhelm C, Lavergne E, Ginefri JC, Combadiere B, Clement O, Darrasse L, Gazeau F (2008) Magn Reson Med 60:1292
104. Shapiro EM, Sharer K, Skrtic S, Koretsky AP (2006) Magn Reson Med 55:242
105. Lux F et al (2011) Angew Chem Int Ed Engl 50:12299
106. Guenoun J, Koning GA, Doeswijk G, Bosman L, Wielopolski PA, Krestin GP, Bernsen MR (2011) Cell Transplant
107. Lin YJ, Koretsky AP (1997) Magn Reson Med 38:378
108. Pautler RG, Silva AC, Koretsky AP (1998) Magn Reson Med 40:740
109. Watanabe T, Frahm J, Michaelis T (2004) NMR Biomed 17:554
110. Lindsey JD, Scadeng M, Dubowitz DJ, Crowston JG, Weinreb RN (2007) NeuroImage 34:1619
111. Jackson SJ, Hussey R, Jansen MA, Merrifield GD, Marshall I, Maclullich A, Yau JL, Bast T. Behav Brain Res
112. Bilgic B, Pfefferbaum A, Rohlfing T, Sullivan EV, Adalsteinsson E (2011) Neuroimage
113. Fukunaga M et al (2010) Proc Natl Acad Sci USA 107:3834
114. Murphy GE, Narayan K, Lowekamp BC, Hartnell LM, Heymann JA, Fu J, Subramaniam S (2011) J Struct Biol 176:268
115. Guerquin-Kern JL, Wu TD, Quintana C, Croisy A (2005) Biochim Biophys Acta 1724:228
116. Quintana C, Wu TD, Delatour B, Dhenain M, Guerquin-Kern JL, Croisy A (2007) Microsc Res Tech 70:281
117. Antharam V, Collingwood JF, Bullivant JP, Davidson MR, Chandra S, Mikhaylova A, Finnegan ME, Batich C, Forder JR, Dobson J (2012) NeuroImage 59:1249
118. Hopp K, Popescu BF, McCrea RP, Harder SL, Robinson CA, Haacke ME, Rajput AH, Rajput A, Nichol H (2010) J Magn Reson Imaging 31:1346
119. Valable S, Barbier EL, Bernaudin M, Roussel S, Segebarth C, Petit E, Remy C (2008) NeuroImage 40:973

ICP-MS for the neurodegenerative and brain sciences

Yonghwang Ha, Olga G. Tsay, and David G. Churchill

Abstract

ICP-MS (inductively coupled plasma mass spectrometry) allows for the determination of "trace" elements and can be usefully related to tissues relating to neurodegenerative diseases. The accurate determination of trace elemental distributions in AD, PD, ALS, etc., allows in part for a better understanding of such diseases. The sensitivity and scope of the ICP-MS technique is therefore important to discuss. The elements detected can be analyzed for both body tissues and fluids. We discuss the practical use of ICP-MS, the instrumental setup; elements and their detection limits, a brief comparison of ICP-MS with other inorganic analysis instruments, sample preparation, and the analysis method are also treated. Next, we discuss metal ion analysis with ICP-MS in the context of neurodegenerative disease. This includes an introduction of neurodegenerative diseases, tissue analysis, fluid analysis and bioimaging of metals in brain tissue samples, and protein analysis application with metals and ICP-MS. The subtopics of (1) isotope dilution analysis, (2) related immunoassay techniques, and (3) hyphenated instrumental applications are also presented. This chapter is meant to be a primer for a synthetic chemist interested in utilizing ICP-MS and related techniques, and is current through 2011.

Keywords

ICP-MS • Alzheimer's disease • Parkinson's disease • Iron • Copper • Zinc

Abbreviations

6-OHDA	6-Hydroxydopamine
AES	Atomic emission spectroscopy
ALS	Amyotrophic lateral sclerosis
APP	Amyloid precursor protein
AS	α-Synuclein
Aβ	Beta amyloid
BBB	Blood–brain barrier
BSA	Bovine serum albumin
Cps	Counts per second
CRC	Collision reaction cell
CSF	Cerebral spinal fluid
cys	Cysteine
CZE	Capillary zone electrophoresis
DOPAC	3, 4-Dihydroxyphenylacetic acid
DRC	Dynamic reaction cell
EDS	Energy dispersive spectroscopy
FI	Flow injection
HPLC	High-performance liquid chromatography
Htt	The huntingtin gene
ICP	Inductively coupled plasma
IDA	Isotope dilution analysis
LA	Laser ablation
LA-ICP-MS	Laser ablation inductively coupled plasma mass spectrometry

Based on chapter published in: Chemical Monthly, Vol 142, No 4, 2011

D.G. Churchill (✉)
Department of Chemistry, Korea Advanced Institute of Science and Technology (KAIST), 373-1 Guseong-dong, Yuseong-gu, Daejeon 305-701, Republic of Korea
e-mail: dchurchill@kaist.ac.kr

W. Linert and H. Kozlowski (eds.), *Metal Ions in Neurological Systems*,
DOI 10.1007/978-3-7091-1001-0_19, © Springer-Verlag Wien 2012

MCI	Mild cognitive impairment
MS	Mass spectrometry
NAA	Neutron activation analysis
NMR	Nuclear magnetic resonance
OES	Optical emission spectroscopy
PE	Polyethylene
PP	Polypropylene
ppm	Parts per million
ppq	Parts per quadrillion
PTFE	Polytetrafluoroethylene
SF-ICP-MS	Sector field inductively coupled plasma mass spectrometry
SN	Substantia nigra
SOD	Super oxide dismutase
ZnT	Zn transporter

Introduction

There are many lines of research relating to neurodegenerative disorders such as Alzheimer's disease, Parkinson's disease, Amyotrophic Lateral Sclerosis (ALS), Huntington's disease, and Creutzfeldt–Jakob disease. Frequently these are related to protein misfolding and aggregation accumulation commonly in, or near, neurons or neurologically related cells. Alzheimer's disease relates to amyloid-β (plaque) and tau proteins (tangles). Parkinson's disease (and Lewy body dementia, etc.) are synucleinopathies involving the abnormal accumulation of α-synuclein in the brain. Further, Huntington's disease is related with Huntingtin protein; Creutzfeldt–Jakob disease is related to the Prion protein. Thus, unwanted protein dimers, trimers, oligomers, plaques, fibrils, and other morphologies of networks may be formed. Many researchers think these phenomena somehow are linked to the *cause or result* of the disease, but there are still many open questions as there is much complexity. To unearth the exact mechanism and chemical species of the disease much more basic research is needed. Other chemical aspects implicated in such disease etiology include oxidative stress, certain toxic environmental chemicals, enzyme malfunctioning, and the simple act of aging. Also, particular metals can bind into misfolded proteins preferentially (e.g., Cu^{2+} into histidine-rich fragments). Some researchers seek to link disease to metal ion concentration. Certain transition metals such as Cu, Zn, and Fe have normal healthful and important roles in catalytic activities or transport in the active site of the enzyme. When they are found abundantly *uncomplexed* and labile, toxic effects can be imparted, e.g., induction of covalent linkage between monomer proteins. Some researchers support that the abundance/activity/function of, e.g., Fe, Cu, Zn (and previously Al) are importantly related to AD and/or PD.

For metal ions in disease, there is a possibility to prepare model systems. There are of course both advantages and technical limitations in studying real living systems as well as model systems. In tissue, neuron samples, cerebrospinal fluid (CSF), e.g., some instruments can be directly analyzed. Specifically, HPLC-MS, IC, ICP-MS, MALDI-MS, ICR-MS, etc., techniques have been developed. In particular, ICP-MS is used broadly and has many advantages. Small sample volumes are allowable; short analysis times are possible (~μs for one metal in one sample); an important hyphenated utility exists with other instruments: HPLC, capillary electrophoresis, SDS–PAGE, laser ablation (LA) systems. ICP-MS has become an essential instrumental technique for determining $[M^{n+}]$ in biological samples (brain, neurons, blood, CSF).

ICP-MS analysis and its practical use

Instrumental setup

Interestingly, ICP-MS is not a very new technique as it was first used commercially in 1983. A crudely estimated 8,000 or more systems are used around the world at the time of writing [1].

To break down the ICP-MS technique, five major topics need to be addressed: the sample injection portion, ICP portion, interfacing, mass separation, and detection (Scheme 1). The sample injection parts are commonly composed of a nebulizer and sample chamber. Connections can be made to many kinds of instruments, including laser ablation systems, liquid chromatography, and capillary electrophoresis. The injector system is designed to analyze liquid samples. For *solids* analysis, extra pure grade acid is used for dissolving the sample. Laser ablation analyzes, e.g., *solid* surfaces, clinical biological tissues, 1- or 2-D gel electrophoresis.

ICP generates plasma from Ar gas and radio frequency power. Its temperature reaches 6,000–10,000 K whereupon, the sample converts into the positive ion (1+) streams (Scheme 2). Positive ion streams are directed to the interface part connecting the ICP (760 Torr) and the mass separation segments (10^{-6} Torr). Regulating devices of the positive ion stream include: the sampler cone, the skimmer cone, and the ion optics portion. Near vacuum is maintained (~1–2 Torr). Cooling protects the instrument from ICP. The ion stream passes through the sampler cone, skimmer cone, and ion optics. The ion optics maintains non-ionic species, such as photons and neutral species, from reaching the detector, which would otherwise cause artifacts [1].

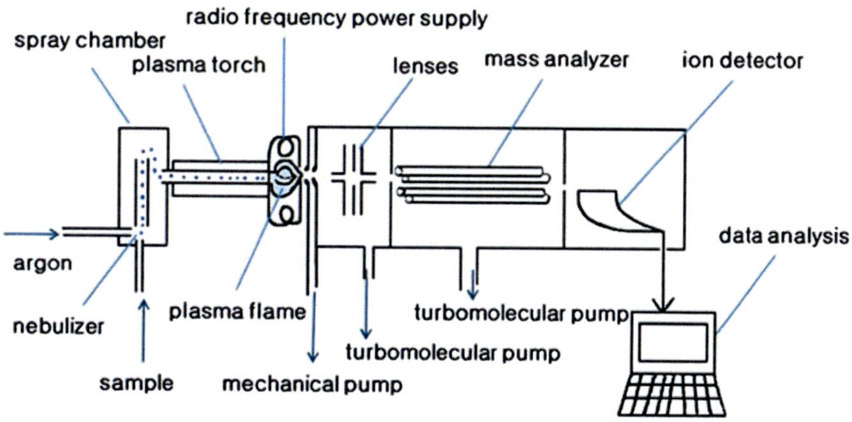

Scheme 1

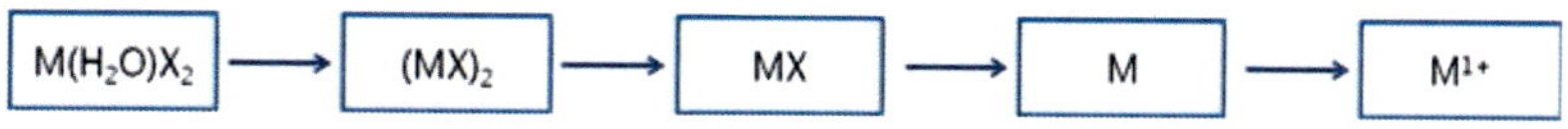

Scheme 2

Positive ion streams contain interfering molecules produced by Ar^+ or O. So, the dynamic reaction cell (DRC) or collision cell separates the analytes from the interferents [2]. Compared with the ICP-OES method, eliminating interference molecules is one of the most important aspects in ICP-MS analysis [3]. There are different hyphenated instruments for mass separation, time of flight (TOF), double focusing sector field (SF), quadrupole, etc. The detector quantifies and amplifies the signal obtained from various concentrations: very low (~ppq) to relatively high (~ppm). Commercially used apparati include the channel electron multiplier, Faraday cup, and discrete dynode electron multiplier [1].

Elements according to their limits of detection and normative values

ICP-MS is engineered to accommodate the analysis of more than 70 elements. Commonly, almost all are *positively* charged elements; some are commonly *negatively* charged elements (e.g., halogens, chalcogenides), etc. [1]. Experimentally, elemental detection is limited; elements are grouped according to the common range of different detection limits, provided below (Table 1) [1]. Most elements (65) can be detected below 1 ppt. For Be, B, Hg, this detection limit is 1 ~ 100 ppt. Si, P, and S have detection limits of 0.1 ~ 1 ppb; Br and I have limits of 1 ~ 50 ppb.

Ions implicated in molecular neurodegeneration such as Cu^{n+}, Zn^{2+}, Fe^{n+} have been analyzed by ICP-MS; these ions are usefully found in the body above the limits of detection. A range of normative values of metal ions are provided below in Table 2. Levels of these elements in, e.g., CSF for healthy human subjects are usually in the range of 20–30 $\mu g\ L^{-1}$ [4], a range appropriate for ICP-MS analysis.

Ions are separated by their mass-to-charge ratio, thus, there is a difference between detecting an ion or a neutral atom. Various mass separators can be used including the TOF, sector field double focusing, quadrupole, etc. To detect neutral atoms, a different strategy is required involving spectroscopic methods, e.g., optical emission spectroscopy (OES), energy dispersive spectroscopy (EDS), nuclear magnetic resonance (NMR) spectroscopy, etc. The elemental precision for neutral atom detection is lower than ion detection [1].

Comparison with other forms of inorganic instrumental analysis

There are several atomic spectroscopic techniques, such as flame atomic absorption, atomic emission spectroscopy (Fig. 1). Flame atomic absorption is a single-element technique. The flame is generated from a graphite furnace (2,500–4,000 K). When the sample enters the flame via the instrumental nebulizer, the ground state atom absorbs at an element-specific wavelength (λ_{abs}) measured and detected using a monochromator, photomultiplier, and a solid-state detector in order to calculate the concentration of element. This technique while being fast, is also relatively inexpensive, and easy to handle without excessive technical training. Flame AAS can be hyphenated with HPLC. However, the

Table 1 Groups of elements (given as their symbols) according to their limits of detection for the ICP-MS technique

Detection limit	Elements (as symbols)	No. of elements
Below 1 ppt	Li, Na, Mg, Al, K, Ca, Sc, Ti, V, Cr, Mn, Fe, Co, Ni, Cu, Zn, Ga, Ge, As, Se, Rb, Sr, Y, Zr, Nb, Mo, Tc, Ru, Rh, Pd, Ag, Cd, In, Sn, Sb, Te, Cs, Ba, La, Hf, Ta, W, Re, Os, Ir, Pt, Au, Tl, Pb, Bi, Ce, Pr, Nd, Sm, Eu, Gd, Tb, Dy, Ho, Er, Tm, Yb, Lu, Th, U	65
1–100 ppt	Be, B, Hg	3
0.1–1 ppb	Si, P, S	3
1–50 ppb	Br, I	2

Adapted from Ref. [1]

Table 2 Trace element concentrations of Cu, Fe, and Zn found for normal human subjects

	Plasma		Whole blood		CSF	
Element (μg/L)	Mean	Error	Mean	Error	Mean	Error
Cu	478	37	859	30	21.7	2
Fe	458	36	454,000	15,100	30.5	9.7
Zn	342	15	5,642	249	24.6	2.5

Adapted from Ref. [4]. Error, standard error of the mean (SEM)

sensitivity here is somewhat low. Detection limits at the ppm level are common, meaning that ICP is superior in sensitivity [1, 5–7].

Atomic emission or OES usually involves the use of an inductively coupled plasma source, 6,000–7,000 K. ICP-OES is multi-elemental. In plasma, electrons in sample atoms are excited to higher energy. The amount of light (wavelength signal λ_{em}) emitted when the excited electrons return to the ground state is then detected. The intensity and integration of the spectral signal gives the concentration of the aforementioned elements. Radial and axial types are used: in the axial type, the plasma is horizontally positioned, so more photons can be procured than for the radial type (improvement by 2–10 times in detection limit) [1]. Because a cone is not used, the tolerance to matrixes is good, and it can more easily be "hyphenated," e.g., with HPLC. Further, it is possible to detect sulfur or phosphorus (see Table 1), because interfering molecules do not need to be removed, when compared to ICP-MS [6].

Sample preparation and analysis

Since ICP-MS can help analyze various sample types such as fluid, tissue, and tissue for bioimaging, care in sample preparation is very important. To protect the sample from metal or element contamination is vital; thus exquisite cleaning of instrumental components such as tubing is very important. For this procedure, high- or extra-pure grade acids (nitric acid) are used and a clean room is required [8]. Plasticware is often used for sampling, transferring, and storage [8]. Glass is generally not used so to prevent leeching of metal ions out of the glass matrix [9]. Usually apparati are made of quartz, Teflon, polyethylene (PE), polypropylene (PP), polysulfide [10], and polytetrafluoroethylene (PTFE) [1].

Fluid analysis

Analyte samples are stored at −20 °C [4]. In some cases, digestion by microwave irradiation is needed. Stainless steel needles are used to sample biological fluids, which is transferred into precleaned polymer tubing (*vide supra*). Samples in Pyrex tubes are placed into steel bombs via Pyrex tubes and sealed with a Teflon lid [11]. In this way, some protein in the buffer system can be directly analyzed [12].

Tissue analysis

Sample containers are cleaned with aqueous HNO_3 for several hours prior to use. Tissues are cut in thin slices on an acid-washed polyethylene surface by an acid-washed quartz knife. Samples can then be homogenized by motor-driven Teflon pestles or dried via a vacuum drier or freeze-drier. In some cases, the samples may be digested in a sealed Teflon "bomb" in a microwave oven in acidic solution. Dilution is made with ultra pure water [13, 14].

Bioimaging analysis

Usually, biological sectioning is performed with a ceramic or stainless steel knife. To obtain samples of micrometer thickness, special instrumental dissectors, microtomes, or cryo-cutting tools are used. Sectioned tissues are mounted and dried on sodium-free glass slides [15–17].

Neurodegenerative diseases and metal ion analysis with ICP-MS

Neurodegenerative diseases

Alzheimer's disease

AD is progressive and irreversible and is a type of dementia that involves eventually grave injury to the brain characterized by extensive neuronal death. Symptoms include memory loss and personality changes. "Mild cognitive impairment" (MCI) is the name given to the earliest detectable stage of AD. Major physiological/pathological

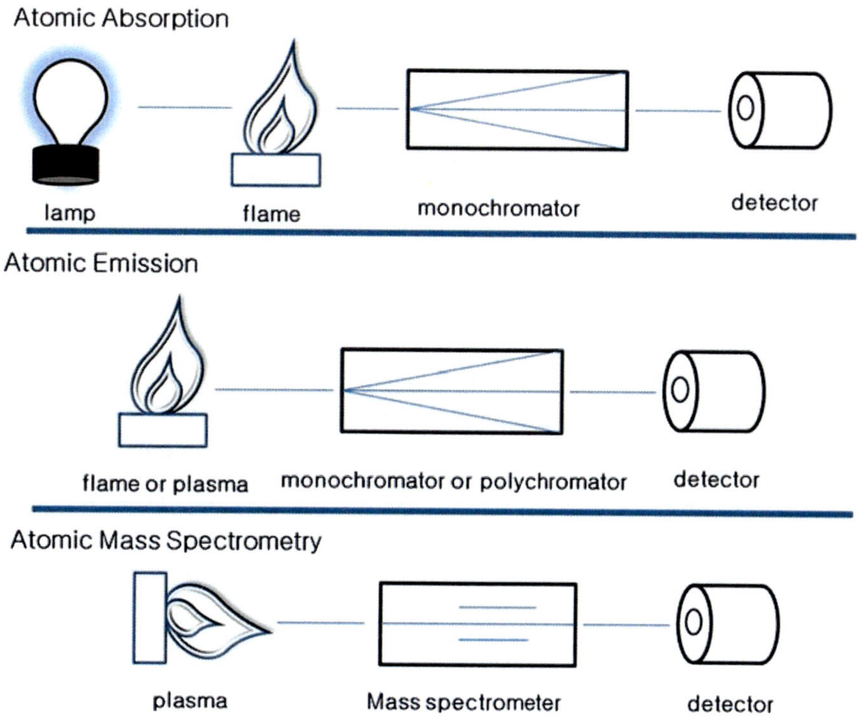

Fig. 1 Schematic illustration of the instrumental setup for atomic absorption and emission, and atomic mass spectrometry. Adapted from Ref. [1]

characteristics include extracellular senile plaques (amyloid) and intra-neuronal neurofibrillary tangles (tau) in the brain. Senile plaques arise from Aβ deposition from excessive scission of the amyloid precursor protein (APP), and neurofibrillary tangles from "hyperphosphorylated" tau.

Metal ions are implicated in disease aetiology (*vide supra*). Metal-catalyzed oxidations relating to transition metal ions of Cu, Fe, and Zn may lead to mitochondrial dysfunction and molecular processes of aging. Copper can enter the cell via ways including the copper transporter 1 (Ctr1), specific for Cu (I), in the intestinal absorptive cell. Inside the cell, copper ions are transferred to certain cellular organelles or molecules including the Golgi complex, superoxide dismutase (SOD), and the mitochondria. Processes are mediated by metallochaperones. These are specialized proteins that can protect metals from scavengers: the trans-Golgi network (Atx1), copper chaperones for SOD (CCS), and cytochrome *c* oxidase (Cox17) in the mitochondria [18]. Genetically, MURR1 mouse (U2af1-rs1), COMMD1 (copper metabolism gen MURR1) were studied; these species were discovered to be important for regulation of copper concentration. COMMD1 can bind to copper ions to induce conformational changes; this relates to the interaction with other regulatory proteins. Metallothioneins (MTs) also participate in copper regulation. Four distinct MTs are known. MT1–MT3 are found primarily in the brain; MT4 exists outside the brain. These molecules are cysteine (cys)-rich and can bind a range of ions (e.g., Cu, Zn). If copper regulation is not efficient, these ions impart a cytotoxic effect. Copper incorporation into the cell sometimes fails in such states as Menkes disease (copper deficiency). In the case of the failure of copper to become excreted *from* the cell, this may signal Wilson's disease (copper accumulation). When Cu(II) levels are not controlled in the cell, Cu (II) can bind to the Aβ protein, which may turn on production of hydrogen peroxide. Cu(II) reduction to Cu(I) occurs through the use of biological reducing agents such as NADH, NADPH, ascorbate, catecholamines, and cholesterol. The Aβ–Cu(I) metal complex reacts with H_2O_2 to give hydroxyl radical and reactive oxygen species [18]. These reactions are known as the Haber–Weiss or Fenton reactions [18–20]. Hydrogen peroxide can be the source of various reactive oxygen species [21] that can induce covalent links between proteins (protein–protein cross-links) and oxidation at sites in the protein backbone [20] (Fig. 2).

Zn^{2+} homeostasis is understood to be highly regulated; but an understanding of many aspects of this regulation are still sought. Various specialized biomolecules such as members of the Zn transporter (ZnT) (more than 10), ZIP (Zn(II)-regulated metal transporters, iron-regulated metal transporter-like proteins) (15 members), and metallothionein (3 isoforms) [18]. Usually Zn^{2+} efflux is related with the action of zinc transporter, Na^+/Zn^{2+} exchanger, and ZIP [22]. Free Zn^{2+} can induce toxic effects. When Zn^{2+} enters the postsynaptic neuronal region through Ca^{2+}–A/K channels, it is observed to generate oxygen species rapidly

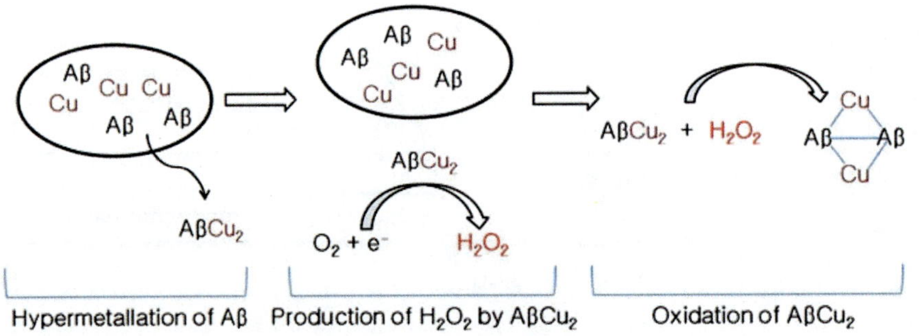

Fig. 2 Formation of cross-linked Aβ using copper. Adapted from Ref. [20]

[23]. These ROS can induce mitochondrial dysfunction and oxidative stress to increase the activity of neuronal nitric oxide synthase and NADPH oxidase, which in turn generates peroxynitrite ($O{=}NOO^-$). This allows mitochondria to decrease the ability of Zn^{2+} sequestration in the cytosol. This "positive feedback" is thought to induce cellular apoptosis or necrosis; eventually neuronal cell death ensues [18]. Zn^{2+} buffering in the cell is regulated by metallothioneins. MTs are composed of 61–68 amino acids, which have a sequence of 20 cysteine (cys) residues that is highly conserved. Such Zn^{2+} sequestration from the cytosole is controlled by mitochondria [24].

The physiological role of iron in the brain is often considered with such topics as (1) embryonic neuronal development, (2) myelin formation, (3) production of neurotransmitters, (4) DNA synthesis, (5) phosphorylation and ATP synthesis [18, 25, 26]. Iron homeostasis is regulated via Transferrin, Transferrin receptors, divalent metal transporters, the ferritin protein, ferroportins, mitochondria, endosomes, etc. Iron trafficking is mediated by Transferrin, Transferrin receptors, cubilin, divalent metal transporter, and the hemoglobin scavenger receptor (CD163). Inside the cell, iron levels are regulated by endosomes and mitochondria. Iron is usually stored in ferritin which can store up to ~4,500 iron ions. This protein is also a highly symmetrically designed container entity of 24 light, and heavy subunits. Iron release is mediated by ferroportin [27]. Iron toxicity largely occurs by oxidative stress based on the Fenton reaction. ROS react with membranes, proteins, and DNA strands alike; the expression of p53 and caspase 3 are stimulated. Cytochrome *c* is released from mitochondria; eventually, cell apoptosis or necrosis ensues [28].

Parkinson's disease

Parkinson's disease is the most common neurodegenerative disorder affecting motor skills: a sufferer may experience muscle rigidity, bradykinesia (Greek etymology: *bradys* = slow, *kinesis* = movement), resting tremor, as well as cognitive impairment. Progressive dopaminergic neuronal cell death occurs in the substantia nigra (SN) region in the brain [29]; SN means black substance. Pathological hallmarks include Lewy bodies and Lewy neuritis, which are eosinophilic and round cytoplasmic inclusion bodies in which the main component protein is α-synuclein filaments. α-Synuclein (AS) is strongly, but not exclusively, linked to Parkinson's disease. Simplistically, AS is to PD as Aβ is to AD. There is also LBD or lewy body dementia. The oligomer form of α-synuclein is known to be more toxic to the neuronal cell than ones in fibrillar form [30]. α-Synuclein is one of the most abundant proteins in the postsynaptic protein. It is composed of ~140 amino acids and usually found in its unfolded form (in the cytosol) [31]. There are many areas of research regarding aggregation of α-synuclein, metal-catalyzed oxidation [32]; certain toxic aldehydes have also been considered as being linked to disease as well (4-hydroxynonenal [33], oxidized cholesterol [34], 3,4-dihydroxyphenylacetaldehyde (DOPAL)) [35]. Long-term exposure to transition metals, or aging, stress can also influence the homeostasis of metals. Uncontrolled iron and copper concentration can generate reactive oxygen species that can allow α-synuclein to cross-link; this may produce aggregates of oligomers and fibrils. Dysregulated free metal ions in neuronal cells can influence mitochondrial dysfunction, effect changes in antioxidant systems of monoamine oxidase type-B (MAO-B), SOD activity, and allow for abnormal interactions with neuromelanin with iron. Some key reactions in mitochondrial respiration involve Fenton chemistry and dopamine oxidation (Scheme 3) [36].

Menkes and Wilson's Diseases

Menkes disease and Wilson's disease are disorders involving atypical *copper* transport and homeostasis, are genetic. They exhibit marked but different neurological manifestations. Mutations in the P-type copper transporting ATPases, ATP7a (MNKP, Menkes protein) exist in the case of Menkes disease; Mutations in ATP7b (WNDP, Wilson protein) are found in the case of Wilson's disease. Significant disruption in copper trafficking results [19, 37]. In Menkes disease, copper blood–brain barrier (BBB) penetration is inhibited leading to copper deficiency in the

Scheme 3

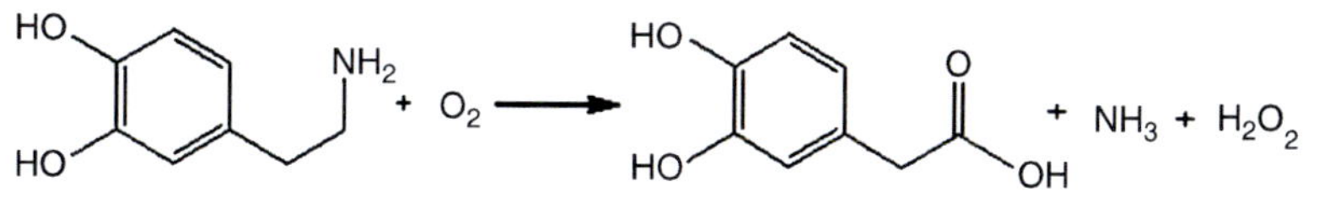

body. Wilson's disease is caused by copper *overload*: whole-body copper accumulation leads to brain/liver copper deficiency. Further, there is holoceruloplasmin biosynthesis and a marked impairment in biliary copper excretion [18, 26].

Amyotrophic Lateral Sclerosis

ALS (sometimes referred to as Lou Gherig's disease) is a progressive and fatal neurodegenerative disease caused by the degeneration of motor neurons. Certain prominent cases include that of Scientist Stephen Hawking. Approximately 10 % of cases are linked by genetic factors; these incidences are termed as cases of familial ALS (fALS). Other cases (the majority) are termed "sporadic" (sALS) [19]. Approximately 20 % of fALS cases are caused by inherited mutations in the gene encoding Superoxide Dismutase 1 (SOD1). SOD1 is homodimeric (2 × 153 AA) and normally functions as an antioxidant by detoxifying the potentially damaging superoxide radical anion. Its important function relates to its structure: each subunit is composed of an eight-stranded "Greek key" β-barrel fold; each one binds one zinc and one copper ion; there is an *intra*subunit disulfide bond as well. Disproportionation of the superoxide anion ($O_2^{\cdot -}$) is effected catalytically at the copper site; here, alternating cycles of reduction and oxidation exist. Copper and Zinc in the enzyme play, not only catalytic roles, but are also involved in preserving correct protein folding and in retaining "negative design." Loss of metal ion (demetallation) may induce protein misfolding and aggregation [19, 38–40].

Prion diseases

Prion diseases, also known as transmissible spongiform encephalopathies, are degenerative neurological diseases that relate to a "contagious" protein and its misfolding. These disorders include the classic Creutzfeldt–Jakob disease, Gerstmann–Sträussler–Scheinker syndrome, fatal familial insomnia as well as Kuru (New Gunea) in humans. Prion diseases are associated with the accumulation of misfolded cellular prion protein (PrP^C); PrP readily binds copper ion. While the exact role of PrP is still unknown, two main biological functions of the prion protein appear to be copper transport and antioxidative activity. Besides copper, an imbalance of other metals (ions) such as iron, manganese, and zinc is also associated with prion diseases [19]. Many investigations of the potential pathogenic role of manganese, including conformational changes of protein and non-homeostatic redox activity, have been made [41, 42].

Huntington's disease

Huntington's disease is a genetic disorder originating from the mutation in gene coding for Huntingtin (*Htt*) [43]. However, the exact function of *Htt*, as with many other proteins mentioned herein, is still unclear, the protein is thought to impact myelin integrity and plays a role, albeit indirectly, in the striatal iron accumulation observed in HD. Proposed mechanisms where iron homeostasis might lead to HD have been previously reviewed [44]. Copper imbalance is also observed in HD patients. A contribution of copper–protein interaction to HD pathogenesis has also been studied [45].

Tissue analysis

In many tissue portions, there is much research and analysis about the concentration of metals in these tissues—the metal concentration can be telling about pathology. There are carefully classified regions of the human brain (Fig. 3) below the broad regions of the cortex and medulla (Fig. 3). Subdivisions of the brain include the hindbrain (Myelencephalon, Metencephalon), Midbrain, Forebrain (Diencephalon, Epithalamus, Third ventricle, Thalamus, Hypothalamus, Subthalamus, Pituitary gland, Telencephalon, White matter, Rhinencephalon, Lateral ventricles, and Cerebral cortex).

Inductively coupled plasma mass spectrometry (ICP-MS) can be used in the assessment of trace element brain tissue to assess regional concentrations. These multi-element analyses help quantity of which metal ions and can be correlated to neurodegenerative disorders. Concentrations of 16 elements (Mg, Al, V, Cr, Mn, Fe, Co, Ni, Cu, Zn, As, Se, Mo, Cd, Hg, Pb) were measured in three regions of the human brain (frontal cortex, frontal white matter, and cerebellum) using ICP-MS by Ejima et al. [46]. These sample analyses from subjects who had died and did not possess incidence of neurological disease were compared with those of the tissues of the patient with FALS. These data and those obtained by other techniques such as neutron activation analysis (NAA) and ICP-AES were in good agreement. Also, the higher concentration of Mn in the samples of the ALS patient than in the controls was observed.

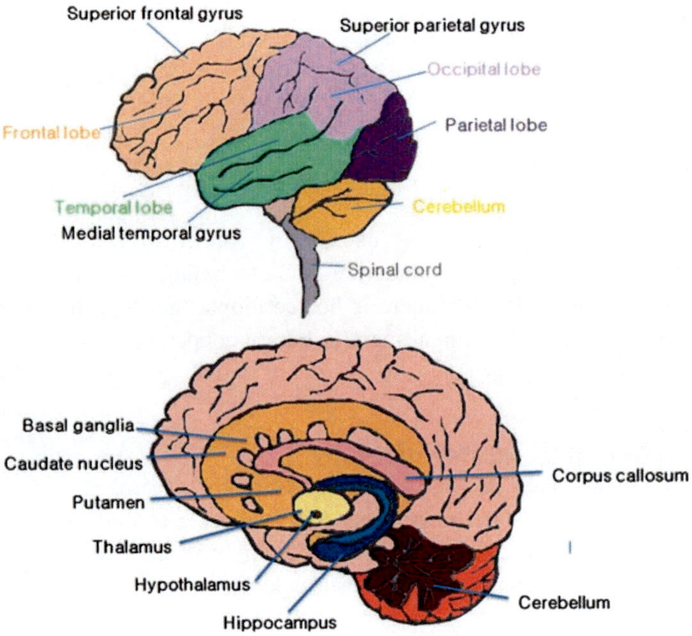

Fig. 3 Some regions of the human brain

The ICP-MS technique has also been widely used in the determination of the concentration of the certain metals in studies relating to AD. The concentration of Zn and Cd have been estimated through analysis of human brain tissue samples from Alzheimer's disease (AD) subjects (brain areas: superior frontal gyrus, superior parietal gyrus, medial temporal gyrus, hippocampus, and thalamus (both hemispheres)) and compared with those for control subjects [47]. Values here are given for all regions of the brain, including both hemispheres. The value ranges of the arithmetic means with standard deviations were reported in ng g^{-1} for Cd and mg g^{-1} for Zn. Concentrations of Cd and Zn in "normal" samples are obtained in the 5–145 ng g^{-1} and 6.4–29.3 mg g^{-1} range with arithmetic means of 26.7 ± 22.2 and 13.8 ± 4.0, respectively. The values for Alzheimer's disease samples were in the 4–104 ng g^{-1} range and 32.1 ± 19.0 ng g^{-1} for Cd and Zn, respectively, the arithmetic mean for Cd and Zn were 7.1–19.4 mg g^{-1} and 13.8 ± 2.7, respectively. Importantly, differences in element concentration are expected when considering brain regions, disease stage, age, and gender.

In a paper by Corrigan et al., 38 different elements were analyzed by NAA and ICP-MS [48, 49]. Brain tissue acquired post-mortem from 11 human subjects previously diagnosed with Alzheimer's disease, and from six controls were used (brain region: temporal lobe and frontal cortex lobe, caudate nucleus and putamen). Elevated tissue concentrations of aluminum, silicon, and calcium were reported (area: cortical areas, basal ganglia) while zinc, selenium, cesium, and cerium were lower. Plaque cores themselves can be successfully analyzed by ICP-MS. Here, it was combined with flow injection (FI) for selective extraction. Concentrations of several plaque metal samples were measured, including Cr, Mn, Ni, Cu, and Pb (0.2–0.8 mg L^{-1}) and Al, Fe, and Zn (2–20 mg L^{-1}) [14].

Furthermore, metal ions of Mo and especially Mn, have attracted much recent attention in the AD [50, 51], Parkinson's disease [43, 52, 53], ALS [13], and prion disease research fields [41, 42]. In one study by Csaszma et al. where a total of 5 brain regions were studied, the mean concentration of Mn was found to be in the range of 1.1–1.3 ppm, whereas that for Mo was between 90 and 150 ppb for dry weight tissue (both control and AD subject). The high concentrations of Mo and Mn 330 ± 42 (ppb) and 2.90 ± 0.07 (ppm), respectively, were found for AD patients in the putamen brain region (dry weight) [50]. Higher concentrations of manganese, as well as Si, Sn, and Al, in the parietal cortex of the AD brain compared to control values were measured by the research group of Srivastava et al. [51]. Concentrations in the non-metal B and other ions of metals such as Na, Te, Cr, Fe showed moderate increases in the same brain region. Changes in Mn, Fe, Cu, and Zn concentration in rat brain where Parkinson's disease was induced by 6-hydroxydopamine (6-OHDA)-induced parkinsonian symptoms were observed in all regions that lay along the dopaminergic pathway [43]. Measurements were performed via ICP-MS at 1, 3, 7, 10, 14, and 21 days after treatment with 6-OHDA. Stable ^{55}Mn, ^{56}Fe, ^{63}Cu, and ^{64}Zn isotopes were measured simultaneously with the procedural detection limits of 8, 70, 27, and 30 pg g^{-1}, respectively. In studies relating to the effects of Mn exposure on neurotoxicity, and the subsequent exacerbation of parkinsonism, concentrations of manganese in the whole brain were measured with an accuracy of an average 94 %; here the analytical detection limit is 0.16 ppb [53]. A study of concentration profiles of Fe and Cu upon Mn exposure in brain tissue for juvenile and adult mice was made [52]. The influence of manganese uptake on the pathogenesis of prion disease was also investigated [42]. Data obtained by ICP-MS indicate the increase in intracellular Mn levels for the mouse brain slice cultures upon chronic Mn exposure, which did not affect levels of Cu, Fe, Zn significantly altered by the aforesaid *manganese* treatment. In an *in vivo* chelation therapy study, decreases of Mn levels by up to 50 % in treated mice compared with untreated controls was determined by ICP-MS [41]. Interestingly, copper, iron, zinc, and cobalt levels remained unchanged.

Concentrations of several metals Rb, V, Mn, Fe, Co, Cu, Zn, and Cd were estimated in formalin-fixed brain tissue from subjects (of Guam) who suffered from ALS; four subjects were confirmed to have Parkinsonism–Dementia Complex (PDC); five served as control subjects [13]. Two metals (Zn and Cd) were reported at different values; for other metals, no significant differences were observed. An increase in the concentration of Zn for both subject groups was observed (gray and white matter). Cadmium gives higher values in ALS, but not in PDC subjects. Alkali metal ions (Li^+, Na^+, K^+, Rb^+, and Cs^+) also can be detected in several AD brain parts [54]. Mg content in AD brain was studied by Andrasi et al. [55]. A decrease in Mn concentration was detected by several instrumental methods including ICP-MS, ICP-AES, and NAA.

Environmental pollutants including heavy metals can enter the food chain and eventually contaminate the human central nervous system (CNS) [56, 57]. Ni and V exposure on nasal and brain areas in dogs was also studied and reported [57]. ICP-MS data showed higher [V] in exposed animals over those in control animals (0.36 ± 0.3 μg g^{-1} vs. 0.19 ± 0.17 μg g^{-1}, respectively).

Fluid analysis

Fluid analysis is very straightforward in ICP-MS. Usually it is possible to analyze the liquid sample directly, e.g., in its neat liquor form. In several studies, the involvement of metal ions in pathogenesis of neurodegenerative diseases have been supported by analysis of body fluids. Certain important fluids (CSF, blood, etc.) relate either directly or indirectly to CNS diseases. ICP-MS widely used for quantification of several metal ions in CSF, urine, serum, and whole blood. In most cases, Sector Field Inductively Coupled Plasma Mass Spectrometry (SF-ICP-MS) was used to avoid several disadvantages of the conventional quadrupole mass filter (Q-ICP-MS): these include signal interference during the experiment originating from plasma, water, reagents, and matrix of the biological samples used [58]. Samples of CSF from subjects of a large group of PD patients ($n = 42$) was analyzed for metal content by the research group of Alimonti in 2007 [8]. A total of 20 metallic elements (Li, Be, Sr, Ba, Zr, V, Cr, Mn, Co, Ni, Mo, W, Cd, Hg, Al, Sn, Sb, Bi, Pb, and Tl) were detected by SF-ICP-MS. Additionally, 6 elements (Si, Mg, Ca, Fe, Cu, and Zn) were quantified by ICP-AES. Researchers represented values from a mixture of transition metals and others statistically for PD subjects and controls and the plot of element discrimination between control and patient: data indicated abnormally low levels of Si, Cr, Fe, Co, Sn, and Pb in the CSF of PD patients compared to values from control samples. Previously, the same research group detected significantly increased concentrations of Pb and V in blood and urine ($p \leq 0.03$, for both metals) and reduced levels of Al, Cd, Hg, and Pb in serum and Cr, Co, Cd, Hg, Pb in CSF ($p \leq 0.003$) [59]. The detection limits for most of the elements range from 0.001 ng ml^{-1} for Cd to 0.05 ng ml^{-1} for Al, and were satisfactory and suitable for the intended purposes. Here, the applied ICP-MS method provided good accuracy; for example, in urine this value ranged from 92.4 % for Ni to 105 % for Hg; in the case of serum from 97.8 % for Mn to 104 % for Al; and in blood it varied between 97.0 % for Cr and 103 % for Pb.

Forte et al. determined concentrations of [Al] and [Mn] in urine, blood, and CSF for 26 PD human cases [60]. Both metals gave no statistical difference in concentration levels for patients and controls in any of the analyzed fluids. One exception was the [Al] in blood serum; this was lower in PD patients and a slight decrease, albeit not statistically relevant, in manganese concentration for whole blood and CSF was also found for PD patients. A similar trend for Al was reported by Bocca et al. in 2006 [61]. Here, concentration values of 3.0 ± 1.8 in PD patients vs. 5.3 ± 3.4ng ml^{-1} for the control group in serum and 11.7 ± 7.05 vs. 20.7 ± 12.6 ng ml^{-1} in urine, respectively, were detected. In a separate study, the analysis of 31 elements in samples of serum for 45 PD patients and 42 controls showed that [Cu], [Al], [Mn], and [Zn] are central ions in separating PD cases from control cases [62]. Concentrations of 31 elements in CSF, blood plasma, and whole blood were determined in living human subjects suffering from (1) multiple sclerosis [63], (2) so-called "Skogholt's" disease (a demyelinating disorder of both the central and the peripheral nervous system involving adult onset) [64], and (3) those from a control group [4]. Through analysis with SF-ICP-MS, a threefold increase of [Cu] and [Fe], and a twofold increase in [Zn] were detected in the CSF of "Skogholt" disease patients compared to those of controls. Other elements such as Rb, Co, P, S, Th, and Se were also found at higher concentration levels in "Skogholt" subjects. Interestingly, the decrease of Mo, Cd, Hg, and Tl levels in "Skogholt" and multiple sclerosis (MS) samples compared to the control group values.

Recently, Isao Hozumi's research group studied biological metal ions, Cu, Fe, Zn, Mg, Mn, in the CSF of patients with ALS (52 people), AD (21 people), and PD (20 people) subjects through the use of ICP-MS [65]. Cu and Zn in AD, Cu, Zn, and Mn in PD, Cu, Zn, Mg, and Fe in ALS were elevated compared to normative values. In particular, Cu ions (18.8 ng/ml) of Parkinson's disease patients showed increased levels over control values (10.2 ng/ml).

Additionally, Mg, Ca, Mn, Fe, Co, Ni, Cu, Zn, Se, Rb, Sr, Mo, Cd, Sn, Sb, Cs, Hg, and Pb were measured from the CSF and plasma of AD patients by the group of Gerhardsson. Quotients of metal contents were assessed between CSF and plasma. The quotients of Mn, Rb, Sb, Pb, and Hg were low, whereas that for Co ($Q_{Co} = 0.09$) was high in AD subjects compared to those of healthy controls ($Q_{Co} = 0.06$) [66].

Bioimaging of metal ions in brain samples and the advent of LA-ICP-MS

In recent years, bioimaging and metal distribution of the brain are researched by way of laser ablation ICP-MS (LA-ICP-MS). LA-ICP-MS is a widely used analytical spectrometric imaging technique; trace metal ion concentration/distribution determinations in brain tissues are made via imaging. Examples of practical applications of LA-ICP-MS in neurodegenerative disease studies have been previously published [67, 68]. In comparison to the tissue or fluid case, it is possible to determine the microstructure of the sample, as well as the metal distribution. Zoriy et al. analyzed human brain tissue for $^{13}C^+$, Cu, and Zn distributions; five adjacent sections were analyzed. The reproducibility was 5.4–6.5 %, 5.8–8.2 %, and 5.1–6.7 %, respectively, for aforementioned elements [15]. Becker et al. investigated the insular region and tissue from brain tumor to detect distribution of Cu, Zn, and certain other elements. The detection limits were 340 ng g^{-1} and 140 ng g^{-1}, respectively [69].

In a Parkinson's disease study of the distribution of several metal ions (iron, copper, and zinc) in a (Parkinson) mouse model brain was compared with control values [67]. The images obtained by LA-ICP-MS clearly show significantly higher iron content in the substantia nigra (SN) brain region in the Parkinson brain in control samples. Through similar research conducted by Matusch et al. [16], quantitative images of Mn, Fe, Cu, and Zn were generated in the Parkinson brain. This "time-course" series of images involved tissue exposure with 1-methyl-4-phenyl-1,2,3,6-tetrahydropyridine (MPTP). The most pronounced changes were observed for Cu and Fe (Fig. 4). There was an increase of Cu (38 %, versus control) in the periventricular zone at 28 days, and of Fe (38 % and 39 %) in the interpeduncular nucleus at a time of 7 and 28 days, respectively.

A report by Hare et al. [70] discussed an observed increase in the concentration of ^{56}Fe in a 6-hydroxydopamine (6-OHDA) lesion mouse brain model within the substantia nigra (SN) compared with control animal subjects.

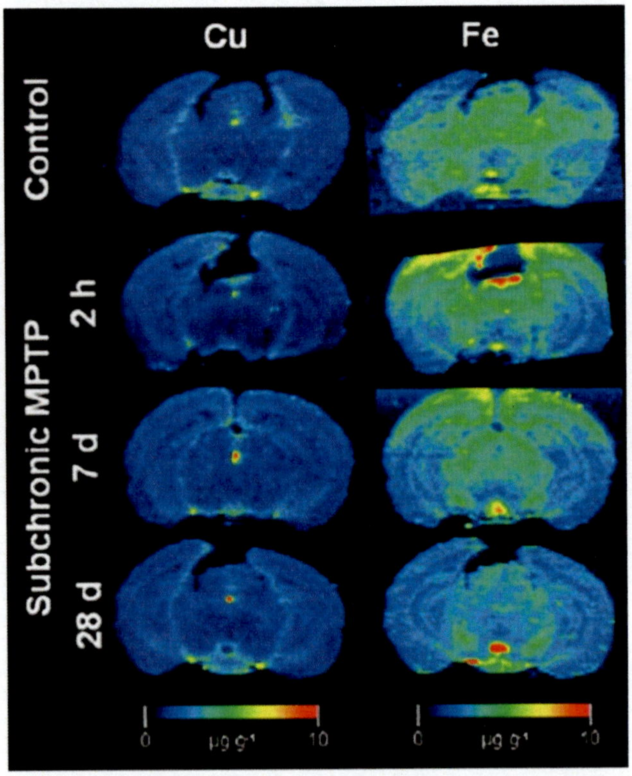

Fig. 4 Images of brain cross-sections and analysis of Cu and Fe content for the control (*top row*), 2 h (*second row*), 7 days (*third row*), and 28 days (*bottom row*) after the last of five daily MPTP injections. Reprinted with permission from Ref. [16] (License from Rightslink/Copyright Clearance Center. Licensee: David G. Churchill, License Date: Nov 18, 2010, License Number: 2551800410899, Type Of Use: reuse in a journal/magazine)

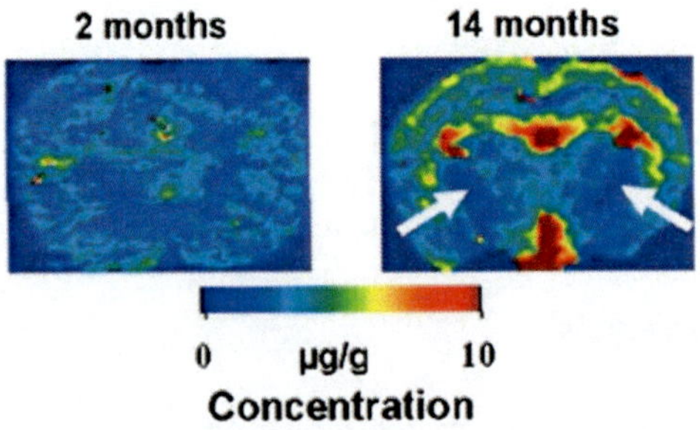

Fig. 5 Images demonstrating distributions of Cu in cross-sections of brains of 2- and 14-month-old mice. Reprinted with permission from Ref. [71]

The quantitative Cu distribution was analyzed for 2- and 14-month-old mice; this was done to probe the notions that (1) neurodegenerative diseases are age-related diseases and that (2) copper may play an important role in the brain in aging [71]. While total Cu content in brain after intravenous (i.v.) administration of ^{67}Cu increased in both cases; however, in the striatum and ventral cortex, Cu showed an approximately 50 % reduction in 14-month-old mice compared to those who were 2 months old (Fig. 5).

Table 3 The contents of Cu, Zn, and Fe in normal people with ICP-MS or related instruments

Reference	Samples	Instrumentation	Units	Cu		Zn		Fe	
				Mean	SD	Mean	SD	Mean	SD
Andrasi et al. [74]	Area occipitalis	ICP-AES	μg/g, dry weight	28	8	75	13	520	60
	Caudatum			37	7	80	7	640	100
	Putamen			42	7	91	20	910	190
	Pallidum			40	10	68	5	830	300
	Thalamus			16	6	66	12	240	40
	Cortex centralis			16	6	102	40	150	30
	Ammon's horn			19	7	86	11	230	100
	Cortex entorhinalis			19	7	54	11	230	100
	Cortex frontalis parasagittalis			17	5	47	14	150	30
	Cortex frontalis basalis			17	5	47	14	150	30
Ejima et al. [46]	Anterior horn in the spinal cord	ICP-MS	μg/g, fresh weight	3.35	1.22	9.06	0.91	15.83	2.5
	Posterior horn in the spinal cord			2.28	0.87	8.28	1.21	12.48	1.62
	Lateral funiculus in the spinal cord			0.77	0.11	5.51	1.26	7.12	1.43
	Posterior funiculus in the spinal cord			0.98	0.33	5.95	0.66	8.36	4.75
	Frontal cortex			3.64	1.44	10.45	1.55	33.82	13.78
	Frontal white matter			4.33	1.14	9.42	3.09	56.22	11.52
	Cerebellum			4.51	1.98	10.54	1.22	29.82	7.83
Lovell et al. [75]	Neuropil	micro-PIXE	μg/g	4.4	1.5	22.6	2.8	18.9	5.3
Panayi et al. [47]	Superior frontal gyrus in left hemisphere	ICP-MS	μg/g, fresh weight			13.4	2.9		
	Superior parietal gyrus in left hemisphere					14.7	5.6		
	Medial temporal gyrus in left hemisphere					13.1	2.2		
	Hippocampus in left hemisphere					15.2	2.4		
	Thalamus in left hemisphere					12.5	3.7		
	Superior frontal gyrus in right hemisphere					13	2.9		
	Superior parietal gyrus in right hemisphere					12.2	1.9		
	Medial temporal gyrus in right hemisphere					15.2	3.5		
	Hippocampus in right hemisphere					14.1	4		
	Thalamus in right hemisphere					14.3	7.8		
Gellein et al. [13]	Gray matter in frontal lobe, formalin-fixed	HR-ICP-MS	μg/g, dry weight	19.8		44		251	
	Gray matter in parietal lobe, formalin-fixed			23.9		64		330	
	Gray matter in occipital lobe, formalin-fixed			22.3		59		313	
	Gray matter in temporal lobe, formalin-fixed			20.9		40		280	
	Gray matter in cerebellum, formalin-fixed			30.1		54		309	
	Gray matter in basal ganglia, formalin-fixed			22.5		43		402	
	White matter in frontal lobe, formalin-fixed			18.9		43		190	
	White matter in parietal lobe, formalin-fixed			1.4		42		185	
	White matter in occipital lobe, formalin-fixed			23		58		245	
	White matter in temporal lobe, formalin-fixed			18.2		31		198	
	White matter in cerebellum, formalin-fixed			19.4		425		198	
Forte et al. [60]	Urine	ICP-AES, ICP-SFMS	μg/L	10.6	4.75	270	157	8.25	5.4
	Serum			1,216	326	705	91.1	1,136	393
	Blood			602	138	4,022	652	324,767	33,896
	CSF			22.5	4.76	32.9	8.85	73.3	72.7
Gellein et al. [76]	CSF, Norway	HR-ICP-MS	μg/L	22.3	2.23[a]	58.6	17.2[a]	74.8	25.2[a]
	Serum, Norway			1,129	124[a]	746	44.3[a]	1,044	76[a]
Alimonti et al. [8]	CSF	ICP-AES, ICP-SFMS	μg/L	21.9	4.77	32.3	11.4	35.5	5.03
Gellein et al. [4]	Plasma	HR-ICP-MS	μg/L	478	37	342	15	458	36
	Whole blood			859	30[a]	5,642	249[a]	454,000	15,100[a]
	CSF			21.7	2[a]	24.6	2.5[a]	30.5	9.7[a]
Hozumi et al. [65]	CSF	ICP-MS	ng/ml	10.2	2.1	5.3	3.3	238	54.7

[a]SEM (standard error of the mean), not SD (standard deviation)

Table 4 The contents of Cu, Zn, and Fe in AD patients with ICP-MS or related instruments

				Cu		Zn		Fe	
Reference	Samples	Instrumentation	Units	Mean	SD	Mean	SD	Mean	SD
Andrasi et al. [74]	Area occipitalis	ICP-AES	μg/g, dry weight	23	2	52	23	380	130
	Caudatum			29	3	49	11	430	90
	Putamen			21	5	43	6	600	250
	Pallidum			27	4	41	4	570	220
	Thalamus			33	18	49	17	310	150
	Cortex centralis			24	5	39	9	370	50
	Ammon's horn			20	3	46	8	310	60
	Cortex entorhinalis			24	6	42	7	380	40
	Cortex frontalis parasagittalis			24	8	46	12	400	30
	Cortex frontalis basalis			19	5	43	11	340	60
Lovell et al. [75]	Senile plaque rim	micro-PIXE	μg/g	22.7	6.5	67	13	52.4	14.5
	Senile plaque core			30.1	11	86.8	21	53.1	13.7
	Neuropil			19.3	6.3	51.4	11	38.8	9.4
Panayi et al. [47]	Superior frontal gyrus in left hemisphere	ICP-MS	μg/g, fresh weight			13.2	2.1		
	Superior parietal gyrus in left hemisphere					12.7	3.1		
	Medial temporal gyrus in left hemisphere					14.6	2.5		
	Hippocampus in left hemisphere					14.8	3.2		
	Thalamus in left hemisphere					12.8	3		
	Superior frontal gyrus in right hemisphere					13.3	2.8		
	Superior parietal gyrus in right hemisphere					13.4	3.1		
	Medial temporal gyrus in right hemisphere					14.7	2.2		
	Hippocampus in right hemisphere					14.3	3.5		
	Thalamus in right hemisphere					14.2	1.1		
Hozumi et al. [65]	CSF	ICP-MS	ng/ml	17.4	10.4	8.4	6.4	238.6	38.7
	CSF, early onset AD			10.3	5.4	3.9	3.3	221.6	16.5
	CSF, late onset AD			20.9	10.7	10.7	6.5	247.2	44.1

Iron and zinc distributions also were investigated [72, 73]. Zinc does not give any changes, whereas Fe gives increased values in the Substantia Nigra, the thalamus, and the CA1 region of the hippocampus. In an Alzheimer's disease study, LA-ICP-MS was also used for metal distribution and also in imaging the localization of Aβ peptide in brain tissue. For the imaging of amyloid plaques, ^{60}Ni and ^{153}Eu radioisotopes were used [17]. Concentrations of metal ions were different in plaque than in non-plaque tissue samples; important plaque: non-plaque ratios were calculated, and some are provided as follows: 3.5:1 for Mg, ~3:1 for Ca, "2:1 for Fe, 3.5:1 for Cu, and 2:1 for Zn."

Additionally, tables are provided regarding the contents of Cu, Zn, and Fe in brain tissues from subjects who were normal or suffered from neurodegenerative diseases (Tables 3, 4, 5, 6, and 7).

Protein analysis applications with metals and ICP-MS

Isotope Dilution Analysis

Isotope Dilution Analysis (IDA) is an important technique that allows the exact concentration of a sample to be obtained by adding an isotopic sample of exact isotope ratio or concentration to an unknown sample (spiking). The experimental isotope ratio of the mixture is then obtained. Advantages here include the fact that results are not influenced by the (1) mode of sample preparation, (2) various instrumental instabilities, (3) signal drift, (4) matrix effects, and (5) sample loss upon equilibrium mixing between the sample and the spike [77]. The two modes for this method are "species specific" and "species unspecific." In the specific case, the sample species is the same as the isotope chemical compound. It is thus necessary to know the exact composition and structure of the sample. The isotope sample can be added at the beginning point of the analytical procedure. The "species unspecific" case allows the sample species to be different from the isotope "spike." This method can be used when the structure and composition of the sample species are not accurately known. Often it is expensive to purchase certain commercially available reagents or even synthesize them. In this case, the isotope "spike" could be added to the sample after the sample preparation and treatment are effected, e.g., performing chromatography immediately before the ionization process [77]. The IDA method does have some challenges related with (1) instruments, (2) spectral interference, (3) detection dead

Table 5 The contents of Cu, Zn, and Fe in PD patients with ICP-MS or related instruments

Reference	Samples	Instrumentation	Units	Cu		Zn		Fe	
				Mean	SD	Mean	SD	Mean	SD
Gellein et al. [13]	Gray matter in frontal lobe, formalin-fixed, dementia complex	HR-ICP-MS	µg/g, dry weight	18.2		72		207	
	Gray matter in parietal lobe, formalin-fixed, dementia complex			40.4		309		530	
	Gray matter in occipital lobe, formalin-fixed, dementia complex			25.9		125		299	
	Gray matter in temporal lobe, formalin-fixed, dementia complex			26.1		98		350	
	Gray matter in cerebellum, formalin-fixed, dementia complex			32.7		206		447	
	Gray matter in basal ganglia, formalin-fixed, dementia complex			24.4		116		268	
	White matter in frontal lobe, formalin-fixed, dementia complex			18.9		89		174	
	White matter in parietal lobe, formalin-fixed, dementia complex			18.1		97		156	
	White matter in occipital lobe, formalin-fixed, dementia complex			20.5		111		186	
	White matter in temporal lobe, formalin-fixed, dementia complex			21.5		72		219	
	White matter in cerebellum, formalin-fixed, dementia complex			21.1		80		179	
Forte et al. [60]	Urine	ICP-AES, ICP-SFMS	µg/L	12.9	5.11	280	144	13.1	6.33
	Serum			994	171	669	118	1,318	481
	Blood			595	128	4,711	892	401,621	92,062
	CSF			23.7	10.5	27.3	10.5	33	29.4
Alimonti et al. [8]	CSF	ICP-AES, ICP-SFMS	µg/L	19.4	7.97	27.7	9.01	28.2	14.6
Hozumi et al. [65]	CSF	ICP-MS	ng/ml	18.8	6.9	14.5	7.6	263.9	112.9

Table 6 The contents of Cu, Zn, and Fe in ALS patients with ICP-MS or related instruments

Reference	Samples	Disease	Instrumentation	Cu		Zn		Fe	
				Mean	SD	Mean	SD	Mean	SD
Gellein et al. [13]	Gray matter in frontal lobe, formalin-fixed	HR-ICP-MS	µg/g, dry weight	25.1		117		303	
	Gray matter in parietal lobe, formalin-fixed			24.4		106		305	
	Gray matter in occipital lobe, formalin-fixed			23.4		56		241	
	Gray matter in temporal lobe, formalin-fixed			24.8		127		357	
	Gray matter in cerebellum, formalin-fixed			29.1		114		342	
	Gray matter in basal ganglia, formalin-fixed			24.1		92		349	
	White matter in frontal lobe, formalin-fixed			18.1		59		169	
	White matter in parietal lobe, formalin-fixed			21.3		84		162	
	White matter in occipital lobe, formalin-fixed			18.4		286		164	
	White matter in temporal lobe, formalin-fixed			16.4		69		217	
	White matter in cerebellum, formalin-fixed			18.7		87		160	
Hozumi et al. [65]	CSF	ICP-MS	ng/ml	19.5	14.8	11.1 + 11.2		282.5	74.9
	CSF, classical type			18.3	9.3	12.7	13	276.8	74.7
	CSF, bulbar type			21	19.8	9.3	8.7	285.9	78.9

time, (4) mass discrimination, etc. Spectral interference occurs from the same atomic mass molecules by H, O, Ar, and so on. A double-focusing spectrometer operating at high resolution helps overcome this problem. For the quadrupole instrument, there is a removable apparatus: a DRC or collision reaction cell (CRC). For DRC's reactive gases such as O_2, CO, NH_3, and CH_4 are used among others; the CRC uses colliding gases, such as H_2 and He [77]. At high counting

Table 7 The contents of Cu, Zn, and Fe in other neurodenerative patients with ICP-MS or related instruments

					Cu		Zn		Fe	
Reference	Samples	Disease	Instrumentation	Unit	Mean	SD	Mean	SD	Mean	SD
Gellein et al. [4]	Plasma	Skogholt	HR-ICP-MS	μg/L	477	23	333	17	440	57
		Multiple sclerosis			487	39[a]	333	16[a]	453	73[a]
	Whole blood	Skogholt			950	56[a]	5,917	291[a]	460,000	14,800[a]
		Multiple sclerosis			944	84[a]	5,897	286[a]	474,000	11,400[a]
	CSF	Skogholt			69.3	7.7[a]	42	4.4[a]	106.2	11.8[a]
		Multiple sclerosis			21.6	2.3[a]	21.5	2[a]	19.8	2.6[a]

[a]SEM (standard error of the mean), not SD (standard deviation)

rates (10^6 + cps) an electron multiplier-type detector has "dead time:" the signal cannot be processed completely. In practice, a high abundance state of isotope can be influenced by "dead time" [77]. Heavier elements may be detected more precisely than lighter elements because of the space-charge effect; heavy isotopes show a higher intensity than light isotopes. This can be corrected through known methodolgy with certified isotopic composition materials [77]. In a report by Jakubowski et al. the investigators tried, via iodine labeling, to quantify bovine serum albumin (BSA), chicken egg white lysozyme and porcine gastric mucosa protein where an iodo-histidine or -tyrosine residue was incorporated into the protein. The stable isotope ^{127}I was detected via ICP-MS. The investigation range was from 0.015 (BSA) to 105 pmol (lysozyme) [78]. So the ICP-MS is essential for IDA. This method can decrease the discrepancy of the matrix effect or sample preparation [77, 79]. Becker et al. performed isotope dilution of $^{63}Cu^+$, $^{65}Cu^+$ in analyzing human brain fluids [80], and studied ^{65}Cu, ^{67}Zn, and ^{54}Fe for tissue of Alzheimer's disease brain [81]. Heumann et al. discussed the possibility of using isotope dilution method with ICP-MS as a routinely for trace elements. Hyphenating, e.g., with liquid chromatography, and coupling of gases can help in the difficulty of stemming from the lack of commercially available isotope-labeled "spike." If the isotope can be generated in a closed system the "spike" will not be lost. Isotope dilution techniques are useful for the accuracy of detecting $MeHg^+$ and Cr^{6+} [82].

Immunoassay techniques

We will now describe how protein quantification can be more thoroughly exploited with ICP-MS. There are techniques for protein quantification via analyzing metal ions tagged with antibodies (Scheme 4); here, the metal is attached covalently to an antibody. Lou et al. reported the use of certain kinds of polymers bearing metal-chelating functional groups; the S–S cross-linked bond in the protein in question allowed for the reaction to polymers. Upon

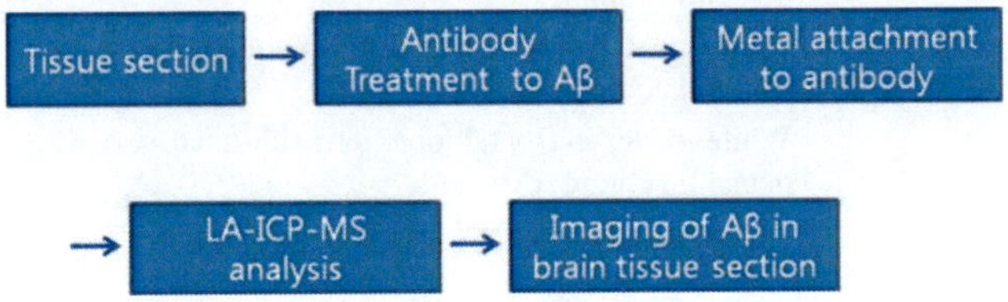

Scheme 4

conjugation with the metal complex, the purified metal ion attached antibodies were analyzed [83] (see Scheme 5).

In 2008, Razumienko et al. compared ICP-MS-based immuno-protein quantification with conventional immunoassays (colorimetric and indirect ELISA). Samples included protein (hPDGF–AA), p53 in cell lines (A431, 293T, KG1a, K562). Whether Tb- or Tm-attached antibodies were used, results of analysis were similar with conventional colorimetric ELISA or the indirect ELISA [84]. Next, specific element-tagged antibody methods were reported by Banarov et al. in 2002. Methods of centrifugal filtration, Protein A affinity, size-exclusion gel filtration, and ELISA with ICP-MS were connected through the use of specific element tagged antibodies. The investigators reported that levels of target protein (0.1–0.5 ng ml^{-1}) can be measured, involving a linear response to over three orders of magnitude. The ICP-MS hyphenated method revealed more precise results obtained more conveniently [85].

Hyphenated instrumental applications

There are related hyphenated instruments whose designs are often considered in the realm of engineering but can also be briefly mentioned here. Another mass separator, the sector-field analyzer can be used and shown to give higher accuracy. But in this case, a prospective high instrument cost, as well as significant maintenance costs are expected. When using instruments of the hyphenated design, it is required to optimize the replicate time, "dwell time," number of sweeps per readings, and other parameters between instruments and

Scheme 5

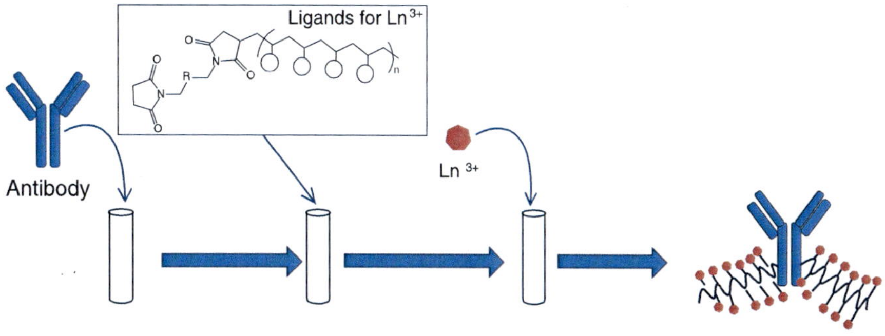

the ICP-MS technique [6]. Helfrich et al. succeeded in coupling the gel electrophoresis technique to that of the ICP-SFMS. They attempted to detect the $^{32}S^{+}$ and $^{31}P^{+}$ of α- and β-caseins so to better understand phosphorylation levels. Results obtained matched well with anticipated values [86].

Conclusion

In this chapter, we summarized the utility of ICP-MS and related techniques for determining trace element concentrations in tissues related to neurodegenerative disease studies. The practical use of ICP-MS was also presented. As discussed, the accurate determination of trace elemental distributions in cases where AD or PD were diagnosed, allow for a better understanding of such diseases and relates to the sensitivity and scope of the technique. Analytes detected involved tissues and fluids. This chapter reviewed the instrumental setup, elements and their detection limits, a brief comparison of ICP-MS with other inorganic analysis instruments, and sample preparation and analysis methods. Also, material about neurodegenerative diseases and metal ion analysis were presented in the context of the ICP-MS technique. The reader should find this chapter has included certain important biophysical, analytical, and clinical studies that underscore the practical use and development of ICP. This chapter while current through the summer of 2011, is not meant to be exhaustive but was meant to be user-friendly. Along with the inclusion of various important studies, it also mentions related techniques, recent improvements, and limitations of current techniques.

Acknowledgment The Molecular Logic Gate Laboratory operated by Professor David G. Churchill received funding from the High Risk High Return Project (HRHRP) (Grant No: N10100016), and the National Research Foundation, (NRF) (Grant No: 2010-0013660 and 2009-0070330). Adaptations of various Figure and Schemes were made carefully by O.G.T.

References

1. Robert T (2007) Practical guide to ICP-MS – a tutorial for beginners. CRC, Boca Raton, FL
2. Tanner SD, Baranov VI, Bandura DR (2002) Spectrochim Acta Part B 57:1361
3. Nam SH, Park Y, Kim JJ, Han SH, Kim WH (2004) Bull Korean Chem Soc 25:447
4. Gellein K, Skogholt JH, Aaseth J, Thoresen GB, Lierhagen S, Steinnes E, Syversen T, Flaten TP (2008) J Neurol Sci 266:70
5. Luo Y, Zhang B, Chen M, Wang J, Zhang X, Gao WY, Huang JF, Fu WL (2010) J Agric Food Chem 58:9396
6. Szpunar J (2000) Analyst 125:963
7. Yaman M (2005) Anal Biochem 339:1
8. Alimonti A, Bocca B, Pino A, Ruggieri F, Forte G, Sancesario G (2007) J Trace Elem Med Biol 21:234
9. Shin JH, Parkin G (1998) Organometallics 17:5689
10. Kempson IM, Skinner WM, Kirkbride KP (2007) J Toxicol Environ Health B Crit Rev 10:611
11. Svantesson E, Pettersson J, Markides KE (2002) J Anal At Spectrom 17:491
12. Deitrich CL, Raab A, Pioselli B, Thomas-Oates JE, Feldmann J (2007) Anal Chem 79:8381
13. Gellein K, Garruto RM, Syversen T, Sjobakk TE, Flaten TP (2003) Biol Trace Elem Res 96:39
14. Beauchemin D, Kisilevsky R (1998) Anal Chem 70:1026
15. Zoriy MV, Becker JS (2007) Int J Mass Spectrom 264:175
16. Matusch A, Depboylu C, Palm C, Wu B, Hoglinger GU, Schafer MKH, Becker JS (2010) J Am Soc Mass Spectrom 21:161
17. Hutchinson RW, Cox AG, McLeod CW, Marshall PS, Harper A, Dawson EL, Howlett DR (2005) Anal Biochem 346:225
18. Kozlowski H, Janicka-Klos A, Brasun J, Gaggelli E, Valensin D, Valensin G (2009) Coord Chem Rev 253:2665
19. Gaggelli E, Kozlowski H, Valensin D, Valensin G (2006) Chem Rev 106:1995
20. Bush AI (2003) Trends Neurosci 26:207
21. Grossi C, Francese S, Casini A, Rosi MC, Luccarini I, Fiorentini A, Gabbiani C, Messori L, Moneti G, Casamenti F (2009) J Alzheimers Dis 17:423
22. Sekler I, Sensi SL, Hershfinkel M, Silverman WF (2007) Mol Med 13:337
23. Sensi SL, Yin HZ, Carriedo SG, Rao SS, Weiss JH (1999) Proc Natl Acad Sci USA 96:2414
24. Frazzini V, Rockabrand E, Mocchegiani E, Sensi SL (2006) Biogerontology 7:307
25. Youdim MBH (2000) Nutrition 16:504
26. Sigel A, Sigel H, Sigel RKO (2006) Neurodegenerative diseases and metal ions. In: Metal ions in life science, p 179
27. Hentze MW, Muckenthaler MU, Andrews NC (2004) Cell 117:285
28. Gaasch JA, Lockman PR, Geldenhuys WJ, Allen DD, Van der Schyf CJ (2007) Neurochem Res 32:1685
29. Goedert M (2001) Nat Rev Neurosci 2:492
30. Pountney DL, Voelcker NH, Gai WP (2005) Neurotox Res 7:59
31. Uversky VN (2007) J Neurochem 103:17
32. Cole NB, Murphy DD, Lebowitz J, Di Noto L, Levine RL, Nussbaum RL (2005) J Biol Chem 280:9678

33. Qin ZJ, Hu DM, Han SB, Reaney SH, Monte DA, Fink AL (2007) J Biol Chem 282:5862
34. Bieschke J, Zhang QH, Bosco DA, Lerner RA, Powers ET, Wentworth P, Kelly JW (2006) Acc Chem Res 39:611
35. Burke WJ, Li SW, Williams EA, Nonneman R, Zahm DS (2003) Brain Res 989:205
36. Sigel A, Sigel H, Sigel RKO (2006) Neurodegenerative diseases and metal ions. In: Metal ions in life science, p 136
37. Strausak D, Mercer JFB, Dieter HH, Stremmel W, Multhaup G (2001) Brain Res Bull 55:175
38. Banci L, Bertini I, Boca M, Girotto S, Martinelli M, Valentine Joan S, Vieru M (2008) PLoS One 3:e1677
39. Ding F, Dokholyan NV (2008) Proc Natl Acad Sci USA 105:19696
40. Li HT, Jiao M, Chen J, Liang Y (2010) Acta Biochim Biophys Sin 42:183
41. Brazier MW, Volitakis I, Kvasnicka M, White AR, Underwood JR, Green JE, Han S, Hill AF, Masters CL, Collins SJ (2010) J Neurochem 114:440
42. Choi CJ, Anantharam V, Martin DP, Nicholson EM, Richt JA, Kanthasamy A, Kanthasamy AG (2010) Toxicol Sci 115:535
43. Tarohda T, Ishida Y, Kawai K, Yamamoto M, Amano R (2005) Anal Bioanal Chem 383:224
44. Rivera-Mancia S, Perez-Neri I, Rios C, Tristan-Lopez L, Rivera-Espinosa L, Montes S (2010) Chem Biol Interact 186:184
45. Fox JH, Kama JA, Lieberman G, Chopra R, Dorsey K, Chopra V, Volitakis I, Cherny RA, Bush AI, Hersch S (2007) PLoS One 2
46. Ejima A, Watanabe C, Koyama H, Satoh H (1996) Tohoku J Exp Med 178:1
47. Panayi AE, Spyrou NM, Iversen BS, White MA, Part P (2002) J Neurol Sci 195:1
48. Corrigan FM, Reynolds GP, Ward NI (1991) Trace Elem Med 8:1
49. Corrigan FM, Reynolds GP, Ward NI (1993) Biometals 6:149
50. Csaszma I, Andrasi E, Lasztity A, Bertalan E, Gawlik D (2003) J Anal At Spectrom 18:1082
51. Srivastava RAK, Jain JC (2002) J Neurol Sci 196:45
52. Moreno JA, Yeomans EC, Streifel KM, Brattin BL, Taylor RJ, Tjalkens RB (2009) Toxicol Sci 112:394
53. Witholt R, Gwiazda RH, Smith DR (2000) Neurotoxicol Teratol 22:851
54. Belavaria C, Andrasi E, Molnar Z, Bertalan E (2005) Microchem J 79:367
55. Andrasi E, Igaz S, Molnar Z, Mako S (2000) Magnes Res 13:189
56. Wang YY, Li S, Piao FY, Hong Y, Liu P, Zhao YF (2009) Neurotoxicol Teratol 31:318
57. Calderon-Garciduenas L, Maronpot RR, Torres-Jardon R, Henriquez-Roldan C, Schoonhoven R, Acuna-Ayala H, Villarreal-Calderon A, Nakamura J, Fernando R, Reed W, Azzarelli B, Swenberg JA (2003) Toxicol Pathol 31:524
58. Bocca B, Forte G, Petrucci F, Senofonte O, Violante N, Alimonti A (2005) Ann Ist Super Sanita 41:165
59. Bocca B, Alimonti A, Petrucci F, Violante N, Sancesario G, Forte G, Senofonte O (2004) Spectrochim Acta Part B 59:559
60. Forte G, Bocca B, Senofonte O, Petrucci F, Brusa L, Stanzione P, Zannino S, Violante N, Alimonti A, Sancesario G (2004) J Neural Transm 111:1031
61. Bocca B, Alimonti A, Senofonte O, Pino A, Violante N, Petrucci F, Sancesario G, Forte G (2006) J Neurol Sci 248:23
62. Ahmed S, Santosh W (2010) PLoS One 5
63. Hopt A, Korte S, Fink H, Panne U, Niessner R, Jahn R, Kretzchmar H, Herms J (2003) J Neurosci Methods 128:159
64. Hagen K, Boman H, Mellgren SI, Lindal S, Bovim G (1998) Arch Neurol 55:1467
65. Hozumi I, Hasegawa T, Honda A, Ozawa K, Hayashi Y, Hashimoto K, Yamada M, Koumura A, Sakurai T, Kimura A, Tanaka Y, Satoh M, Inuzuka T (2011) J Neurol Sci 303:95
66. Gerhardsson L, Lundh T, Londos E, Minthon L (2011) J Neural Transm 118:957
67. Becker JS, Zoriy M, Matusch A, Wu B, Salber D, Palm C, Becker JS (2010) Mass Spectrom Rev 29:156
68. Becker JS (2010) Int J Mass Spectrom 289:65
69. Becker JS, Zoriy M, Becker JS, Dobrowolska J, Matusch A (2007) J Anal At Spectrom 22:736
70. Hare D, Reedy B, Grimm R, Wilkins S, Volitakis I, George JL, Cherny RA, Bush AI, Finkelstein DI, Doble P (2009) Metallomics 1:53
71. Wang LM, Becker JS, Wu Q, Oliveira MF, Bozza FA, Schwager AL, Hoffman JM, Morton KA (2010) Metallomics 2:348
72. Becker JS, Matusch A, Palm C, Salber D, Morton KA, Becker S (2010) Metallomics 2:104
73. Becker JS, Lobinski R, Becker JS (2009) Metallomics 1:312
74. Andrasi E, Farkas E, Scheibler H, Reffy A, Bezur L (1995) Arch Gerontol Geriatr 21:89
75. Lovell MA, Robertson JD, Teesdale WJ, Campbell JL, Markesbery WR (1998) J Neurol Sci 158:47
76. Gellein K, Roos PM, Evje L, Vesterberge O, Flaten TP, Nordberg M, Syversen T (2007) Brain Res 1174:136
77. Rodriguez-Gonzalez P, Marchante-Gayon JM, Alonso JIG, Sanz-Medel A (2005) Spectrochim Acta Part B 60:151
78. Jakubowski N, Messerschmidt J, Anorbe MG, Waentig L, Hayen H, Roos PH (2008) J Anal At Spectrom 23:1487
79. Vogl J (2007) J Anal At Spectrom 22:475
80. Becker JS, Sela H, Dobrowolska J, Zoriy M, Becker JS (2008) Int J Mass Spectrom 270:1
81. Becker JS, Zoriy M, Pickhardt C, Przybylski M, Becker JS (2005) Int J Mass Spectrom 242:135
82. Heumann KG (2004) Anal Bioanal Chem 378:318
83. Lou XD, Zhang GH, Herrera I, Kinach R, Ornatsky O, Baranov V, Nitz M, Winnik MA (2007) Angew Chem Int Ed 46:6111
84. Razumienko E, Ornatsky O, Kinach R, Milyavsky M, Lechman E, Baranov V, Winnik MA, Tanner SD (2008) J Immunol Methods 336:56
85. Baranov VI, Quinn Z, Bandura DR, Tanner SD (2002) Anal Chem 74:1629
86. Helfrich A, Bettmer J (2007) J Anal At Spectrom 22:1296

Index

W. Linert and H. Kozlowski (eds.), *Metal Ions in Neurological Systems*,
DOI 10.1007/978-3-7091-1001-0, © Springer-Verlag Wien 2012